5th International Conference on Materials and Manufacturing Engineering (ICMME-2020)

IOP Conference Series: Materials Science and Engineering
Volume 954

Kanchipuram, India
7 – 8 September 2020

Editors:

T. Rajmohan
A. Nandakumar
K. Palani

ISBN: 978-1-7138-4782-3
ISSN: 1757-8981

Printed from e-media with permission by:

Curran Associates, Inc.
57 Morehouse Lane
Red Hook, NY 12571

CURRAN ASSOCIATES INC. proceedings.com

Some format issues inherent in the e-media version may also appear in this print version.

Institute of Physics
Dirac House, Temple Back
Bristol BS1 6BE UK

Phone: 44 1 17 929 7481
Fax: 44 1 17 920 0979

techtracking@iop.org

Additional copies of this publication are available from:

Curran Associates, Inc.
57 Morehouse Lane
Red Hook, NY 12571 USA
Phone: 845-758-0400
Fax: 845-758-2633
Email: curran@proceedings.com
Web: www.proceedings.com

TABLE OF CONTENTS

IOP Conf. Series: Materials Science and Engineering **954** (2020) 011001 doi:10.1088/1757-899X/954/1/011001

PREFACE

This proceeding comprises selected papers from the 5th **International Conference on Materials and Manufacturing Engineering (ICMME-2020)** at SRI CHANDRASEKHARENDRA SARASWATHI VISWA MAHAVIDYALAYA (SCSVMV), Kanchipuram, India. This book covers a wide range of topics within the materials and manufacturing disciplines. The content includes technical papers and review articles mainly focused in synthesis, development and characterization of new materials and machining-associated studies on various materials.

This book covers some of the recent advancements in the areas such as optimization techniques, computational fluid dynamics, tribology, alternate fuels, turbomachinery, renewable energy, thermal engineering and metal forming process. The collective expertise of the editorial team in a variety of aspects in the mechanical engineering field to manage the editorial process efficiently for providing a high-quality review process. Also, this book will be providing a technological and scientific platform for researchers, scientists, engineers and academicians seeking advancements in the area of materials and manufacturing engineering.

The conference is initially planned on 19-20 March 2020, but with the outbreak of COVID-19, it is rescheduled on 7-8 September 2020. The conference is organized through online virtual mode using Internet Communication Technologies (ICT) via Google meet and WebEx. Through the ICT technology , the majority of participants were actively participated in the conference. It is worth the option of virtual participation that allows expanding the geography significantly and increasing the number of participants. More than 350 participants participated in the conference and 87 papers from different regions of India and other countries are selected for presentation during the conference sessions.

After the conference, all the papers were thoroughly reviewed by the experts and finally 50 papers were accepted for publication in the *IOP Conference Series: Materials Science and Engineering*.

We thank all the reviewers for their time and effort in reviewing the papers even in the COVID-19 situation , as well as the authors, who took into account the reviewers' recommendations and improved their papers to ensure the publication high quality in *IOP Conference Series: Materials Science and Engineering* Series.

Dr.T.Rajmohan,
Associate Professor & Head,
Convenor – ICMME-2020.
Department of Mechanical Engineering,
Sri Chandrasekharendra Saraswathi Viswa Mahavidyalaya (SCSVMV)
Kanchipuram-631561,
Tamilnadu, India.
E.mail: rajmohanscsvmv@yahoo.com

IOP Conf. Series: Materials Science and Engineering **954** (2020) 011001 doi:10.1088/1757-899X/954/1/011001

GUEST EDITORS

Dr. T. RAJMOHAN
Associate Professor & Head,
Department of Mechanical Engineering,
Sri Chandrasekharendra Saraswathi Viswa Mahavidyalaya (SCSVMV)
Kanchipuram-631561,
Tamilnadu, India.
E.Mail: rajmohanscsvmv@yahoo.com

Dr. A. NANDAKUMAR
Assistant Professor,
Department of Mechanical Engineering,
Sri Chandrasekharendra Saraswathi Viswa Mahavidyalaya (SCSVMV)
Kanchipuram-631561,
Tamilnadu, India.
E.Mail: nandakumar.a@kanchiuniv.ac.in

Dr. K. PALANI
Assistant Professor,
Department of Mechanical Engineering,
Sri Chandrasekharendra Saraswathi Viswa Mahavidyalaya(SCSVMV)
Kanchipuram-631561,
Tamilnadu, India.
E.Mail: palani.k@kanchiuniv.ac.in

IOP Conf. Series: Materials Science and Engineering **954** (2020) 011001 doi:10.1088/1757-899X/954/1/011001

COMMITTEE

Chairman	: Prof. Dr. G. Sriram, Dean, Engg., & Tech., SCSVMV
Convenor	: Dr. T. Rajmohan, Associate Professor & Head, SCSVMV
Organizing Secretary	: Dr. A. Nandakumar & Dr. K. Palani, SCSVMV
Technical Committee	: Dr. J. Paulo Davim, Univ. of Aveiro, Portugal
	Dr. Sridar Idapalapati, Nanyang Tech. Univ., Singapore
	Dr. B.V.R. Chowdari, Nanyang Tech. Univ., Singapore
	Dr. Kittichai Sojiphan, KingMongkut's Univ. of Tech.,Thailand
	Dr. Nikhil Ranjan Dhar, BUET, Bangladesh
	Dr. X. G. Hu, Hefei Univ. of Tech., China
	Dr. Mohd Faizul Bin Mohd Sabri, Univ. of Malaya, Malaysia
	Dr. G. Nagarajan, Professor, Anna University, Chennai
	Dr. P. Ramkumar, IIT, Madras
	Dr. I. A. Palani, IIT, Indore
	Dr. Uday Shanker Dixit, IIT, Guwahati
	Dr. S. Jayavel, IIITDM, Chennai
	Dr. J. Srinivas, NIT, Rourkela
	Dr. Vishal Santosh Sharma, NIT, Jalandhar.
	Dr.–Ing. M. Duraiselvam, NIT, Trichy
	Dr. A. Velayudham, Scientist 'F', CVRDE, DRDO
	Dr. K. Palanikumar, Sri Sairam Inst. Tech., Chennai
	Dr. P. Kuppan, VIT, Vellore
	Dr. R. Ramanujam, VIT, Vellore
	Dr. M. Natarajan, VIT, Vellore
	Dr. M. Velu, VIT, Vellore
	Dr. R. Ramesh, SVCE, Sriperumbudur
	Dr. S. Ranganathan, GMRIT, AP
	Dr. T. Senthilvelan, PEC, Pondicherry
	Dr. K. Pitchandi, SVCE, Sriperumbudur
	Dr. M. Vetrivel, SCSVMV
	Dr. S. Arumugam, SCSVMV
	Dr. A. Arunpremnath, SCSVMV
	Dr. K. Venkatesh, SCSVMV
	Dr. R. Vinayagamoorthy, SCSVMV
	Dr. A. Tamilarasan, SCSVMV
	Dr. D. Vijayan, SCSVMV
	Dr. S. Vijayabhaskar, SCSVMV
	Dr. K. Mohan, SCSVMV

IOP Conf. Series: Materials Science and Engineering **954** (2020) 011001 doi:10.1088/1757-899X/954/1/011001

KEYNOTE SPEAKER

Dr. Hortense LE FERRAND
School of Materials Science and Engineering
Nanyang Technological University (NTU)
Singapore

IOP Conf. Series: Materials Science and Engineering **954** (2020) 011001 doi:10.1088/1757-899X/954/1/011001

SNAP SHOTS IN ICMME-2020

KEYNOTE ADDRESS

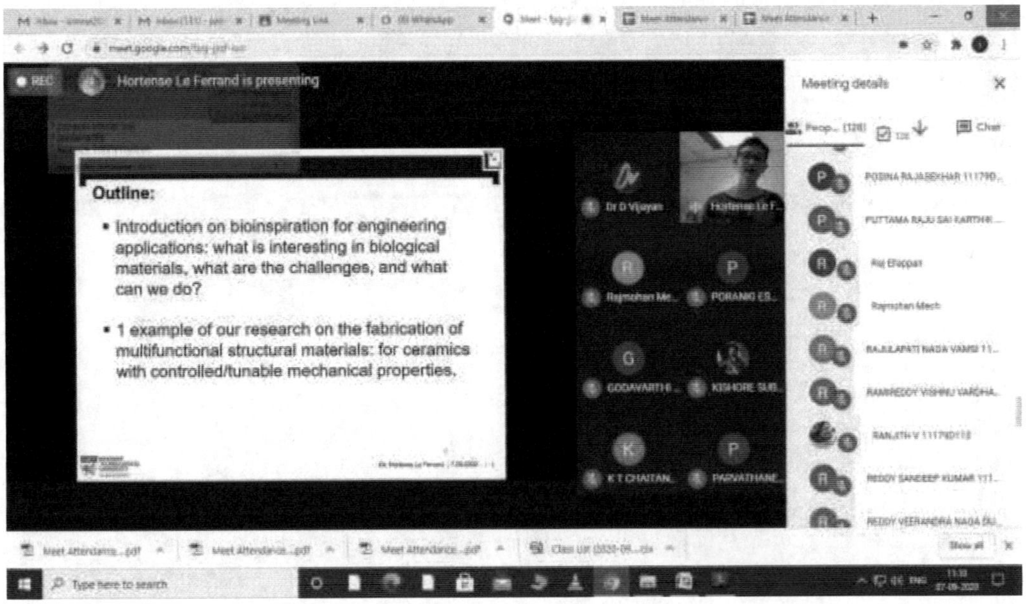

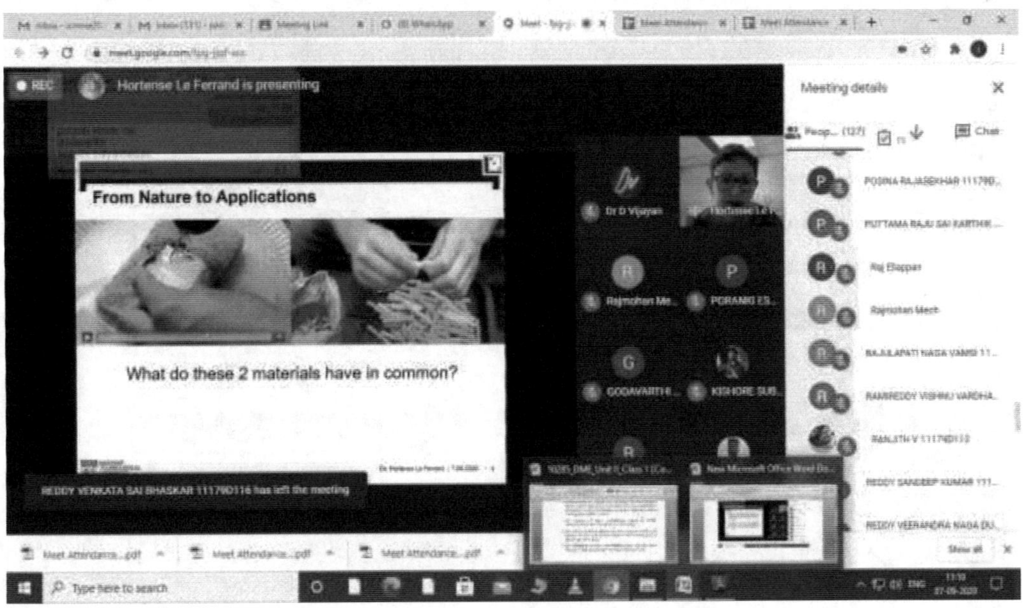

IOP Conf. Series: Materials Science and Engineering **954** (2020) 011001 doi:10.1088/1757-899X/954/1/011001

CONFERENCE PRESENTATIONS

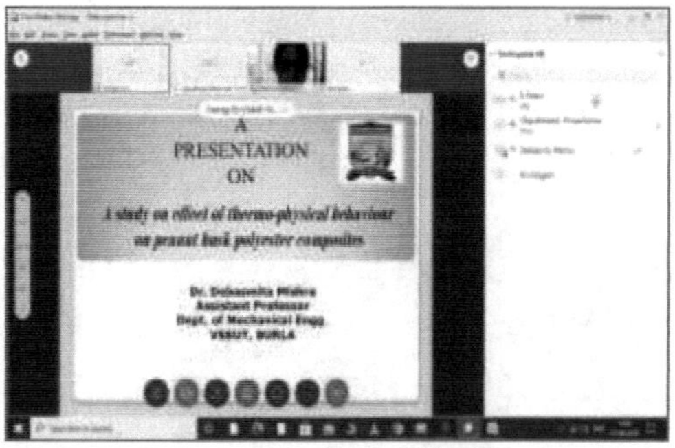

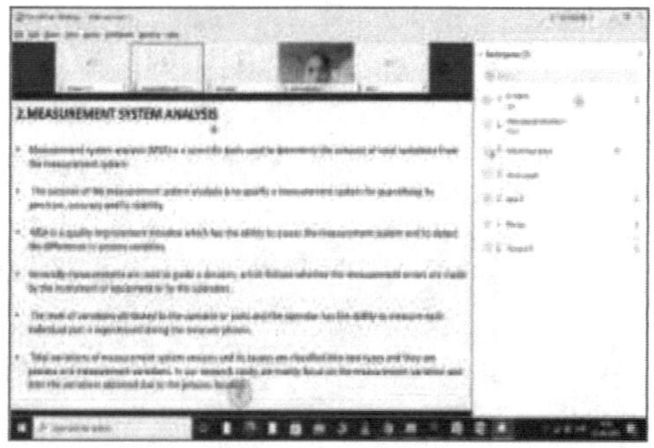

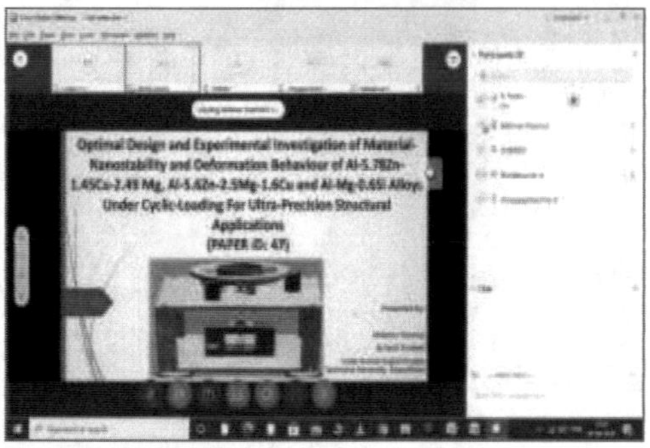

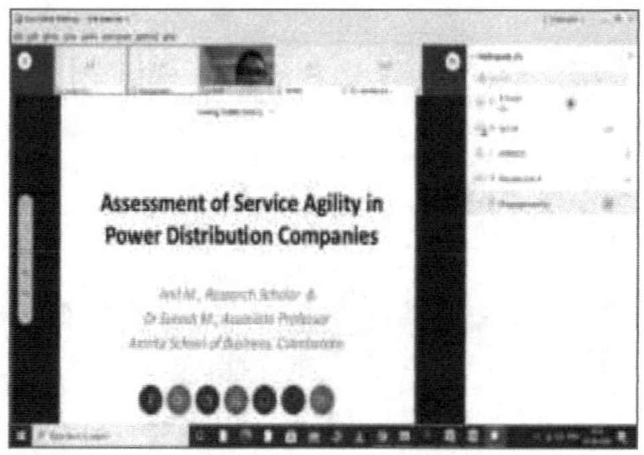

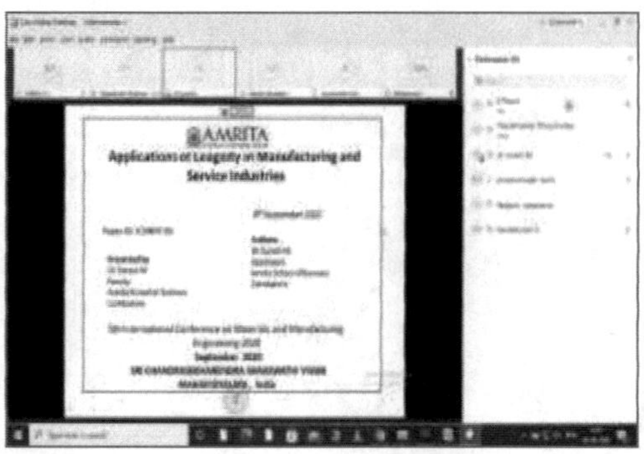

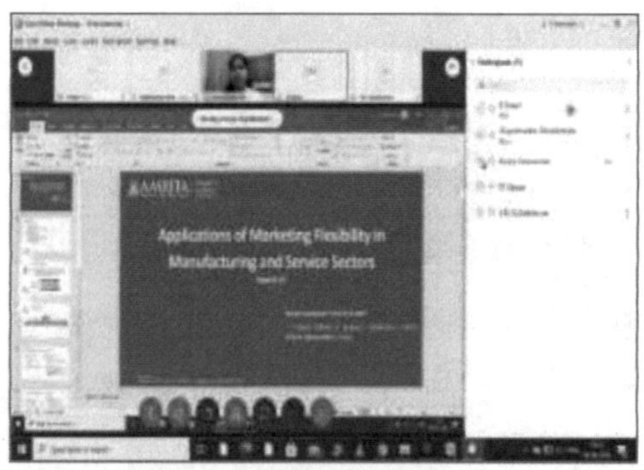

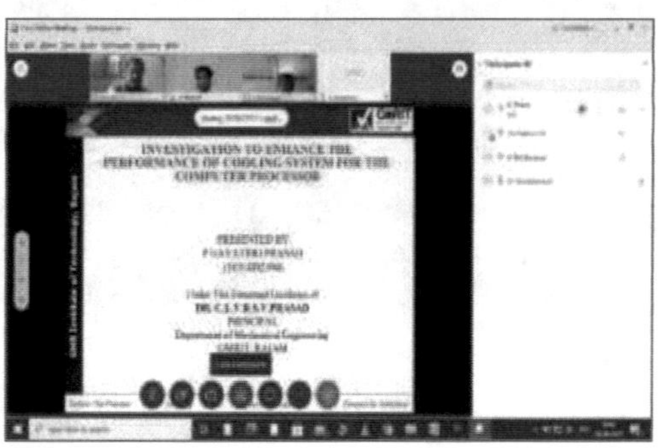

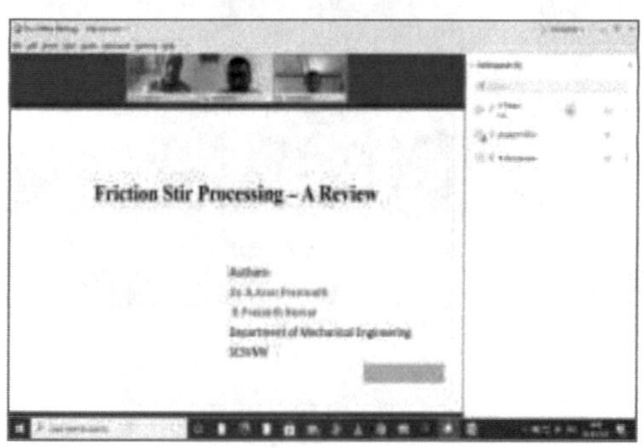

IOP Conf. Series: Materials Science and Engineering **954** (2020) 011002 doi:10.1088/1757-899X/954/1/011002

Peer review declaration

Abstract

All papers published in this volume of *IOP Conference Series: Materials Science and Engineering* have been peer reviewed through processes administered by the Editors. Reviews are conducted by expert referees to the professional and scientific standards expected of a proceedings journal published by IOP Publishing.

- **Type of peer review:** Single-blind with two level review system
  (i) The conference coordinators reviewed the submitted papers relevance to the conference topics and checked whether the papers meet the IOP format requirements.
  (ii) The independent reviewers evaluated the submitted papers according to the following criteria as the relevance to meet the study, novelty of work, quality and scientific knowledge, paper structure for IOP format and adequate references etc., and then all submitted papers were checked by English-professors.
- **Conference submission management system:**
  All the papers were submitted via EasyChair link and conference Email.
- **Number of submissions received:** 87 submissions received
- **Number of submissions sent for review:** 80 submissions were sent for review
- **Number of submissions accepted:** 49 papers accepted
- **Acceptance Rate (Number of Submissions Accepted / Number of Submissions Received X 100):** 56.32
- **Average number of reviews per paper:** 2
- **Total number of reviewers involved:** 20
- **Any additional info on review process:** -Nil-
- **Contact person for queries:**

Dr.T.Rajmohan,
Associate Professor & Head,
Convenor – ICMME-2020.
Department of Mechanical Engineering,
Sri Chandrasekharendra Saraswathi Viswa Mahavidyalaya (SCSVMV)
Kanchipuram-631561,
Tamilnadu, India.
E.mail: rajmohanscsvmv@yahoo.com

IOP Conf. Series: Materials Science and Engineering **954** (2020) 012001 doi:10.1088/1757-899X/954/1/012001

Predictive Maintenance using Machine Learning Based Classification Models

Anisha Chazhoor[1]*, Mounika Y[1], Vergin Raja Sarobin M[1], M V Sanjana[1], Yasashvini R[1]

[1]School of Computer Science and Engineering, Vellore Institute of Technology, Kelambakkam - Vandalur Road, Rajan Nagar, Chennai, Tamil Nadu- 600127 India.

Email: anisha.chazhoor2017@vitstudent.ac.in

Abstract. Machine learning facilitates predictive maintenance due to the advantages it holds over traditional methods of maintaining semi-conductor devices such as preventive and breakdown maintenance. Several predictive models using machine learning on the Semiconductor Manufacturing process dataset (SECOM) will be applied in this paper. The dataset contains the information related to semiconductor manufacturing process, with the attributes corresponding to signals collected from semiconductor devices. Due to the high-dimensionality of the data and class imbalance problem in the SECOM dataset, it poses several challenges related to data pre-processing, which is an essential step incorporated in this work while applying various machine learning models. Comparison and analysis of various predictive machine learning classification models were carried out based on the performance metrics like, accuracy and Receiver Operating Characteristic (ROC) curve.

1. Introduction

Large volumes of data are generated and shared every single day. Analyzing such large volume of data to make meaningful results is a taxing job and manually impossible. Developing a method to analyze data to find patterns and to achieve meaningful interpretations is necessary for any application domain for making informed decisions. But it is possible to apply various supervised or unsupervised machine learning algorithms such as Random Forest, Logistic Regression, Support Vector Machine and clustering algorithm to do better predictions [1]. The efficiency and time taken by each algorithm can be compared to reach the expected output. SECOM dataset consists of 591 features which are the signals generated from the semiconductor device to sense anomalies. It contains only 104 fail cases. Thus, it is an example of class imbalance problem. The dataset is not very large as it has only 1567 examples. But it has high dimensionality. It could hence cause problems like the curse of dimensionality.

Hence, to address these problems in the dataset, data pre-processing [2-3] is required, which is an essential pre-requisite for the application of machine learning models. Approaches like Synthetic Minority Oversampling Technique (SMOTE) analysis, correlation, feature selection etc., could be applied to remove initial hurdles like class imbalance, correlated features etc. After this step, machine learning models are applied and compared in order to select the best possible prediction approach according to the needs. The organization of the paper is as follows. The review of related work to predictive maintenance models in semiconductor manufacturing process and handling high-dimensional, class-imbalanced datasets is presented in Section II. Then a description of various predictive machine learning models is given in Section III. A series of experimentation and results are presented in Section IV. Finally, Section V concludes the findings of our paper regarding the predictive models.

2. Related Works

In one approach, Anghel et al., [1] compare and contrast two different predictive models and apply it to SECOM dataset. Pre-processing is performed, in which null and redundant data is removed. Over-sampling is performed to overcome the problem of class imbalance. Next, feature selection is performed using two approaches: one with Multivariate Adaptive Regression Splines (MARS) algorithm and another with Support Vector Machine (SVM) algorithm. After the important features are selected, Gradient Boosting Trees (GBT) is applied on MARS and Neural Networks using Tensor Flow is applied on SVM. They conclude that the approach involving NN is more effective [2].

In another paper, Kerdprasop et al., have developed a unique algorithm for pre processing of the SECOM dataset called the MeanDiff algorithm, which creates clusters on the basis of pass or fail cases, compare value differences and then rank features according to the calculated value differences. Also, columns with more than 55% null values and redundant data are removed. The performance of the clean data on Naïve Bayes, Decision Tree, k-Nearest Neighbors (k-NN) and Logistic regression algorithms are compared. It is concluded that though Naïve Bayes provides the best performance, the false positives are very high. On the other hand, decision tree provides low false positives and false negatives [3]. In another well-known approach, Verdier et al., replace the Euclidean distance method with the Mahalanobis method. The machine learning algorithm used is k-NN. While the traditional method used involves Hotelling-T2 model, this method involving Hotelling-T2 has several disadvantages. For example, it works only on Gaussian data which is not always relevant, as mostly real-life data is non-Gaussian. Thus, the proposed method eliminates the need to convert data to Gaussian data [4].

In the research work, Munirathinam et al., [5] have worked on methods based on machine learning techniques to build an accurate model for fault detection. Feature selection techniques that range from removing features with constant value and features containing missing values (above 55%) to statistical analysis such as chi-square and Principal Component Analysis (PCA) are used. To overcome problems like over fitting, more computational power and less prediction accuracy, techniques like subject matter expert knowledge, correlation analysis and variable component analysis have been used. The main disadvantage is that by duplicating of existing examples (over sampling), it makes the model prone to over fitting. To increase the prediction probability of rare classes, deep learning is required. A deep belief network is a deep learning technique that supports both supervised and unsupervised learning. The nodes and the number of layers constituting a deep belief network are different and it depends on the dataset used. It is difficult to obtain global optimization using gradient descent.

A new technique called particle swarm optimization is used to improve the performance and to carry out global optimization. The objective of the proposed technique is to exhaust the search space of a problem with the intention of finding the parameters which maximize the required objective. This algorithm imitates the particles behavior in a swarm. The position of a particle x is determined by its location in the kth step and its velocity in the (k+1)th step. The optimal position can be determined from the input parameters, which are the position of the particles after each step [6]. Computing the distances between the data points is computationally expensive and time consuming. To reduce the computation time, many fast k-NN search algorithms like the a Principal Axis search Tree (PAT) algorithm, Lower Bound search Trees (LBT), Modified principal axis search tree (MPAT), and Orthogonal Search Trees (OST) were proposed. The performances of all the above-mentioned algorithms were compared to find the best k-NN search algorithm. By conducting experiments on various datasets, it is seen that the OST algorithm performs well for most cases. For a data set from real images, the LAI algorithm is a good choice. If a data set with a huge number of dimensions and the pre-processing time is important, the MPAT method becomes the best choice. The disadvantage of this technique is that it is highly influenced by the number of dimensions, number of data points, and data distribution of a data set [7].

A multiple classifier machine learning approach for Predictive Maintenance (PdM) is proposed. Susto et al., propose the creation of dynamical decision rules to be adopted for the generation of quantitative indicators, associated with the possible problems that could damage the system in question. The relationship between these factors and the operational costs and failure risk could also be determined. This paper proposes training of multiple classification modules, to provide different trade-offs between the performance determining factors. The information gathered will be used to make an operational cost-based decision system for maintenance. This approach is suitable for semiconductor datasets, given their size, because this method works well for high-dimensional and censored datasets. This method is proved to have been working better than classical preventive maintenance (PvM) approaches and single SVM classifier [8].

SECOM dataset, being a high-dimensionality dataset, originally requires a two-step framework consisting of the dimensionality reduction and machine learning algorithm (like SVM) steps, for its analysis. But this demands high storage (of original data) and processing capacity (in the pre-processing stage) from the system, which increases the complexity. This paper proposes a unique algorithm called Incremental Projection Vector Machine (IPVM), which combines new dimensionality reduction feature in incremental format combined with feed-forward NN training simultaneously. Thus, when new samples are input in the system, the task of creating SVD becomes easier as one need not compute the full rank version on the entire dataset, instead just update the existing version. The results show that this method is far better than the two stage learning combinations like (SVD, BP). Hence, this approach is highly suitable for high-dimensional, large sample data like the SECOM dataset [9]. The SECOM dataset has class imbalance issues along with high dimensionality. The class imbalance is handled with SMOTE analysis. The algorithms used are Naive Bayes, k-NN and decision trees with the metrics Chi-square, Information gain metrics and Relief [10]. Not all data collected and features are relevant. So simple data manipulation or data transformation is required. This is often achieved through pre-processing steps like standardization, normalization, signal enhancement (smoothening and sharpening), principle component analysis and multidimensional scaling.

McCann et al., investigate various feature selection techniques and study how accurate they are in recognizing the causal effects in the SECOM semi-conductor manufacturing. After using these feature selection techniques along with simple algorithms like Naïve Bayes, it is concluded that a hybrid approach involving the appropriate feature extraction techniques and the existing business improvement techniques be developed to improve performance [11]. The semi-conductor manufacturing process is based on the concept of multi-stage manufacturing system. It refers to the system that involves more than one workstation to manufacture the final product akin to the model adopted by several automobile industries. Cascade Quality Prediction Method (CQPM) focuses on analysing the complex variable relationship in multi-stage manufacturing using a combination of multiple principle component analysis and iterative Dichotomiser 3 algorithm to extract rules. Ten-fold cross validation scores were calculated to compare the predicted models. It is observed that CQPM performs better than the other models. From the relatively low value of geometric mean, it is evident that the probability of misclassification in negative class is still high [12].

In the research work [13], Narul Afiqah A. Majid et al., presents the use of mechatronics technology that can help to implement predictive maintenance in mechanics by combining intelligent and predictive maintenance instrument. This paper uses the above concept and shows the choice of the appropriate strategy in the vibration of diagnostic process of the mechanical system. It discusses the role of mechatronics as a prediction tool behind the use of signature analysis of rotary machines. It proposes a method called Vibration Fault Simulation System (VFSS), a simulation system for detecting the unique vibration faults signatures of a rotating machine. Each time, the VFSS system is tested on various types of faults such as vibration, imbalanced rotor disc or ball, mechanical looseness, etc and signature of each fault is noted. The paper concludes that vibration is the most reliable parameter to be monitored for predictive maintenance purposes.

In the research work [14], Tayaba Abbasi et al., discusses the use of Recurrent Neural Network (RNN) using long short-term memory (LSTM) to carry out predictive maintenance of Air booster compressor (ABC) motor, an essential equipment in the oil and gas industry. Two most popular training algorithm i.e., Levenberg Marquardt and Bayesian Regularization were compared for training the network. Levenberg Marquardt has proven to be the fastest, generating lower root mean square error. The optimal number of epochs and hidden layers are 750 and 15 respectively. The main advantage of this algorithm is that it does not require any prior expertise or feature engineering. It can also perform multistep ahead prediction with high prediction accuracy. In the research work[15], Xiang Li et al., uses Random Forest Regression model to implement IoT equipment maintenance predictive model. This machine learning algorithm was chosen since it performs well on the data compared to other algorithms. The "equipment life prediction system" runs under the Ubuntu system. The algorithm successfully passes the unit testing.

IOP Conf. Series: Materials Science and Engineering **954** (2020) 012001 doi:10.1088/1757-899X/954/1/012001

This paper combines the actuality of equipment maintenance forecasting work, based on the engine life prediction data, and systematically discusses the construction processes and presents a Python implementation for the same.

3. Machine Learning Models

Random forest is a type of ensemble learning, which uses the approach of generating multiple trees simultaneously and vote for the most accurate tree classification is presented in Algorithm1.

Algorithm 1: Random forest

```
for i←0 to number of trees, n
    for i←0 to number of nodes, l
        Select k out of m features.
        Calculate the node d using best split point based on highest information gain.
        Split the node into daughter nodes.
    end for
end for
for i←0 to n
    Use generated rules to predict output
    Calculate votes for each predicted output.
end for
return highest voted predicted target as the final prediction.
```

Logistic regression is one of the simplest and most efficient algorithms to implement binary dependent variable problems. It is a statistical approach with the main objective being to calculate the weights associated with the variables is explained in Algorithm 2.

Algorithm 2: Logistic regression

Randomly generate *w=(w[0],w[1],..,w[n-1],1)*

while $\nabla E_{in} > \epsilon$

  for *i*←0, 1, 2...

    Compute gradient :

  $\nabla E_{in} = -\frac{1}{N}\sum_{n=1}^{N}\frac{y_n x_n}{1+e^{y_n w^T(t) x^n}}$

    Compute gradient

    $w(t+1) \leftarrow w(t) - \eta \nabla E_{in}$

  end for

end while

return w

Algorithm 3: Decision tree

Set training instances to root

Current *node*←*root*

for *i*←*c*

  Calculate information gain for every attribute

  $E(S) = \sum_{i=0}^{c} -p_i \log_2 p_i$

  where c is the number of classes and p_i is the probability of randomly selecting class i.

  $E(T,X) = \sum_{c \in X} P(c)E(c)$ where E(T,X) is the average entropy of the child attributes.

  Information Gain=E(S)-E(T,X)

end for

Find feature with greatest information gain.

Set this feature to be the splitting criterion at the current node.

if *information gain*←0

  return current *node*←*leaf*

end if

Decision tree classifies the dataset into classes based on the attribute value. It utilizes the concept of attribute selection. There are several methods of performing attribute selection, with the most popular ones in practice being using information gain and Gini index. The given algorithm is implemented for information gain methodology is explained in Algorithm 3. The multilayer perceptron as its name suggests has multiple layers with the lower most layer being the input layer followed by a number of hidden layers and the output layer at the top. All neurons in one layer are fully connected to the ones in adjacent layer. It utilizes the back-propagation method which is shown in Algorithm 4.

Algorithm 4: Multi-layered perceptron

Procedure MLP
1: Take input layer, forward propagate the patterns of training data and generate an output.
2: Calculate error, ϵ
3: Minimize ϵ using cost function f
4: Call Back_propagate()
5: Find derivative with respect to each weight *w[i]* in the network, and update model.
Procedure Back_propagation
1:**for** every node in the output layer
2: calculate error signal
3: **end for**
4: **for** all hidden layers
5: **for** every node in layer
6: Calculate signal error
7: Update each node's weight in the network
8: **end for**
9: **end for**

4. Implementation and Results

The dataset has been fitted to several machine learning models such as k-NN, decision tree, logistic regression and multi-layer perceptron. Univariate analysis has also been implemented. As the number of features is very high, several dimensionality reduction techniques such as principal component analysis and Linear Discriminant Analysis have been implemented. Accuracy and cross-validation scores have been noted. Boosting algorithms such as AdaBoost have been implemented and the accuracy scores are compared with k-NN algorithm. The efficiency of the models can be judged using their ROC curves. The first step in pre-processing is to find the ratio between the pass and fail cases. As we can see in the Figure 1, the ratio of pass: fail is almost 14: 1. This makes it a case of class imbalance which needs to be dealt separately.

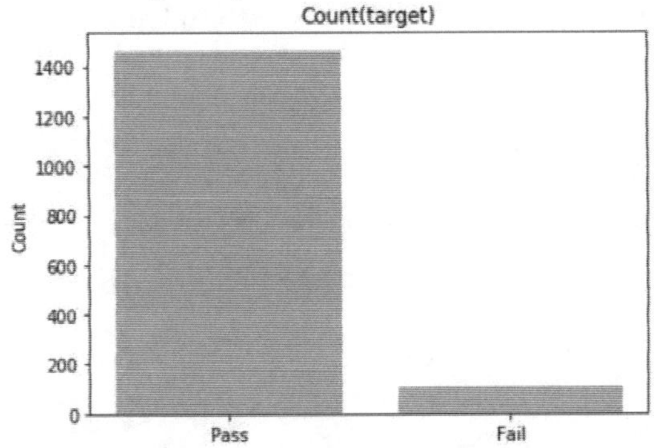

Figure 1. Count of pass and fail cases

The models that have been implemented are k-NN, logistic regression, MLP, random forest, AdaBoost and decision trees. Various Dimensionality reduction methods like PCA, LDA are used. The percentage of accuracy scores have been noted and compared simultaneously in Table 1. It could be observed that Logistic Regression with False Discovery Rate is the best performing algorithm and Random Forest with LDA is the least performing algorithm with respect to accuracy.

Table 1. Comparison of various algorithms with PCA and LDA

Algorithm	With PCA	With LDA	With False Discovery Rate
Random Forest	93.87	85.20	94.13
MLP	91.63	89.03	91.32
Logistic Regression	90.56	88.77	94.64
AdaBoost	91.07	89.03	91.58

Decision trees have also been implemented with dimensionality reduction techniques and the accuracy scores have been noted. The comparative study of the algorithms has been shown in Table 2. The dimensionality reduction with respect to decision trees is in terms of pruning to avoid over fitting. All values are in %.

Table 2. Comparison of accuracy various algorithms with and without dimensionality reduction

Algorithm	Without Dimensionality	With Dimensionality
Random Forest	76.588	94.13
MLP	90.051	91.63
Logistic regression	88.52	94.64
AdaBoost	92.09	92.602
Decision trees	86.48	93.622

The class imbalance problem is solved by SMOTE analysis. The accuracy before and after SMOTE sampling are recorded in Table 3. All the accuracy values are in %. The cv is set as 5.

Table 3. Comparison of mean cross validation scores and accuracy scores after SMOTE sampling on various algorithms

Algorithm	Mean cross validation score	Using SMOTE
Random Forest	92.40	94.3877
MLP	77.82	88.9285
Logistic regression	81.70	84.693
k-NN	92.46	55.357

The dependencies of the attributes with each other and their influence in determining the predicted output can be visualized using a correlogram as given below in Figure 2. PCA has been used as a pre-processing method with the intention of performing dimensionality reduction. The Explained Variance Ratio is defined as the fraction of variance of the current principal component with the total variance. As one can notice in the graph, the Explained Variance Ratio increases largely at the initial components and stabilizes after around 150-200 components. The said graph is shown as Figure 3.

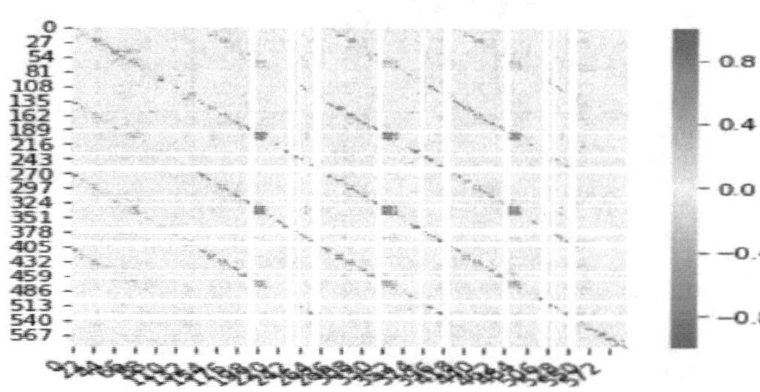

Figure 2. Correlogram showing dependencies between attributes

IOP Conf. Series: Materials Science and Engineering **954** (2020) 012001 doi:10.1088/1757-899X/954/1/012001

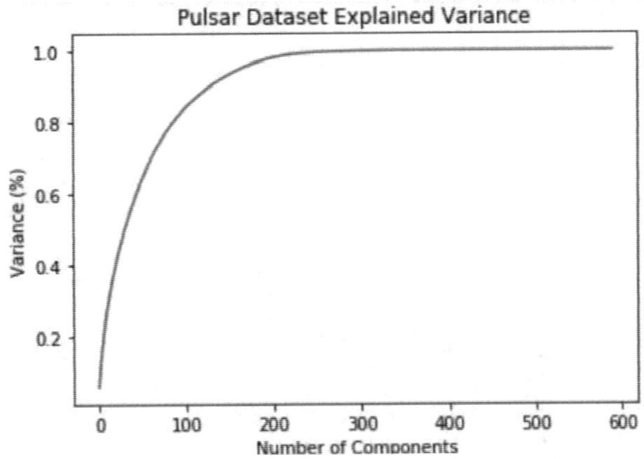

Figure 3. Explained variance ratio for PCA analysis

The ROC curves have been implemented for four machine learning models. The Area Under the ROC Curve(AUC) values describe the efficiency with which a certain parameter can be used to distinguish between two groups. Initially, the ROC curve has been implemented for the simplest algorithm, logistic regression. The results are shown in Figure 4.

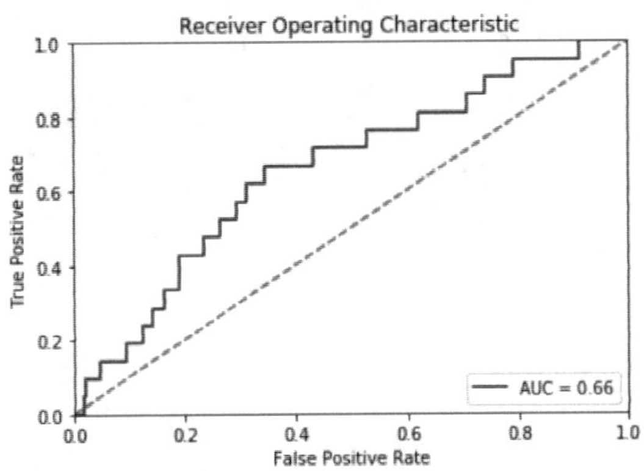

Figure 4. ROC curve for logistic regression

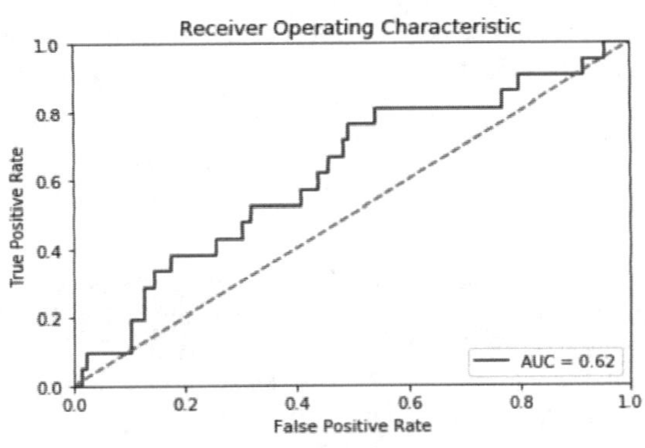

Figure 5. ROC curve for MLP

The ROC curve of MLP has been implemented in the Figure 5 given below. The ROC curve for k-NN has been implemented in Figure 6.

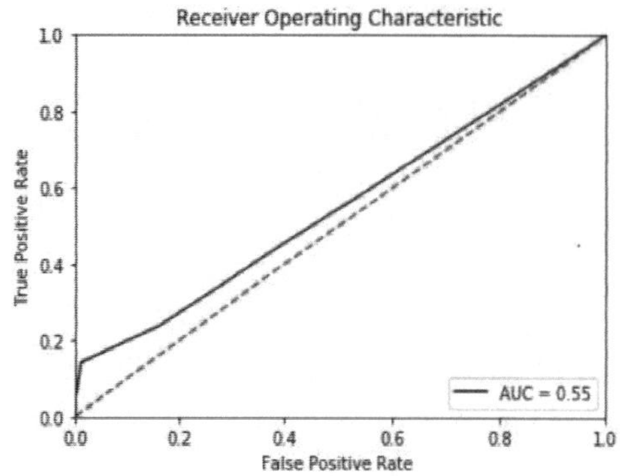

Figure 6. ROC curve for k-NN

The ROC curve for random forest is given in Figure 7.

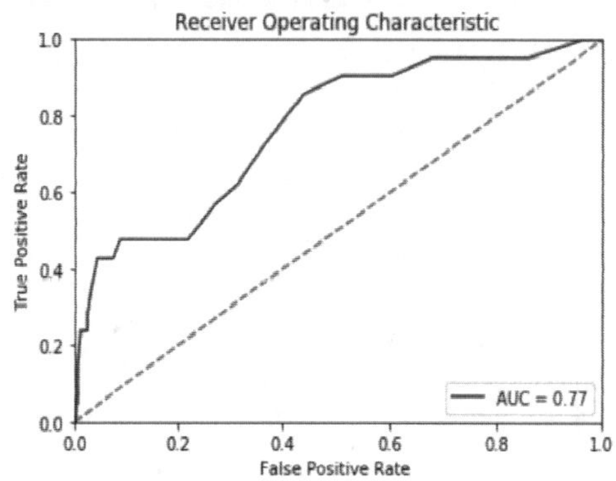

Figure 7. ROC curve for random forest

5. Conclusions

With the extensive comparative study performed, it can be concluded that, random forest performs poorly on SECOM dataset due to class imbalance problem. Logistic regression performs considerably better with a commendable accuracy with PCA. The model fitting most accurately to the dataset is Logistic Regression with False Positive Rate. Due to the class imbalance problem, certain algorithms do not perform up-to the mark and special hybrid algorithms like evolutionary machine learning could be integrated. SMOTE technique helped in the enhancement of the accuracy in Random Forest algorithm. The mean cross-validation score is the highest when cross-validation technique is used with k-NN algorithm.

6. References

[1] Gondalia, A., Dixit, D., Parashar, S., Raghava, V., Sengupta, A. and Sarobin, V.R., 2018. IoT-based healthcare monitoring system for war soldiers using machine learning. Procedia computer science, 133, pp.1005-1013.

[2] Anghel, I., Cioara, T., Moldovan, D., Salomie, I. and Tomus, M.M.,*2018 IEEE 16th International Conference on Embedded and Ubiquitous Computing (EUC) (pp. 29-36).*

[3] Kerdprasop, K. and Kerdprasop, N., *2011. Int. J. Mech, 5(4), pp.336-344.*

[4] Verdier, G. and Ferreira, A., *2010,IEEE Transactions on Semiconductor Manufacturing, 24(1)*, pp.59-68.

[5] Munirathinam, S. and Ramadoss, B., *2016,IACSIT International Journal of Engineering and Technology, 8(4), pp.273-285.*

[6] Kim, J.K., Han, Y.S. and Lee, J.S., *2017. Concurrency and Computation: Practice and Experience, 29(11), p.e4128.*
[7] Liaw, Y.C., *2011. ,International Conference on Image Processing, Computer Vision, and Pattern Recognition (IPCV) (p. 1).*
[8] Susto, G.A., Schirru, A., Pampuri, S., McLoone, S. and Beghi, A., *2014, IEEE Transactions on Industrial Informatics, 11(3), pp.812-820.*
[9] Zheng, Q., Wang, X., Deng, W., Liu, J. and Wu, X., *2010, December, Australasian Joint Conference on Artificial Intelligence (pp. 132-141). Springer, Berlin, Heidelberg.*
[10] Bach, M. and Werner, A., *2017, September, International Conference on Information Systems Architecture and Technology (pp. 182-194). Springer, Cham.*
[11] McCann, M., Li, Y., Maguire, L. and Johnston, A., *2010, Causality: Objectives and Assessment (pp. 277-288).*
[12] Arif F., Suryana, N., Hussin, B., *2013.,Ieri Procedia, 4, pp.201-207*
[13] Nurul Afiqah A. Majid, Asan G.A Muthalif, *2017, IOP Conf. Ser.: Mater. Sci. Eng. 260 012006*
[14] Tayaba Abbasi *, King Hann Lim, Ke San Yam,*2019 IOP Conf. Ser.: Mater. Sci. Eng.495 012067*
[15] Xiang Li,Li Wei , Jianfeng He, *2018, IOP Conf. Ser.: Mater. Sci. Eng.466 012001*

Passive flow control strategies for low Reynolds number airfoils

Rajesh Senthil Kumar T[1*], Mohini Priya Kolluri[1], Gopal Subramaniyan V R[1], Sripathi A D[1]

[1]Department of Aerospace Engineering, Amrita School of Engineering, Coimbatore, Amrita Vishwa Vidyapeetham, India.
[1*]Corresponding author: t_rajesh@cb.amrita.edu

Abstract. This study focuses on ameliorating the aerodynamic performance of an airfoil in low Reynolds number regime by implementing a passive flow control technique to suppress the effects created by the bubble bursting which holds significance in enhancing the aerodynamic performance of SUAVs which fall under this regime. A numerical study is conducted for the flow over E387 airfoil at a Reynolds number 10^5 to investigate the effects of the implementation of the passive flow control configuration on the laminar separation bubble dynamics. It is noted that orientation, thickness, inlet location, and configuration of the passive flow control passage had imposed significant effects on lift and drag coefficients generated by the airfoil. A significant drag reduction was achieved when the airfoil is implemented with a slot oriented at 30° that has a converging exit which resulted in the rise of the airfoil's aerodynamic efficiency.

1. Introduction

In low Reynolds number regime, even due to a slight adverse pressure gradient, the formation of the laminar separation bubble is encountered on an airfoil. The bursting phenomenon of the separation bubble is detrimental to the aerodynamic performance of the airfoil used in SUAVs that fly in this regime. With an increase of incidence or a reduction in speed, the separated shear layer may fail to reattach and the short bubble may burst to form either a long bubble or an unattached free shear layer [1]. O'Meara *et al.* [2] observed that with an increase in the angle of attack, the laminar separation point is moved forward without any significant change in the length of the bubble. Hence, various active and passive flow control techniques are being adopted to improve the performance of the airfoils or wings in low Reynolds numbers regime. A numerical study by L. Huang *et al.* [3] shows that in low Reynolds numbers blowing on NACA0012 at downstream locations decreased the drag and a suction jet near the leading edge increased the lift. A. V. M. Kumar *et al.* [4] studied that constant and periodic heat sources at the lower nose of the NACA0012 have reduced drag. Sang Hoon Kim *et al.* [5] utilized a single synthetic jet to get an increase in lift and decrease in drag when the velocity of the synthetic jet is thrice the free stream velocity at higher angles of attack. The laminar separation bubble was eliminated by synthetic jet at 0° angle of attack in an experiment conducted by M. Gul *et al.* [6]. A numerical study done by Xie *et al.* [7] on S809 airfoil with slot had increased the lift and slightly decreased the drag produced at higher angles of attack. Based on the research conducted above on flow control techniques, this study focusses on enhancing the aerodynamic performance of an airfoil in low Reynolds number regime by implementing a passive flow control technique. A numerical study is conducted to investigate the aerodynamic performance of E387 airfoil when implemented with a passive flow control configuration in 10^5 Reynolds number.

2. Numerical setup

The S809 airfoil implemented with a slot from the Xie *et al.* [7] paper is validated at a Reynolds number 10^6 and E387 airfoil implemented with various slot passages is considered for the numerical study at 0° angle of attack and Reynolds number 10^5 while the chord of the airfoil is considered as 0.305m which falls under the range of typical chord length of SUAVs. A structured C-H mesh domain having dimensions 19.67 times the airfoil chord length is generated in ANSYS ICEM CFD 2019 R1 with appropriate cell growth rate and biasing of nodes toward the airfoil surface ensuring that wall y^+ is less than 1 as shown in figure 1. ANSYS FLUENT 2019 R1 employing transient pressure-based solver with double precision is used for performing numerical simulations. Transition – SST (4 equations) turbulence model is used for predicting boundary layer physics and laminar separation bubble dynamics accurately. The no-slip wall boundary condition is set for the airfoil surfaces, the inlets of S809 and E387 airfoils are specified with velocities 47.8929ms^{-1} and 4.8755ms^{-1} respectively, both their outlets are specified with 0 Pa gauge pressure and the turbulent intensity is set to 0.11%. All the residual values are set to 0.0001 and the simulations are run for a flow time of 4s.

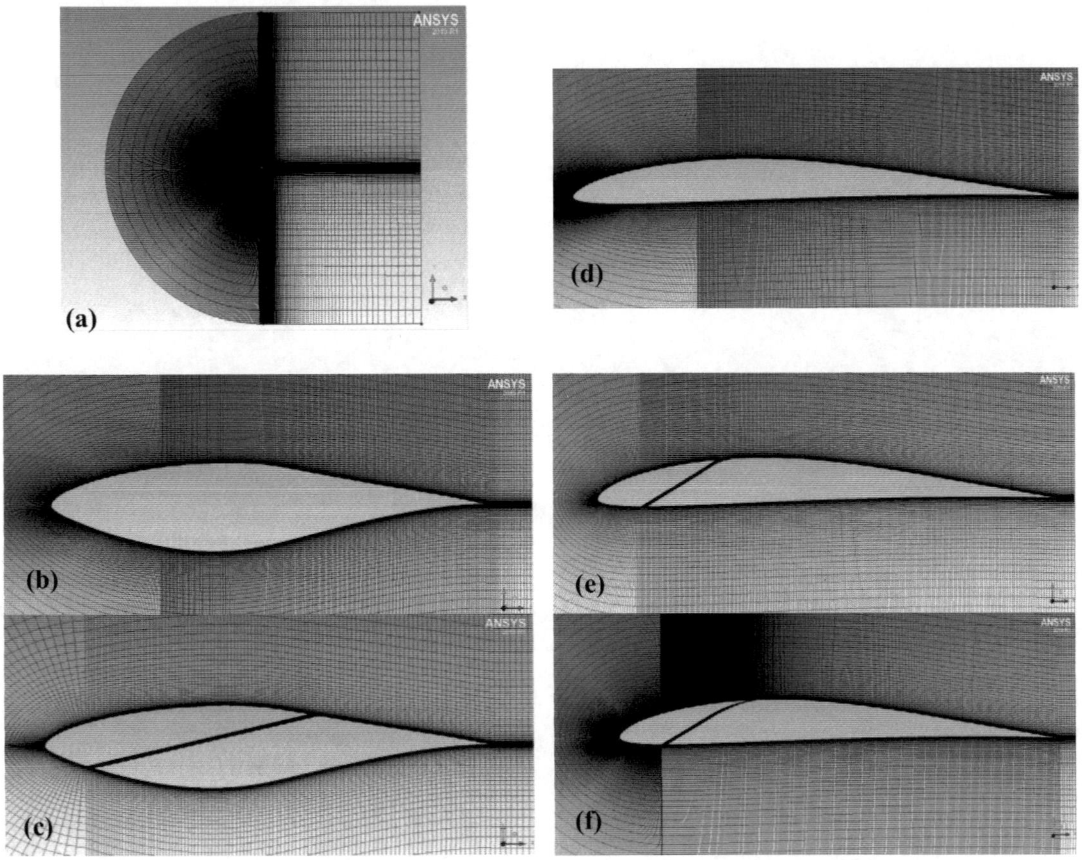

Figure 1. The Computational Structured C-H Mesh for the flow field around (b) S809 airfoil (c) S809 airfoil with slot, (d) E387 airfoil, (e) E387 airfoil with 0.002 m thick slot oriented at 30°, and (f) E387 airfoil with the same slot having a converging exit in ANSYS ICEM CFD.

3. Numerical Validation

Numerical simulation results of S809 airfoil is validated with experimental results [8] and are compared with the results obtained by Xie *et al.* [7]. It is observed that the numerical results that are run with Transition SST turbulence model closely matched with the experimental results following a similar trend of a dip and rise in lift coefficient at higher angles of attack compared to the results from Xie *et al.* [7]

paper which used SST-k-ω turbulence model (2 equations eddy viscosity) as shown in figure 2. Numerical simulations of S809 airfoil implemented with slot suggest that upon addition of the slot there resulted in a rise in the drag and fall in the lift. Numerical simulation results and experimental results [9] of E387 airfoil are observed to be closely matching with each other as shown in figure 3.

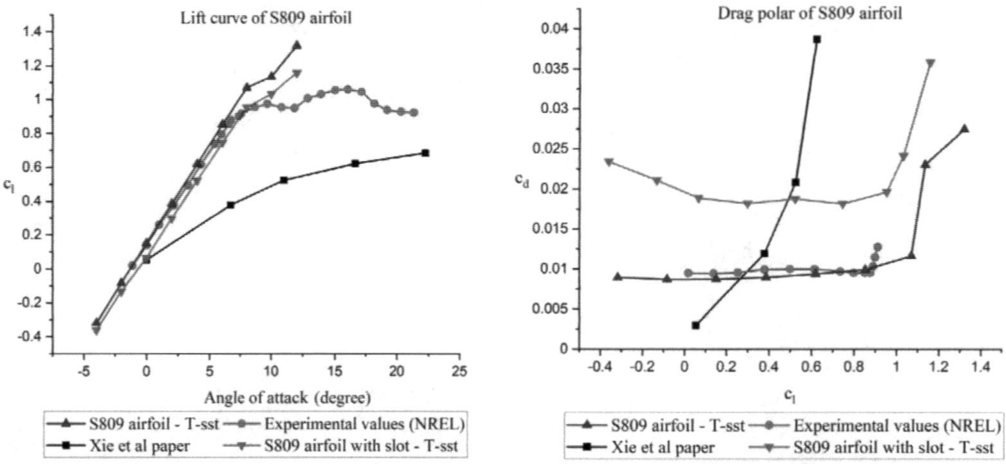

Figure 2. The plots showing Lift curve and Drag polar for the flow past S809 airfoil for $R_e = 10^6$ obtained from numerical simulation results, experimental results [8], Xie *et al.* [7] paper and numerical simulations obtained for S809 airfoil with slot.

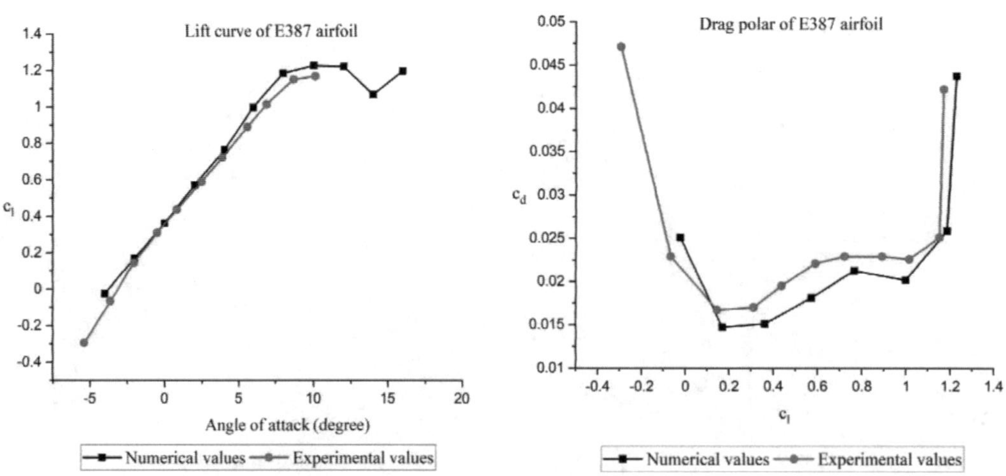

Figure 3. The plots showing Lift curve and Drag polar obtained from numerical simulations and experimental data [9] of the flow past an E387 airfoil at $R_e = 10^5$.

4. Results and discussion

Numerical simulations are performed at 10^5 Reynolds number for the flow past E387 airfoils implemented with 0.003 m thick slots oriented at 12.97°, 30°, 60° and 90° w.r.t chord line with all their inlet locations at 0.1c from the leading edge on the airfoil's pressure surface at 0° angle of attack. The E387 airfoil with 12.97° slot orientation, the slot cuts through the laminar separation bubble while bubble bursting takes place for the airfoils with remaining slot orientations. The numerical results suggest that there is a loss in lift coefficient upon addition of the slot of any orientation to the airfoil.

With increasing slot orientation from 30° to 90°, both the lift and drag coefficients generated by the airfoils kept decreasing as shown in table 1. The airfoil with the slot oriented at 90° has very little flow passing through the slot and it produced the least amount of both drag and lift coefficients among the airfoils. Although the airfoil with the slot oriented at 30° generated the highest drag coefficient, it generated the least amount of loss in the lift coefficient as compared to that of the airfoils with other slot orientations and hence this slot orientation is considered for the subsequent simulations.

Table 1. Time-averaged Lift and Drag coefficients generated for the flow past E387 airfoils with 0.003 m thick slots oriented at 12.97°, 30°, 60° and 90° respectively at α=0°, t = 4 s and $R_e = 10^5$.

Slot orientation	12.97°	30°	60°	90°	No slot
c_l	0.29762	0.34961	0.29762	0.28561	0.37630
c_d	0.01948	0.02478	0.01728	0.01208	0.01574
c_l / c_d	15.278	14.109	17.223	23.627	23.903

Numerical simulations are performed at 10^5 Reynolds number for flow past E387 airfoil implemented with 0.001 m and 0.002 m thick slots oriented at 30° with their inlet location at 0.1c from the leading edge on the airfoil's pressure surface at 0° angle of attack. The slot exit geometries for both airfoils are made smooth for facilitating gradual turning of the flow to obey the Coanda effect. The 0.001 m thick slot passage is too slender to let any flow into it. Thus, results obtained for this slot thickness are almost identical with the results of an airfoil without any slot passage. The airfoil with the 0.002 m thick slot is considered for the subsequent simulations as it has relatively improved the aerodynamic efficiency of the airfoil as compared to that of 0.003 m thick slot as shown in table 2.

Table 2. Time-averaged Lift and Drag coefficients generated for the flow past E387 airfoils implemented with slots oriented at 30° having various thicknesses at α=0°, t = 4 s and $R_e = 10^5$.

Slot thickness	0.003 m	0.002 m	0.001 m	No slot
c_l	0.34961	0.34358	0.34114	0.37630
c_d	0.02478	0.02145	0.01562	0.01574
c_l / c_d	14.109	18.232	23.914	23.903

Numerical simulations performed on the E387 airfoil implemented with a 0.002 m thick slot oriented at 30° which is shifted rightward from slot inlet position 0.1c to 0.15c from the leading edge on the airfoil's pressure surface, suggest that there is a slight increment in the lift coefficient as compared to the airfoil whose slot is not shifted, but the lift coefficient generated is still around 6.8% lower than that of the airfoil without slot. Although reduction of slot thickness has reduced the drag coefficient, the shifting of the slotted passage only increased the drag coefficient even further as compared to the airfoil whose slot is not shifted, which resulted in a decrease in aerodynamic efficiency as shown in table 3.

Numerical simulations were performed for the E387 airfoil implemented with a 0.002 m slot oriented at 30° whose exit is modified into a converging passage to facilitate the acceleration of flow at the slot exit. Despite this modification generated a lower lift coefficient as compared to the airfoil without slot, a 13.43% reduction in drag coefficient was achieved as shown in table 3, which resulted in a considerable improvement in aerodynamic efficiency of the airfoil.

When the E387 airfoil is implemented with a slot which is shifted rightward to the inlet position 0.15c from the leading edge on the pressure surface of the airfoil and the slot's exit is modified into a converging passage, the numerical simulation results suggest that the lift coefficient generated is nearly same as that of the airfoil with a converging exit alone and the drag coefficient generated is nearly same as that of the airfoil without slot as shown in table 3.

Table 3. Time-averaged Lift and Drag coefficients generated for the flow past E387 airfoils with shifted slot with converging exit, shifted slot, slot with converging exit and for the flow past E387 airfoil with unmodified slot, all slots are oriented at 30° and have thickness 0.002 m at α=0°, t = 4 s and $R_e = 10^5$.

	Shifted slot	Converging slot exit	Shifted slot with converging exit	Unmodified slot	No slot
c_l	0.35063	0.32475	0.32434	0.34348	0.37630
c_d	0.02338	0.01363	0.01568	0.02145	0.01574
c_l / c_d	17.192	27.414	20.951	18.232	23.903

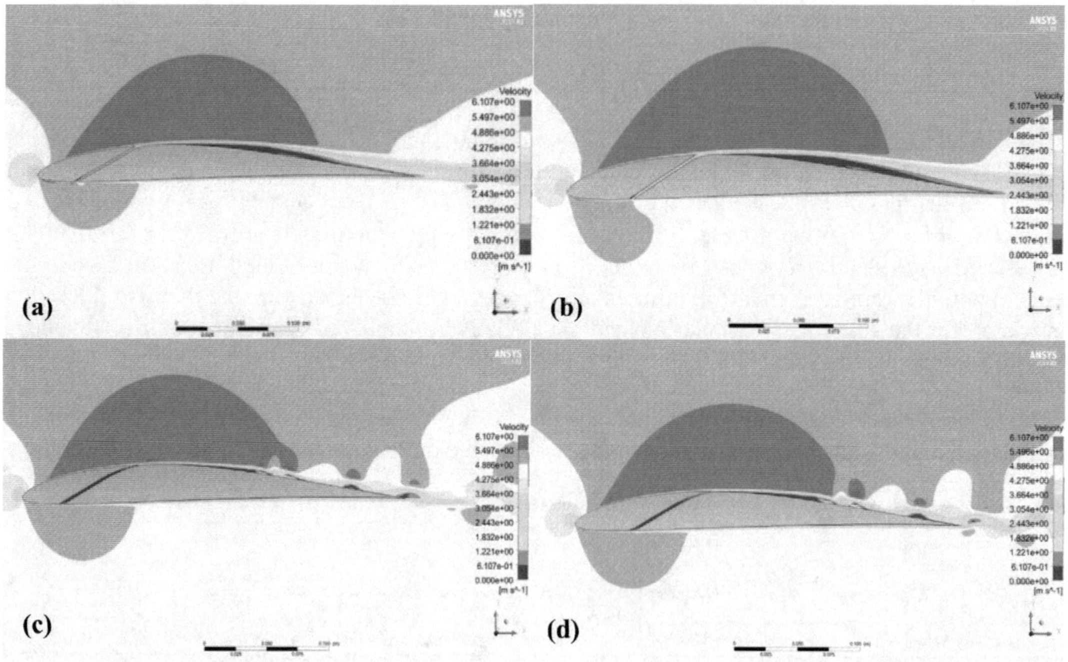

Figure 4. Computed Velocity contours obtained from numerical simulations for the flow past (a) E387 airfoil implemented with a 0.002 m thick slot with its inlet located at 0.1c and oriented at 30°, (b) when the slot inlet is shifted to 0.15c, (c) with a converging slot exit, and (d) when the slot inlet is shifted to 0.15c and with a converging exit at α=0°, t = 4 s and $R_e = 10^5$.

5. Conclusions

1. Numerical simulations are executed for the E387 airfoils implemented with various passive flow control configurations to investigate their aerodynamic performances. The orientation, thickness, inlet location, and configuration of the passive flow control passage had imposed significant effects on the lift and drag coefficients generated by the airfoil. The loss in the lift coefficient is encountered upon the addition of any slot configuration to the airfoil.
2. The airfoil implemented with a slot oriented at 30° generated the highest lift coefficient among airfoils with other slot orientations.
3. While a 0.001 m thick slot is too slender to let any flow into it, a 0.002 m thick slot has improved the aerodynamic efficiency of the airfoil than that of a 0.003 m thick slot.
4. The gradual turning of the flow coming out of the slot was facilitated by smoothening of the slot exit geometry to obey the Coanda effect.

IOP Conf. Series: Materials Science and Engineering **954** (2020) 012002 doi:10.1088/1757-899X/954/1/012002

5. Upon modifying the slot's exit into a converging passage, a maximum of 13.43% reduction of drag coefficient leading to a considerable rise in aerodynamic efficiency of the airfoil by 14.69% was achieved.

References

[1] Mohsen Jahanmiri 2011 *Laminar Separation Bubble: Its Structure, Dynamics and Control* (Göteborg: Chalmers)

[2] O'Meara, M. M., Mueller and T. J. 1987 Laminar Separation Bubble Characteristics on an Airfoil at Low Reynolds Numbers *AIAA J.* **25** pp 1033–41

[3] L. Huang, P. G. Huang, R. P. LeBeau and T. Hauser 2004 Numerical Study of Blowing and Suction Control Mechanism on NACA0012 Airfoil *J. Aircr.* **41** p 1005

[4] A. V. M. Kumar, Rajendrakumar. N., Gogineni, P. C., Gopan, N., and Rajesh Senthil Kumar T. 2019 *Effect of temperature on the aerodynamics of airfoil in low Reynolds number flow AIP Conf. Proc.* vol 2134 (Thiruvananthapuram: AIP) p 020010

[5] Sang Hoon Kim and Chongam Kim 2009 Separation control on NACA23012 using synthetic jet *Aerosp. Sci. Technol.* **13** pp 172–182

[6] M Gul, O Uzol and I S Akmandor 2014 *An Experimental Study on Active Flow Control Using Synthetic Jet Actuators over S809 Airfoil J. Phys.: Conf Ser.* vol 524 (Ankara: IOP) p 012101

[7] Yonghui Xie, Jianhui Chen, Huancheng Qu, Gongnan Xie, Di Zhang and Mohammad Moshfeghi 2013 Numerical and Experimental Investigation on the Flow Separation Control of S809 Airfoil with Slot *Math. Probl. Eng.* **2013** Article ID 301748

[8] Dan M. Somers 1997 *Design and Experimental Results for the S809 Airfoil* NREL/SR-440-6918 (Pennsylvania: Airfoils, Incorporated.; NREL)

[9] Gregory A. Williamson, Bryan D. McGranahan, Benjamin A. Broughton, Robert W. Deters, John B. Brandt, and Michael S. Selig 2012 *Summary of Low-Speed Airfoil Data* vol 5 (Urbana-Champaign: Department of Aerospace Engineering, University of Illinois at Urbana-Champaign)

IOP Conf. Series: Materials Science and Engineering **954** (2020) 012003 doi:10.1088/1757-899X/954/1/012003

Applications of marketing flexibility in manufacturing and service sectors

Kavya Gopakumar [1] and M.Suresh[2]
[1, 2] Amrita School of Business, Coimbatore, Amrita Vishwa Vidyapeetham, India

Email: [1] kaavya93@gmail.com; [2]drsureshcontact@gmail.com

Abstract Most businesses are now faced with extreme pressures that relate to advances in globalization, rapidly evolving technology, as well as changes in the demands and behaviors of consumers. Marketing flexibility is therefore the key strategy for a business which helps to ensure customer satisfaction, achievement of success, differentiation among the competitors and superior performance under the environmental fluctuations. Unlike the plenty of studies pertaining to manufacturing and operational flexibility, there are only little efforts made to investigate and research on the topic marketing flexibility. Hence the concept of marketing flexibility is relevant for the prevailing situation and moreover it is undoubtedly an important topic for researchers that help in investigating firm performance and the sustainability of competitive advantage. This article presents an extensive study about marketing flexibility and its practical as well as theoretical applications through various approaches.

Keywords: *Flexibility; Marketing flexibility; Agility; Flexibility in services; Flexibility in manufacturing*

1. Introduction

It was found that most of the studies in the field of flexibility were concerned about operational and manufacturing facets rather than flexibility in the field of marketing. Also, these works also explored the versatility of marketing in broader strategic context. [1] define flexibility as "Flexibility is the degree to which an organization has a variety of managerial capabilities and the speed at which they can be activated, to increase the control capacity of management and improve the controllability of the organization."

In this fast-moving volatile, uncertain, complex and ambiguous world, nothing is static. This is applicable to the businesses as well varying from various domains – finance, marketing, human resources etc. Organizations that lacks flexibility was found not doing well in their business as expected compared to the competitors in the same field. Flexibility concept is still under the stage of studying wherein lots of researches and studies have to be performed in order to understand its dynamics and applicability. Organizations that have adapted marketing flexibility can face demand fluctuations both short and long term and also cope with the threats that they may face in this ever-changing business environment. Marketing flexibility helps the business to succeed and survive in such challenging situations. It acts as a key to the growth strategy of the organization.

2. Individual Approaches

2.1 Corporate Entrepreneurship

The research performed by [2] deals with a study of selected managers in South Africa working in different industries. It was found that there is a positive relationship between different market orientation, flexibility, and job satisfaction factors. The authors suggest that companies should develop their business strategies by promoting their marketing, flexibility and job satisfaction orientation.

According to [3], there have been changes in the way of doing business after the introduction of the Liberalization, Privatization, and Globalization (LPG) phenomenon. LPG has eroded boundaries which led to more challenging competition and providing importance and priorities to the customers. The research explains the lessons that are to be learnt from the corporate world by closely examining various firms such as Nokia vs Micromax (Developmental (NPD), Distribution, and Price Flexibility), Wal-Mart/Big Bazaar Vs Subhiksha (Promotional, Distribution, Logistics, and IT Flexibility) to name few.

2.2 New Product Development

In order to develop complex value-constellation systems that are capable of optimizing the satisfaction and advantages of all involved parties, the flexibility of marketing programs must be extended to all involvement, interaction and implementation processes [4]. The author explains a theoretical approach of the study that has few limitations. The first limitation is that the model built and addressed is merely a preliminary attempt to identify the main areas of the marketing system that need versatility to improve the co-creation process of value. The next limitation is that the flexibility research of participatory marketing processes has focused primarily on the advantages of this approach. The third limitation is that the developed model needs to be tested and improved through empirical research.

In order to cope with the market changes and performance, two distinct marketing skills are necessary to be adopted by the organisations - marketing flexibility and marketing planning capabilities. The authors [5] use view based on resource and literature exploring strategic flexibility to understand more about the paradox that is identified for planning the flexibility which is essential in achieving superior organizational responsiveness and a productive New Product Development (NPD).

2.3 Arts industry

Location advantages along with product development, advertising and pricing strategies have become strategic tool for the arts industry [6]. The author studies how location and other marketing mix factors can affect the financial flexibility of Performing Arts Organisations (PAOs) and their success.

The article published in Tribune Business News [7] demonstrates how well marketing flexibility can be applied in the field of arts. There have been so many customisations by the customers regarding their ticket buying options based on the varieties of the concerts that they are attending. In the last decade, the orchestra has made several changes to tackle subscriber-drain- more flexible exchange policies and encourage listeners to create their own concert packages.

2.4 Automobile industry

The case study of Tata Nano has been studied and explores to establish a significant role in recovering the sales profits of this car. The study done by [8] reported that inflexible marketing was one of the main causes of Nano's initial poor sales. Tata Motors recognised this issue and fixed it by producing the revised version of Nano 2012, which clearly means that Tata Motors could incorporate high flexibility in marketing. This is very clear when it was observed that the sales improved drastically by a thriving margin immediately after the launch of Nano 2012 bringing back. Hence Tata Nano could reach its maximum potential by the effective use of marketing flexibility as the situation demanded.

The study [9] concludes with the finding that automotive organisations were forced to adopt business-environment volatility and market dynamism to have the appropriate level of flexibility in their marketing mix. This author explains the importance of why marketing flexibility needs to be practised in the companies by taking the example of market leaders who controlled the industry for long time. The research also shows that even the market leaders today are having the fear of losing their positions in the market and they consider factors like innovation and inventions are essential to retain the same. The constantly fluctuating fuel prices, declining customer loyalty, increased knowledge and constant switching between the petrol and diesel variants have increased risks not only for individual companies but also for the entire market. Marketing flexibility if integrated on all these critical issues will ensure that both feasibility and relevance are retained by adjusting to the various rising situations.

2.5 Telecommunication industry

In the Indian Telecom Sector, the series of different events has turned the industry into a more competitive and dynamic environment [10]. It was observed that telecom organisations could manage and retain their customers by applying flexibility in various marketing activities such as branding, point of purchase, product creation, product differentiation, product planning and development, pricing strategies and loyalty programs. The author also suggests to use marketing and flexibility interaction matrix to develop flexibility integration model by considering other business functions like finance, supply chain and CSD.

2.6 Finance

[11] found that both initial public offering firms and seasoned equity offering firms are succeeded by a more aggressive marketing promotion in the initial two years following the offering. The link between marketing investment and firm value is moderated by the strategic flexibility of competitors with respect to issuers. It was also found that the only those issuers with less strategic flexibility could benefit more from the competition compared to that of firms that have adopted an aggressive marketing strategy.

The research in the field of commodity production done by [12] revealed that when two spot market prices are negatively correlated, the commodity producer would hedge more both financially and operationally. It was also observed that the marketing flexibility impact was more significant when the correlation is negative. The research concludes that the commodity producer should make operational and financial decisions simultaneously.

IOP Conf. Series: Materials Science and Engineering **954** (2020) 012003 doi:10.1088/1757-899X/954/1/012003

2.7 Methods to measure marketing flexibility

The research conducted by [13] identified the factors that should be utilized to maximize flexibility. It is suggested that the marketing-based flexibility is needed to improve the planning in organisations. A conceptual model of marketing flexibility orientation was developed based on the dynamic capability and contingency theories [14].

[15] developed a marketing flexibility measurement scale (AUTOFLEX) for automobile companies which includes a 26-item scale of six different dimensions including price, customer orientation etc. The study also paves ways for the future research on marketing flexibility in automotive field. The first limitation that can be observed in this scale is that, only a hand-picked number of automobile organisations operating in four-wheeler department were taken for the study. The other limitation that can be observed is that the study has taken only the company's perspective and not the customer's point of view which is also an important factor in deciding the dimensions and the factors. The third limitation is that the developed scale cannot be used for other segments of the automobile industry though these segments are similar in many aspects.

[16] developed a scale - FLEXMARK to measure and test the marketing flexibility in the organisation. This scale is an important indicator of organisational performance and can be used to measure organisational flexibility. The development of scale revealed that 59 attributes and 12 dimensions were relevant including new product development and Innovation, price measure, channel partner, customer satisfaction, structure of sales and marketing department, employee audit, training needs etc. Though the scale is not yet tested for reliability and validity, it is first of its kind in the field of flexibility which can be used by managers while taking strategic decisions. There is no particular industry or sector identified for the application of this scale. Hence further studies can be done to customise or standardise the scale as per the specific industry or sector needs and demanded flexibilities.

The research has come up with a 26-item AUTOFLEX scale that showed good reliability with the support of strong proof of content, construct, and nomological validity [17]. The scale could also establish its correlation with measure of market orientation which was essential for the nomological validity. The AUTOFLEX scale has adopted the concept of marketing flexibility that helps to make decisions according to the customer preference changes. The degree of marketing flexibility and various ways to enhance it by finding areas which are missing can be found and rectified by the automobile managers by using this scale.

3. Strategic flexibility

[18] define market focused strategic flexibility as "The firm's intent and capabilities to generate firm-specific real options for the configuration and reconfiguration of appreciably superior customer value propositions". The authors propose an integrative model with a proper approach in place that act as a mediator of market-focused strategic flexibility in marketing strategy frameworks. Organizations should use of strategic flexibility to combat business uncertainties. Strategic flexibility refers to a firms ' ability to acknowledge, adjust and adapt to changes in the environment. Environmental dynamism is one among many important factors that influences strategic flexibility. It was found that there is positive and significant relationship between strategic flexibility and exploitative innovation performance [19].

The managers should be capable enough to understand when they have to stop or reverse resource commitments and act accordingly [20]. The authors show that in a vicious cycle of insensitivity, self-serving interpretation and inaction, managers can become unconsciously trapped. Changes in today's demand are a "contextual" approach to marketing intelligence and planning. Another perspective of strategic flexibility is being studied in the research conducted by [21]. Strategic flexibility enables a company to decide whether the emergence of new market, new technology, political rules/regulations and public opinion can be controlled or not.

Classical business borders are beginning to blur because of changing contexts. Product expansion and product extension are not considered to be going hand in hand. Instead, they are complementary as they are the parts of a dynamic process. Marketing and flexibility in development have both a positive effect on the competitive success of an organization and on product innovation in a material way [22].

4. Coordination flexibility

From a marketing perspective, strategic flexibility helps a firm to redefine strategies, and to reconfigure and redeploy resources with respect to a firm's products and markets. The research conducted by [23] and [24] exemplifies the same concept. The research by [25] is an attempt to bridge the gap in the coordination flexibility from the view of collective cross-functional connections and flexibility in the usage of resources. This research explores the need of intra organisational coordination flexibility (IOCF) for marketing. IOCF deals with the readiness to accommodate deviations from the standards kept according to the strategies and to adapt to them and plan their cross functional activities more effectively and efficiently. This is considered as a critical factor for succeeding in the business.

5. Retail business

[26] establishes a systematic approach on how the expected total sales can be predicted over a certain period of time when the days designated for sales campaign are provided. The [26] says that when N campaign days were allocated over a period of M days, the total sales may not incur any additional cost when the marketing flexibility is introduced. The study is applying ABC analysis to define a Good Sales Day (GSD) and using this, the values of four expected daily campaigns were calculated that can be used to maximise the expected total sales.

As per [27], "A flexible organizational structure is a must for organization to be viewed as an open system. An informal organic and centred enterprise with a flat hierarchy, loose job descriptions, self and focused teams and cross ñ functional involvement meets this requirement quite well." The research urges the need to study how temporary flexibility can affect the attitude of retailer employees with respect to the flexible labour hours. The relation between flexibility and enterprise can have a major impact on the competitiveness and efficiency of the enterprise.

There was another research conducted to understand how small businesses and environmental uncertainty are related. It was observed that responsiveness was considered as an unavoidable factor that the small businesses with higher levels of market orientation would consider sustaining in the changing industries [28]. The aspects of technological turbulence in the field of competitive advantage and innovative ideas for product supply were also explored in the research. These factors were closely related to the higher degrees of market orientation.

6. Technology/E-Commerce/Online industry

The research conducted by [29] concluded that the different characteristics of product novelty are associated in a different way to role flexibility of R&D/marketing. Hence, careful measures should be taken while planning the product development in such cases and ensure that the integration between R&D and marketing are done effectively.

The research done by [30] presents a comprehensive view of flexibility. The ways by which online businesses accommodate technology, human resources, operations, marketing, finance and management define how they can become versatile in the industry. It is essential to concentrate on customisation and personalisation of products and services offered online as it shows how much the firms are competitive and flexible in accordance with the fluctuating customer demands. As per the research, the author suggests having flexibility in the field of market research, products/services offered, pricing, distribution and public relations which are essential components of marketing field. As the environment is facing complex transformations very frequently, an improvement in the strategic decisions' quality along with online business flexibility is really necessary to sustain in the market.

A new generation of mobile devices has significantly increased the internet's reach, resulting in the rapid growth in the field of m-commerce [31]. Along with the traditional method of accessing internet via PCs, mobile access seems to attract more people due to its versatility, making it possible to access the Internet from anywhere. This has paved its way to the popularisation of m-commerce over the existing e-commerce. The model developed by [31] discusses the relevance of optimal strategy in allocation of promotional budget between PC and mobile promotions.

7. Key issues in further research

Not only is the subject of flexibility relevant to current practice, but it is also an important topic for scholars when researching how effectively a firm works and competitive advantage required for its sustainability. Managers face significant challenges when handling external change and adapting to it [32]. The article also mentions about the economic crisis that has occurred in many countries including Europe which has undoubtedly affected the need for flexibility in firms. The other factor impacting the marketing flexibility is the changes in the consumer behaviour. By the implementation of strategic marketing flexibility in the firms, they can achieve sustainable competitive advantage.

The major challenge faced by managers in the field of flexibility is the implementation of flexibility. The managers should be able design portfolios to increase the flexibility or should consider developing strategic options that are creative and differentiable by interacting with the customers. It was also suggested that further research is needed to address how innovations can affect the market orientation and flexibility alignment.

8. General applications

[33] find that niche markets are unstable. Because of this, chances of catastrophic failures are really high and this shows the importance of market flexibility in order to survive among the competitors for long term. The author illustrates how strategic flexibility can be incorporated in niche marketing – via systematic contingency planning and portfolio development. First of all, an organisation should face an adverse unavoidable mishap, overcome its shock and continue its business as before in order to survive in long term. This equips the organisation to face any problems and its consequences in the future – both short term and long term. Any company that over invests in a single resource is unlikely to possess required flexibility, and hence it could fail.

In order to develop complex value-constellation processes capable of enhancing the optimal benefits of all stakeholders, marketing system versatility will significantly improve the value-creation process through engagement, interaction and implementation [34]. Flexibility, within specific limits determined by the expertise, resources and productivity objectives of the business, represents the main feature of participatory marketing programs, enabling interactions with and among staff, consumers and other stakeholders [35].

The research done by [36] highlights critically the gap in the relation between technology and strategic management process that organizations follow. The field of flexibility and organizational strategy is evolving and there must be proper understanding how it affects the enterprise on different management processes leading to increased competition and better performance. The author also illustrates how versatility has developed as a philosophy and takes the form of a strategic catalyst to improve the performance of the organizations.

9. Labor market flexibility

[37] says that to secure flexibility in uncertain and changing environment, firm can ensure flexibility in two ways: Flexibility on the extensive margin and flexibility on the intensive margin. The former is the employer's ability to change the number of employees and the latter is changes in working time, or in the range of tasks that employees perform. Countries with more stable labour markets should have a competitive advantage precisely when changes are more important namely those industries that suffer from high variance shocks [38].

10. Futuristic research directions

In this paper, 36 articles from 1992 to 2019 have been collected through various trusted websites, which helped in exploring the studies conducted in the field of marketing flexibility and its applications in various segments varying from finance, retail business, flexibility scale development etc. The approaches, real time case studies, applications, advantages as well as disadvantages have been summarised.

Companies operating in a complex and rapidly changing world need to build marketing flexibility to achieve sustainable competitive edge. It was observed that marketing flexibility has been applied in numerous fields. The case of Tata Nano is such an example which showed how the application of marketing flexibility helped Tata in reaching its maximum potential sales. There have been studies in

IOP Conf. Series: Materials Science and Engineering **954** (2020) 012003 doi:10.1088/1757-899X/954/1/012003

the automotive industry. Marketing flexibility scale has been developed by taking many factors into consideration which can be used to measure organisational performance and organisational flexibility.

In order to apply marketing flexibility, there must be continuous efforts to be put to explore and enhance the number of options keeping in mind the various ways to expand the current business by the firms. The scale developed for the measurement of marketing flexibility needs to be enhanced further in such a way that it is more precise and can be used in every segment irrespective of the types.

11. Conclusion

Flexibility guarantees the firm's competitive edge in normal circumstances, while it ensures stability in the event of turbulent times. The diversification of strengths, such as participating in multiple product markets, developing the ability to use multiple distribution channels, customisation and personalisation of products, proper resource utilisation and designing operating procedures to deal with environmental change, can be used to adapt the marketing flexibility in the organisations.

Through implementing steps to improve organizational marketing flexibility, business will definitely benefit by achieving both long-term sustainability and short-term profits. There is a need of more systematic studies and research in the field of marketing flexibility. Efforts are needed to be put in by the managers to understand the gap and act accordingly. This topic needs more research and if done properly, the businesses can achieve success in no time.

References

[1]. Sharma, M. K., & Jain, P. K. (2010). Revisiting flexibility in organizations: exploring its impact on performance. *Global Journal of Flexible Systems Management*, *11*(3), 51-68.

[2]. Adonisi, M., & Van Wyk, R. (2012). The influence of market orientation, flexibility and job satisfaction on corporate entrepreneurship. *International Business &Economics Research Journal,* 11(5), 477-486.

[3]. Singh, N. (2014). Marketing flexibilities: Lessons from the corporate. In *The Flexible Enterprise* (pp. 333-345). Springer, New Delhi.

[4]. Gurău, C. (2010). Marketing flexibility for new product development. *Marketing From Information to Decision*, (3), 177-183.

[5]. Asseraf, Y., Luis, L. F., &Shoham, A. (2017). The "Marketing Flexibility–Marketing Planning" Paradox and New Product's Performance: An Abstract. In *Marketing at the Confluence between Entertainment and Analytics* (pp. 1389-1389). Springer, Cham.

[6]. Lai, C. A., & Poon, J. P. (2009). Location, marketing, and the financial flexibility of nonprofit performing arts organizations in second tier cities. *Journal of Nonprofit& Public Sector Marketing*, *21*(2), 160-183.

[7]. Dobrin, P. (2008, Jul 30). The philadelphia inquirer peter dobrin column: Orchestra marketing membership flexibility. *McClatchy - Tribune Business News.*https://www.inquirer.com/philly/columnists/peter_dobrin/20080730_Orchestra_marketing_membership_flexibility.html

[8]. Singh, N., &Shalender, K. (2014). Success of Tata Nano through marketing flexibility: a SAP–LAP matrices and linkages approach. *Global Journal of Flexible Systems Management*, *15*(2), 145-160.

[9]. Shalender, K., & Singh, N. (2015). Marketing flexibility: Significance and implications for automobile industry. *Global Journal of Flexible Systems Management*, *16*(3), 251-262.

[10]. Sharma, A., Medudula, M. K., &Patro, S. (2015). Marketing flexibility interaction matrix and consumer clusters preference criteria in telecommunication sector. *Global Journal of Flexible Systems Management*, *16*(3), 295-307.

[11]. Kurt, D., & Hulland, J. (2013). Aggressive marketing strategy following equity offerings and firm value: the role of relative strategic flexibility. *Journal of Marketing*, *77*(5), 57-74.

[12]. Civelek, I. (2014). Commodity production with marketing flexibility and financial hedging. *Current Topics in Management*, *17,*119-128.

[13]. Mishra, R., & Mishra, O. N. (2018). A hybrid PCA-AHP-Multi-grade Fuzzy approach to assess marketing-based flexibility. *Marketing Intelligence & Planning*. 36(2), 213-229.

[14]. Panomjerasawat, J., Jhundra-indra, P., &Muenthaisong, K. (2015, July). Marketing flexibility orientation and marketing performance: the conceptual framework. In *Allied Academies International Conference. Academy of Marketing Studies. Proceedings* (Vol. 20, No. 2, p. 17). Jordan Whitney Enterprises, Inc.

[15]. Shalender, K., Singh, N., & Sushil. (2017). AUTOFLEX: marketing flexibility measurement scale for automobile companies. *Journal of Strategic Marketing*, *25*(1), 65-74.

[16]. Singh, N. (2011, December). Flexmark: scale for testing flexibility in the marketing system. In *11th Global Conference on Flexible Systems Management, IIM Kozhikode* (pp. 09-12).

[17]. Shalender, K. (2017). Marketing flexibility measurement scale: conceptualization, development and nomological validation. *Journal of Strategic Marketing*, *25*(7), 618-628.

[18]. Johnson, J. L., Lee, R. P. W., Saini, A., &Grohmann, B. (2003). Market-focused strategic flexibility: Conceptual advances and an integrative model. *Journal of the Academy of Marketing Science*, *31*(1), 74-89.

[19]. Cingöz, A., &Akdoğan, A. A. (2013). Strategic flexibility, environmental dynamism, and innovation performance: An empirical study. *Procedia-Social and Behavioral Sciences*, *99*, 582-589.

[20]. Shimizu, K., &Hitt, M. A. (2004). Strategic flexibility: Organizational preparedness to reverse ineffective strategic decisions. *Academy of Management Perspectives*, *18*(4), 44-59.

[21]. Logman, M. (2008). Contextual intelligence and flexibility: understanding today's marketing environment. *Marketing Intelligence & Planning*. 26(5), 508-520.

[22]. Direction, S. (2018). Unlocking the strategic value of flexibility: Functionally specialized routes to product innovation. 34(8), 13-14.

[23]. Grewal, R., &Tansuhaj, P. (2001). Building organizational capabilities for managing economic crisis: The role of market orientation and strategic flexibility. *Journal of marketing*, *65*(2), 67-80.

[24]. Sanchez, R. (1995). Strategic flexibility in product competition. *Strategic management journal*, *16*(S1), 135-159.

[25]. Dewsnap,B., Micevski,M., & Cadogan,J.W. (2012).Coordination flexibility in sales and marketing: indepth insights into an important new concept *(2012).AMA Winter Educators' Conference Proceedings*, American Marketing Association *Winter 2012, 116-117.*

[26]. Nour, S., Sumita, U., & Yoshii, J. (2015). Development of enhanced marketing flexibility by optimally allocating sales campaign days for maximizing total expected sales. *Global Journal of Flexible Systems Management*, *16*(1), 87-95.

[27]. Chaudhuri, R. (2008). Mapping flexibility in retail strategy. *Global Journal of Flexible Systems Management*, *9*(1), 11-20.

[28]. Didonet, S., Simmons, G., Díaz-Villavicencio, G., & Palmer, M. (2012). The relationship between small business market orientation and environmental uncertainty. *Marketing Intelligence & Planning*. 30(7), 757-779.

[29]. Jin, Z. (2001). The nature of NPD and role flexibility of R&D/marketing in a fast growing high-tech setting. *R&D Management*, *31*(3), 275-285.

[30]. Ozer, M. (2002). The role of flexibility in online business. *Business horizons*, *45*(1), 61-69.

[31]. Yoshii, J., &Sumita, U. (2016). Development of marketing flexibility for e-Commerce by assessing impact of mobile devices on sales with multiple classes of customers. In *Managing Flexibility* (pp. 273-295). Springer, NewDelhi.

[32]. Combe, I. (2012). "Marketing and flexibility": debates past, present and future. *European Journal of Marketing*. 46(10), 1257-1267.

[33]. Hamlin, R., Henry, J., & Cuthbert, R. (2012). Acquiring market flexibility via niche portfolios: The case of Fisher & Paykel Appliance Holdings Ltd. *European Journal of Marketing*, *46*(10), 1302-1319.

[34]. Gurău, C. (2009). Marketing flexibility in the context of the service-dominant logic. *The Marketing Review*, *9*(3), 185-197.

[35]. Călin, G. (2012). Strategies for increasing marketing flexibility: An application of the service dominant logic. *MB School. Annals of Faculty of Economics*, University of Oradea, Faculty of Economics, 4(1), 685-689.

[36]. Sharma, M. K., Sushil., &Jain, P. K. (2010). Revisiting flexibility in organizations: exploring its impact on performance. *Global Journal of Flexible Systems Management*, *11*(3), 51-68.

[37]. Beatson, M. (1995). Labour market flexibility. *Employment Department, Research Series*, (48).

[38]. Cunat, A., & Melitz, M. J. (2012). Volatility, labor market flexibility, and the pattern of comparative advantage. *Journal of the European Economic Association*, *10*(2), 225-254.

IOP Conf. Series: Materials Science and Engineering **954** (2020) 012004 doi:10.1088/1757-899X/954/1/012004

INVESTIGATIONS OF FLEXURAL AND HARDNESS BEHAVIOUR OF GRAPHENE COMPOSITES

B. Vijaya Ramnath[1,*], S. Gowtham[1], J. Melvin antony[1], D. Kumaran[1], G. Lokeshwaran[1]

[1]Department of Mechanical Engineering, Sri Sairam Engineering College, West Tambaram,
Chennai-44, Tamil Nadu, India.

[1,*] Corresponding author mail ID: vijayaramnathb@gmail.com

Abstract. In this paper, it was investigated about the flexural and hardness behaviour of graphene reinforced with hemp fiber composites. Hemp fibres with graphene of two different laminates are prepared. Epoxy resin of grade LY556 along with hardener HY 951 is blended with graphene composites through a hand layup method. The mechanical properties of graphene reinforced NaOH treated and untreated hemp fibre laminates are tested for their efficiency. Results indicate that Graphene when it is reinforced with hemp fibre, the mechanical properties namely flexural and hardness behaviour are enhanced to a greater extent. These composites have various academic and engineering fields.

1. Introduction

Qiaoling Wang et al. [1] studied and regenerating process of ischiadic nerve of rats, and procures the basics for the clinical application of nano materials and combined to isolate the extracellular matrix of ischiadic nerve of rats with graphene oxidized. By using Hummer's Method, graphene oxidized is generated by the addition of potassium permanganate to a solution of graphite, sodium nitrate and sulphuric acid. To scan and detect the functional groups in graphene oxidized of sample Fourier IR spectrum and by using the pellet method, the microcosmic morphological appearance of graphene oxidized was observed by using the scanning electron microscope. Graphene oxidized/decellularized scaffolds were prepared by the use of oscillation mixing method and operation bridging of injured ischiadic nerves was conducted [2]. Comparisons among groups were analysed by variance, and the comparison of two means was detected. Graphene oxidized could combine with the biomaterial-decellularized scaffold to repair the injury of ischiadic nerve and regenerating process of injured nerve. This provided new thoughts and theoretical & experimental bases for nano materials to be applied to clinic treatment. Azraai et al. [3] experimented that, by adding graphene nano platelets is added with the epoxy grouts and following test namely tensile, compressive and flexural test was carried out to study the effect of graphene-epoxy based composites. By comparing graphene based and epoxy grout, the results showed same strength was observed on both the cases due to the weak interface in graphene nano platelets and epoxy composites.

By performing the tensile and flexural test, the results showed Graphene based epoxy grouts is slightly lower than ones of neat epoxy grout [4]. The graphene has improved the strength, ductility of the grout, hence reducing its brittle behaviour. Wu et al. [5] investigated about the effect of mechanical properties and strain induced on crystallization behaviour on natural rubber/ graphene oxide composites. In order to analyse the oblique size of graphene oxidized, to fabricate the natural rubber/graphene oxide nano composites we use various sizes of graphene oxidized sheets. Adding graphene oxidized can increase the modulus of natural rubber. Dynamic mechanical measurement and swelling ratios (Qf/Qg) indicate that it has stronger interfacial interaction with Natural Rubber. XRD shows more effective in accelerating the strain-induced crystallization of natural rubber. The strong interfacial interaction facilitates the stress transfer and strain induced crystallization, both of which lead to the improved modulus.

Rutkowski et al. [6] observed that by applying pressure methods sintering of alumina-graphene material takes place. By making different combinations of graphene with alumina followed by processing in hot pressing or spark plasma sintering. The microstructural properties are studied under Scanning Electron Microscope (SEM). The various measurements of mechanical and elastic properties were carried by ball-on- disk method. Vijaya Ramnath et al. [7] determined the mechanical properties of intra-layer abaca–jute–glass fiber reinforced composite. The composite is made up of five layers with three layers of jute and abaca enclosed by two layers of glass fibers. The composites are manufactured with three different fiber orientations and the compositions are varied in three different proportions. The fabricated composite samples are tested to investigate their various mechanical properties.

Xusheng Du et al. [8] studied that the formation of graphene/epoxy composites was made by simple thermal reduction method from functionalized graphene sheets as a raw material. The graphene incorporated with epoxy showed better fracture energy to withstand and reduction of thermal expansion coefficient to lesser value. Partially cured G/E composites were used as interleaves in carbon fiber/epoxy (CF/E) composites and co-cured an increase in interlaminar fracture energy to a greater extent. The damage sensing capacity of the graphene and its electrical response method was demonstrated, where the change in electrical resistance increased almost linearly with crack increment. Rajesh Kumar Prusty et al. [9] investigated the effect of graphene oxide in glass fiber/epoxy composites at elevated temperature environments. Present investigation shows the effect of environmental temperature on the mechanical response of glass fiber/epoxy (GE) composite loaded with a range of graphene oxide content. Presence of GO in the composite significantly strengthens the fiber/matrix interface as well as the matrix. Good dispersion of GO is achieved up to 0.5 wt%.

Exposure to elevated temperatures deteriorates the flexural properties of GE as well as GO modified composites but the extent of degradation remains higher in case of GO-GE composites [10]. The storage modulus increases with increasing GO content up to 0.5%. Therefore, GO incorporation may not always strengthen polymeric composites rather it depends on the environmental temperature. Due to high thermal conductivity of graphene- polymer composites, it can be used as an interfaced link for the applications [11]. This paper talks majorly about modulating the graphene-polymer interfacial thermal transport. By using molecular dynamics simulations, the analysis showed that there was a covalent functionalization of graphene that could considerably reduce the graphene-polymer interfacial thermal resistance [12]. To reduce the thermal resistance, we can use butyl groups and also it is most effective.

Jijun Tang et al. [13] studied the properties of Graphene Oxide/Epoxy Resin Composites. The graphene oxide was obtained by pressurized oxidation method using natural graphite as raw materials. The epoxy resin composites were prepared by casting. The glass transition temperature decreased and the damping capacity is improved. Graphene oxide significantly enhances the toughness of the composites while it does not decrease the strength of the composites. As the graphene oxide content is 0.5 weight, the impact intensity of the composites reaches the largest value when compared with pure epoxy resin matrix, and the damping capacity is improved. Vijaya Ramnath and et al. [14] investigated the Shear and Hardness of Abaca based Hybrid composite fabricated by hand layup process. Abaca fiber has more strength than other fibers like kenaf, banana and sisal, the composite with this fiber can be suitable replacement material for automotive applications. The properties like double shear and hardness were evaluated and the result shows that the double shear properties and hardness of the hybrid composites [GFRP +Abaca+Raffia] is higher than other two combinations.

2. Materials used

2.1 Hemp fiber

In this work, Hemp fiber has been used as reinforcement that can be used for various applications like paper, textiles, clothing, biodegradable plastics, paint, insulation, biofuel, food, and animal feed. Normally, the hemp stalks would be water-retted first before the fibres were beaten off the inner herd by hand known as scrutching which is nowadays replaced by crushing rollers and brush rollers Many commercial products are made as composite sink, rope, jewellery and bio fuels.

2.2 Graphene

Graphene is an allotrope of carbon in the form of a two-dimensional hexagonal lattice, it is about 200 times stronger than the commercial steel. It efficiently conducts heat and electricity and is nearly transparent.

2.3 Epoxy and Sodium Hydroxide

In this work epoxy is used as resin matrix which is largely petroleum derived. The resin used for this study is Epoxy Resin with brand name of LY 951 which is manufactured by fibre Glass Development Corporation Company. It has superior flexural strength and adhesive characteristics and fatigue resistance. Also, sodium hydroxide (NaOH) also known as lye and caustic soda is an inorganic compound used in this work to treat the hemp fibers.

3. Methodology

The conventional methods like retting, water retting are used for chemical treatments. After treatment drying and combing of fibers were done to make it ready for fabrication.

3.1 Hand layup Method

It is an easy and simple method of fabricating natural fiber composites. Even though the production is very slow, it suitable for all types of fibers. In this work, the layup process involves manipulating each ply into shape by hand and then firmly stuck to the previous layer or mold surface leaving no air pocket between plies. This can produce high quality complex features, has relatively low start-up costs, and is highly adaptable to new parts and design changes. However, it is far from perfect, as production rates can be low and the costs of both materials and labour are sometimes high.

4. Testing of composites

The following tests were performed on the fabricated composites as per ASTM standard.

4.1 Flexural Test

Flexural tests are generally used to find the flexural modulus or flexural strength of a material. The material is laid horizontally over two points of contact and then a force is applied to the top of the material through either one or two points of contact until the sample fails. The maximum recorded force is the flexural strength of that particular sample. The specimen is prepared as per ASTM D 790 standard.

IOP Conf. Series: Materials Science and Engineering **954** (2020) 012004 doi:10.1088/1757-899X/954/1/012004

4.2 Hardness Test

Hardness is a measure of how resistant solid matter is to various kinds of compressive force. The Rockwell method measures the permanent depth of indentation produced by a force/load on an indenter.

5. Result and Discussion

The following section gives result of mechanical test carried out in this work namely flexural and hardness.

5.1 Flexural Test

The results of flexural test for treated and untreated s specimens are furnished in Table 1.

Table 1. Flexural Test values for various specimens

SPECIMEN	FLEXURAL LOAD
HEMP FIBER + GRAPHENE (untreated)	0.42 kN
HEMP FIBER (NaOH) + GRAPHENE (treated)	0.49 kN

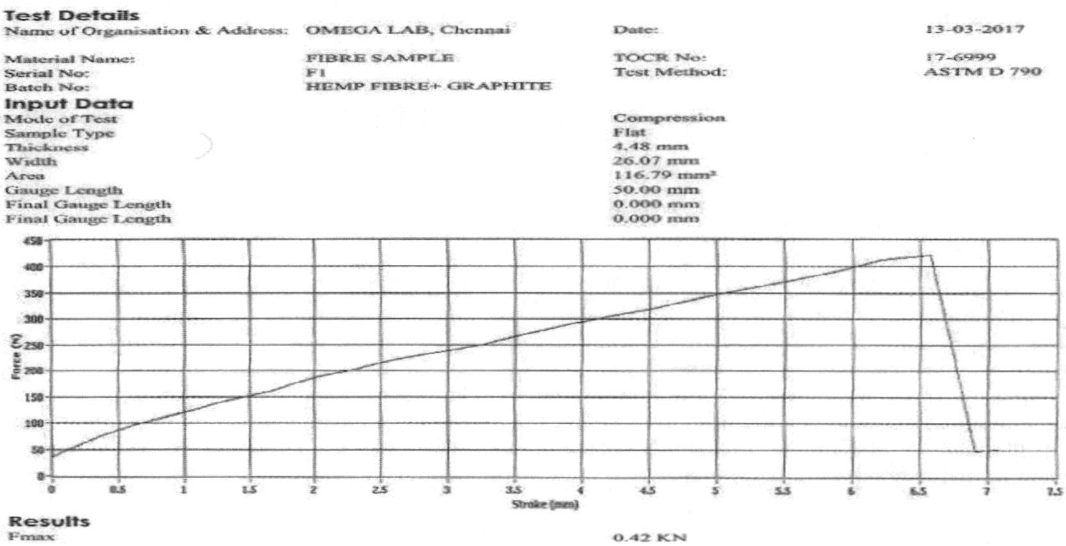

Figure 1. Result of Flexural Test (untreated)

It is seen that flexural load capacity is high for specimen 2 which has been treated with NaOH and high potential graphene. It is noted that fiber treatment increases flexural load carrying capacity which is nothing but flexural strength of the fabricated composites. The Figure 1 represents the results of specimen at various stages of flexural test of untreated fiber and the Figure 2 represents the results of specimen at various stages of flexural test of treated fiber.

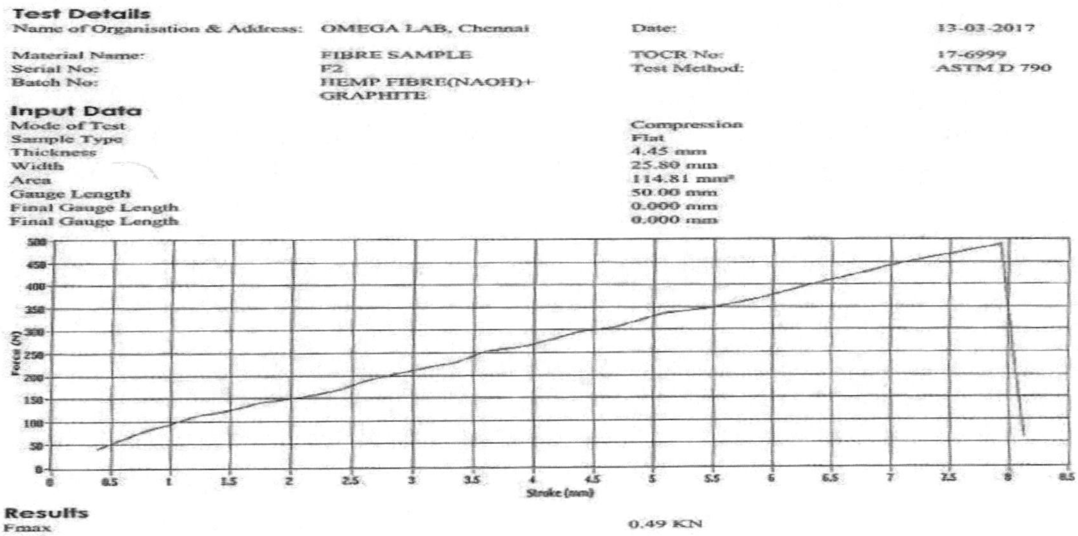

Figure 2. Result of Flexural Test (treated)

5.2 Rockwell Hardness Test

The hardness test results of treated and untreated specimens are tabulated. The test has been performed with 1/16 ball indenter with 100 Kg load. The results indicated that hardness have been partially increases by treating the fiber.

Table 2. Rockwell hardness of different specimens

SPECIMEN	ROCKWELL HARDNESS NUMBER
HEMP FIBER + GRAPHENE	32 HRB
HEMP FIBER (NaOH) + GRAPHENE	35 HRB

6. Conclusion

In this work hand lay process is used to fabricate hemp fiber reinforced natural fiber composites. Two types of composites namely treated and untreated laminate were fabricated. Flexural and hardness properties were evaluated. It was concluded that fiber treatment improves flexural and hardness of composites.

7. References

[1] Qiaoling Wang, Jinlong Chen, Qingfei Niu, Xiumei Fu, Xiaohong Sun, Xiaojie Tong, "The application of graphene oxidized combining with decellularized scaffold to repair of sciatic nerve injury in rats", Saudi Pharmaceutical Journal 25 (2017) 469–476.

[2] B. Vijaya Ramnath, S. Rajesh, C. Elanchezhian, A. Santosh Shankar, S. Pithchai Pandian, S.Vickneshwaran, and R. Sundar Rajan, "Investigation on Mechanical Behaviour of Twisted Natural Fiber Hybrid Composite Fabricated by Vacuum Assisted Compression Moulding Technique", Fibers and Polymers 2016, Vol.17, No.1, 80-87.

[3] S. N. A. Azraai, K. S. Lim, N. Yahaya, N. M. Noor, "Characterization of Mechanical Properties of Graphene Modified Epoxy Resin for Pipeline Repair", International Journal of Chemical, Molecular, Nuclear, Materials and Metallurgical Engineering Vol:10 (1) 2016.

[4] Sharavanana, B. VijayaRamnath, C. Elanchezhianb, S. Rajesh, "Experimental investigation of tensile and impact behaviour of hemp flax hybrid composite", International Journal of Mechanical and Production Engineering Research and Development", 8, (3) 2018, 549-556.

[5] X. Wu, T. F. Lin, Z. H. Tang, B. C. Guo, G. S. Huang, "Natural rubber/graphene oxide composites: Effect of sheet size on mechanical properties and strain induced crystallization behaviour", express Polymer Letters Vol.9, No.8 (2015) 672–685.

[6] Pawel Rutkowski, Piotr Klimczyk, Lucyna Jaworska, Ludoslaw Stobierski, Dariusz Zientara1, Karolina Tran, "Mechanical properties of pressure sintered alumina-graphene composites", Materia£Y Ceramiczne /Ceramic Materials/, 68, 2, (2016), 168-175.

[7] B. Vijaya Ramnath, V.M. Manickavasagam, C. Elanchezhian, C. Vinodh Krishna, S. Karthik, K. Saravanan "Determination of mechanical properties of intra-layer abaca–jute–glass fiber reinforced composite", material and design 60 (2014) 643-652.

[8] Xusheng Du, Helezi Zhou, Weifu Sun, Hong-Yuan Liu, Guangnan Zhou, Huamin Zhou, Yiu-Wing Mai, "Graphene/epoxy interleaves for delamination toughening and monitoring of crack damage in carbon fibre/epoxy composite laminates", Composites Science and Technology 140 (2017) 123-133.

[9] Rajesh Kumar Prusty, Sohan Kumar Ghosh, Dinesh Kumar Rathore, Bankim Chandra Ray "Reinforcement effect of graphene oxide in glass fibre/epoxy composites at in-situ elevated temperature environments: An emphasis on graphene oxide content" Composites: Part A 95 (2017) 40–53.

[10] B. Vijaya Ramnath, R. Sharavanan, M. Chandrasekaran, C. Elanchezhian, R. Niranjan Raja, R. Sathyanarayanan, and S. Junaid Kokan, "Experimental Determination of Mechanical Properties of Banana Jute Hybrid Composite", Fibers and Polymers 2015, Vol.16, No.1, 164-172.

[11] Kaliappan, Dr. R. Kesavan, Dr. B. VijayaRamnath, "Investigation on effect of fibre hybridization and orientation on mechanical behaviour of natural fibre epoxy composite", Bulletin of material science, Vol.44, 773-782, 2017.

[12] B. VijayaRamnath, R. Niranjan Raja, R. Sathyanarayanan, C. Elanchezhian, A. Rajendra Prasad, S. Junaid Kokan, V.M. Manickavasagam "Evaluation of mechanical properties of abaca–jute–glass fibre reinforced epoxy composite" material and design 51 (2013) 357-366.

[13] Jijun Tang, Haijun Zhou, Yunxia Liang, Xinlan Shi, Xin Yang, and Jiaoxia Zhang "Properties of Graphene Oxide/Epoxy Resin Composites" Journal of Nanomaterials Volume 2014, Article ID 696859.

[14] B Vijaya Ramnath, C Elanchezhian, V.M Manickavasagam, S Gowri Prasad, S Arvindh Swamy and R Keshav Raj, "Experimental Investigation on Shear and Hardness of Abaca based Hybrid Composites", MATE Web of conference 74, 000039 (2016).

IOP Conf. Series: Materials Science and Engineering **954** (2020) 012005 doi:10.1088/1757-899X/954/1/012005

OPTIMISING THE PARAMETERS AND INCREASING THE HARDNESS OF DUPLEX STAINLESS STEEL BY ANNEALING PROCESS IN VALVE MANUFACTURING INDUSTRIES

T.Varun Kumar[1], K.Karthick[1], K.Dhanushkaran[1], D. Jagadeesh[1]
[1]P.A. College of Engineering and Technology, Pollachi - 642002, Tamil Nadu, India

E-Mail: ercrazyvaun@gmail.com

Abstract: Duplex stainless steels (DSS) are designed to have a phase assemblage consisting of approximately 50% austenite and 50% ferrite. Such an assemblage makes these materials very attractive by providing an excellent combination of high hardness, toughness, and corrosion resistance. This project discuss about the increasing of hardness in duplex stainless steel (DSS), by suitable annealing process. Increasing of austenite increases the hardness of the material and the increasing of ferrite increases the ductile nature of the material. It is too crucial to obtain the suitable result by suitable match of annealing process. Thus the Design of Experiments were done and heating temperature of 1240^{0}C with a holding time of 4.5hrs and cooled to a temperature of 1043.9^{0}Cprovides the maximum hardness value of 311.235 BHN.

1. Introduction

The presence of austenite content in the material is responsible for hardness of the material, so the study had been done to increase the austenite content of the material. Heating temperature, cooling temperature, holding time is the three major parameters which alter the hardness when increased or decreased. A cumulative value with the increased hardness is to be found by using DOE. Annealing heat treatment method changes chemical and sometimes physical properties of a material to decrease the ductility nature and developing the hardness of the material, enhances to work longer. It is reason to heat the material above the recrystallization temperature and maintaining a suitable amount of time with suitable temperature, then cooling it into the required temperature. Heating temperature, cooling temperature and holding time are important process parameters for obtaining fine microstructure. Several articles have recommended that the steels should be heated into the temperature region between the upper and lower critical temperatures to get a fine microstructure. Parametric constraints for annealing (ASTM A995/A995M-Gr.5A (CE3MN)), the references were studied to determine the influenced process parameter and their ranges. Paramount mechanical properties can be obtained by controlling the influenced process parameters. The influenced parameters like Heating temperature, Holding time and cooling temperature were varied while the other input variables were kept constant. Based on this concept the experimental methodologies were done and the results were discussed below.

2. Experimental methodology

Based on the journals referred and experiments conducted, it was observed that the heating temperature above 1140°Chave appreciable improvement in the mechanical properties due to needle ferrite formation hence, heating temperature above 1140°C is taken into account. Lower shot velocity results in porosities, whereas increased holding time also leads to improvement of hardness. Therefore, holding time from 3.5 to 4.5hrs was taken into account. Cooling temperature, when the specimen is heated in the furnace there will be a austenite in the specimen so when the cooling is done there will be a formation of the ferrite content which gives ductility to the material so when the cooling temperature is increased above 1045°C there will be a less ferrite formation, which results in a increased hardness.

The process parameter range considered for conducting the experiments are listed below,

Heating temperature: $1090 \leq 1140 \geq 1240$

Holding time: $3.5 \leq 4 \geq 4.5$

Cooling temperature: $995 \leq 1045 \geq 1095$

The structure of the selected design of experiments should be such that the experimental investigation complies with full factorial design. This means that every possible combination between the levels of the various factors is analyzed. For the 3^3full factorial design, heating temperature, Holding time and cooling temperature were selected as factors, and each was run at three levels and the experiments were carried out by varying the process parameter based on the full fractional approach. Heating temperature, holding time and cooling temperature are the major three input parameters which are taken into account for this project work. The process parameters and the change in hardness according to the change in process parameters were studied and their changes are plotted in the graph.

2.1 Heating Temperature v/s Hardness

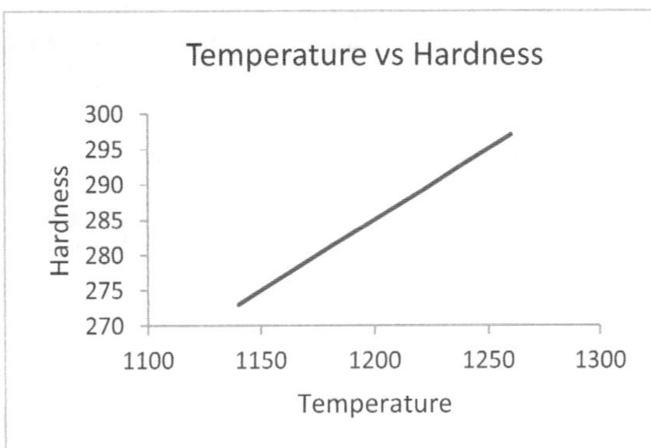

Figure 1. Heating Temperature v/s Hardness

2.2 Hardness v/s Time

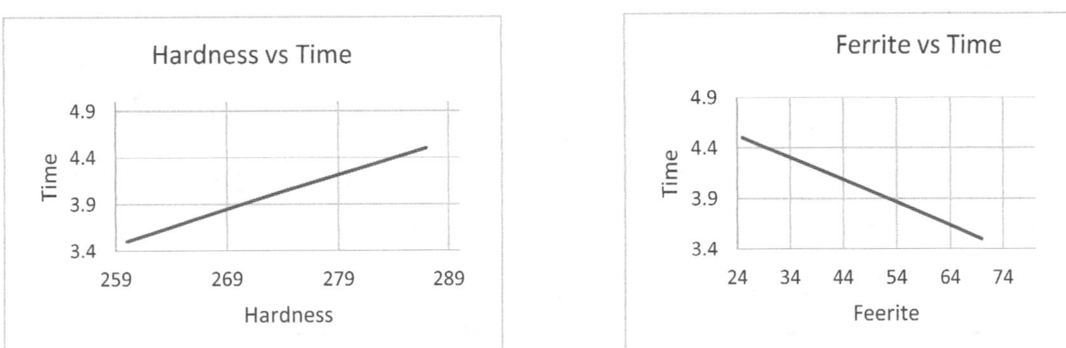

Figure 2. Hardness v/s Time

IOP Conf. Series: Materials Science and Engineering **954** (2020) 012005 doi:10.1088/1757-899X/954/1/012005

2.3 Cooling Temperature v/s Time

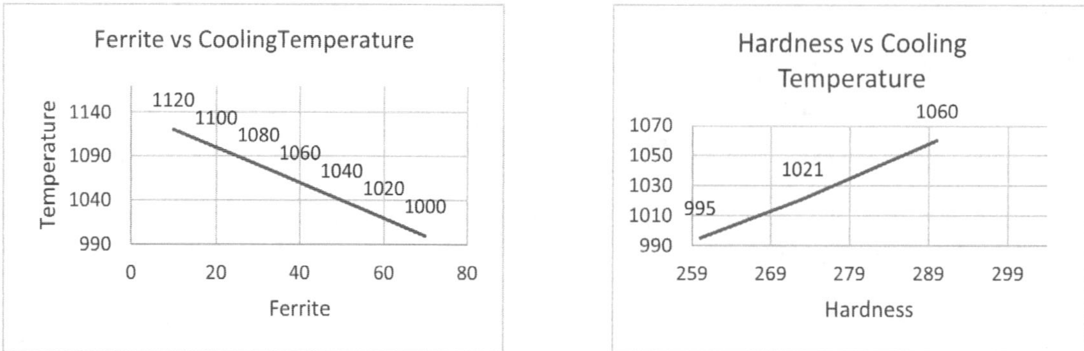

Figure 3. Cooling Temperature v/s Time

Table 1. Brinell Hardness Value

Std	Run	Factor 1 A:Heating temperature DEGREE	Factor 2 B:Holding time hrs	Factor 3 C:Cooling time DEGREE	Response 1 Hardness BHN
4	7	1190	4	1045	276
5	11	1190	4	1045	278
3	12	1190	3.5	1045	276
14	1	1190	4	1045	278
12	2	1240	4	1045	290
7	3	1240	4.5	995	304
11	4	1140	3.5	995	260
8	8	1190	4.5	1045	291
15	9	1190	4	1045	279
9	10	1190	4	995	261
13	14	1140	4	1045	275
10	15	1140	4.5	1095	285
6	5	1240	3.5	1095	301
1	6	1190	4	1045	280
2	13	1190	4	1095	282

By collaborating the above values, different combination of parameters were obtained and according to the parameters, hardness test is conducted in brinell hardness machine and the best 15 results are plotted above. According to the hardness value obtained from the brinell hardness machine and by the predicted data on the hardness and according to the change in the process parameters the DOE will be conducted.

3. Results and Discussion

3.1 Statistical Analyses

The Table 3.1 statistical analysis shows that F- values attained 112.5, 112.5 and 220.5 for heating temperature, holding time and cooling time respectively. The parameters attained highest value will influence the responses utmost. From the Table 1, it has been observed that the F statistical value attained about 44.34 for cooling time. From the statistical analysis, it is confirmed that cooling time influences the hardness utmost compare to heating temperature and holding time.

Table 2. Statistical Analyses for Hardness

Parameters	Sum of	df	Mean	F-value	p-value
A-HEATING TEMPERATURE	112.5	1	112.5	22.62421	0.005075
B-HOLDING TIME	112.5	1	112.5	22.62421	0.005075
C-COOLING TIME	220.5	1	220.5	44.34345	0.001152
AB	33.33333	1	33.33333	6.70347	0.04889
AC	0.333333	1	0.333333	0.067035	0.806025
BC	75	1	75	15.08281	0.011599
AÂ²	112.509	1	112.509	22.62603	0.005074
BÂ²	149.4321	1	149.4321	30.05141	0.002755
CÂ²	51.58597	1	51.58597	10.37415	0.023442

HARDNESS = +2024.95049 - 7.10300 Heating Temperature - 151.38235 Holding Time + 4.88482 Cooling Time + 0.200000 Heating Temperature * Holding Time + 0.000200 Heating Temperature * Cooling Time - 0.300000 Holding Time * Cooling Time + 0.002624 Heating Temperature^2 + 30.23529 Holding Time^2 - 0.001776 Cooling Time^2.

3.2 Interactive and direct effects of Heating temperature, Holding Time and cooling temperature on annealing process

The effect of each parameter on responses can be analyzed graphically using the interaction surface plots. It shows the relationship of process parameters on responses in way of linear and non-linear function. The interaction plots helps to understand the effects of variation considering one process parameter simultaneously with constant other two parameter. The interaction and directs effect figures obtained for the response hardness is shown in

IOP Conf. Series: Materials Science and Engineering **954** (2020) 012005 doi:10.1088/1757-899X/954/1/012005

Figures 6.1(a-c). There are several observations made from the interaction direct effect figures are discussed in the following. It has been observed that increase in heating temperature increases the hardness drastically because, the metal after heating above 1140°C there will be the needle ferrite formation while cooling which results in the increase of hardness. While considering the cooling temperature, the metal which is heated to a certain temperature will be 100 percent austenite and while cooling there will be the formation of ferrite content. So while increasing the cooling temperature above 1045°C there will be less ferrite which increases the hardness of the material. However, an F- statistical value reveals that the process parameters holding time and cooling temperature also have a considerable influence on hardness. At maximum heating temperature, the improved results can be obtained by using the following. Mechanical properties are influenced by the increased heating temperature with holding time and cooling temperature. From the table below DOE for the different combination is obtained, from this the heating temperature of 1240°Cwith holding time of 4.5hrs and cooling temperature of 1040.3°Cyields the maximum results of 310.9483BHN with a desirability of 0.986.

Table 3. Result Analyses for Hardness

Number	Heating temperature	Holding time	Cooling time	Hardness	Desirability	Desirability	
1	1240	4.5	1040.333	310.9483	0.986032	1	Selected
2	1240	4.5	1048.635	311.5447	0.985673	1	
3	1240	4.498528	1044.663	311.2085	0.985217	1	
4	1239.854	4.499998	1046.422	311.335	0.985119	1	
5	1240	4.5	1037.727	310.7107	0.984954	1	
6	1240	4.499999	1050.451	311.6424	0.984854	1	
7	1239.569	4.499993	1044.501	311.0577	0.98299	1	
8	1240	4.5	1054.185	311.8068	0.982365	1	
9	1240	4.495273	1042.704	310.8799	0.982153	1	
10	1240	4.495153	1045.614	311.0898	0.982106	1	
11	1240	4.494442	1048.682	311.2474	0.980814	1	
12	1239.39	4.5	1039.45	310.5594	0.980622	1	
13	1240	4.498445	1055.221	311.7625	0.980212	1	
14	1240	4.5	1057.455	311.91	0.979309	1	
15	1239.047	4.499999	1046.89	310.9531	0.978389	1	
16	1240	4.490207	1047.967	310.9764	0.97733	1	
17	1240	4.499999	1029.349	309.7829	0.977317	1	
18	1238.509	4.5	1048.469	310.7763	0.973568	1	
19	1240	4.476397	1052.481	310.5041	0.963926	1	
20	1240	4.497986	1068.305	311.8831	0.962031	1	

IOP Conf. Series: Materials Science and Engineering **954** (2020) 012005 doi:10.1088/1757-899X/954/1/012005

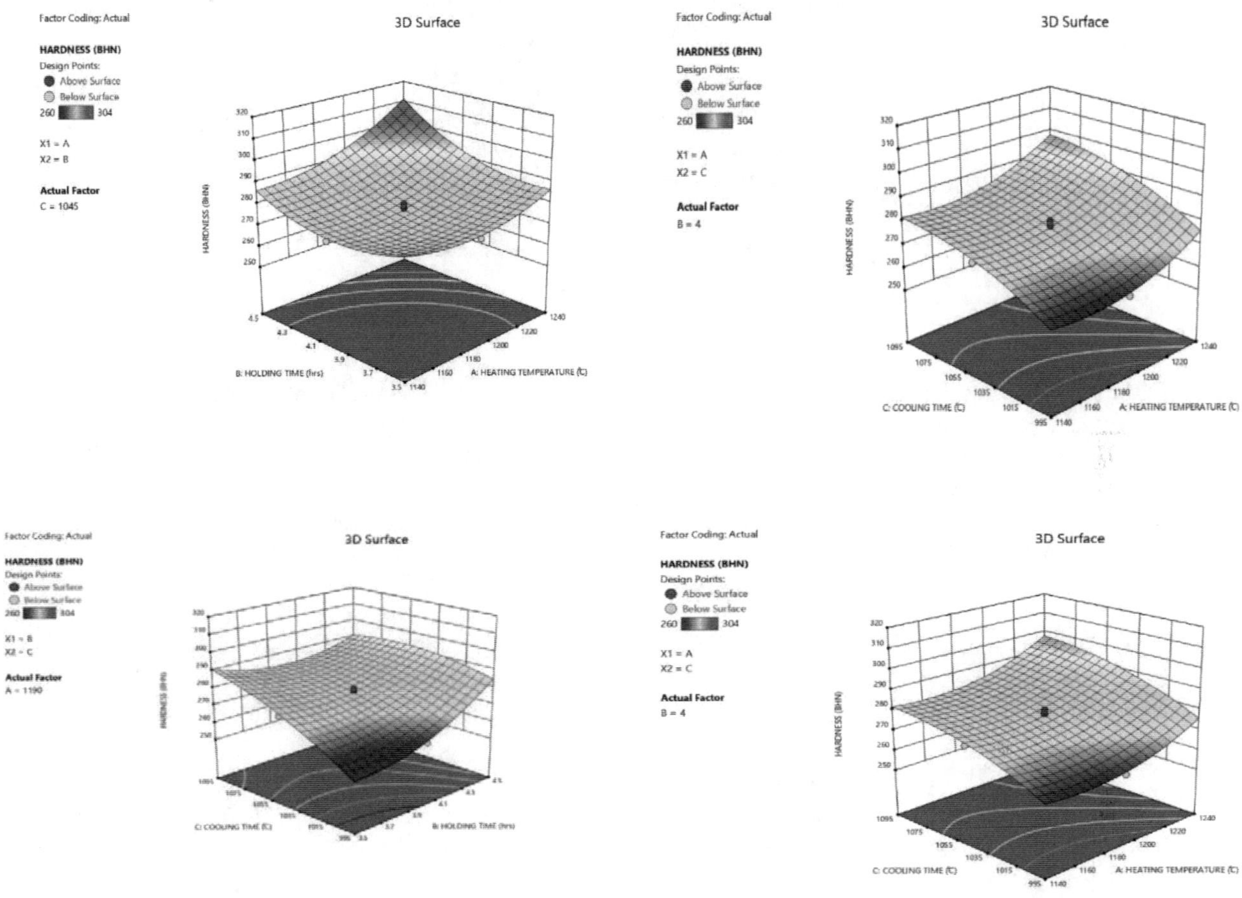

Figure 4. a-c Interactive surface plots for hardness

3.3 Microstructure for Astm A995/A995M

A metallurgical microscope is used to study the changes in the morphology of the dendritic structure of the annealing in ASTM A995/A995M Duplex Stainless steel under the selected process parameters. Microstructures of 100X magnification were shown in Figure 5.With the increment of Heating temperature, a significant elimination in micro-porosity and other defects were noted through the quantitative metallographic analysis as discussed earlier. The casted ASTM A995/A995M microstructure region shows two different colors with white and black. The white region is Austenite and a black area is Ferrite, thus the ferrite changes into needle like structure while the temperature increases which results in the increased hardness. The dendritic branches act as the load carrying member. Mechanical properties of the alloy were increased by the refinement of dendritic branches. The alloy samples of ASTM A995/A995M at a heating temperature of 1240^0c with a holding time of 4.5 hrs and a cooling temperature of 1043.9080c has the increased austenite content which results in improved hardness.

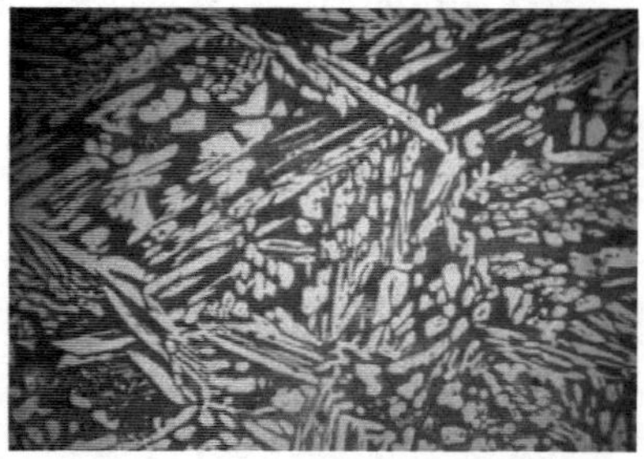

Figure 5. Microscopic view of DSS in ×100 Zoom

3.4 Validation

The validation was done to observe the deviations of experimental values and predicted values. The deviation between the experimental and predicted value gives a details of residuals (e). The Figure 6 shows the deviation of experimental values and predicted values; higher the deviations of data points from mean line higher the residuals. The figure shows the data points closer to the mean line; thus the residuals become lesser ad increases the accuracy of prediction. The Table 2 shows the deviation in prediction of hardness for desirability analysis results; it is observed from the table, the 'e' falls less than 2 for hardness thus the observation of lower residuals reveals that the predicted value is highly acceptable and accurate.

Table 4. Validation

Sl.no	Injection pressure (A)	Shot Velocity (B)	Furnace Temperature (C)	Brinell hardness (MH) (Predicted)	Brinell hardness (MH) (Experimental)	Residual (MH)
	Kgf/cm^2	m/s	°C	BHN	BHN	e
1.	208.981	0.6	677.83	125.48	124.23	1.25

4. Conclusion

Thus from the above chapters we can able to conclude that the austenite percentage in the duplex stainless steel is the only factor which determines the hardness of the material. Since this project is to improve hardness of the material, so the increase of austenite content in the material will automatically increases the hardness. Thus from this research work, on conducting DOE and various experiments regarding hardness, the heating temperature of 12400C with holding time of 4.5hrs and cooled up to 970^{0}C provides the maximum hardness of 311 without changing the material. So we suggest these parameters to the valve manufacturing industries to improve hardness of their product.

5. References

[1] Eric Johnson, Yoon-Jun Kim, L. Scott Chumbley and Brian Gleeson 2004. Initial phase transformation diagram determination for the CD3MN cast duplex stainless steel, *Elsevier Ltd*, Vol.**50**, pp.1351-1354.

[2] P. I. Popoola1, O. S. Fatoba1, O. M. Popoola and S. L. Pityana 2016. The influence of heat treatment and process parameters optimization on hardness and corrosion properties of laser alloyed X12CrNiMo steel, *Springer Science*, Vol.**8**, pp.579-589.

[3] S.M. Mahbobur Rahman, Kazi Ehsanul Karim, and MD. Hasan Shahriar Simanto 2016. Effect of heat treatment on low carbon steel: an experimental investigation, *Applied mechanics and materials* Vol. **860**, pp 7-12.

[4] Maoyu ZHAO and Qianwang CHEN 2013. Parameters optimization of low carbon low alloy steel annealing process, *Acta Metall. Sin. (Engl. Lett.),* Vol.**26**, No.2, pp.122-130.

[5] Yoon-Jun Kim, L. Scott Chumbley and Brian Gleeson 2008. Continuous cooling transformation in cast duplex stainless steels CD3MN and CD3MWCuN, *Journal of materials engineering and performance*, Vol.**17**, pp.234-239.

[6] J.O. NILSSON, P. KANGAS, T. KARLSSON and A. WILSON 2000. Mechanical properties, micro structural stability and kineticsof s-phase formation in 29Cr-6Ni-2Mo-0.38N super duplex stainless steel, *Metallurgical and materials transactions*, Vol.**31**, pp.35-45.

[7] Marcelo Aquino MARTORANO, Caio Fazzioli TAVARES and Angelo Fernando PADILHA 2012. Predicting delta ferrite content in stainless steel castings, *ISIJ international*, Vol.**52**, No.6, pp.1054-1065.

[8] Rastee D. Koyee, Uwe Heisel, Rocco Eisseler and Siegfried Schmauder 2014. Modelling and optimization of turning duplex stainless steels, *Journal of Manufacturing Processes*, Vol.**16**, pp.451-467.

[9] Jan Nachazel, Petr Martinek, Pavel Podany and Ivana Polakova 2015. Effect of heat treatment on the microstructure of duplex stainless steel, *Applied mechanics and advanced engineering research*, Vol.**751**, pp.9-14.

[10] Sun Jianliang, Li Shuo, Qiu Chouwu and Peng Yan 2018. Numerical and experimental investigation of induction heating process of heavy cylinder, *Applied thermal engineering*, Vol.**134**, pp.341-352.

[11] Cheng-Jiang Huang, David J. Browne and Shaun McFadde 2005. A phase-field simulation of austenite to ferrite transformation kinetics in low carbon steels, *Elsevier Ltd*, Vol.**54**, pp.11-21.

[12] M. Rosso, I. Peter and D. Suani 2013. About Heat treatment and properties of duplex stainless steels, *Journal of achievements in materials and manufacturing engineering*, Vol.**59**, pp.25-36.

[13] Jithin M, Anees Abdul Hameed, Ben Jose and Anush Jacob 2015. Influence of heat treatment on duplex stainless steel to study the material properties, *International journal of scientific and technology research*, Vol.**4**, issue.2, pp.291-293.

IOP Conf. Series: Materials Science and Engineering **954** (2020) 012005 doi:10.1088/1757-899X/954/1/012005

[14] Varun Kumar T, Manoj Mohan Prasad V, Santhosh D, Prasanth C, Ranjith K 2020. Evaluation of mechanical and interfacial properties of carbon fiber reinforced polymer (CFRP) composite materials, *Material Today,* Proc.21, 477-482.

SMEs Perception towards Adoption of Measurement System Analysis Strategy: A Case Study in Pump Manufacturing Industry

T.Varun Kumar[1*], G.Ramya[2], P.Kowsalya[2],B.Shanmuga priya[3] and P.Tharani[2]

[1]Professor, P.A.College of Engg and Tech, Pollachi - 642002,Tamil Nadu, India.
[2]Student, P.A.College of Engg and Tech, Pollachi - 642002,Tamil Nadu, India.

E mail: ercrazyvarun@gmail.com and priyadaddu87@gmail.com

Abstract. Every industry from large scale to small scale is nowadays seeking for their continuous improvement. It is necessary to compare the performance of the measurement processes with the standard systems. In a typical scenario, one of the two processes represents an attempt to improve the other process, and one must determine if the attempt has been successful. Since repeatability and reproducibility (R&R) studies are often used to determine the adequacy of each measurement process separately, it is suitable to compare the results of these studies to determine if there is any difference in the variability of the two processes. In this paper, we consider R&R studies with two factors of parts and operators. We consider situations where the operator factor can be either random or fixed. Each operator makes repeated measurements on each part in a sample of process output and the different operators are used for each measurement process and the parts used for measurements are the same. A ratio commonly used to determine the adequacy of a single measurement process is used to compare the variability of the two measurement processes. Two statistical tests a result for differences in their ratio depends on their confidence intervals are proposed and compared using computer simulation. Finally, our study enables delegates to assess the performance of new and existing measurement systems.

Keywords: Measurement System Analysis (MSA), Analysis of Variance (ANOVA), Small Medium Enterprises (SMEs), Precision to Tolerance Ratio (PTR).

1. INTRODUCTION

Nowadays, every production industry faces difficult competition due to their increased customer demand [1]. Every production industry has to actively perform their improvement to satisfy the customers. In our study, quality is one of the important factors, not only for our research; generally for the entire production unit it plays a major role. This paper presenting a case study of measurement system analysis applied in small and medium scale pump industry [2]. Measurement system analysis is a system which plays a major role in lean six sigma. This measurement system is different from the traditional approach; consist of the measurement method, measurement process, measured part and the measurement environment [3]. Sources of variations in measurement are one of the main sources of variation in quality. In production industries, quality problems are solved by identifying and correcting inaccurate data and measurement process. The ability to make right decisions depends on the availability of measurement process. In our research, we consider Gage R&R studies with two main factors one is parts and the other is operators [4]. Each operator makes the repeated measurements on each part in the samples and also three different operators are used for each measurement process and the parts to be measured is the same component. Outcome of the results of simulation are used to provide the recommendations for the effective instruments. Measurement system analysis is recommended to fulfil the requirements of the client and also it is used to ensure that company owns an acceptable measurement system [5].

2. MEASUREMENT SYSTEM ANALYSIS

Measurement system analysis (MSA) is a scientific tools used to determine the amount of total variations from the measurement system [6]. The purpose of the measurement system analysis is to qualify a measurement system for quantifying its precision, accuracy and its stability. MSA is a quality improvement initiative which has the ability to assess the measurement system and to detect the differences in process variables.

Generally measurements are used to guide a decision, which follows whether the measurement errors are made by the instrument or equipment or by the operators [7]. The level of variations attributed to the operator or parts and the operator has the ability to measure each individual part is experienced during the measure phases and is presented in Figure 1. Total variations of measurement system analysis and its causes are classified into two types and they are process and measurement variations. In this study, the focus is mainly on the measurement variation and also, the variations obtained due to the process.

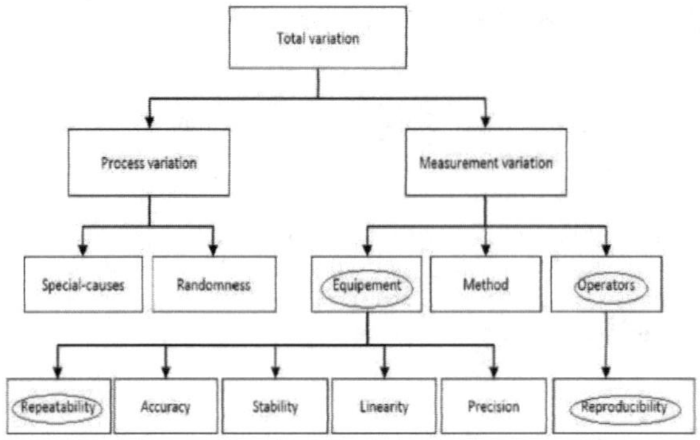

Figure 1. Total Measurement Variations

2.1 Bias

The difference between the average value of all measurements and true value is called as bias.

2.2 Stability

Stability is the capacity of a measurement system to produce the same values over time when measuring the same sample. It statistically monitors the state of measurement system over a period of time and shows the variations in the readings obtained due to wear and tear as it gets into their use.

2.3 Linearity

Linearity is the difference in bias values over the range of instrument capability. The difference in the accuracy values of a gage through the expected operating range of the gage. It may be classified as a good linearity and bad linearity.

2.4 Repeatability

Repeatability is the variation in measurements obtained with measurement instrument when used several times by same appraiser while measuring the identical characteristics on the same part. It also states that when repeated measurements are made of same conditions such as same operator, same units, same setup and same environmental conditions.

2.5 Reproducibility

Reproducibility is the variations in the average of measurement made by the different operators using the same measuring instrument on the same identical part. It is caused by the measurement made by the appraisers.

3. GAGE R&R

In MSA gage R&R plays a major role and it is the combined estimate of measurement system of repeatability and reproducibility [8]. This paper based on R&R analysis by using Analysis of Variance [ANOVA] method. In ANOVA method we could separate the repeatability and reproducibility interprets easily. In the Gage R&R, generally there are two to three operators are involved to measure the process outputs [9]. In our study, we utilize 3 operators to participate and they were to measure 10 parts randomly and 3 trials taken by each operator.

This paper is based on gage R&R analysis to know Precision to Tolerance (P/T ratio) and Analysis of Variance (ANOVA) method using Minitab software. According to the Automotive Industry Action Group (AIAG) standard [10], the least count of the measuring instrument should be 1/10th of the tolerance. AIAG standard says that precision to tolerance ratio is below 10% (acceptable), average between 10% to 30% ratio is (conditionally acceptable) and if it is above 30 % (not acceptable). Table 1 shows the AIAG standard tolerance assumption criteria.

Table 1. AIAG Standard Tolerance Assumption Criteria

Classification	Criteria		
	Accepted	**Caution**	**Rejected**
%contribution	<1%	1-10%	>10%
%Tolerance	<10%	10-30%	>30%
NDC	>10	4-9	< 4

4. EXISITING CASE STUDY

A measurement system of gage R & R study was done in small and medium scale pump industry [11]. The manufacturing industry produces the monoblock pumps, submersible pumps, etc. Here we concentrate on the single section of monoblock pump and process studies are made and data's are collected. The data were collected from 3 operators; they measured 10 parts (outer diameter of stamping). Each part was measured 3 times at different interval of times, and finally there were 90 readings were collected, which has shown in the table (II, III and IV). Here we take the stamping component shown in Figure 2 is taken for our measurement and the measured dimensions of the parts to be 105mm within the tolerance of ± 0.9 mm.

Figure 2. Stamping Component and calibration made on Vernier calliper with least count 0.05 mm

4.1 Beginning of the Project

According to the Automotive Industry Action Group (AIAG) standards, for measurement process requires three operators measuring 10 items in three times at different intervals. All inspections had to be measured under conditions of the same instrument, with same setup at the same environmental condition, and the sample items had the same characteristics [12]. Each operator is measuring the each sample item three times in random sequence. For evaluating the repeatability of the measurement system, here the operator was for measuring is the same person.

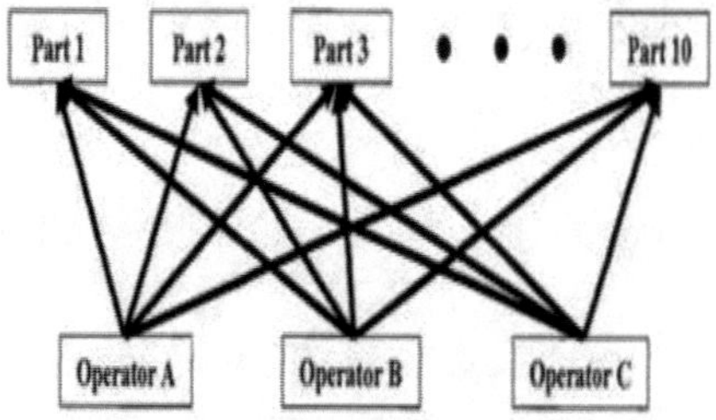

Figure 3. Gage R&R Methodology of Reading

The above Figure 3 shows the gage R&R methodology [13], that how the three appraisers are taking the measurement reading for the 10 sample parts without knowing each other's result. After all this data's are collected, then the measurement data would be entered in software (worksheet made) program and analyzed using excel software. All the mathematical calculation using respective formulas are in the worksheet and finally, the results would be assessing the capability of the measurement system.

4.2 Readings before MSA System

The initial measurements were obtained by the operators with a Vernier calliper (Least Count=0.05mm). The values are in the Table 2 for operator A, Table 3 for operator B and Table 4 for the operator C.

Table 2. Data Collected From Operator A

	TRIAL 1	TRIAL 2	TRIAL 3
PART 1	105.20	105.10	105.20
PART 2	105.30	105.20	105.30
PART 3	105.50	105.50	105.40
PART 4	105.50	105.50	105.50
PART 5	105.50	105.60	105.60
PART 6	105.80	105.70	105.70
PART 7	105.90	105.80	105.90
PART 8	105.90	105.80	105.90
PART 9	105.90	105.90	105.80
PART10	105.70	105.60	105.70

Table 3. Data Collected From Operator B

	TRIAL 1	TRIAL 2	TRIAL 3
PART 1	105.10	105.00	105.10
PART 2	105.30	105.20	105.30
PART 3	105.40	105.50	105.50
PART 4	105.40	105.60	105.60
PART 5	105.60	105.60	105.60
PART 6	105.70	105.80	105.70
PART 7	105.70	105.80	105.80
PART 8	105.90	105.80	105.90
PART 9	105.90	105.80	105.90
PART10	105.80	105.60	105.60

Table 4. Data Collected From Operator C

	TRIAL 1	TRIAL 2	TRIAL 3
PART 1	105.00	105.00	105.10
PART 2	105.30	105.30	105.30
PART 3	105.50	105.50	105.50
PART 4	105.60	105.50	105.50
PART 5	105.60	105.60	105.60
PART 6	105.80	105.70	105.70
PART 7	105.80	105.80	105.80
PART 8	105.90	105.90	105.90
PART 9	105.90	105.90	105.90
PART10	105.60	105.60	105.60

The readings are made by the three different operators and the individual readings are noted separately on the Tables 2, 3 and 4 respectively. Total 90 readings were obtained from the measurement system and these readings were made at different interval time [14]. One thing is should be kept, that while measuring the parts, it should be randomly picked and the appraiser does not remains the existing reading. This data's were used by the mathematical formulas to find the average, ranges and precision to tolerance ratio (P/T Ratio). This all values are obtained by mathematical symbols or framing formulas in the worksheets were by directly uploading the data's and obtain the values of the final reading. Table 5 shows the Gage R&R Study of reading using the vernier calliper with 0.05 mm least count.

Table 5. MSA- Gage R&R study for Vernier caliper (Least Count 0.05 mm)

MEASUREMENT SYSTEM ANALYSIS -GAUGE REPEATABILITY & REPRODUCIBILITY STUDY

Product:	Pump	Instrument:	Digital Caliper					
Date:	22.08.2014	Least count:	0.05mm	Specification:	USL:	105.9 mm	LSL:	105 mm
Process:	Stamping	Industry:	JAY PUMPS Pvt.Ltd-CBE					

Sample	Operator A					Operator B					Operator C				
Parts	Trial 1	Trial 2	Trial 3	X bar	Range	Trial 1	Trial 2	Trial 3	X bar	Range	Trial 1	Trial 2	Trial 3	X bar	Range
1	105.20	105.10	105.20	105.1666667	0.1	105.10	105.00	105.10	105.0666667	0.1	105.00	105.00	105.10	105.0333333	0.1
2	105.30	105.20	105.30	105.2666667	0.1	105.30	105.20	105.30	105.2666667	0.1	105.30	105.30	105.30	105.3	0
3	105.50	105.50	105.40	105.4666667	0.1	105.40	105.50	105.50	105.4666667	0.1	105.50	105.50	105.50	105.5	0
4	105.50	105.50	105.50	105.5	0	105.40	105.60	105.60	105.5333333	0.2	105.60	105.50	105.50	105.5333333	0.1
5	105.50	105.60	105.60	105.5666667	0.1	105.60	105.60	105.60	105.6	0	105.60	105.60	105.60	105.6	0
6	105.80	105.70	105.70	105.7333333	0.1	105.70	105.80	105.70	105.7333333	0.1	105.80	105.70	105.70	105.7333333	0.1
7	105.90	105.80	105.90	105.8666667	0.1	105.70	105.80	105.80	105.7666667	0.1	105.80	105.80	105.80	105.8	0
8	105.90	105.80	105.90	105.8666667	0.1	105.90	105.80	105.90	105.8666667	0.1	105.90	105.90	105.90	105.9	0
9	105.90	105.90	105.80	105.8666667	0.1	105.90	105.80	105.90	105.8666667	0.1	105.90	105.90	105.90	105.9	0
10	105.70	105.60	105.70	105.6666667	0.1	105.80	105.60	105.60	105.6666667	0.2	105.60	105.60	105.60	105.6	0
			XA bar	105.5966667				XB bar	105.6				XC bar	105.59	
			RA		0.09			RB		0.11			RC		0.03

1	Average of all ranges (R Bar)	0.07667
2	Diff.among operators (R diff)	0.01
3	sigma (o) repeatability for n=3 trials	0.04468
4	sigma (o)reproducibilibity, n=3 oprs	0.00777
5	sigma gage (o)2	0.00206
6	Precision Tolerance ratio (P/T)	0.30232

As per the MSA Standard learned from AIAG and industrial practices ,the (P/T) ratio should be 0.1 % to 0.3%

The primary purpose of this worksheet methodology is the graphical approach to determine whether the measurement process is adequate to measure the manufacturing process variations. From the worksheet, we have formulated the table with mathematical signs and by entering the iteration values, it just evaluate and give the final gage results. This table is simple and easier compared to the mathematical formulas by applying and finding the results for each and every iteration.

From the result, the ranges and the average values are not within the control limits by using the vernier calliper with least count of 0.1mm and precision to tolerance ratio (P/T ratio) is 30% where the system is conditionally accepted and need some improvements according to the AIAG standard reference manual. Therefore, the system clearly identifies the reason for the lack of control in process variability [15]. This problem can be remedied, by changing the system which has the ability to detect the variation within the parameters by increasing the discrimination of measurements in the gage R&R [16].

4.3 ANOVA Method

MINITAB is one of the statistical software, reliable and easy to use and it has been used by many successful companies for their deployment of quality control functions. In MINITAB software by using ANOVA method, we can evaluate the repeatability and reproducibility of the measurement system [17]. Analysis of the variance (ANOVA) is a statistical method that tests the contribution of control factors and interactive effects between the different factors. To conduct MSA [18], for our test, there are 3 operators to participate and to measure 10 parts randomly and 3 trials taken by each operator. While doing this, we must ensure that the parts are given to the appraisers in a random order which they do not remember their earlier result. They also should not see other appraisers measuring the parts. Once the measurements are completed, the data is analysed using the Minitab software.

The gage R & R study (crossed) can be made for each operator [19], measured each part by using Minitab software is analysed. Result of the collected data illustrated in figure 5 would be interpreted in the form of percentage of contribution [20]. If this percentage of contribution is below 10% the system would be acceptable. If the contribution is in the range of 10% to 30%, the measurement systems seem to be conditionally acceptable depending on the application, cost of gage and cost of repairs, etc. If it is beyond 30%, the measurement system was no fit for the system and need some improvements. Here, our study works affirms that the current measuring system could not produce the precise and accurate data. The graphical results shows that the part to part variation of the percentage contribution are higher in the current system and the appraisers are measuring the system of range and average values are not within the control limits. From the results, it clearly shows the remedial actions to be made for further improvement of product variations among the appraisers and the process. The Figure 4 shows the graphical results of calibration of vernier calliper with least count of 0.05 mm as the standard measuring instrument.

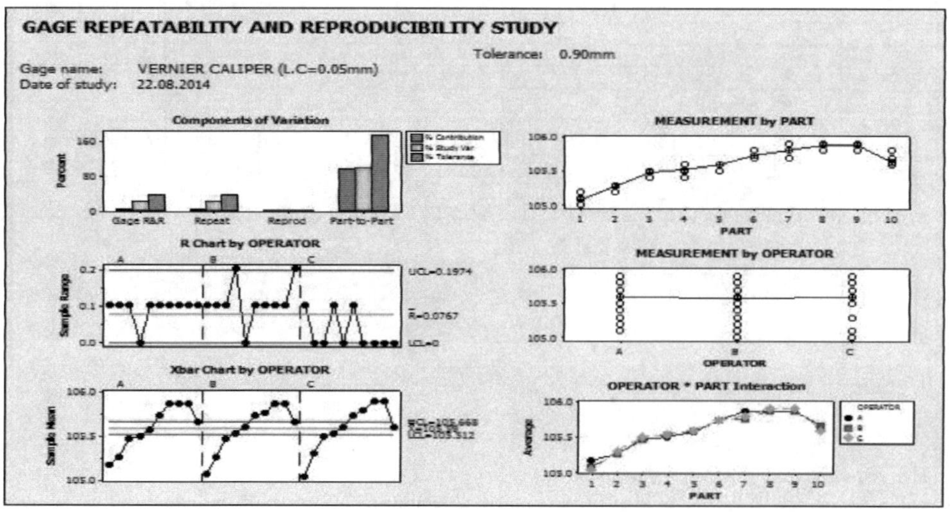

Figure 4. Graphical results for vernier calliper (least count=0.05 mm)

4.4 Gage R&R Study- ANOVA Method: Initial Data Results

By the outcome of the ANOVA results, we could clearly see the variations and the percentage of contributions are higher. Here, the numbers of distinct categories (NDC) are not up to the acceptable condition. NDC indicate the number of groups within the process data that the measurement system concern. According to the AIAG standard assumptions, the NDC should be more than 10 is good and acceptable system. If it is between 4 to 9 categories, the group seems to be some good and bad and further some measures to be made. If NDC is < 4 it is useless for certain decision making. According to our existing measuring system, using the vernier calliper instrument with least count 0.1mm here the number of distinct categories is 6 and the results obtained in the current measuring system clearly shows further improvements should be made for quality system. Figure 5 shows the Gage R&R study for the two way ANOVA method [21].

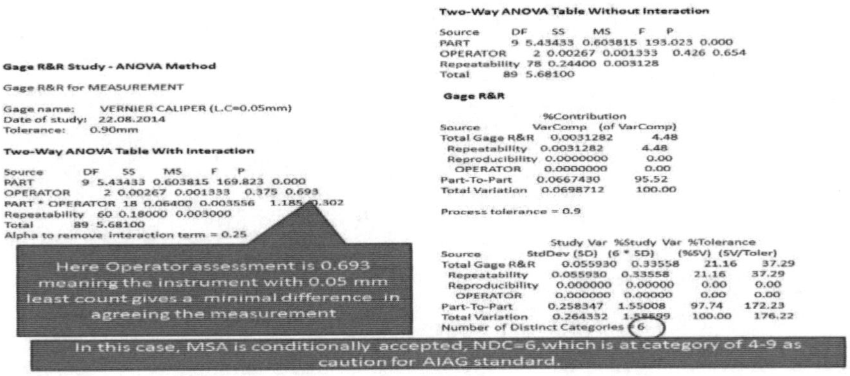

Figure 5. Gage R&R study – Two way ANOVA Method

5. OUTCOME OF THE PROJECT

This problem can be remedied, by changing the system which has the ability to detect the variation within the parameters by increasing the discrimination of measurements. The decision of replacing the vernier calliper (least count=0.05 mm) with a vernier calliper (least count=0.02 mm). This measurement system will have adequate discrimination due to apparent resolution of vernier with 0.02 mm least count. With this small relative measurement the average ranges are seems to be within the control limits and process variations are reduced. The measurements made by the vernier calliper (least count=0.02 mm) with gage Repeatability and reproducibility [22] study are shown in Tables 6, 7 and 8 respectively.

Table 6. Data Collected From Operator A

	TRIAL 1	TRIAL 2	TRIAL 3
PART 1	105.00	105.00	105.10
PART 2	105.25	105.30	105.28
PART 3	105.40	105.42	105.40
PART 4	105.45	105.48	105.50
PART 5	105.58	105.56	105.56
PART 6	105.72	105.73	105.74
PART 7	105.79	105.80	105.78
PART 8	105.85	105.85	105.85
PART 9	105.88	105.89	105.90
PART10	105.58	105.60	105.59

Table 7. Data Collected From Operator B

	TRIAL 1	TRIAL 2	TRIAL 3
PART 1	105.10	105.00	105.10
PART 2	105.30	105.30	105.30
PART 3	105.42	105.40	105.43
PART 4	105.50	105.48	105.50
PART 5	105.55	105.55	105.55
PART 6	105.73	105.74	105.74
PART 7	105.80	105.80	105.78
PART 8	105.80	105.85	105.85
PART 9	105.85	105.89	105.90
PART10	105.59	105.59	105.59

Table 8. Data Collected From Operator C

	TRIAL 1	TRIAL 2	TRIAL 3
PART 1	105.00	105.00	105.10
PART 2	105.25	105.30	105.30
PART 3	105.43	105.45	105.45
PART 4	105.50	105.50	105.50
PART 5	105.55	105.56	105.56
PART 6	105.74	105.74	105.74
PART 7	105.80	105.80	105.80
PART 8	105.85	105.85	105.85
PART 9	105.90	105.90	105.90
PART10	105.59	105.60	105.59

After Suggested measuring instrument with Vernier calliper least count (0.02 mm) again 90 readings are made by the three different operators and the individual readings are noted separately on the table (VI, VII and VIII). These readings were entered in the worksheet we designed and the values are obtained from the measurement system. This data's were used by the mathematical formulas to find the average, ranges and precision to tolerance ratio (P/T Ratio). Table 9 shows the Gage R&R Study of reading using the vernier calliper with 0.02 mm least count.

Table 9. MSA- Gage R&R study for Vernier caliper (Least Count 0.02 mm)

MEASUREMENT SYSTEM ANALYSIS -GAUGE REPEATABILITY & REPRODUCIBILITY STUDY

Product: Pump Instrument: Vernier caliper Study by: Mr.T.Varun kumar.
Date: 22.08.2014 Least count: 0.02mm Specification: USL: 105.9 mm LSL: 105 mm
Process: Stamping Industry: JAY PUMPS Pvt.Ltd-CBE

Sample	Operator A					Operator B					Operator C				
Parts	Trial 1	Trial 2	Trial 3	X bar	Range	Trial 1	Trial 2	Trial 3	X bar	Range	Trial 1	Trial 2	Trial 3	X bar	Range
1	105.00	105.00	105.10	105.03333	0.1	105.10	105.00	105.10	105.1	0.1	105.00	105.00	105.10	105	0.1
2	105.26	105.30	105.28	105.28	0.04	105.30	105.30	105.30	105.3	0	105.28	105.30	105.30	105.3	0.02
3	105.40	105.42	105.40	105.40667	0.02	105.42	105.40	105.44	105.4	0.04	105.42	105.44	105.44	105.4	0.02
4	105.46	105.48	105.46	105.46667	0.02	105.50	105.48	105.50	105.5	0.02	105.50	105.50	105.50	105.5	0
5	105.58	105.56	105.56	105.56667	0.02	105.54	105.56	105.56	105.6	0.02	105.56	105.56	105.56	105.6	0
6	105.72	105.72	105.74	105.72667	0.02	105.74	105.74	105.74	105.7	0	105.74	105.74	105.74	105.7	0
7	105.76	105.80	105.78	105.78	0.04	105.80	105.80	105.80	105.8	0	105.80	105.80	105.80	105.8	0
8	105.84	105.86	105.84	105.84667	0.02	105.80	105.84	105.84	105.8	0.04	105.84	105.86	105.84	105.8	0.02
9	105.88	105.86	105.88	105.87333	0.02	105.86	105.88	105.88	105.9	0.02	105.90	105.90	105.90	105.9	0
10	105.58	105.60	105.60	105.59333	0.02	105.60	105.58	105.60	105.6	0.02	105.60	105.60	105.60	105.6	0
			XA bar	105.55733				XB bar	105.6				XC bar	105.6	
			RA		0.032			RB		0.026			RC		0.016

1	Average of all ranges (R Bar)	0.02467
2	Diff.among operators (R diff)	0.01333
3	sigma (o) repeatability for n=3 trials	0.01437
4	sigma (o)reproducibilibity, n=3 oprs	0.00777
5	sigma gage (o)2	0.00027
6	Precision Tolerance ratio (P/T)	0.10893

Here, the MSA is excellent.As P/T ratio=10%, this is because the instrument with 0.02 mm least count used as per the standard.

IOP Conf. Series: Materials Science and Engineering **954** (2020) 012006 doi:10.1088/1757-899X/954/1/012006

5.1 ANOVA Method for Improved Results

The use of ANOVA method is to conform and provide further insight to the data. Although the values can be calculated manually, most of them will prefer computer program to generate the results. This method will provide information concerning the causes of measurement system or gage variation, if the repeatability is large compared to the reproducibility the instruments need some maintenance, or the gage need to be redesigned etc. At the other hand, if the reproducibility is large compared to repeatability then the appraisers needs to be better trained to use and read gage instrument [23]. Using Minitab software, graphical results for the measured data from vernier calliper (least count=0.02 mm) are shown in the Figure 6. The graphical results clearly shows the components of variations are within the limits and the appraisers are measuring the system of range and average values are within the control limits. The measurement of parts variation is rectified by the improved systems [24]. The result shows, the remedial suggestion makes an excellent quality improvement in their process.

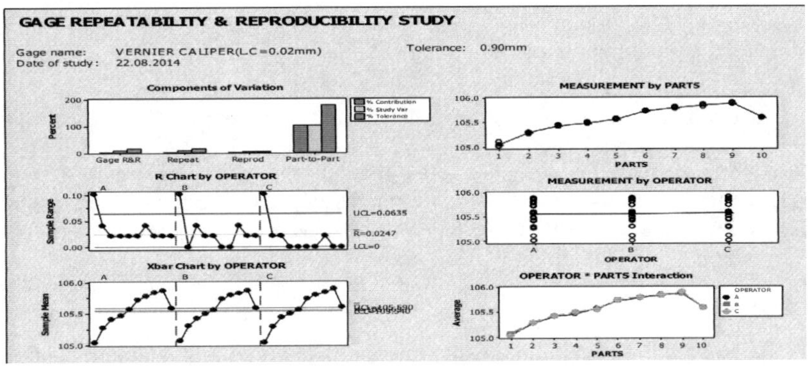

Figure 6. Graphical Results for Vernier Caliper (Least Count=0.02 mm)

5.2 Gage R&R Study - ANOVA Method: Improved Data Results

The Suggested measuring instrument results shows the excellent performance in MSA. The results are obtained by Minitab gage R&R (crossed) analysis using the ANOVA method [25]. The output in Figure 7 shows, that the percent of operator assessment is 0.043 meaning that the instrument using with the least count of 0.02 mm gives a minimal difference in agreeing the measurement. Also, the number of distinct categories is equal to 17, where by the AIAG standard reference manual, if NDC is > 10, then the system is excellent and acceptable.

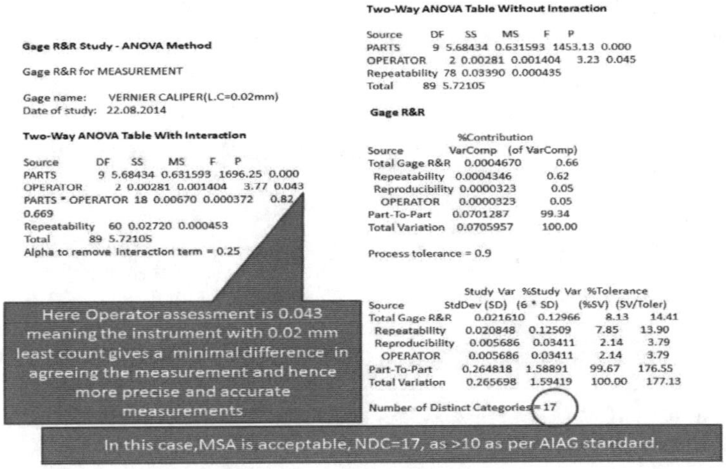

Gage R&R Study - ANOVA Method

Gage R&R for MEASUREMENT

Gage name: VERNIER CALIPER(L.C=0.02mm)
Date of study: 22.08.2014

Two-Way ANOVA Table With Interaction

Source	DF	SS	MS	F	P
PARTS	9	5.68434	0.631593	1696.25	0.000
OPERATOR	2	0.00281	0.001404	3.77	0.043
PARTS * OPERATOR	18	0.00670	0.000372	0.82	0.669
Repeatability	60	0.02720	0.000453		
Total	89	5.72105			

Alpha to remove interaction term = 0.25

Two-Way ANOVA Table Without Interaction

Source	DF	SS	MS	F	P
PARTS	9	5.68434	0.631593	1453.13	0.000
OPERATOR	2	0.00281	0.001404	3.23	0.045
Repeatability	78	0.03390	0.000435		
Total	89	5.72105			

Gage R&R

Source	VarComp	%Contribution (of VarComp)
Total Gage R&R	0.0004670	0.66
Repeatability	0.0004346	0.62
Reproducibility	0.0000323	0.05
OPERATOR	0.0000323	0.05
Part-To-Part	0.0701287	99.34
Total Variation	0.0705957	100.00

Process tolerance = 0.9

Source	StdDev (SD)	Study Var (6 * SD)	%Study Var (%SV)	%Tolerance (SV/Toler)
Total Gage R&R	0.021610	0.12966	8.13	14.41
Repeatability	0.020848	0.12509	7.85	13.90
Reproducibility	0.005686	0.03411	2.14	3.79
OPERATOR	0.005686	0.03411	2.14	3.79
Part-To-Part	0.264818	1.58891	99.67	176.55
Total Variation	0.265698	1.59419	100.00	177.13

Number of Distinct Categories = 17

Here Operator assessment is 0.043 meaning the instrument with 0.02 mm least count gives a minimal difference in agreeing the measurement and hence more precise and accurate measurements

In this case,MSA is acceptable, NDC=17, as >10 as per AIAG standard.

Figure 7. Gage R&R study – Two way ANOVA Method Improved Results

6. INTERPRETING THE RESULTS

Minitab uses the two way analysis of Variance (ANOVA) procedure to calculate the variance in the components, and it uses the components to estimate the % variation due to the measuring system. These % variations are appearing in the gage R&R table [25]. The % contribution is based on the estimates of the variance components and each value in the varComp is divided by the total variation and multiplied by 100.

%Study Var for gage R&R = (0.12966/1.59419) * 100 ≈ 8.13%.

By comparing the measurement system variation with the % tolerance, then Minitab calculates % tolerance compare with the measurement system variation to specification. From the Minitab, it multiplies the resulting proportion by 100 and reports it as the percent Tolerance.

% Tolerance for gage R&R is (0.12966/8) * 100= 14.41%.

The %Study Var results indicate that the measurement system should be less than 10% of the overall variation in this study.

Total Gage R&R, %Study Var = 8.13

The number of distinct categories represents how many separate groups of parts the system can distinguish. If the number <4 , the system cannot discriminate between parts, if the number is 4 to 9 then the parts can be divided into two groups(pass/fail). According to AIAG standard the system is acceptable when the number of distinct categories >10. In this case, the number of distinct parts= 17, Here the system is Excellent and acceptable. From the graphical representation, we understand that the most variations are generated by part to part.

7. CONCLUSION

In our study, the measurement system comparison is made before and after the suggested improvement. A result clearly shows that the suggested measurement system analysis is excellent and acceptable after replacing the existing instrument of vernier calliper with least count 0.02mm. The purpose of doing the measurement system analysis is to identify and to reduce the measurement variation in a system. In this study the suggested measuring instrument of (vernier calliper with LC=0.02mm) contributes very little to overall variations as noticed in data representation and in graphical representation using ANOVA in Minitab software. Results show the percentage of contribution of Gage R&R would be acceptable according to AIAG standard. Variations are mostly obtained from the parts, not from instruments, operators, nor methods. With this study, the Company knew that its measurement system was giving accurate and precise information and hence the measurement systems could be acceptable and reliable. If the measurement systems are not satisfactory, operators can know what the cause of problem and this system helps the operators to determine which function should be fixed to improve the measurement system. Any modern companies of SMES [18], in future would go for quality audit systems [20] such as lean six sigma, ISO [21], Agile or TQM implementation [22] etc., this system plays a basic key factor to entering into the process.

ACKNOWLEDGMENT

The authors are thankful for the anonymous reviewers for their invaluable comments which helped them to improve their paper.

REFERENCES

[1] Jiménez, J.d.B. and Lorente, J.J.C. (2001), "Environmental performance as an operations objective", *Int J Oper & Prod Mgt, **21*** (12) 1553-1572

[2] T.Varun Kumar, Ganesh Babu, Saravanan Muthusamy, (2016), "Assessing the awareness of the agile manufacturing for organizational change in Indian small manufacturing firms: An empirical investigation", *J Organ Change Mgt, **29*** (5) 713-73

[3] Colotla, I., Shi, Y. and Gregory, M. (2003), "Operation and performance of international manufacturing networks", *Int J Oper & Prod Mgt, **23*** (10) 1184.

[4] Nagarur, N. (1992), "Some performance measures of flexible manufacturing systems", *Int J Prod Res **30*** (4) 799-809.

[5] Kumar, M.T.V., Babu, B.G., and Saravanan, M. (2016), "A framework for assessing the awareness of the agile manufacturing environment in Indian SMEs", *Int J Adv Engg Tech* ***7*** (2) 201-208

[6] Montgomery, D. C., Runger, G. C., (1993), "Gauge Capability and Designed Experiments Part I: Basic Methods", *Qual Engg* ***6*** (1) 115-135.

[7] Ismail, T.H. (2007), "Performance evaluation measures in the private sector: Egyptian practice", *Manag Audit J* ***22*** (5) 503-513.

[8] Kumar, V. T., Babu, G. B., Saravanan, M., & Murugan, M. (2014), "Agile manufacturing in small and medium scale enterprises", *Inter J Innov Res Sci Engg Tech* ***3*** (2) 253–260.

[9] Brignall, S. and Ballantine, J. (1996), "Performance measurement in service businesses revisited", *Inter J Serv Indus Mgt* ***7*** (1) 6-31.

[10] Chen, K. S., Wu, C. H., Chen, S. C., (2008), "Criteria of Determining the P/T Upper Limits of GR&R in MSA", Quality & Quantity, Vol.42, No.1, pp.23-33.

[11] Hong, P., and Jeong, J. (2006). "Supply chain management practices of SMEs: from a business growth perspective", *J Enter Inform Mgt* ***19*** 292-302

[12] Varun Kumar, T., Ganeshbabu, B. and Saravanan, M. (2014), "Effective performance of an user friendly dynamic tool design in small and medium scale enterprises", *Appl Mecha Mater* ***591*** 215-218.

[13] R.K. Burdick, A.E. Allen, G.A. Larson, (2002), "Comparing variability of two measurement processes using R&R studies", *J Qual Tech* ***34*** (1) 97–105.

[14] Wang, F.K., Chien, T.W., (2010), "Process-oriented basis representation for a multivariate gauge study", *Comp & Indus Eng* ***58*** 143-150.

[15] Chen, K. S., Wu, C. H., Chen, S. C., (2008,) "Criteria of Determining the P/T Upper Limits of GR&R in MSA", *Qual & Quant* ***42*** (1) 23-33.

[16] Al-Refaie, A., Bata, N., (2010), "Evaluating measurement and process capabilities by GR&R with four quality measures", *Measur* ***43*** (6) 842-851.

[17] Burdick, R. K. and Larsen, G. A. (1997). "Confidence Intervals on Measures of Variability in R&R Studies", Journal of Quality Technology 29, pp. 261–273.

[18] P. Tsai, (1989), "Variable gauge repeatability and reproducibility study using the analysis of variance method", *Qual Engg* ***1*** 107-115.

[19] Saraph J.V., Benson, P.G. and Schroeder, R.G, (1989), "An instrument for measuring the critical factors of quality management", *Deci Sci* ***20*** (4) 810-829.

[20] R. R. Smith, S. W. McCrary, and R. N. Callahan, (2007), "Gauge repeatability and reproducibility studies and measurement system analysis: A multi-method exploration of the state of practice", *J Qual Tech* ***23*** (1) 1-11.

[21] Ahire, S.L. (1996). An empirical investigation of quality management in small firms. *Prod Invent Mgt J* 44-50.

[22] Tsekouras, K., Dimana, E. & Skuras, D. (2002), "Adoption of a quality assurance scheme and its effect on firm performance: A study of Greek firms implementing ISO 9000", *Tot Qual Mgt* ***6*** (1) 827-841.

[23] Lupan, R., & Bacivarof, I. C. (2005), "A Relationship between Six Sigma and ISO 9000:2000", *Qual Engg* ***17*** (4) 719-725.

[24] Zhang, Z.H, Waszink, A., & Wijngaard, J. (2000). An instrument for measuring TQM implementation for Chinese manufacturing companies. *Inter J Qual Reliab* ***17*** (7) 730-755.

[25] Varun Kumar T, Manoj Mohan Prasad V, Santhosh D, Prasanth C, Ranjith K. (2020), "Evaluation of mechanical and interfacial properties of carbon fiber reinforced polymer (CFRP) composite materials. *Mater Today Proc* ***21*** 477-482.

IOP Conf. Series: Materials Science and Engineering **954** (2020) 012007 doi:10.1088/1757-899X/954/1/012007

A review on Friction Stir Welding in Aluminium Alloys

Stephen Leon.J[1*], Bharathiraja.G[1] and Jayakumar.V[2]

[1]Department of Mechanical Engineering, Saveetha School of Engineering,
Saveetha Institute of Medical and Technical Sciences, Chennai, Tamilnadu, India.
[2]Department of Mechanical Engineering, Amrita School of Engineering,
Amrita Vishwa Vidyapeetham, Chennai, Tamil Nadu, India
*E-mail: stephenleonj@gmail.com

Abstract. Joining of aluminium alloys is always a challenging task due to its easily oxidising property in its molten stage. Although its application is huge in the building of aerospace structure, 2XXX and 7XXX series alloys are categorised as non-weldable aluminium alloys. Recent development in the application of friction stir welding process in the joining of these non-weldable alloys simplified the fabrication processes. In this review article, significance of friction stir welding comparing with other solid state metal joining process is addressed. Recent developments and application of this relatively new metal joining process in various industries are also discussed. Particular emphasis has been given to the mechanism responsible for the formation of this solid state joining technique. Apart from this, major process parameters that influence the formation of weld and weld defects caused by the improper selection of process parameters are also explained in detail.

1. Introduction

Aluminium became metal of choice for many applications due to its most economical and attractive characters like light weight, good appearance, fabric ability and corrosion resistance [1]. Aluminium has poor strength in its pure form. Its mechanical properties are improved by alloying with iron, silicon, manganese and magnesium to form non heat treatable alloys (1xxx, 3xxx, 4xxx and 5xxx series). Heat treatable high strength aluminium alloys (2xxx, 6xxx and 7xxx series) are developed by alloying pure aluminium with copper, magnesium silicate and zinc [2]. Depending on its physical and mechanical properties theses alloys are used in various fields like, airframes, engines, and ultimately, for missile bodies, fuel cells, and satellite components. In industries arc welding is one of the widely practiced contemporary manufacturing processes, which is carried out by heating the metal pieces to their melting point and fusing them together. The characters like forming aluminium oxide in molten stage, hydrogen solubility, thermal expansion and shrinkage during solidification makes aluminium alloys cannot be joined using conventional arc welding technique [3].

1.1. *Weldability of aluminium alloys*

MIG and TIG welding are the commonly used fusion joining methods in the fabrication of different aluminium alloys. Weldability of different alloys is shown in Fig.1. Fig.1 reveals that some alloy series have major problems in fusion welding. Fusion welding is carried in a temperature more than the solidus temperature of the base metal. Although oxide formation during fusion welding is completely avoided in TIG and MIG welding, there were many other problems associated with the thermal properties of aluminium.

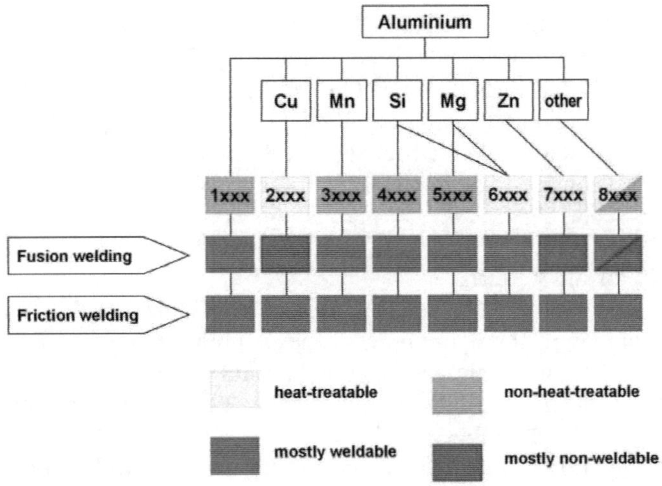

Figure 1. weldability chart of aluminium alloys

Thermal conductivity of aluminium is much higher (3 times higher than steel). As fusion welding is carried out with higher heat input, heat affected zone experiences high temperature rise during the joining process [4]. Higher temperature rise in heat affected zone eradicates the mechanical properties and results poor strength in that zone. Apart from this, thermal expansions of aluminium alloys are about twice that of steel [5]. Solidification shrinkage in aluminium weld metal is about 6% by volume and it can be the main cause for distortion, especially in thicker welds. Inclusion of copper in Aluminium alloy 2xxx series results formation of solidification cracks in fusion welding. The cooper content which causes hot cracking and poor solidification in the microstructure and porosity in the fusion zone made the 2xxx series impossible to join using fusion welding [6].

1.2. Solid state welding

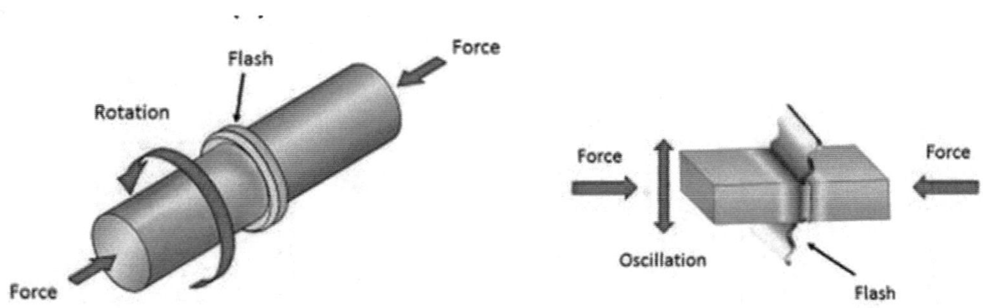

Figure 2 Fiction welding [7]

The major drawbacks of obtained casting structure in fusion welding like hot cracking, porosity and other macro and micro structural defects can be avoided in friction welding as the entire process is carried out below the solidification temperature of the base metal. In this process material is plasticised using the friction heat generated between the moving and stationary surfaces [8]. Once the material is plasticised a lateral force called "upset is applied to fuse the plasticised metal parts. As the entire joining process is carried out below the melting point of base metal, it does not require filler material. Defects linked with melting and solidification is completely eliminated in this solid metal joining process. Based on the relative motion between the moving and stationary surfaces they are classifies as spin welding and linear friction welding.

Friction stir welding (FSW) is an upgraded version of conventional friction welding process. It is invented by Wayne Thomas at The Welding Institute (TWI), Cambridge, in 1991. In FSW, the frictional heat generated between the rotating non consumable tool and the stationary workpiece is used as a major heat source to join the workpiece which is assisted by the heat generated through the plastic deformation of material during welding. Due to inherent potentials of FSW, it has become a process of choice for producing high-quality joints in both similar and dissimilar materials in just a short time span. FSW exhibits comparatively better weld quality in all aluminium alloys [9]. Sound joints obtained through friction stir welding rapidly replacing all fastening methods like riveting in many industries. In this review article, advantages of friction stir welding over other aluminium joining techniques are discussed. The currents state of understanding and developments of fiction stir welding are reviewed in detail.

2. Significance of Friction stir welding

FSW has now established itself to be a remarkable solid state welding technique to effectively join similar and dissimilar aluminum alloys. The process does not require any consumables (filler material, fluxes and shielding gas, etc.) for joining, produces no harmful emissions, safe to humans and is, therefore, considered to be an energy-efficient, environment-friendly, and clean material joining process, as summarised in Table1 and Fig.3

Table 1. Metallurgical and environmental benefits of FSW

Metallurgical benefits	Environmental benefits
• Solid phase process • Low distortion • Good dimensional stability and repeatability • No loss of alloying elements • Excellent mechanical properties in joint area • Fine recrystallized microstructure • Absence of solidification cracking • Post FSW formability	• No shield gas required • Minimal surface cleaning • No harmful emission • Eliminate solvent requirement • Consumable material saving such as rugs, wire or any other gases • Only 2.5 % of energy needed comparing with laser weld.

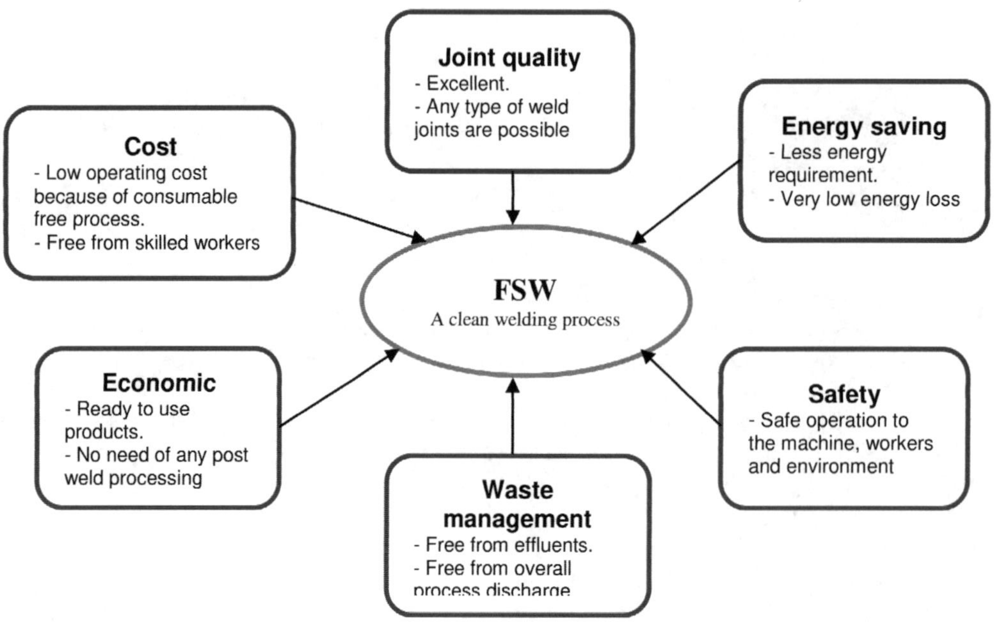

Figure 3. Significance of FSW

2.1. Industrial developments and applications

Friction stir welding is the latest joining process has a huge number of applications in automobile industries. The Ford Motor Co. has produced thousands of Ford GT automobiles with a central tunnel assembled by FSW that isolates the fuel tank from the interior counterparts (Fig. 4). Mazda patented and employed plunge friction spot welding (PFSW) that is a spin-off technology of friction stir spot welding (FSSW) for producing the rear-door structure of almost 100,000 automobiles [10].

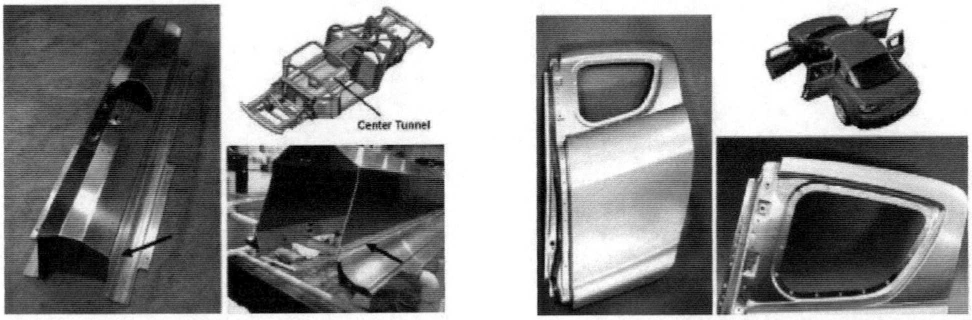

Figure 4. Tunnel assembly in Ford and spot welded door in Mazda [10]

In this method, even though there is a pull-out hole left on the spot weld, the fatigue life and the strength of the joint are satisfactory. The AUDI R8 used a three-sided re-entrant probe to accomplish the welding of tailored blanks for the center-tunnel closing panel in the AUDI R8. The same company has employed the FSSW process for overlap welding of side members in the BMW 5 series. Honda Motor Corporation has employed dissimilar FSW of aluminum alloy with steel to produce a component to be used in the front structure of a Honda Accord [11].

Figure 5. Ship deck plate construction [12]

Figure 6. FSW lap joint in Eclipse 500 jet [10]

In the ship building industries, Japan manufacturers are the leaders in employing the FSW process to construct the sections made of aluminum alloys. Fig. 5 shows ship deck plate assembly constructed using friction stir welding. In the aerospace industry, the friction stir (FS) welded lap joints are used as the rivet replacement technology to join the circumferential and longitudinal internal stiffeners and to attach doublers at door and window cutout locations [12]. As another application, Fig. 6 illustrates the Eclipse 500 business class jet constructed using FSSW. Better quality, stronger joints, and lighter structure at a reduced cost for assembly and production cycle time are some of the advantages of using FSSW in the Eclipse jet assembly [10].

Apart from normal conventional joints most challenging longitudinal and circumferential joints are successfully done in rocket fuel tank (Falcon 9) [13]. Fig. 7 illustrates the fuel tank assembly of a rocket fabricated by the FSW process in the Boeing Company. They replaced the tungsten inert gas (TIG) conventional fusion welding process with FSW, and obtained a staggering 99% reduction in the welding costs. Other few examples of major industries that use Friction stir welding as joining technique to manufacture their products are listed in Table 2.

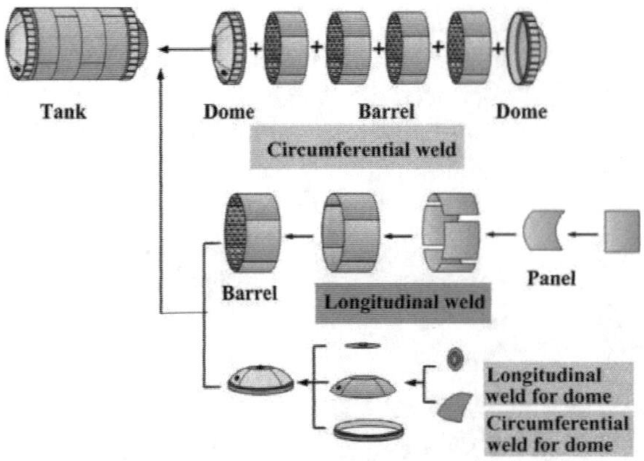

Figure 7. FSW in rocket fuel tank [13]

Table 2. Friction stir welding applications in industry.

Application	Industries
Heat exchangers	Marine Aluminum, Norway
Shipbuilding	Marine Aluminum, Norway
Delta II rockets	Boeing, United States
Shipbuilding	SAPA, Sweden
Automotive components	SAPA, Sweden
Laser system housings	General Tool, United States
Motor housings	Hydro Aluminum (formerly Marine Aluminum), Norway
Suburban trains	Alstom, Germany, and Hydro Marine
Automotive components	Showa, Japan
Train bodies	Hitachi, Japan
Automotive components	Tower Automotive, United States
Aircraft structure	Eclipse, United States
Commercial shipbuilding	Advanced Joining Technologies, United States
Space shuttle external tank	Lockheed Martin, United States
Shipbuilding	Friction Stir Link, United States
Automotive components	Honda Motor Company, Japan

3. Overview of macroscopic processes during FSW

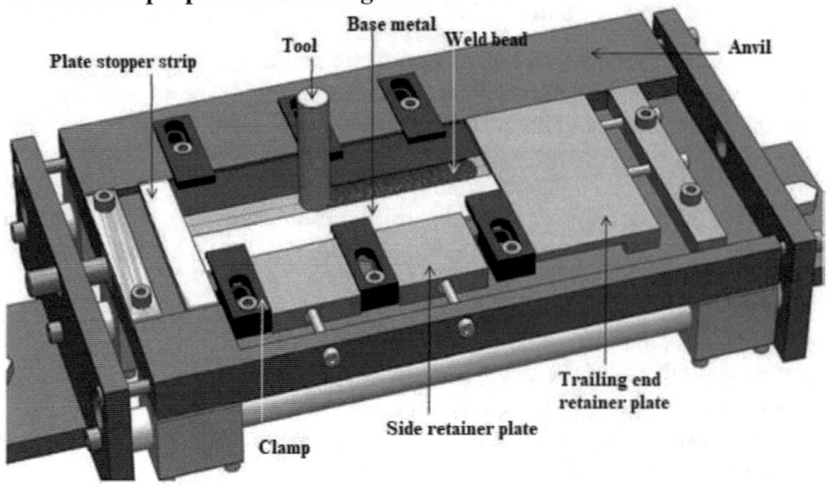

Figure 8. Solid model of FSW fixtures

The BMs to be welded are rigidly held with the help of specially designed clamps. The clamps are designed in such a way that they do not protrude after being mounted on the anvil so as to avoid their obstruction with the welding tool. After placing the BM on the anvil, clamps are tightened to firmly hold the BM from top (Fig. 8). Unlike other conventional fusion welding techniques friction stir welding does not depend on external source for the heat generation during the process [14]. The frictional heat developed between the tool and the workpiece is the major heat source for the entire joining process. Frictional heat developed along the tool matrix interface completely depends on the relative velocity between the contacting surfaces. From the operational point of view, based on the relative motion between the tool and workpiece this processes is further divided into three different sequences of operations as shown in Fig. 9.

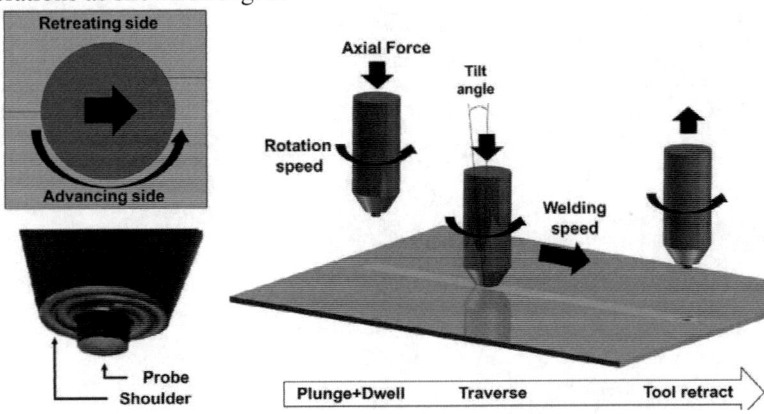

Figure 9. Sequence of operations in FSW [15]

Plunge and Dwell – Rotating tool allowed to plunge into the stationary workpiece in the weld line till the tool shoulder touches the top surface of the workpiece (Fig. 9). Frictional heat is generated in the tool/matrix contact surface. Rotating tool allowed to stay in the same position till the temperature is increased enough to plasticise the material. In this stage entire tool pin (probe) is plunged into the base metal. The rotating tool pin inside the base metal and rotating shoulder over the top surface of the base metal not only produce heat but also it stir the plasticised material under the shoulder to flow from the advancing side to the retreating side and develops sound joint along the weld line.

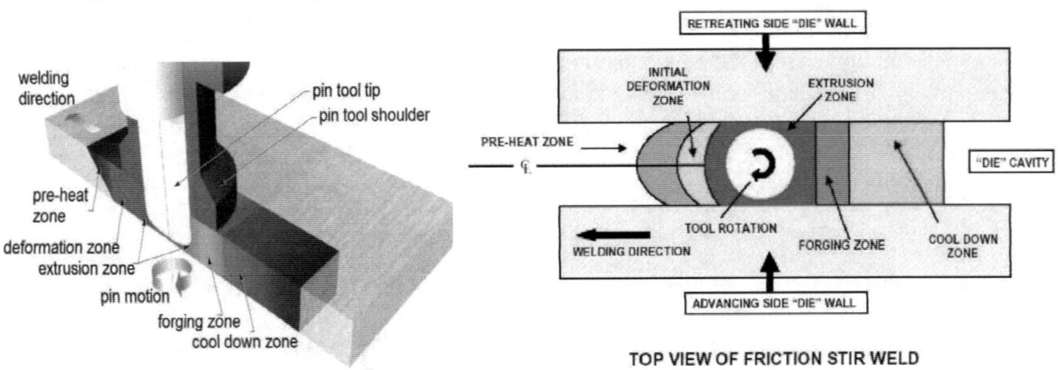

Figure 10. Processes in material joining during FSW [16]

Traverse – Rotating tool allowed to slide along the weld line. This sequence of operation is also called as weld stage as the joining of metal happens along the weld line in this stage. Tool travel speed (revolutions / mm of linear movement) is the key factor for the sound join inn this stage. Feed rate of tool is selected based on the thermal properties of material to be joined [17]. Optimum feed rate results defect free joints. As the entire joining process is done in lower temperature quality of joint is very sensitive to the selected weld parameters. Material along the weld line preheated in front of the tool, plasticised under the shoulder, extruded from the front of the tool and forged at the back side of the tool as shown in Fig. 10

Retract – Once the welding is done the rotating tool is pulled upwards. This is a finishing stage. One of the draw of friction stir welding. During the upward motion of the tool, tool probe moves the material upwards which leaves key hole at the end point of the weld line. A new technique of self-refilling friction stir welding (SRFSW) relying on non-consumable joining tool has been developed to repair the keyhole left at the end of process [18].

4. Structural zone characteristics of FSW

Based on the post weld changes in the microstructure of joined base metal obtained weld joint divided into two different zones (Fig. 11).

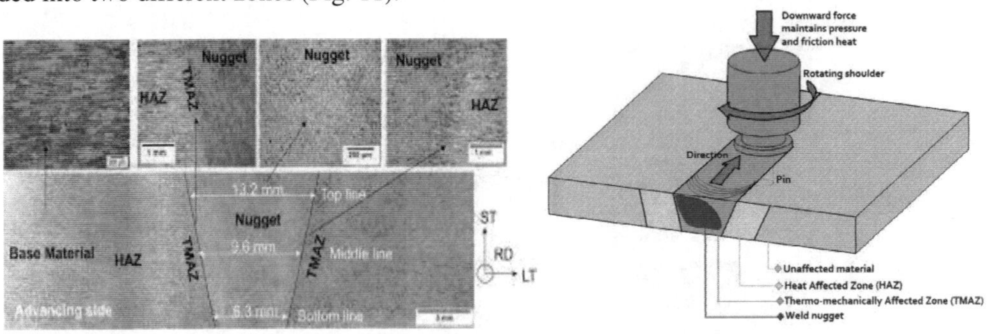

Figure 11. Microstructural view of different zones in FSW joint [19]

4.1. Weld nugget or Stir zone (SZ)

This zone refers to a region which is affected by the rotational and traverse movement of the tool pin. The size of this region depends on the diameter of the tool pin. This region experiences grain refinement due to the dynamic recrystallization. Grains in SZ are equiaxed, and are extremely smaller in size compared to BM. Sometimes this region displays onion ring-type structure depending upon BM and process conditions. Hardness values are usually lower than the BM of heat-treatable alloys [20].

4.2. Thermo-mechanically affected zone (TMAZ)

This zone exist between SZ and heat affected zone. This represents the entire deformed zone under the shoulder other than the nugget zone. Material in this region experiences thermal cycles as well as mechanical deformation due to the rotary motion of the tool shoulder. Thermal effects on this region are comparatively lesser than the stir zone which indicates lower degree of deformation and partial recrystallization of grains. In this region, hardness reaches to the minimum value due to partial precipitate dissolution and grain growth caused by attainment of high temperature at some local area in TMAZ [21].

4.3. Heat affected zone (HAZ)

This zone can be observed in all metal joining processes. In friction stir welding this region represents the area which is not mechanically deformed but affected by the thermal cycles developed during the joining process. When this region subjected to the temperature more than the recrystallization temperature of the material, grain are coarsened and the density of the strengthening agents in the alloy is reduced. Normally this region is the low strength region in Friction stir welded joint [22]. When this region experiences lesser temperature than recrystallization temperature during the process, then the low strength region shifts from HAZ to TMAZ in a joint. Hardness value of this region is always lesser than the base meatal.

4.4. Unaffected zone

Region adjacent to the HAZ away from joint is the BM that does not deform. However, it experiences thermal cycle but not to the extent to bring significant changes in microstructure. Chart shown in Fig. 12 summarises the microstructural evolution in different zones.

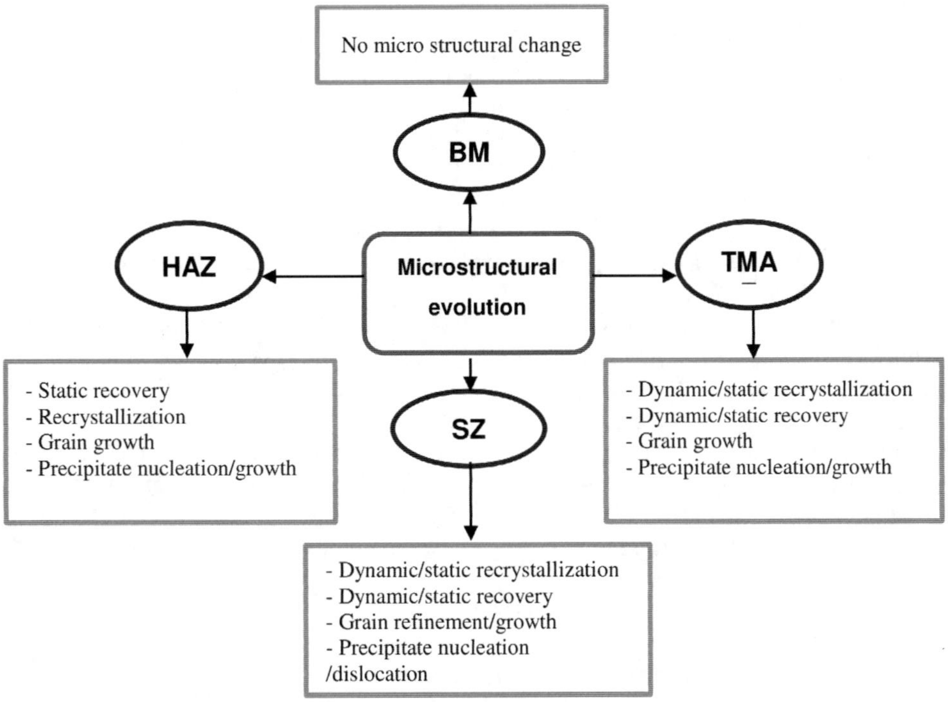

Figure 12. Microstructure evolution during FSW at different zones

5. Effects of process parameters:

Optimum range of process parameters for a friction stir welding process depends on resulting joint efficiency and production rate. Although the process is affected by so many direct and indirect parameters, the key factors and their effects are listed in Table. Selection of these parameters for a joining process depends on the thermal and mechanical properties of base metal and the heating and cooling boundary conditions.

Table 3. Process parameters and its effects

Parameter	Effects
Tool rotational speed	Heat generation, mixing material and breaking oxide layer
Welding speed	Effective heat input and weld appearance
Downward force	Frictional heat generation
Tool tilt angle	Enhancing material flow and weld appearance

5.1. Tool rotational speed

Tool rotation speed is responsible for the heat generation as well as uniform material flow in the joining process. Increase in process temperature is achieved through the increase in heat supply. The simple way to increase heat supply is by increasing the tool rotational peed. McClure et al [23] recorded 40℃ increases in temperature with an increase in tool rotational speed form 300 to 650 rpm. But the temperature raise did not increased considerable at higher speed. Swaminathan [24] observed only 20℃ temperature rise when the tool rotational speed increased form 900 rpm to 1200 rpm. This indicates that increase in tool rotational speed beyond a limit does not have much influence on heat generation. Lower rotational speed causes defects due to the insufficient material flow in stir zone and higher tool rotational seed results excess heat and eradicates post weld properties in heat affected zone. Therefore, tool rotational speed has to be optimised based on the optimum heat supply.

IOP Conf. Series: Materials Science and Engineering **954** (2020) 012007 doi:10.1088/1757-899X/954/1/012007

5.2. Weld speed

Forward motion of the tool assists the movement of the plasticised material under the tool shoulder from advancing side to the back of the immersed pin. Apart from that, effective heat input to the material to be joined depends on the weld speed. High weld speed drastically reduces the effective heat supply during welding [25]. Reduction in heat supply reduces the process peak temperature and results insufficient material flow around the tool pin in stir zone. Insufficient material flow leads to tunnel defect in stir zone. Selection of higher welding speed increases torque in the tool and develops excess stress in tool pin [26]. As tool pin is completely immersed in the material the excess stress on tool pin results tool pin breakage. Decrease in weld speed not only leads to the excess heat supple but also increases production time.

Simultaneous action of heating and stirring of material in stir zone leads to grain refinement in stir zone. Recrystallization of material in stir zone is affected by the increase in tool rotational speed and decrease in welding speed. Refined grains in weldment ensure better weld quality. Tool rotation and traversing speed must be selected to ensure efficient welding. Lower rotational speed combined with higher traversing speed results in colder weld whereas, higher rotational speed along with lower traversing speed results in hotter weld. Being easily controllable weld parameters many researches (Table 4) has been done to optimise the tool rotational speed with respect to the tool feed rate.

Table 4. Literature survey on optimum process parameters

Material	Analysed parameter levels		Optimum outcome value		Reference
	Tool traverse speed (rpm)	Tool rotational speed (mm/min)	Tool traverse speed	Tool rotational speed	
AA1050	800 to 1600	100 to 300	1200	100	[27]
AA2024-T3	800 to 1600	35 to 140	1200	35	[28]
AA2024-T4	560 to 1800	11 to 45	900	35	[29]
AA2024- T6	1100 to 1900	10 to 70	1500	40	[30]
AA4047	500 to 1100	45 to 55	1100	55	[31]
AA5083	500 to 1400	16 to 40	950	28	[32]
AA6061 –T6	900 to 1120	20 to 40	1120	40	[33]
AA7075-T6	630 to 1250	20 to 60	1000	20	[34]

5.3. Downward force and plunge depth

Vertical force exerted to the workpiece through the tool is one among the easily controllable parameter. Proper selection of vertical force ensures quality weld even if there are tolerable thickness variations in the materials to be joined. As vertical force is one among the key factor on frictional heat generation during the process, higher vertical force enables high speed welding [35]. Plunge depth is the depth of penetration of shoulder surface in the workpiece as shown in Fig. Plunge depth increases heat generation as well as it controls the forging of flow material in the stir zone [36]. Axial force can be increased through increasing plunge depth which results higher heat generation. Higher plunge depth affects grain growth in stir zone and affects ductile strength of the joint [37]. Lower plunge depth reduces vertical force which in turn reduces heat supply and leads to improper material flow around the tool [38]. Therefore, appropriate plunge depth is important for producing good-quality joints by ensuring adequate forging pressure required to consolidate flowing material properly as well as full penetration of the tool inside BM.

5.4. Tool tilt angle

Inclination of tool rotational axis with respect to the base metal surface is tool tilt angle as shown in fig. It has its own influence on frictional heat generation, material flow and consolidation of flowing material around the tool [39]. Peak process temperature rise is directly proportional to the tool tilt angle [40]. Tool tilt angle helps to prevent excess spill out of material during the joining process. Rajendran et al. [41] identified that tool tilt angle controls the size of the stir zone which indicates its

IOP Conf. Series: Materials Science and Engineering **954** (2020) 012007 doi:10.1088/1757-899X/954/1/012007

effects on material flow during the process. Tilting the tool by some degree in such a way that the front of tool is higher than its rear, helps in adequate forging of the plasticized material behind the tool to complete the joint.

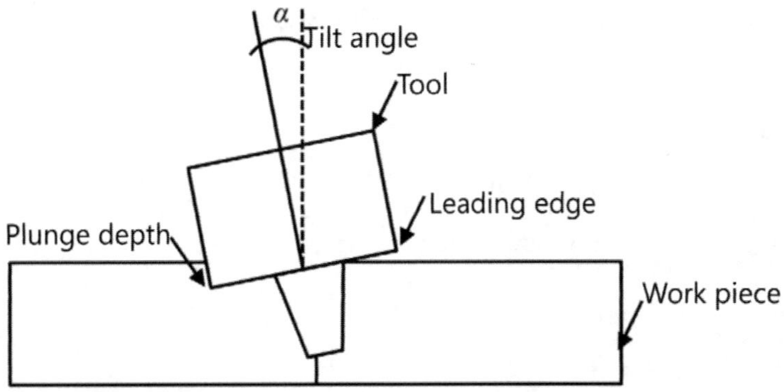

Figure 13. FSW tool with plunge depth and tilt angle.

Apart from this direct process influencing parameters, other parameters like tool geometry, base metal material properties, workpiece thickness, heat dissipating boundary conditions and tool material properties also have its own indirect effect on weld quality. The chart shown below (Fig.14) explains the parameters and its effects in the joining process.

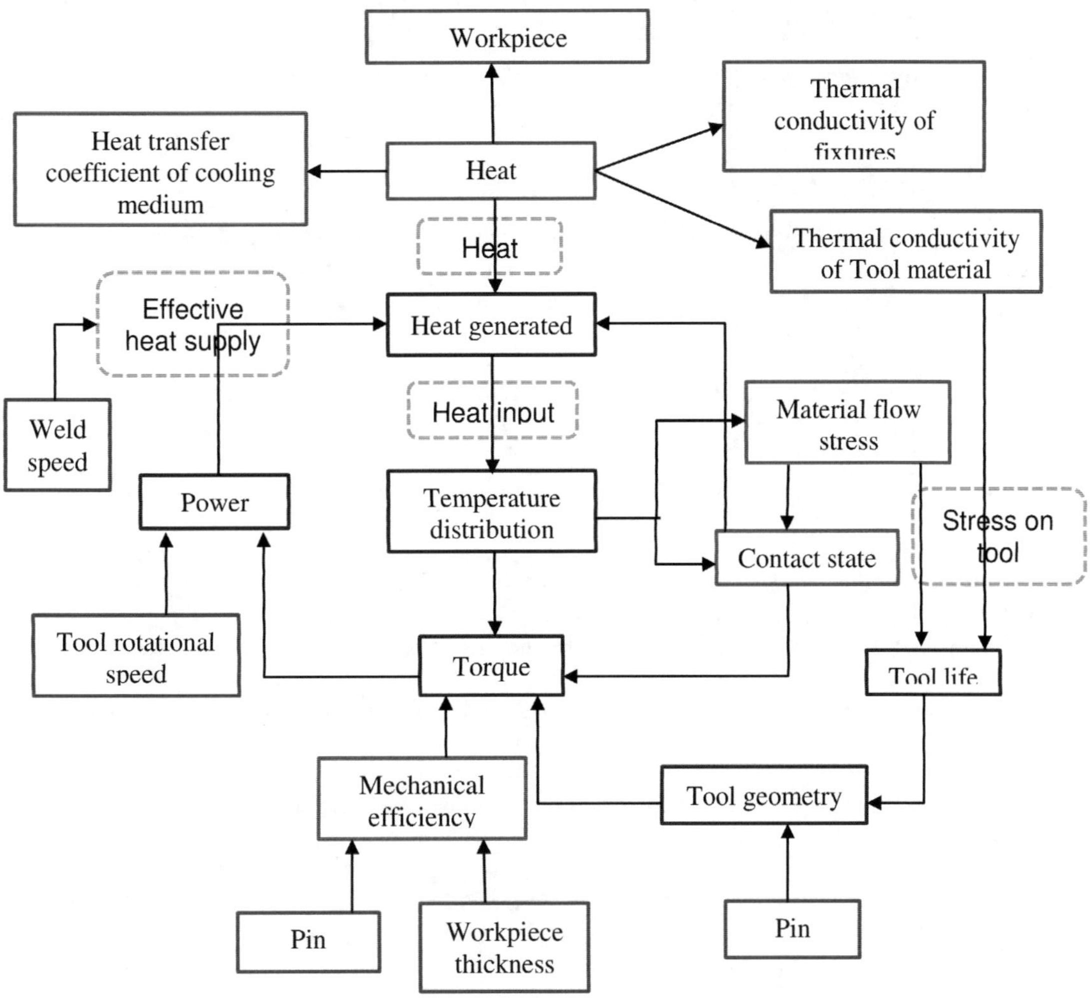

Figure 14. Process parameters and its effects on joining process in FSW

IOP Conf. Series: Materials Science and Engineering **954** (2020) 012007 doi:10.1088/1757-899X/954/1/012007

6. Weld defects and causes

Improper selection of process parameter leads to insufficient/excess heat supply during the process which in turn causes insufficient/excess flash of material flow in the joint. It creates weld defects in friction stir welding. Various commonly observed weld defects and its causes are given in Table 5.

Table 5. FSW defects and causes

Defect	Macroscopic appearance	Location	Causes	Ref
Tunnel	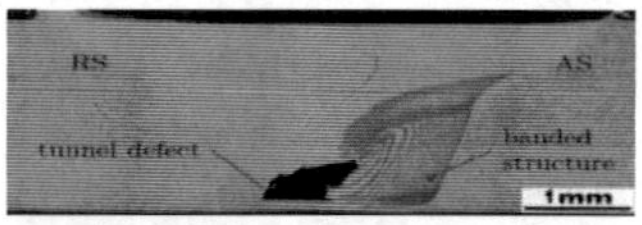	Between SZ and TMAZ in the advancing side	- Low tool rotational speed. - High feed rate. - Insufficient plunge depth. - Improper pin offset.	[42]
Kissing bond	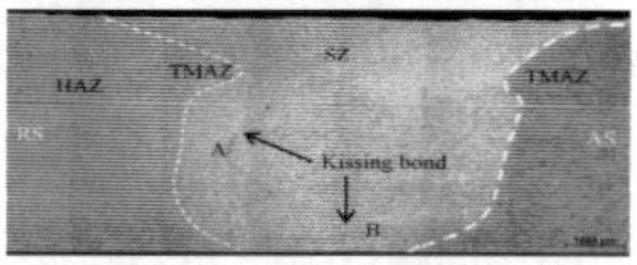	Advancing side of the SZ	- Insufficient material flow. - Improper removal of oxide layer form faying surface. - Insufficient heat input	[43]
Void	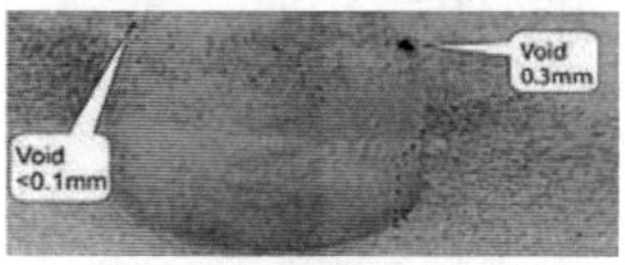	Under the weld surface or the advancing and retreating sides of the weld	- Inadequate forging pressure. - High tool traverse velocity.	[44]
Incomplete root penetration	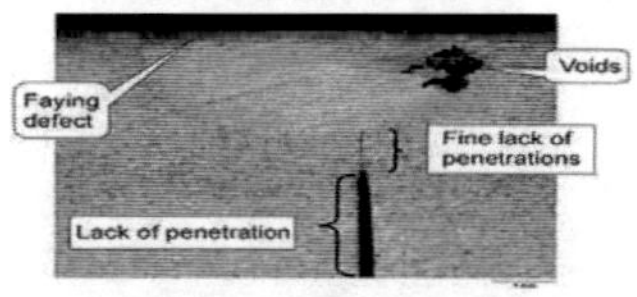	Below the SZ at the interface of the faying surface	- Local variations in the plate thickness. - Improper tool design. - Inappropriate plunge depth.	[44]
Hooking		Both retreating and advancing sides of TMAZ	- Improper tool design. - Inappropriate tool tilt angle. - Insufficient heat supply	[45]
Excess flash	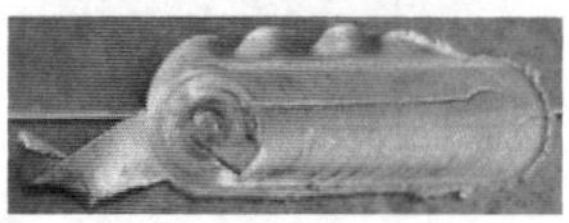	Outer edges of the weld joint	- High tool rotation speed. - Low weld speed. - High vertical force.	[46]

7. Conclusions

Temperature dependent shear strength of the aluminium alloy is the key factor that decides the material flow and defect free joint formation in friction stir welding. Heat generation and heat flow during the process depends on the chemical composition of base metal, surrounding environmental influence on cooling through heat dissipating boundaries. For the chosen process parameters there always be a maximum amount of temperature rise and for any chosen combination of parameters process peak temperature cannot be increased more than the material solidus temperature. Heat energy developed during the process cannot be completely utilised for the formation of stir zone. Optimised processing conditions necessitate that the material reaches a processing temperature which ensures thermal softening takes place and this also means that a degree of slip must occur between the FSW tool and the workpiece material. Tool pin has to be designed in such a way that it increases the material strain rate in the stir zone and decreases the clagging of material with the tool surface. Clagging of base material with tool results weld defects and reduces joint strength. Optimum heat generation during the process is a function of a number of quantitative factors like,

- Thermal conductivity of the tool material.
- Initial temperatures of the tool and base metal.
- The contribution of heat generation through plastic deformation of material in SZ.

- Frictional heat generated along the tool/matrix interface.
- The ability to retain constant temperature gradient during the process by maintaining constant balance between heat input and heat loss through the heat dissipating boundaries

Optimum heat supply conditions depend on so many other environmental factors like spindle construction, clamping and backing bar employed to produce the joint. In order to understand the relationship between temperature rise and material flow during the process every single parameter that directly and indirectly influences the process should be taken into consideration. Hence, for any aluminium alloy the set of process parameters cannot be fixed as constant and it can only be defined as an optimum range based primarily on the geometry of the joint, construction of the workpiece restraint system and design of the FSW machine

References

[1] C.-H. Ng, S. N. . Yahaya, and A. A. . Majid, "Reviews on aluminum alloy series and its applications," *Acad. J. Sci. Res.*, vol. 5, no. 12, pp. 708–716, 2017, doi: 10.15413/ajsr.2017.0724.

[2] J. Joel and M. Anthony Xavior, "Aluminium Alloy Composites and its Machinability studies; A Review," *Mater. Today Proc.*, vol. 5, no. 5, pp. 13556–13562, 2018, doi: 10.1016/j.matpr.2018.02.351.

[3] I. Kwee, W. De Waele, and K. Faes, "Weldability of high-strength aluminium alloy EN AW-7475-T761 sheets for aerospace applications, using refill friction stir spot welding," *Weld. World*, vol. 63, no. 4, pp. 1001–1011, 2019, doi: 10.1007/s40194-019-00732-1.

[4] J. Stephen Leon and V. Jayakaumar, "Numerical modeling of thermal field during friction stir welding using non-circular pin," *Caribb. J. Sci.*, vol. 53, no. 1, 2019.

[5] M. M. Atabaki, M. Nikodinovski, P. Chenier, J. Ma, M. Harooni, and R. Kovacevic, "Welding of Aluminum Alloys to Steels: An Overview," *J. Manuf. Sci. Prod.*, vol. 14, no. 2, pp. 59–78, 2014, doi: 10.1515/jmsp-2014-0007.

[6] M. M. Moradi, H. Jamshidi Aval, and R. Jamaati, "Microstructure and mechanical properties in nano and microscale SiC-included dissimilar friction stir welding of AA6061-AA2024," *Mater. Sci. Technol. (United Kingdom)*, vol. 34, no. 4, pp. 388–401, 2018, doi: 10.1080/02670836.2017.1393976.

[7] Y. S. Sato, H. Yamanoi, H. Kokawa, and T. Furuhara, "Microstructural evolution of ultrahigh carbon steel during friction stir welding," vol. 57, pp. 557–560, 2007, doi: 10.1016/j.scriptamat.2007.04.050.

[8] W. Li, A. Vairis, M. Preuss, and T. Ma, "Linear and rotary friction welding review," *Int. Mater. Rev.*, vol. 61, no. 2, pp. 71–100, 2016, doi: 10.1080/09506608.2015.1109214.

[9] J. Stephen Leon and V. Jayakumar, "An investigation of analytical modelling of friction stir welding," *Int. J. Mech. Prod. Eng. Res. Dev.*, vol. 9, no. 1, 2019, doi: 10.24247/ijmperdfeb201918.

[10] W. J. Arbegast, "Friction stir welding after a decade of development - Its not just welding anymore," *Weld. J.*, 2006.

[11] G. Sangalli, G. V. B. Lemos, D. Martinazzi, C. R. de L. Lessa, A. B. Beskow, and A. Reguly, "Towards Qualification of Friction Stir Welding to AA5083-O and AA5052-O Aluminum Alloys," *Mater. Res.*, vol. 22, no. 5, pp. 1–8, 2019, doi: 10.1590/1980-5373-mr-2019-0349.

[12] J. Martin and S. Wei, "Friction stir welding technology for marine applications," *Frict. Stir Weld. Process. VIII*, pp. 219–226, 2016, doi: 10.1007/978-3-319-48173-9_24.

[13] G. Wang, Y. Zhao, and Y. Hao, "Friction stir welding of high-strength aerospace aluminum alloy and application in rocket tank manufacturing," *J. Mater. Sci. Technol.*, vol. 34, no. 1, pp. 73–91, 2018, doi: 10.1016/j.jmst.2017.11.041.

[14] S. L. Campanelli, G. Casalino, C. Casavola, and V. Moramarco, "Analysis and comparison of friction stir welding and laser assisted friction stir welding of aluminum alloy," *Materials (Basel).*, vol. 6, no. 12, pp. 5923–5941, 2013, doi: 10.3390/ma6125923.

[15] A. Magalhães, J. De Backer, and G. Bolmsjö, "Thermal dissipation effect on temperature-controlled friction stir welding," *Soldag. e Insp.*, vol. 24, pp. 2–10, 2019, doi: 10.1590/0104-

9224/SI24.28.

[16] M. J. Lasley, "a Finite Element Simulation of Tempetrature and Material Flow in Friction Stir Welding," *Engineering*, no. April, 2005.

[17] Stephen Leon J, Bharathiraja G, and Jayakumar V, "Analytical and experimental investigations of optimum thermomechanical conditions to use tools with non-circular pin in friction stir welding," *Int. J. Adv. Manuf. Technol.*, vol. 107, no. 11–12, 2020, doi: 10.1007/s00170-020-05341-7.

[18] L. Zhou *et al.*, "New technique of self-refilling friction stir welding to repair keyhole," *Sci. Technol. Weld. Join.*, vol. 17, no. 8, pp. 649–655, 2012, doi: 10.1179/1362171812Y.0000000058.

[19] B. Malard, F. De Geuser, and A. Deschamps, "Microstructure distribution in an AA2050 T34 friction stir weld and its evolution during post-welding heat treatment," *Acta Mater.*, vol. 101, pp. 90–100, 2015, doi: 10.1016/j.actamat.2015.08.068.

[20] G. Sorger, T. Sarikka, P. Vilaça, and T. G. Santos, "Effect of processing temperatures on the properties of a high-strength steel welded by FSW," *Weld. World*, vol. 62, no. 6, pp. 1173–1185, 2018, doi: 10.1007/s40194-018-0612-8.

[21] H. Nami, H. Adgi, M. Sharifitabar, and H. Shamabadi, "Microstructure and mechanical properties of friction stir welded Al / Mg 2 Si metal matrix cast composite," *Mater. Des.*, vol. 32, no. 2, pp. 976–983, 2011, doi: 10.1016/j.matdes.2010.07.008.

[22] J. S. Leon and V. Jayakumar, "Transient heat input model for friction stir welding using non-circular tool pin," *FME Trans.*, vol. 48, no. 1, 2020, doi: 10.5937/fmet2001137L.

[23] M. Z. H. Khandkar, J. A. Khan, and A. P. Reynolds, "A thermal model of the friction stir welding process," *ASME Int. Mech. Eng. Congr. Expo. Proc.*, vol. 5, pp. 115–124, 2002, doi: 10.1115/IMECE2002-33817.

[24] G. Swaminathan and S. Sathiyamurthy, "Experimental study of mechanical and metallurgical properties of friction stir welded dissimilar aluminum alloys," *Int. J. Mech. Prod. Eng. Res. Dev.*, vol. 8, no. 1, pp. 1049–1058, 2018.

[25] M. Al-Moussawi and A. J. Smith, "Defects in Friction Stir Welding of Steel," *Metallogr. Microstruct. Anal.*, vol. 7, no. 2, pp. 194–202, 2018, doi: 10.1007/s13632-018-0438-1.

[26] X. He, F. Gu, and A. Ball, *A Review of Numerical Analysis of Friction Stir Welding*, no. March. Elsevier Ltd, 2014.

[27] E. T. Akinlabi, R. M. Mahamood, S. A. Akinlabi, and E. Ogunmuyiwa, "Processing parameters influence on wear resistance behaviour of friction stir processed Al-TiC composites," *Adv. Mater. Sci. Eng.*, vol. 2014, 2014, doi: 10.1155/2014/724590.

[28] P. Carlone and G. S. Palazzo, "Influence of Process Parameters on Microstructure and Mechanical Properties in AA2024-T3 Friction Stir Welding," *Metallogr. Microstruct. Anal.*, vol. 2, no. 4, pp. 213–222, 2013, doi: 10.1007/s13632-013-0078-4.

[29] V. Shokri, A. Sadeghi, and M. H. Sadeghi, "Effect of friction stir welding parameters on microstructure and mechanical properties of DSS–Cu joints," *Mater. Sci. Eng. A*, vol. 693, no. February, pp. 111–120, 2017, doi: 10.1016/j.msea.2017.03.054.

[30] C. Rajendran, K. Srinivasan, V. Balasubramanian, H. Balaji, and P. Selvaraj, "Mechanical properties and microstructural characteristics of friction stir welded AA2014-T6 aluminium alloy joints," *J. Mech. Behav. Mater.*, vol. 28, no. 1, pp. 169–185, 2020, doi: 10.1515/jmbm-2019-0019.

[31] S. K. Panda, P. Chandrasekhar, and S. Singh, "Finite Element Simulation of Friction Stir Butt Welding of AA4047 Aluminium Plates," *IRACST – Eng. Sci. Technol. An Int. J.*, vol. 2, no. 6, pp. 2250–3498, 2012.

[32] J. Kundu and H. Singh, "Friction stir welding of AA5083 aluminium alloy: Multi-response optimization using Taguchi-based grey relational analysis," *Adv. Mech. Eng.*, vol. 8, no. 11, pp. 1–10, 2016, doi: 10.1177/1687814016679277.

[33] N. Haldar, S. Datta, and R. Kumar, "Experimental studies on friction-stir welding of AA6061 using Inconel 601 tool," *J. Brazilian Soc. Mech. Sci. Eng.*, vol. 40, no. 9, 2018, doi: 10.1007/s40430-018-1378-z.

[34] J. Langari and F. Kolahan, "The effect of friction stir welding parameters on the microstructure, defects, and mechanical properties of AA7075-T651 aluminium alloy joints," *Sci. Iran.*, vol. 26, no. 4 B, pp. 2418–2430, 2019, doi: 10.24200/sci.2018.5700.1434.

[35] A. Forcellese, M. Simoncini, and G. Casalino, "Influence of process parameters on the vertical forces generated during friction stir welding of AA6082-T6 and on the mechanical properties of the joints," *Metals (Basel).*, vol. 7, no. 9, 2017, doi: 10.3390/met7090350.

[36] J. W. Kwon *et al.*, "Influence of tool plunge depth and welding distance on friction stir lap welding of AA5454-O aluminum alloy plates with different thicknesses," *Trans. Nonferrous Met. Soc. China (English Ed.*, vol. 22, no. SUPPL.3, pp. s624–s628, 2012, doi: 10.1016/S1003-6326(12)61775-0.

[37] S. Rathee, S. Maheshwari, A. N. Siddiquee, and M. Srivastava, "Effect of tool plunge depth on reinforcement particles distribution in surface composite fabrication via friction stir processing," *Def. Technol.*, vol. 13, no. 2, pp. 86–91, 2017, doi: 10.1016/j.dt.2016.11.003.

[38] V. Malik, N. K. Sanjeev, H. S. Hebbar, and S. V Kailas, "Investigations on the Effect of Various Tool Pin Profiles in Friction Stir Welding Using Finite Element Simulations," *Procedia Eng.*, vol. 97, pp. 1060–1068, 2014, doi: 10.1016/j.proeng.2014.12.384.

[39] J. Stephen Leon and V. Jayakumar, "Effect of tool shoulder and pin cone angles in friction stir welding using non-circular tool pin," *J. Appl. Comput. Mech.*, vol. 6, no. 3, 2020, doi: 10.22055/JACM.2019.29340.1585.

[40] H. M. Kuhbanani, H. Yasemi, and H. A. Derazkola, "Effects of Tool Tilt Angle and Plunge Depth on Properties of Polycarbonate FSW Joint," vol. 7, no. 4, pp. 41–55, 2019.

[41] M. Ilangovan, S. R. Boopathy, and V. Balasubramanian, "Effect of tool pin profile on microstructure and tensile properties of friction stir welded dissimilar AA6061- AA5086 aluminium alloy joints," *Def. Technol.*, 2015, doi: 10.1016/j.dt.2015.01.004.

[42] A. Tongne, M. Jahazi, E. Feulvarch, and C. Desrayaud, "Banded structures in friction stir welded Al alloys," *J. Mater. Process. Technol.*, vol. 221, pp. 269–278, 2015, doi: 10.1016/j.jmatprotec.2015.02.020.

[43] N. Zhou, D. Song, W. Qi, X. Li, J. Zou, and M. M. Attallah, "Influence of the kissing bond on the mechanical properties and fracture behaviour of AA5083-H112 friction stir welds," *Mater. Sci. Eng. A*, vol. 719, pp. 12–20, 2018, doi: 10.1016/j.msea.2018.02.011.

[44] M. F. Xavier and V. Jayabalan, "Journal of Materials Processing Technology Tool travel speed effects on the microstructure of friction stir welded aluminum – copper joints," *J. Mater. Process. Tech.*, vol. 217, pp. 105–113, 2015, doi: 10.1016/j.jmatprotec.2014.11.007.

[45] F. Gharavi, F. Fadaeifard, K. A. Matori, R. Yunus, and N. K. Othman, "Mechanical and structural evaluation of friction stir welded 6061 aluminium alloy lap joints at different welding speeds," *Kov. Mater.*, vol. 54, no. 5, pp. 351–361, 2016, doi: 10.4149/km_2016_5_351.

[46] N. Soni, S. Chandrashekhar, K. A., and C. V.R., "Defects Formation during Friction Stir Welding : A Review," *Int. J. Eng. Manag. Res.*, vol. 7, no. 3, pp. 121–125, 2017, doi: 10.13140/RG.2.2.19381.93921.

INFLUENCE OF BORON CARBIDE ADDITION ON PARTICLE SIZE OF COPPER ZINC ALLOYS SYNTHESIZED BY POWDER METALLURGY

L.Rajeshkumar[a*], R.Suriyanarayanan[a], K.Shree Hari[a], S.Venkatesh Babu[a], V.Bhuvaneswari[a], M.P.Jithin Karunan[b]

[a]Department of Mechanical Engineering, KPR Institute of Engineering and Technology, Coimbatore, Tamilnadu – 641407, India

[b]Department of Manufacturing Engineering, National Institute of Technology, Tiruchirappalli, Tamilnadu - 620015, India

*Corresponding author e-mail: lrkln27@gmail.com

Abstract. Material systems with the combination of metals and non-metals forming an alloy systems has numerous advantages when compared with the monolithic materials. Synthesizing such materials systems can easily be done by many solid state processing methods. Copper Zinc (Cu-Zn) forms a ductile material system finding its applications in various conductive and tribological applications. Nano composites finds lot more applications due to the high specific area occupied by the particles. Focus of the current project is on the study of influence of Boron Carbide (B_4C) powders addition into the Cu-Zn matrix while synthesizing them through high energy ball milling. B_4C powders are added to the matrix at 10 % by volume. Particle size of the powder are determined by Scherer and Williamson-Hall synthesis methods after characterizing the milled powders by X-Ray diffraction (XRD). As the B_4C powders are brittle, maximum volume fraction of these powders are expected to influence particle size of Cu-Zn alloy powders. Powder morphology is analysed by scanning electron microscopy (SEM). Influence of B_4C powder particles upon grain size of Cu-Zn alloy powder blends was very minimum due to the tendency of particles to get saturated with respect to particle size after repeated milling.

Keywords: *Ball milling, XRD, SEM, particle size*

1. INTRODUCTION

Metal matrix composites (MMCs) are constitutive materials comprising of more than one material whose properties can be customized as per the requirements by making suitable modifications in the content of constituents [1]. MMCs can be claimed to be reliable only when there is a reasonable return of performance for the cost incurred in processing of the material. Tailorable properties are readily achieved only by the use of such composite materials and it has been in the limelight for the past few decades [2, 3]. Copper, Aluminium and Magnesium are the most commonly used base metals as far as MMCs are concerned. Copper is majorly used due to its meritorious characteristics like better electrical and thermal conductivity and enhanced ductility [4]. Boron Carbide (B_4C) is one among the hardest

IOP Conf. Series: Materials Science and Engineering **954** (2020) 012008 doi:10.1088/1757-899X/954/1/012008

material and ranks between 9 and 10 in Mohr's scale of hardness. It is characterized by better chemical and thermal stability [5]. As boron carbide possess inherent hardness and better mechanical properties, it could be positively used in various punitive industry applications including wear resistant components, vehicles and body armors in military vehicles and elevated temperature nozzles. Usage of boron carbide is supplemented by many other better characteristics such as low specific gravity and elevated melting point [6-9].

The technology of manufacturing MMCs is in competition with other modern material technologies, for example powder metallurgy [10]. Powder metallurgy (P/M) technique is a solid state processing method of composite materials mainly used to manufacture the particulate reinforced composites [11]. Various classical components with intricate shapes and medium sizes can be fabricated through this technique easily [12-14]. Common methodology of powder metallurgy involves three steps: first is blending and synthesis of precursor powder particles known as mechanical alloying, second is to densify the alloyed powders by applying pressure over them when placed inside a closed die called compaction and the third is to consolidate them by reducing their density further through heating called sintering [15]. Mechanical alloying is carried out by ball milling, densification process in a die and plunger and sintering is done in a muffle furnace. PM technique renders numerous merits when compared to the typical material processing methods including accommodation of shape complexity, better dimensional stability, less energy consumption during manufacturing and optimum material usage [16, 20-22]. As the PM manufactured components were greatly influenced by the compaction and densification, particle size of the precursor powders play a salient role in the powders compressibility [17-19]. Due to this fact, the mechanical alloying of the powder particles were carried out using a high energy ball mill. Unlike normal particle reduction methods such as pulverization, ball milling utilizes the enormous energy from the steel balls generated due to their impacts with the powder particles which results in a large scale forging thus minimizing the powder particle size [23, 24]. Here XRD is utilized to assess the various constituents present within the milled powders with the help of the diffraction spectra obtained by passing X-rays through them and could also be used in determination of particle size based on the peak properties. SEM is a widely used microscopic imaging method that renders a high quality image through the sputtering of electrons that were focused over the surface to be examined and is used to reveals the internal characteristics of blend powders in micro level [25].

In the current work copper and zinc powders were blended with various weight fractions of boron carbide powders and were mechanically alloyed for various milling durations. Influence of addition of a hard ceramic particle in a ductile-ductile system is analyzed with respect to the particle size of the blended composite powders. After milling, the particle size of the milled powders was calculated by XRD and the dominant process responsible for the variation of particle size was determined by SEM.

2. EXPERIMENTAL DATA

2.1. *Materials*

Copper, Zinc and Boron Carbide are purchased from Coimbatore metal ltd, Coimbatore, with standard particle size of 325 mesh. The manufacturing of Copper, Zinc, Boron Carbide composites by means of mechanical alloying in a high energy ball mill. Copper is mixed with the Zinc (Cu - 90%, Zn -10%), then Copper-Zinc is reinforced with Boron Carbide in 10 % by weight.

Table 1. Material properties

PROPERTIES	COPPER	ZINC	BORON CARBIDE
Poisson ratio	0.34	0.25	0.20
Young's modulus (GPa)	120	108	450
Shear modulus (GPa)	48	43	180

Melting point (°C)	1085	420	2350
Mohr's hardness	9.0	2.5	3.0

The materials are added to the vial of ball milling setup with the ratio between balls and powders (BPR) as 10:1. Toluene is used as the process control agent (PCA) as per the requirement. The speed ratio of sun and the planet wheels, to which the vial is attached, were set as 1:4. The planetary gears rotate in self-contradicting direction within a regular interval of time of 15 seconds between each change in rotation direction. The speed of rotation of the vials were maintained at 120 rpm [26]. These process standards were maintained for various milling durations of 10, 20 and 30 hours. In order to dissipate the heat from the vials generated due to friction, they were set to be idle for 10 minutes after a continuous run for 75 minutes.

2.2. Characterization by XRD and SEM

The main objective of the XRD technique is to characterize the powders for their particle size based on the peaks obtained. In the current work a total of ten samples were characterized by XRD, which were collected at various milling time periods, like 10, 20 and 30 hours. Usual procedures were adopted to carry out XRD analysis. About 60 gm of powders were taken from every sample and evaluated. The powders were closely packed in a glass slit of size 12 mm × 12 mm × 2 mm. The measurements of XRD were carried out using Schimadu XRD machine with K_a value of 0.154 nm.

The SEM produces images from the samples through scanning the specimen with the support of a focused electron beam. The milled powders were evaluated using a FE-SEM model Carl Zeiss Microscope to identify the prevailing particle size variation phenomenon after each milling cycle. Current study helps to converge to a result that the particle size cannot be downsized beyond steady state.

2.3. Analysis of particle size by Williamson-Hall synthesis

The values obtained from the XRD analyses were used to calculate the particle size and strain induced on the surface of the powders. The calculations were based on the Williamson-Hall equation, as given in Equation (1).

$$\beta\cos\theta = \frac{k\lambda}{t} + 4\varepsilon\sin\theta \tag{1}$$

where *k* is the shape factor (value = 0.5), λ is the wavelength of the X-rays (value 0.154 nm). A plot between β cos θ (*y*-axis) and 4 sin θ (*x*-axis) would render a straight-line equation in which the slope (m) could be equated to the lattice strain directly and the intercept (c) could be used to calculate the particle size (*t*) by using the following relation: $c = k\lambda/t$.

2.4. Analysis of particle size by Scherer Equation

Similar to the Williamson-Hall synthesis, the particle size (t) of the powders were evaluated from the X-ray peak broadening data in orthogonal direction to the powder surface by using Scherer equation. It is given by:

$$t = \frac{0.9\lambda}{\beta\cos\theta} \tag{2}$$

where θ is the Bragg angle, λ is the wavelength of X-ray radiation (1.54 Å), and β is the full width at half maximum of the considered X-ray peak (FWHM) in radians. Since Scherer's equation gives only the particle size of the powder and neglects the lattice strain induced during variation of particle size, it is less sensitive when compared with other synthesis equations [27].

3. RESULTS AND DISCUSSION

3.1. Particle Size reduction

Generally copper is a ductile metal while zinc is partially brittle in nature upto certain temperature. When these elements were combined to form an alloy system they were expected to behave more like brittle system. Due to X-ray peak broadening there is a raise in powder deformation as the milling hours increase thereby increasing the strain induced on the surface responsible for the reduction of particle size of the powders. Purchased copper and zinc particles were observed to have micron sized particles which got reduced directly to nano scale after a long milling cycle of 10 hours. This is due to the fact that the powder particles were imparted with defects in larger magnitudes over their surface which shifted the particle size to even lower values. From the figures, it could be observed that the X-ray peaks experienced a mild shift towards their right which was solely because of the homogenous suspension of zinc atoms into the abundant copper matrix. Yet this phenomenon has to be verified by taking XRD of purchased zinc powder. When the unreinforced copper-zinc powders were milled for various durations, it could be observed that there is a continuous reduction in particle size due to the presence of partially brittle zinc. Minimum particle size was obtained during the highest duration of milling of 30 hours. Figure 1 shows the XRD peaks of copper-zinc alloys for various milling durations.

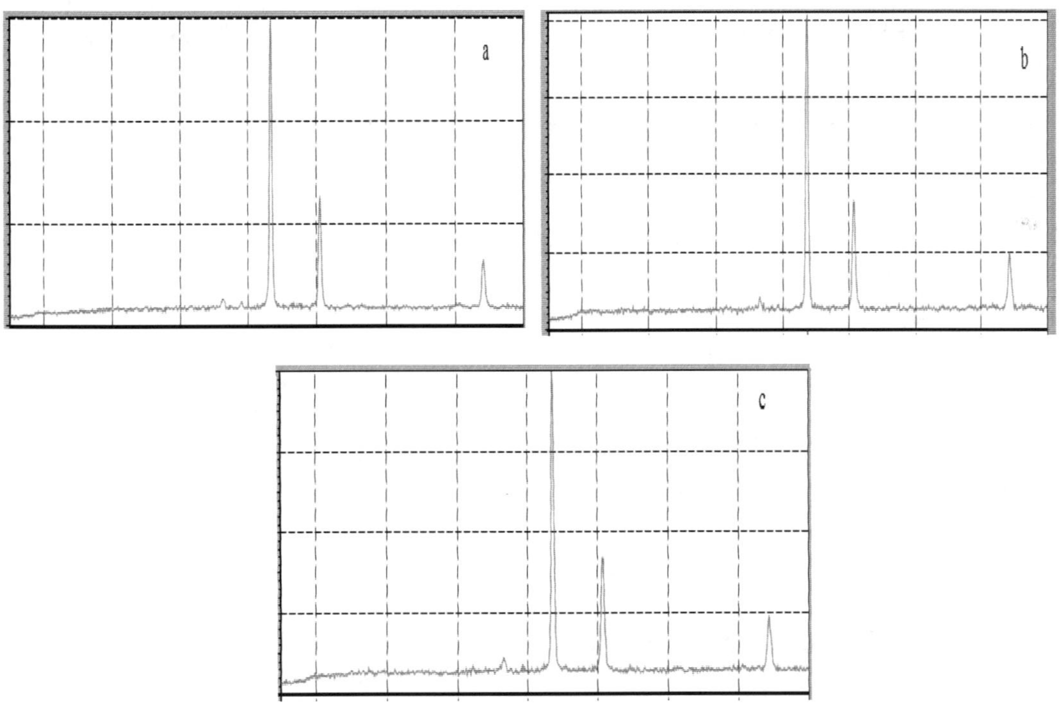

Figure 1 XRD peaks of Cu-Zn alloy for various milling duration (h) – a) 10h b) 20h c) 30h

On contrary, when 10 % B_4C particles were reinforced to Cu-Zn system, the behaviour of composite powders was slightly different. The particle size of the powders decrease at 30 hours after increasing during 20 hours of milling. This could be attributed to the higher hardness of the B_4C powders which could have acted as additional milling agent for reducing the powder particle size but due to higher amount of surface defects the particles developed a tendency to

IOP Conf. Series: Materials Science and Engineering **954** (2020) 012008 doi:10.1088/1757-899X/954/1/012008

stick to each other thereby increasing the average particle size of the composite powders. This is the reason for higher particle size of B_4C reinforced composites when compared with unreinforced alloy [27]. Yet the tendency of particles to undergo further reduction could be studied by possibly increasing the volume fraction of B_4C powders. Table 2 shows the variation of particle sizes after various milling durations calculated by Williamson-Hall synthesis and Scherer's equation.

Table 2 Variation in particle size of milled powders

Milling duration (hours)	Particle size by Williamson-Hall equation (nm)		Particle size by Scherer's equation (nm)	
	Cu-Zn	Cu-Zn + 10 % B_4C	Cu-Zn	Cu-Zn + 10 % B_4C
10	14.16	27.75	24	17.5
20	13.87	38.54	29.12	19.58
30	9.72	31.53	22.35	19.14

Figure 2 shows the XRD peaks of Cu-Zn alloys reinforced with 10 % B_4C composites milled for various milling durations.

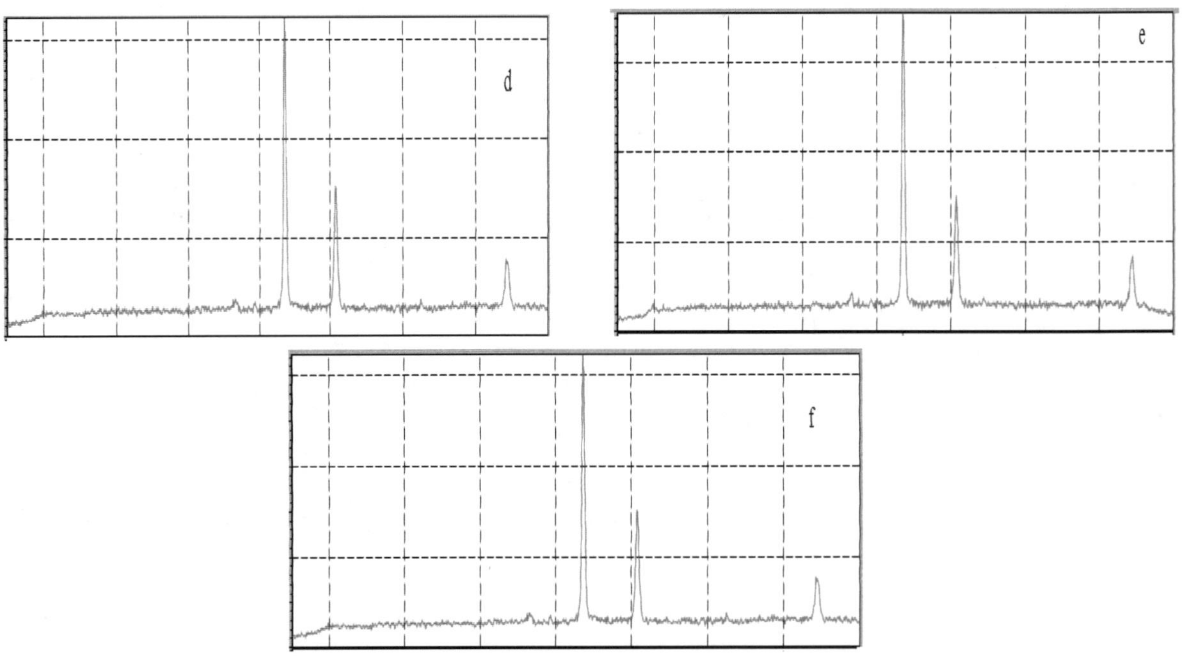

Figure 2 XRD peaks of Cu-Zn-10 % B_4C composites for various milling duration (h) – a) 10h, b) 20 h, c) 30 h

3.2. *SEM analysis*

From the SEM micrographs the dominant process of milling could be observed and inferred. Figure 3a, b & c shows the SEM micrographs of Cu-Zn alloy powders milled for 10, 20 and

30 hours respectively. From the images it could be seen that fracture has dominated resulting in reduction of particle size of the powders. Irregular grain boundaries depicts that ductile copper has undergone repeated deformation due to the impact of hardened steel balls on the surface of the powders. Since zinc is also partially brittle, it could also been an additional milling agent. Yet evidences to substantiate the above claim is not abundant in the current study.

Similarly figure 4 a, b & c shows the SEM micrographs of Cu-Zn composites reinforced with 10 % B4C. Images illustrates that in spite of undergoing repeated fracture, the particle sizes remain almost at larger sizes than unreinforced alloy. This could be possibly due to the fact that B4C is a material of high hardness which might have caused too much deformation in Cu-Zn system which reached saturation.

Figure 3 – SEM micrographs of milled Cu-Zn alloys a) 10 h, b) 20 h, c) 30 h

After this the particle size could not be reduced further and the particles has established a tendency to agglomerate with each other resulting in increased particle size. This phenomenon is called cold welding in mechanical alloying [3]. As a result of cold welding of particles, their size was observed to be larger than the unreinforced alloy. Nevertheless, cold welding could not be claimed to be predominant as the particle size of the composite powders continuously reduce. Hence it could be concluded that the particle size had reached saturation within the process itself such that impossibility of further reduction increased the particle size when compared with the unreinforced alloy.

4. CONCLUSION

Copper zinc alloys were reinforced with boron carbide powders and the composites were synthesized using mechanical alloying. Conducted experiments and the characterization techniques rendered the following conclusions as results of the work.

- Unreinforced copper zinc powder particles reached the size of nano scale at 10 hours of milling and continuously reduced thereafter. Minimum particle size obtained was

9.72 nm by Williamson-Hall synthesis equation and 22.35 nm by Scherer's equation after 30 hours of milling.

- From SEM image it could be concluded that reduction of particle size of unreinforced Cu-Zn alloy was due to continuous fracture of powder particles induced because of prolonged milling cycles.
- Particle size of Cu-Zn composite powders containing 10 % B_4C also reached nano scale and the minimum particle size was obtained as 31.53 nm by Williamson-Hall synthesis equation and 19.14 nm by Scherer's equation after 30 hours of milling.
- SEM image of Cu-Zn with 10 % B_4C depicts that particle size reduction was due to fracture despite the fact that average particle size of B_4C reinforced composites was greater than unreinforced alloy.
- Increased particle size of B_4C reinforced composites could be due to the cold welding that might have occurred due to acute reduction of particle size of composite powders in presence of highly brittle and hard B_4C particles.

Figure 4 – SEM micrographs of milled Cu-Zn + 10 % B_4C a) 10 h, b) 20 h, c) 30 h

REFERENCES

[1] Barick, Prasenjit, Dulal Chandra Jana, and Natarajan Thiyagarajan. "Effect of particle size on the mechanical properties of reaction bonded boron carbide ceramics." *Ceramics International* 39, no. 1 (2013): 763-770.

[2] Janoš, Pavel, Jakub Ederer, Věra Pilařová, Jiří Henych, Jakub Tolasz, David Milde, and Tomáš Opletal. "Chemical mechanical glass polishing with cerium oxide: Effect of selected physico-chemical characteristics on polishing efficiency." *Wear* 362 (2016): 114-120.

[3] L Rajesh kumar, KS, Amirthagadeswaran. "Corrosion and wear behaviour of nano Al2O3 reinforced copper metal matrix composites synthesized by high energy ball milling." *Particulate Science and Technology* 38, no. 2 (2020): 228-235.

[4] Ahmed, Afzaal. "Deposition and analysis of composite coating on aluminum using Ti–B4C powder metallurgy tools in EDM." *Materials and Manufacturing Processes* 31, no. 4 (2016): 467-474.

[5] Manjunatha, B., H. B. Niranjan, and K. G. Satyanarayana. "Effect of amount of boron carbide on wear loss of Al-6061 matrix composite by Taguchi technique and Response surface analysis." In *IOP Conference Series: Materials Science and Engineering*, vol. 376, no. 1, p. 012071. IOP Publishing, 2018.

[6] Dias, Alexandre Nogueira Ottoboni, Aline da Silva, Carlos Alberto Rodrigues, Mírian de Lourdes Noronha Motta Melo, Geovani Rodrigues, and Gilbert Silva. "Effect of high energy milling time of the aluminum bronze alloy obtained by powder metallurgy with niobium carbide addition." *Materials Research* 20, no. 3 (2017): 747-754.

[7] Zhou, Yanchun, and Mohamed N. Rahaman. "Effect of redox reaction on the sintering behavior of cerium oxide." *Acta materialia* 45, no. 9 (1997): 3635-3639.

[8] Ghasali, Ehsan, Masoud Alizadeh, Morteza Niazmand, and Touradj Ebadzadeh. "Fabrication of magnesium-boron carbide metal matrix composite by powder metallurgy route: comparison between microwave and spark plasma sintering." *Journal of Alloys and Compounds* 697 (2017): 200-207.

[9] Izu, Noriya, Toshio Uchida, Ichiro Matsubara, Toshio Itoh, Woosuck Shin, and Maiko Nishibori. "Formation mechanism of monodispersed spherical core–shell ceria/polymer hybrid nanoparticles." *Materials Research Bulletin* 46, no. 8 (2011): 1168-1176.

[10] Hayun, Shmuel, Amir Weizmann, Moshe P. Dariel, and Nahum Frage. "The effect of particle size distribution on the microstructure and the mechanical properties of boron carbide-based reaction-bonded composites." *International Journal of Applied Ceramic Technology* 6, no. 4 (2009): 492-500.

[11] Venkatesh, B., and B. Harish. "Mechanical properties of metal matrix composites (Al/SiCp) particles produced by powder metallurgy." *International Journal of Engineering Research and General Science* 3, no. 1 (2015): 1277-1284.

[12] Prakash, K. Soorya, P. M. Gopal, D. Anburose, and V. Kavimani. "Mechanical, corrosion and wear characteristics of powder metallurgy processed Ti-6Al-4V/B4C metal matrix composites." *Ain Shams Engineering Journal* 9, no. 4 (2018): 1489-1496.

[13] Čapek, Jaroslav, and Dalibor Vojtěch. "Microstructural and mechanical characteristics of porous iron prepared by powder metallurgy." *Materials Science and Engineering: C* 43 (2014): 494-501.

[14] Wu, Jie, Lei Xu, Zhengguan Lu, Bin Lu, Yuyou Cui, and Rui Yang. "Microstructure design and heat response of powder metallurgy Ti2AlNb alloys." *Journal of Materials Science & Technology* 31, no. 12 (2015): 1251-1257.

[15] Jung, Hye Jin, Youngku Sohn, Hong Gye Sung, Hyung Soo Hyun, and Weon Gyu Shin. "Physicochemical properties of ball milled boron particles: dry vs. wet ball milling process." *Powder Technology* 269 (2015): 548-553.

[16] Avcıoğlu, Suna, Merve Buldu, Figen Kaya, Cem Bülent Üstündağ, Erol Kam, Yusuf Ziya Menceloğlu, Hayri Yılmaz Kaptan, and Cengiz Kaya. "Processing and properties of boron carbide (B4C) reinforced LDPE composites for radiation shielding." *Ceramics International* 46, no. 1 (2020): 343-352.

[17] Guo, Hao, and Zhongwu Zhang. "Processing and strengthening mechanisms of boron-carbide-reinforced aluminum matrix composites." *Metal Powder Report* 73, no. 2 (2018): 62-67.

[18] Gu, Junfeng, Ji Zou, Peiyan Ma, Hao Wang, Jinyong Zhang, Weimin Wang, and Zhengyi Fu. "Reactive sintering of B4C-TaB2 ceramics via carbide boronizing: Reaction process, microstructure and mechanical properties." *Journal of Materials Science & Technology* 35, no. 12 (2019): 2840-2850.

[19] Nieto, Andy, Hanry Yang, Lin Jiang, and Julie M. Schoenung. "Reinforcement size effects on the abrasive wear of boron carbide reinforced aluminum composites." *Wear* 390 (2017): 228-235.

[20] Jain, Shobhit, R. S. Rana, and Prabhash Jain. "Study of microstructure and mechanical properties

of Al-Cu metal matrix reinforced with B4C particles Composite." *International Research Journal of Engineering and Technology* 3, no. 1 (2016): 499-504.

[21] Karakoç, Halil, Şener Karabulut, and Ramazan Çıtak. "Study on mechanical and ballistic performances of boron carbide reinforced Al 6061 aluminum alloy produced by powder metallurgy." *Composites Part B: Engineering* 148 (2018): 68-80.

[22] Kozień, Dawid, Piotr Jeleń, Maciej Sitarz, and Mirosław M. Bućko. "Synthesis of boron carbide powders from mono-and polysaccharides." *International Journal of Refractory Metals and Hard Materials* 86 (2020): 105099.

[23] Meignanamoorthy, M., and M. Ravichandran. "Synthesis, properties and microstructure of sintered and hot extruded boron carbide reinforced AA8079 (Al-Cu-Fe-Si-Zn) matrix composites." *Materials Research Express* 5, no. 11 (2018): 116508.

[24] Izu, Noriya, Woosuck Shin, Ichiro Matsubara, and Norimitsu Murayama. "The effects of the particle size and crystallite size on the response time for resistive oxygen gas sensor using cerium oxide thick film." *Sensors and Actuators B: Chemical* 94, no. 2 (2003): 222-227.

[25] Gao, Xin, Hongyan Yue, Erjun Guo, Shaolin Zhang, Longhui Yao, Xuanyu Lin, Bao Wang, and Enhao Guan. "Tribological properties of copper matrix composites reinforced with homogeneously dispersed graphene nanosheets." *Journal of materials science & technology* 34, no. 10 (2018): 1925-1931.

[26] L.Rajesh kumar, K S Amirthagadeswaran. "Variations in the properties of copper-alumina nanocomposites synthesized by mechanical alloying." Materiali in Technologjie, 53(1) (2019): 57-63

[27] Sivasankaran, S., K. Sivaprasad, R. Narayanasamy, and P. V. Satyanarayana. "X-ray peak broadening analysis of AA 6061100− x− x wt.% Al2O3 nanocomposite prepared by mechanical alloying." *Materials Characterization* 62, no. 7 (2011): 661-672.

IOP Conf. Series: Materials Science and Engineering **954** (2020) 012009 doi:10.1088/1757-899X/954/1/012009

DESIGN AND OPTIMIZATION OF STATIC CHARACTERISTICS FOR A STEERING SYSTEM IN AN ATV

L.Rajeshkumar[a*], V.Bhuvaneswari[a], B.Pradeepraj[a], M.Praveen pauldurai[a], C.Palanivel[a], S.Sandeep kumar[a]

[a]Department of Mechanical Engineering, KPR Institute of Engineering and Technology, Coimbatore, Tamilnadu 641407, India

[*]Corresponding Author email: lrkln27@gmail.com

Abstract. The suspension and steering system of an ATV (All-Terrain Vehicle) plays a vital role in vehicle dynamics for off road conditions by load transfer and providing stability. Incorrect geometry and additional mass on the front axle affect the stability and performance of the vehicle in dynamic condition. This can be optimized by varying the geometric, design and mass characteristics of the wheel assembly and suspension system parameters. CAD model of the wheel assembly and steering gear box had been developed using SOLIDWORKS, with respect to geometric parameters obtained on suspension and steering system analysis using LOTUS software with various structural and geometric constraints. Finite element analysis of the wheel assembly was done using ANSYS WORKBENCH for static and basic dynamic loading conditions. Then topology analysis has been done to reduce the additional mass of the wheel assembly without compromising its geometric and load transfer characteristics.

1. Introduction

Performance analysis of an off-road vehicle with respect to the environment of its operation is termed as terra mechanics. Driving a vehicle in free snow, desert sand and soft mud conditions can be included in off-road driving conditions. Such rugged terrain conditions may cause much threat towards driving the vehicle in those conditions and hence the behaviour of vehicle in such tough conditions depends upon the interaction of the vehicle with such harsh environments and conditions. It is therefore important to be empathetic over the vehicular mechanics in off-road conditions. In order to meet the operating necessities of the vehicle in off-road terrains, choice of design factors and configuration of the vehicle becomes significant [1]. During the past few years focus is being diverted towards proposing and developing a wheeled and tracked vehicle, using theoretical and analytical approaches, for operating in unpredictable and rough driving conditions.

Any off-road vehicle configuration can be specified by their operating power, dimensions and overall kerb weight. Some of the factors involved in selecting the off-road vehicle parameters include operation condition, cost of manufacturing and maintenance of the vehicle, operational environmental impact, reliability, economy and above all the need for its development. Virtual approach serves better for initially confining the operational environment and need of the vehicle. One such design is the design of an all-terrain vehicle (ATV) for the event of SAE BAJA.

Based on the need for the event vehicle configurations are taken as: Track width = 1250 m, Wheelbase = 1450 m, Ground clearance = 400 mm [2, 3]. Weight of the vehicle affect the overall performance of the vehicle. That also increases the mileage of the vehicle. The weight is reduced with the optimization of the components is done by the analysis using the software. The geometry of the systems will help in handling and maneuverability of terrain vehicles. Normally there are few issues in the handling of the vehicle system like direction control of the terrain vehicle and the control of the vehicle by posing resistance to the external instabilities [4]. Some of the parameters for the geometry are camber, caster, toe angle, roll center and roll axis. Caster is geometrical alignment of vertical axis of steering with respect to the steered wheel taken along the longitudinal direction. Caster angle is adjusted to optimize the handling characteristics of a vehicle. There are two types of caster one is positive caster and the other is negative caster. Positive caster provides the rate of inherent ability of steering wheel to center itself with respect to the caster of the vehicle so as to mechanically follow the path of steering axis. Such geometry renders better vehicular control in terms of direction and stability. Negative caster results in lighter and easier steering but reduced stability when driving in a straight line [5-7].

Camber can be defined as the angle between the vehicle vertical axis and the steered wheel vertical axis when viewed in the lateral direction from front or rear of the vehicle. There are two types: one is positive camber and the other is the negative camber [8, 9]. Positive camber stabilizes your car. Negative camber alters the handling qualities particularly in the cornering. By this we can maximize the contact patch area. There are two terms to be known for the cornering one is slip angle and other is force of cornering. Stability and direction control of all-terrain vehicles mostly depend on the aforementioned parameters [10]. If a side force is not induced along the direction perpendicular to the plane of wheel rotation, then it might be subjected to a force at an angle of α will act on the wheel due to the path of contact of the tire with the surface and is presented in Figure 1. The angle α is called slip angle and the force induced along the path of contact of tire at zero camber is called cornering force [11-13].

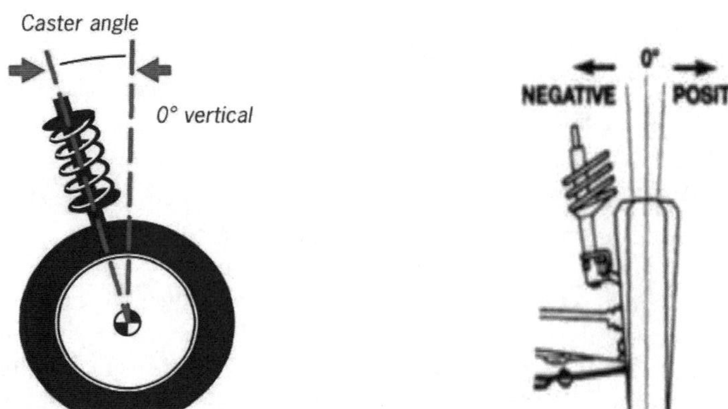

Figure 1. Schematic of Caster and Camber [10]

Toe angle denotes the tire or wheel geometry angle and represents how much the wheels are turned around the vertical axis when pointed straight-on in top view and is presented in Figure 2. Toe angle is also defined as the direction of pointing of tyres with respect to the longitudinal center axis of the chassis when the view is from the top of the vehicle [14].

IOP Conf. Series: Materials Science and Engineering **954** (2020) 012009 doi:10.1088/1757-899X/954/1/012009

Figure 2. Representation of various Toe angles

2. Geometry of the Model

2.1. Ackermann Steering geometry

Ackermann steering geometry is defined as the arrangement of wheels according to the geometry in such a way that it caters the needs of the inner or outer wheels to rotate at different speeds when the vehicle is turning in a curve at various radii and speed. During design of the steering mechanism scrub in tires induced by cornering should be as minimum as possible. These factors should ensure that tires of the vehicle, while turning, should be devoid of sliding but undergo only pure rolling. Hence the wheels should be placed in such a way that they follow a curved path at different curve radius with a common radius of curvature [15]. In order to satisfy this condition a simple relationship between steering angle and vehicle motion direction has to be adopted as shown in equation 1.

$$\text{Cot}\ \varphi - \text{Cot}\ \theta = w/b \qquad (1)$$

Where, θ and φ is the angle between inside and outside front wheel with the common center point, w is track width and b is wheel base. Substituting the values of φ as 33.84° and θ as 57.81°, we get

$$\text{Cot}\ 33.84 - \text{Cot}\ 57.81 = 1250/1450$$

$$0.8620 = 0.8620$$

Thus, Ackermann condition is satisfied by the current design of steering assembly and is presented in Figure 3. The geometry is achieved with the help of lotus software as shown in Figure 4. To match the instantaneous center for our chassis design, number of iterations are made.

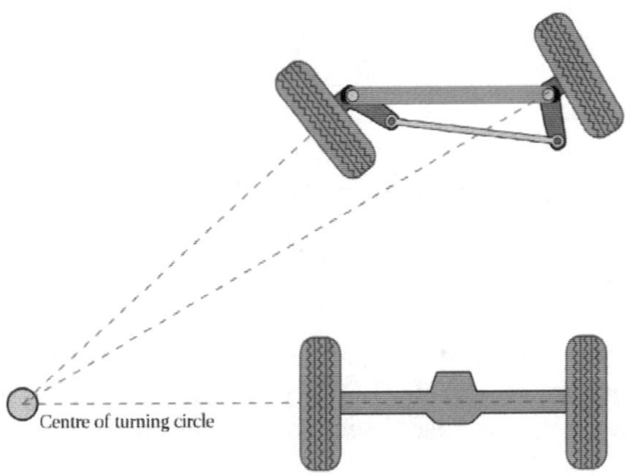

Figure 3. Ackermann Steering geometry

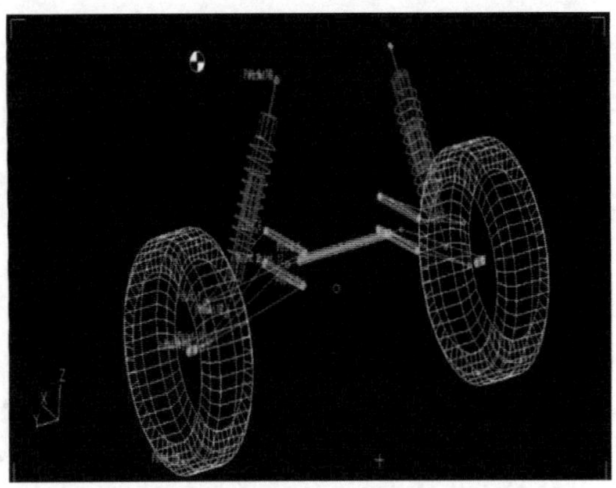

Figure 4. Steering geometry model

IOP Conf. Series: Materials Science and Engineering **954** (2020) 012009 doi:10.1088/1757-899X/954/1/012009

2.2. Instantaneous Center

The instantaneous center is called roll center. The roll center of vehicle denotes the point of corner forces acting in the suspension of the vehicle towards the body of the vehicle [16-18]. Suspension geometry is the major deciding factor of the geometric center placement in a vehicle and it is usually calculated by the concept of instantaneous center method. The SAE's definition of the force-based roll center is, "The point in the transverse vertical plane through any pair of wheel centers at which lateral forces may be applied to the sprung mass without producing suspension roll" [19, 20]. The roll center for the geometry shown in Figure 5.

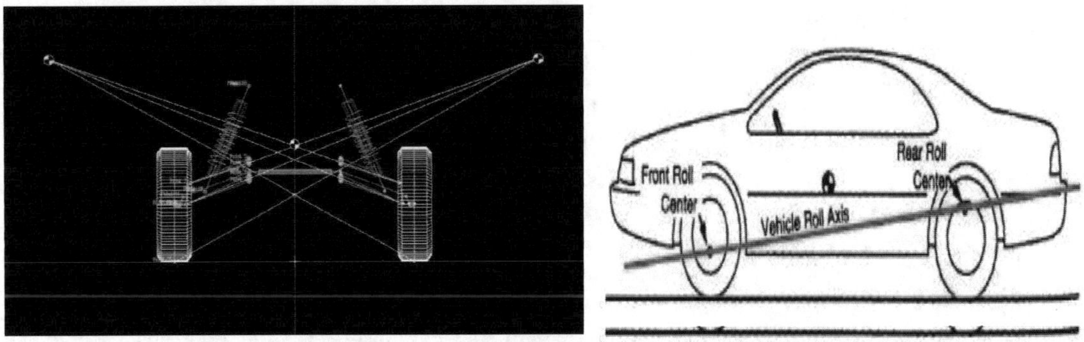

Figure 5. Geometry of a) Roll center, b) Roll axis

Roll axis could be defined as the line connecting the roll center at the front and rear and it is an imaginary line in side view. If the roll center height in front is lower than the rear then it is called anti dive, it will happen during the breaking. If front is higher than the lower then it is called anti squad, it will happen during the acceleration. If both the roll center height is same then it is neutral roll axis [21 -23]. Roll axis shows the mass shift from vehicle CG based on the inclination as shown in figure 5 b. It will be helpful during the turning of the vehicle to have the more grip in the front to turn in the curved path and to avoid the slip of the tires since it is an off-road vehicle which have the loose sand.

3. Design and Optimization of the component

The wheel assembly components were designed based on obtained geometry parameters obtained from lotus analysis. Then the deformation and stress analysis were done for the components using ANSYS, and optimization of the components carried out to reduce un-sprung mass without affecting load transfer characteristics.

3.1. Hub Design

The basic hub design was done based on considering parameters like PCD of rim, PCD of brake disk, Distance of wheel pivot point from wheel center (obtained from LOTUS), Bearing and stub axle size, Bolt size and strength. The analysis was done for various load acting on the wheel assembly for static condition of the vehicle.

3.1.1. Load calculations

Considering,

Radius of wheel, R = 292.1 mm

Acceleration due to gravity, $g = 10\ m/s^2$
Acceleration of vehicle, $a = 2\ m/s^2$
Mass of vehicle, $M_1 = 160$ kg
Mass of driver, $M_2 = 80$ kg
PCD of wheel, $D = 110$ mm
PCD of brake disk, $d_1 = 65$ mm
Brake disc diameter, $d_2 = 150$ mm

Assuming, the height of bump, $h = 100$ mm

Bump force [17], $F_B = (M * a * \sqrt{(2Rh-h^2)}) / (R-h)$
$= ((160+80) * 2 * \sqrt{(2*292.1 * 100 - 100^2)}) / (292.1 - 100)$
$= 550$ N

Thrust Force [17, 18], $F_T = (M_1 + M_2) * a$
$= (160+80) * 2$
$= 480$ N

Torque in Rim PCD [19], $T_R = F_T * (D / 2)$
$= 480 * (110 / 2)$
$= 26400$ N-mm

Braking force at caliper, $F_\mu = 3000$ N

Torque on brake disc PCD, $T_B = F_\mu * ((d_2 - d_1) / 2)$
$= 3000 * ((150 - 65) / 2)$
$= 127500$ N-mm

Load acting on one side of wheel due to mass of vehicle,
$F = (M * g) / 4$
$= ((160+80) * 10) / 4$
$= 600$ N

Force acting on steering arm of knuckle = 150 N

Force distribution of load acting on wheel due to mass of vehicle, upper mount = 96 N

Lower mount = 384 N

The material for all the wheel assembly components were chosen as Al6061-T6, because it has less mass and high ultimate tensile strength considerably. The properties of the selected materials is shown in Table 1.

Table 1. Material properties of Al6061-T6 [20]

PROPERTY	VALUES
Density	2703 kg/m^3
Modulus of Elasticity	70 GPa
Poisson's ratio	0.33
Bulk modulus	68.627 GPa
Modulus of rigidity	26.316 GPa
Ultimate Tensile Strength	310 MPa

The basic hub design has a mass of 0.50425 kg, for the material of Al 6061-T6 as shown in Figure 6(a). The hub was analyzed in FEA method by meshing the components into smaller parts (as shown in Figure 6(b) and then subjected to various load and torque acting on wheel assembly using ANSYS.

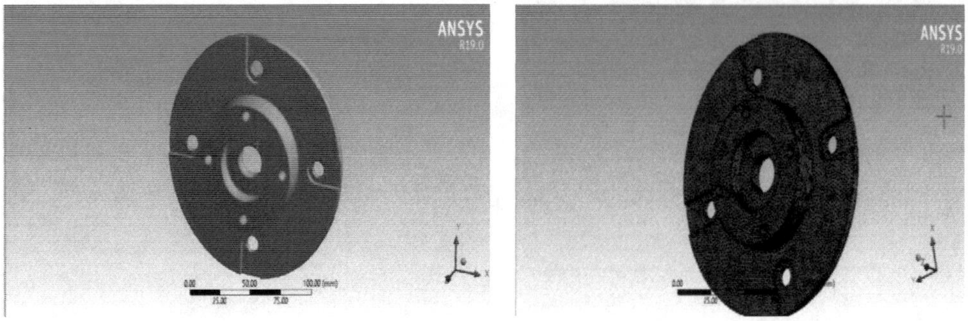

Figure 6. Modelling of hub a) Mass property, b) Meshing

Then total deformation and equivalent stress acting on the hub have been solved and verified for optimum characteristics. Figures 7(a) and 7(b) show the various loads acting on the hub and its deformation for the applied loads. Figure 8 shows the stresses acting on the hub.

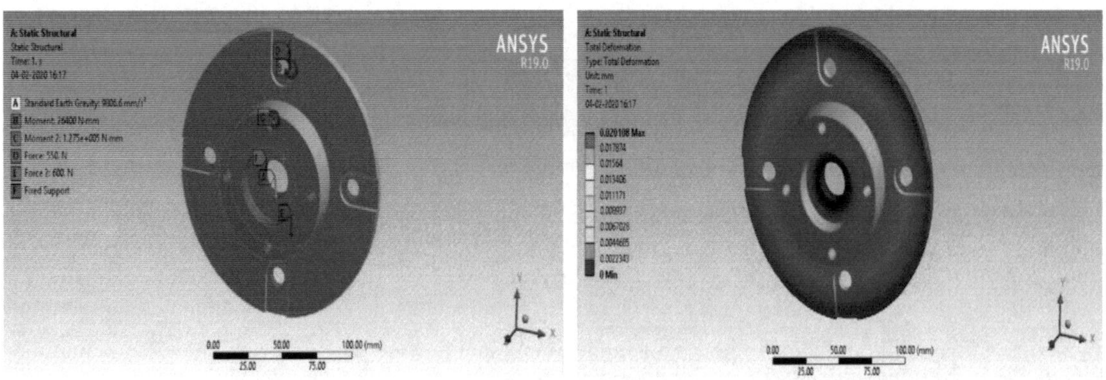

Figure 7. (a) Various load acting on hub and (b) Deformation of hub

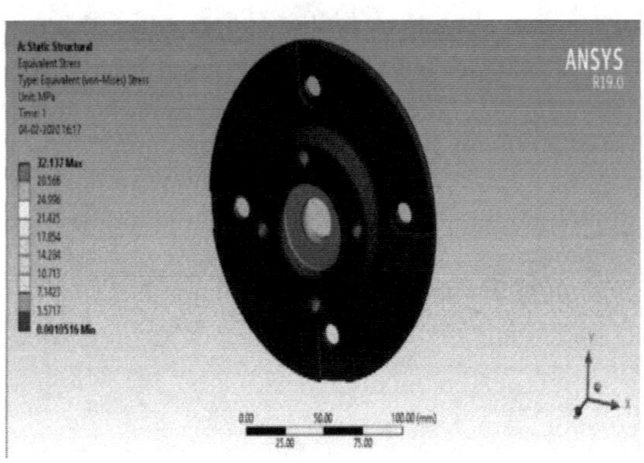

Figure 8. Stress acting on hub

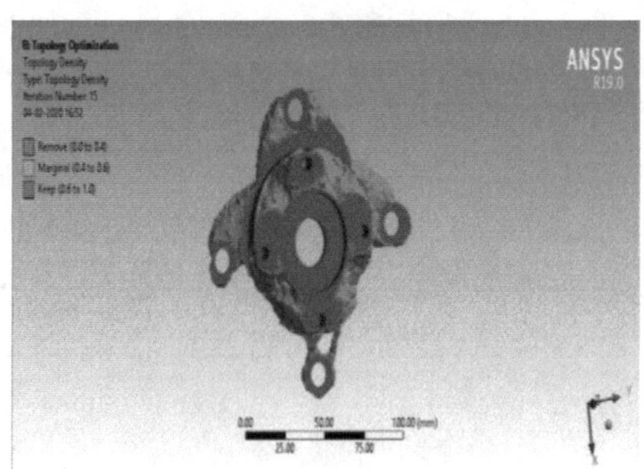

Figure 9. Topology optimization

IOP Conf. Series: Materials Science and Engineering **954** (2020) 012009 doi:10.1088/1757-899X/954/1/012009

After analyzing the hub design, the components were optimized using topology optimization tool on ANSYS as shown in Figure 9. The optimization was done to reduce mass of the component without affecting the load transfer and geometric characteristics of hub. Then the optimized hub has been given boundary structure by modifying the hub model using SOLIDWORKS. Then the hub has been again analyzed using ANSYS for verifying load and stress characteristics. Figures 10(a) and 10(b) depict the mass property and meshing of optimized hub.

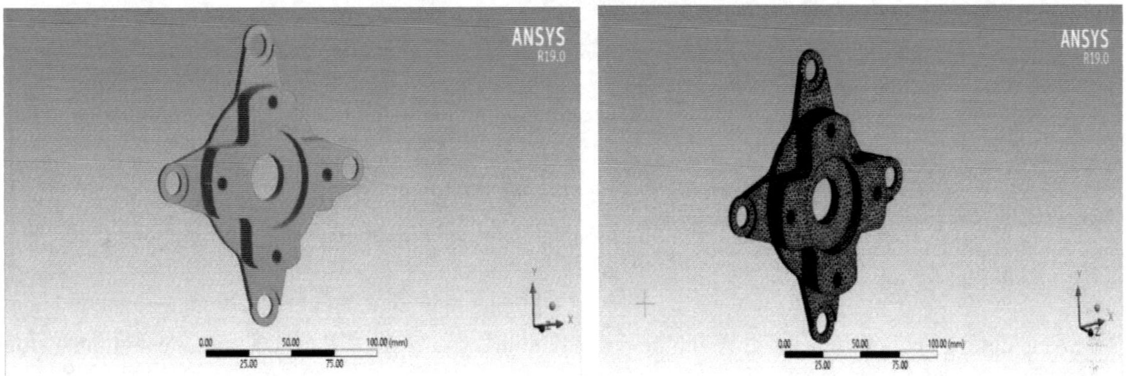

Figure 10. Optimized hub (a) Mass property and (b) Meshing

The mass of the optimized hub has a mass of 0.22177 kg which is about 44% of the basic model hub. Based on the above design factors and considerations and load acting, corresponding deformation of the modified hub is shown in Figures 11(a) and 11(b) respectively.

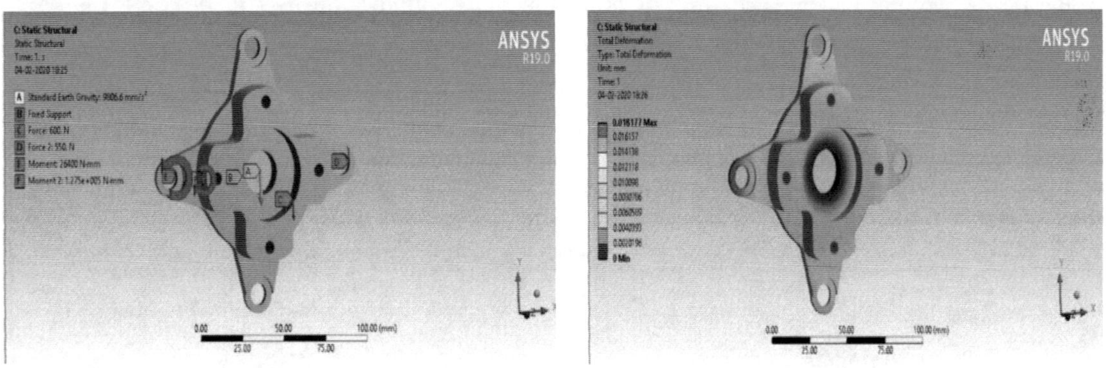

Figure 11. (a) Load acting on optimized hub and (b) Deformation of the optimized hub

The value of stresses acting on the modified hub remains almost the same as that of the currently used hub. This shows that the design of hub has been suitably modified without compromising the load acting on the hub. Figure 12 shows the stresses acting on the optimized hub.

IOP Conf. Series: Materials Science and Engineering **954** (2020) 012009 doi:10.1088/1757-899X/954/1/012009

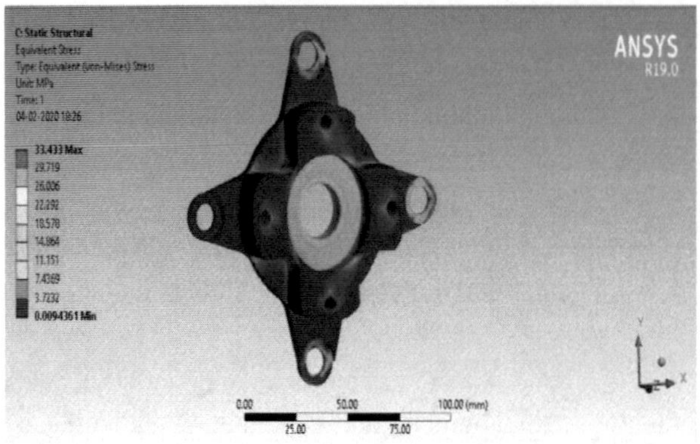

Figure 12. Stress acting on the optimized hub

3.2. Knuckle Design

The design and optimization of the knuckle is similar to that of the hub. The designing of the knuckle is based on the parameters like Upper and lower arm mounting points obtained from LOTUS, Stub axle diameter, Steering tie rod mounting point obtained from LOTUS, Brake caliper mounting point, Clearance between brake disc. CAD model of knuckle is shown in Figure 13.

The knuckle was analyzed using ANSYS by meshing the components into smaller components and applying loads on the mounting and required points of the knuckle. The mass of the non-optimized knuckle is about 0.31549 kg. The deformation and stress characteristics of the knuckle for the loading condition was found and were optimized using topology optimization tool in ANSYS. Figure 14 shows the ANSYS model of steering knuckle.

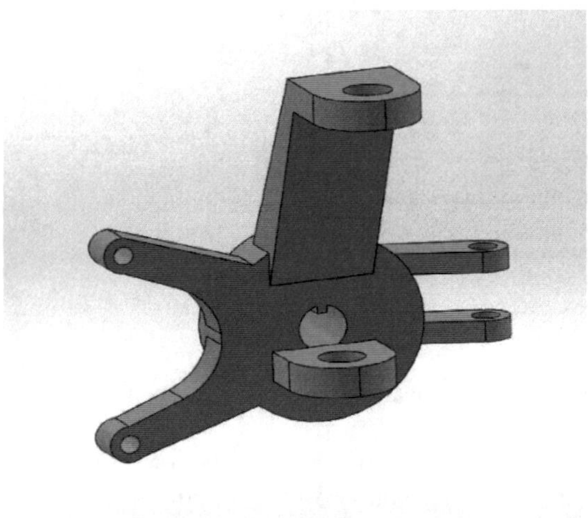

Figure 13. Knuckle CAD model

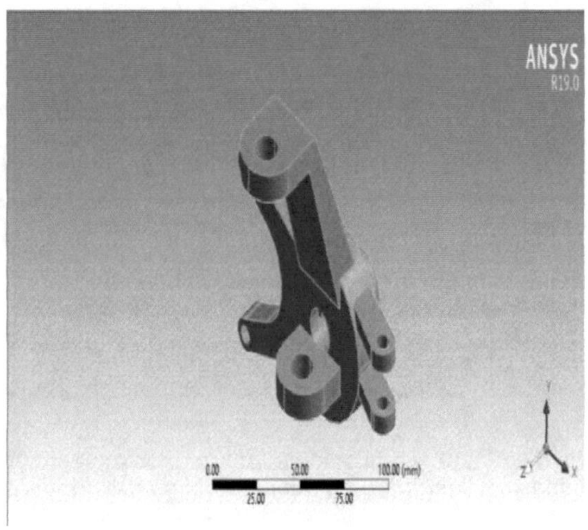

Figure 14. ANSYS model of knuckle

Figures 15(a) and 15(b) show the meshing of modelled knuckle and the loads acting on it. Corresponding deformation of the knuckle and stresses acting on it is illustrated in Figures 16(a) and 16(b) respectively.

IOP Conf. Series: Materials Science and Engineering **954** (2020) 012009 doi:10.1088/1757-899X/954/1/012009

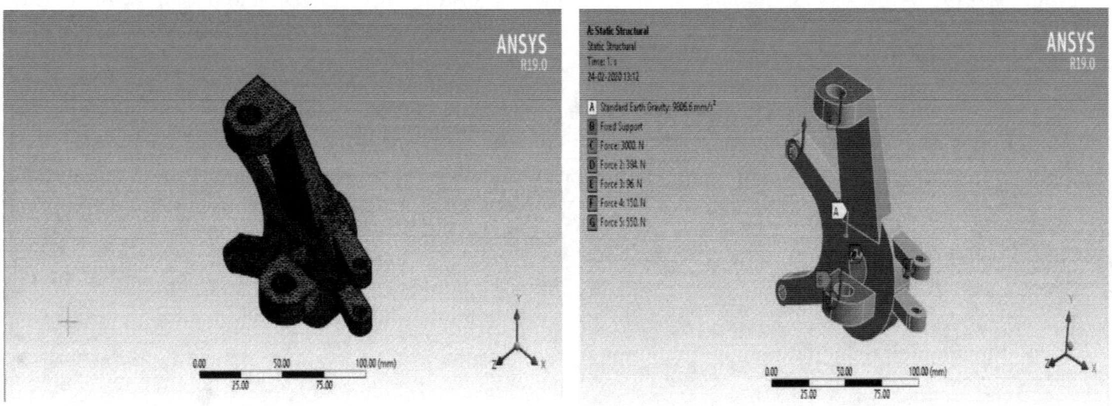

Figure 15. (a) Meshing of knuckle model and (b) Loads acting on the knuckle

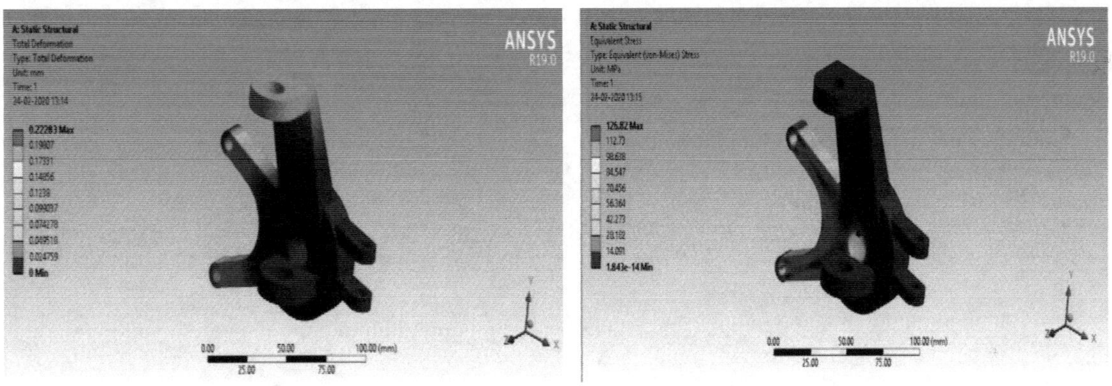

Figure 16. (a) Deformation of the knuckle and (b) Stress on the knuckle

Knuckle is then subjected to topology optimization which is shown in Figure 17 for reducing its mass without compromising the stresses acting on it. The optimization of the knuckle was done and suitable boundary model have been developed by SOLIDWORKS as shown in Figure 18 and analysis have been done for the same loading conditions to verify the load transfer and stress characteristics.

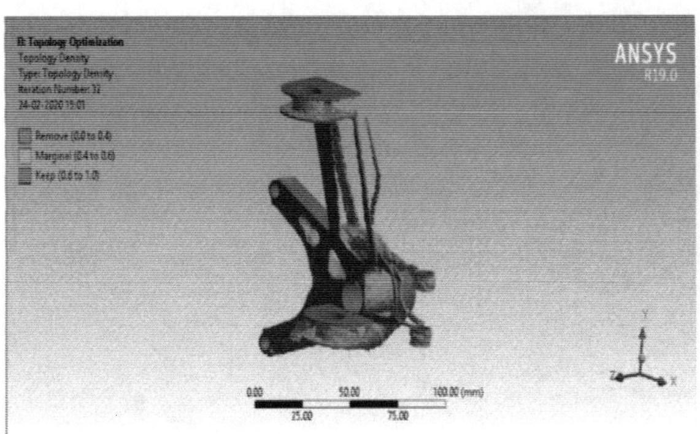

Figure 17. Topology optimization

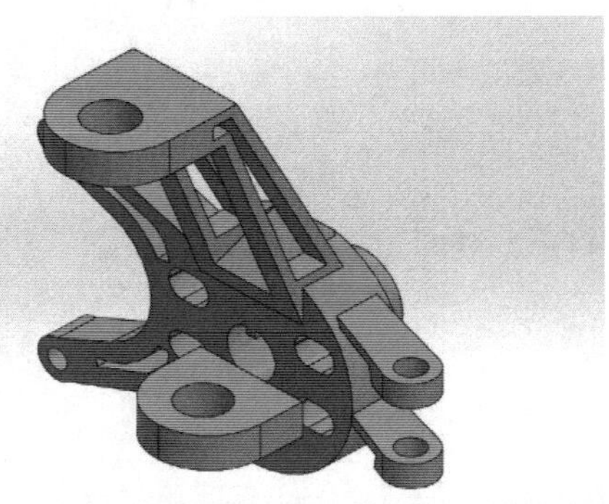

Figure 18. Knuckle CAD model

Figures 19(a) and 19(b) show the ANSYS model of optimized knuckle and its meshing. Load acting, deformation and stresses on the optimized knuckle is shown in Figures 20(a) and 20(b) and Figure 21 respectively.

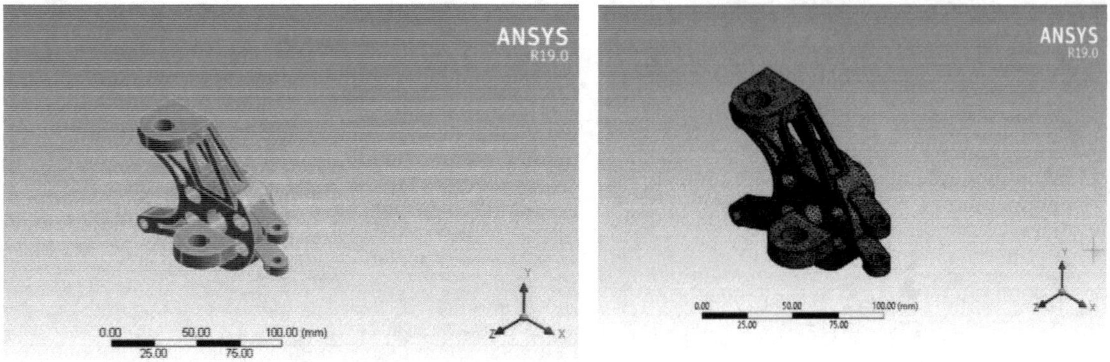

Figure 19. Optimized knuckle (a) Mass property and (b) Meshing

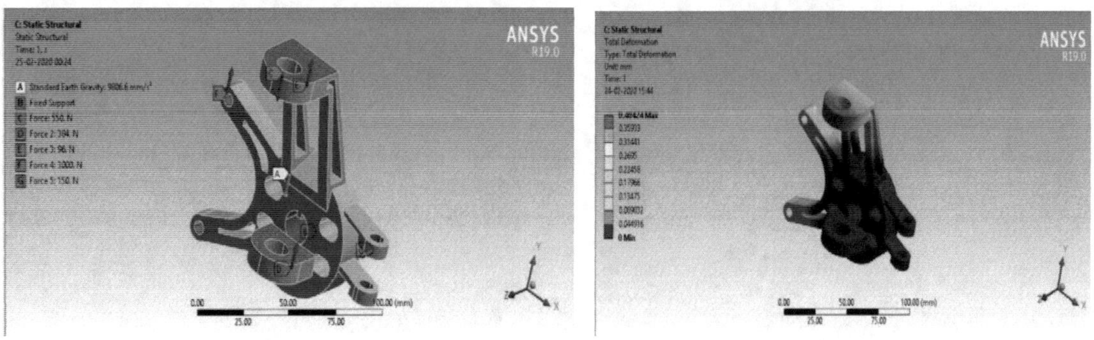

Figure 20. (a) Loads acting on the knuckle and (b) Deformation of optimized knuckle

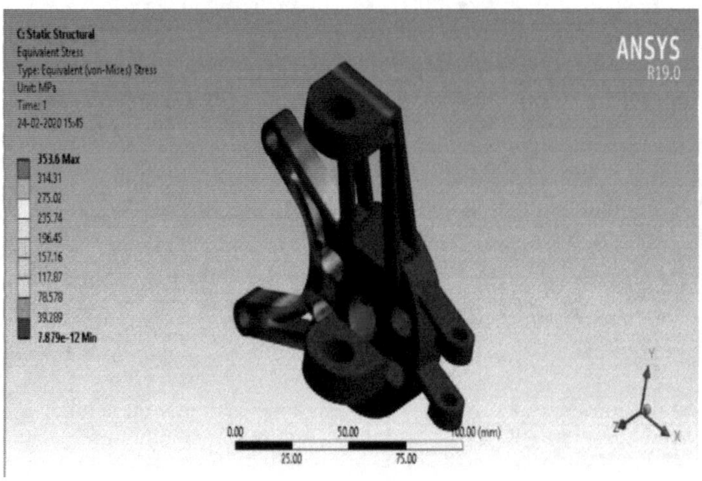

Figure 21. Stress acting on the knuckle

4. Conclusion

The optimization of both the geometric and physical characteristics of the suspension and the steering system has been achieved in static condition. The performance characteristics of the vehicle have been recorded for the static condition and the un-sprung mass has been reduced considerably to reduce the steering effort and improve load transfer using various software.

References

[1] Shaisundaram, V & Loganathan, Karikalan & Vignesh, V & Tamilmani, R & Akash, Mushfiqur. (2018). Design and Analysis of Steering Knuckle Component For Terrain Vehicle. International Journal of Recent Trends in Engineering and Research. 4. 10.23883/IJRTER.2018.4092.MYRIK.
[2] Sanjay Yadav, Ravi Kumar Mishra, Varish Ansari, & Shyam Bihari Lal 2016, 'Design and Analysis of Steering Knuckle Component', International Journal of Engineering Research And Technology, doi: 10.17577/ijertv5is040818
[3] Mutha, A, Thosar, S, & Ghodmare, N 2017, 'Design and optimization of a steering knuckle of FSAE car', Lecture Notes in Mechanical Engineering, doi: 10.1007/978-981-10-1771-1_50
[4] Rajesh kumar L, S Sivalingam, K Prashanth, K Rajkumar (2020), "Design, Analysis and Optimization of Steering Knuckle for All Terrain Vehicles", AIP Conference Proceedings, Vol. 2207, pp. 020003. https://doi.org/10.1063/5.0000044
[5] Dusane, S V., Dipke, MK, & Kumbhalkar, MA 2016, 'Analysis of Steering Knuckle of All Terrain Vehicles (ATV) Using Finite Element Analysis', in IOP Conference Series: Materials Science and Engineering, doi: 10.1088/1757-899X/149/1/012133
[6] Saravanakumar A, Rajeshkumar L, Balaji D and Karunan M P J, "Prediction of Wear Characteristics of AA2219-Gr Matrix Composites Using GRNN and Taguchi-Based Approach" Arabian Journal for Science and Engineering (2020), https://doi.org/10.1007/s13369-020-04817-8
[7] Krishna, MMR 2001, 'Finite element shape optimization of a steering knuckle for a heavy truck-A case study', in SAE Technical Papers, doi: 10.4271/2001-01-0634
[8] Jagwinder Singh (2015), 'STATIC STRUCTURAL ANALYSIS OF SUSPENSION ARM USING FINITE ELEMENT METHOD', International Journal of Research in Engineering and Technology, doi: 10.15623/ijret.2015.0407064
[9] Zeng, Z & Zhang, Y 2012, 'Topological optimization of the steering knuckle in a formula SAE racing car', Qiche Gongcheng/Automotive Engineering
[10] Bhardwaj, S, Ashok, B, Lath, U, & Agarwal, A 2018, 'Design and Optimization of Steering Upright to Reduce the Weight Using FEA', in SAE Technical Papers, doi: 10.4271/2018-28-0081
[11] Liu, YH, Liu, SM, He, WT, & Yu, QH 2015, 'Effect of resistance wall design on the drop forging for steering knuckle', Suxing Gongcheng Xuebao/Journal of Plasticity Engineering, doi: 10.3969/j.issn.1007-2012.2015.03.007
[12] Shuaib, M, Haleem, A, Kumar, L, Rohan, & Sharma, D 2019, 'Design and analysis of steering knuckle joint', in Lecture Notes in Mechanical Engineering, doi: 10.1007/978-981-13-6469-3_38
[13] Madhusudhanan, S, Rajendran, I, & Raguganeshkumar, B 2017, 'Fatigue Analysis of Steering Knuckle using Finite Element Simulation: Technical Note', International Journal of Vehicle Structures and Systems, doi: 10.4273/ijvss.9.3.02
[14] L Rajesh kumar, KS, Amirthagadeswaran. "Corrosion and wear behaviour of nano Al2O3 reinforced copper metal matrix composites synthesized by high energy ball milling." Particulate Science and Technology 38, no. 2 (2020): 228-235. doi: 10.1080/02726351.2018.1526834

[15] Joshi, T & Joshi, V 2016, 'Design and Analysis of Suspension and Steering Components of F1 Prototype .', Imperial Journal of Interdisciplinary Research IJIR
[16] Wang, H & Yu, L 2012, 'Strength analysis of steering knuckle arm on veyron car', in Advanced Materials Research, doi: 10.4028/www.scientific.net/AMR.562-564.627
[17] Tang, GT, Shi, BQ, & Zhang, WM 2013, 'Structural response surface optimization design of steering knuckle arms under multiple working conditions based on fuzzy evaluation', Beijing Keji Daxue Xuebao/Journal of University of Science and Technology Beijing
[18] L.Rajesh kumar, K S Amirthagadeswaran. "Variations in the properties of copper-alumina nanocomposites synthesized by mechanical alloying." Materiali in Technologjie, 53(1) (2019): 57-63
[19] Madhusudhanan, S, Arun Kumar, K, Praveen Kumar, V, & Mahendran, T 2016, 'Fatigue analysis of automative steering knucle', Mechanics and Mechanical Engineering
[20] Wenxing, L 2012, 'Study on optimization design method of drilling fixture of the steering knuckle of vehicles', in Advanced Materials Research, doi: 10.4028/www.scientific.net/AMR.580.33
[21] Gowtham, V, Ranganathan, AS, Satish, S, John Alexis, S, & Siva Kumar, S 2016, 'Fatigue based design and analysis of wheel hub for Student formula car by Simulation Approach', in IOP Conference Series: Materials Science and Engineering, doi: 10.1088/1757-899X/149/1/012128
[22] Muhammad Aizzat, Z, Hairi, Z, & Saiful Amri, M 2016, 'Dynamic Curvature Steering Control for Autonomous Vehicle: Performance Analysis', IOP Conference Series: Materials Science and Engineering
[23] Baskara Sethupathi, P, Chandradass, J, Praketh, G, Kumar Asuthkar, A, & Vinay Kumar, M 2018, 'Design and fabrication of four wheel steering system for light motor vehicles', in IOP Conference Series: Materials Science and Engineering, doi: 10.1088/1757-899X/402/1/012185

IOP Conf. Series: Materials Science and Engineering **954** (2020) 012010 doi:10.1088/1757-899X/954/1/012010

Assessment of Service Agility in Power Distribution Company

M Anil [1] and M.Suresh[2]
[1, 2] Amrita School of Business, Coimbatore, Amrita Vishwa Vidyapeetham, India

Email: [1] anilmkseb@gmail.com; [2]drsureshcontact@gmail.com

Abstract Power Distribution Company (DISCOM)s are basically engaged in providing electricity to power the various customer needs of the people in its area of operation. The availability of quality power free from voltage sags, voltage swells, dips etc. without any interruption contributes to customer delight in the sector. Unfortunately, the network constraints brought in by overhead, bare conductors to convey power from generating stations to transmission Sub-stations and from there to the end consumers is a serious cause for outages to customers and create dissatisfaction among them. This paper attempts to develop assessment framework of service agility for a case DISCOM in India using multi-grade fuzzy approach. The service agility attributes are identified and classified using Importance Performance analysis (IPA). This assessment framework may helps managers of DISCOM for understanding their current service agility level and focus on the weaker attributes to enhance their service performance. This assessment framework can help DISCOM's managers to understand their current level of service agility and focus to improve their service performance.

Keywords: *Service agility; Power Distribution Company; Service Quality; Energy Management; Multi-grade Fuzzy; Importance Performance analysis*

1. Introduction

Unlike traditional production-oriented organisations, service delivery organisations differ in their overall functioning. In addition to Service Quality and Product Quality, the employee behaviour and the ability of employees and customers to co-create a service delivery process in the service-scape play an important role in the satisfaction levels derived out of service delivery process by both the service provider and the service recipient alike. In a product-service continuum, if we arrange various types of services available in the scenario, the services of a Power Distribution Company (DISCOM) to its customers take a position more towards the services end than to the product end of the continuum. DISCOMs are basically engaged in providing electricity to power the various customer needs of the people in its area of operation. The availability of quality power free from voltage sags, voltage swells, dips etc. without any interruption contributes to customer delight in the sector. Unfortunately, the network constraints brought in by overhead, bare conductors to convey power from generating stations to transmission Sub-stations and from there to the end consumers is a serious cause for outages to customers and create dissatisfaction among them. Though network hardening measures like converting bare overhead network to aerial bunched cable (ABC) system or underground (UG) cable system offer a solution to the issue, the prohibitively high cost of implementation thwarts the progress of such measures in a power distribution utility; with limited resources for sourcing capital outlay from the routine operational profits obtained by the utilities, in the strict regulatory regime in vogue, especially after meeting the Corporate Social Responsibilities of the utilities.

The functional aspects of a power distribution company include complex networks, feeding from multiple sources, inadvertent access to bare (uninsulated) energised network to the public etc. which

IOP Conf. Series: Materials Science and Engineering **954** (2020) 012010 doi:10.1088/1757-899X/954/1/012010

contribute to serious issues as regards the safety of general public and to the service provider employees of the utility. The rate of occurrence of accidents, in spite of the deployment of personal protective equipments (PPEs) by the employees, contract workers etc.; are on the rise. The members of the general public are also not saved from chances of accidents, mostly due to unintended contact with the bare conductor systems of network with foreign conducting objects or due to snapping of bare conductors of the network and consequent falling to the ground and remaining energised in spite of the various protective systems and protocols to render such snapped conductors de-energised in the event of contact with the earth. Standardisation of procedures and construction practices go a long way in providing safety to employees and public. Inculcating a safe work culture through training and re-training of the employees, contractors' personnel and imparting safety awareness propaganda to the public are useful means to reduce the causation of accidents to the public and employees alike. An umpteen number of customers are installing solar photo-voltaic generation on their roof-tops and getting their status converted to power producers (PROSUMERs) creating "grid-tied" solar photo-voltaic systems in distributed de-centralised mode and the consequent bi-directional power flows in the network is another potential causative factor of accidents to the employees of the utility, engaged in maintenance activities of the network. The organisation, Case Company advises its employees to create permanent identification marks within the network to identify the presence of such PROSUMERs and to switch off their solar generation while undertaking repairs and maintenance in such parts of the distribution network.

2. Literature Review

The agile methods in production are more popular in the industrial manufacturing parlance [1]; the applicability of leanness, agility and leagility has been brought to the service industry [2;3;4], of late. Healthcare service delivery [5;6;7] and office management were the popular early areas of application of these techniques in service sector[8]. Later on, logistics, supply chain management and tourism and transportation sectors were subjected to agility assessment and consequent improvement in wastage reduction, resource utilisation effectiveness and finesse in service delivery process were brought into these areas. The lean service publications were very few for the period from 1993 to 2001 and the trend slowly took off to the level of 30 annual publications by the year 2011 [9]. Agility in service delivery process is very important in service intensive industries [10;11;12;13]. The business of power distribution is a service intensive industry as the people element in this sector is critical and can influence the level of satisfaction to the customers in a great way. No previous study involving fuzzy logic based Importance Performance Analysis (IPA) was observed to be made in the business of a Power Distribution Company (DISCOM) upon survey of available literature.

Unlike other infrastructural sector industries, power distribution is more visible a service and the various processes involved create a multitude of customer touch points or interaction points which the service delivery personnel have to manage effectively and skilfully to create value for the customers. Even a very slight aberration on the part of the service delivery agent of the power distribution company is sufficient to create a strong dent in the customer relationship of the company, which needs to be maintained for a life time as the service provider- customer relationship often prolong for quite a long time and often transcends to generations of customers.

Importance- Performance Analysis (IPA) has been used as an effective technique in evaluating service attributes based on the comparative analysis of attribute importance vis-à-vis attribute performance [14]. In importance- performance analysis, attribute performance and attribute importance are both independent variables [15] and they are plotted in an X-Y scatter plot with attribute performance on X-axis and attribute importance on Y-axis. In the scatter plot thus obtained, the mean values of attribute performance and attribute importance are utilised to divide the plot area into four quadrants with attributes in the North-West quadrant (Quadrant I) need more attention as they are more important, but have low performance levels. The quadrant on the north-east portion (Quadrant II) consists of those attributes having high importance as well as performance levels. The south-west and the south-east quadrants (Quadrants III & IV respectively) consist of those attributes with low importance. The level of performance for attributes in south-east quadrant (Quadrant IV) is high; which is an unnecessary or undesired situation and the performance levels for these attributes are redundant. The critical attributes are those in Quadrants I & II. Those attributes in quadrant I are highly important, but the performance levels are less. Hence, these attributes are to be given more care and performance levels of these attributes are to be increased.

3. Research Methodology

The study utilised multi-grade fuzzy logic to assess the agility index of the power distribution company. The various attributes for the study were enlisted from the first author's own experience of having worked in a power distribution utility for a long period spanning from 1991 to 2020 and after considering inputs from colleagues and accomplices. The conceptual model for agility assessment with two enablers, five criteria and thirty eight attributes was developed as shown in Table 1. Questionnaire based responses of five experts in the field of power distribution were collected personally by the author on a 10-point scale, after giving a brief description about the attributes, criteria and enablers. Weightages were worked out from the responses of the experts, to obtain the importance against each attribute, criterion and enabler. Ratings were obtained by interviewing five operational level Engineers and performance levels were assessed from the weightages obtained by interviewing the experts.

Mean of the attribute level performance and importance were worked out to obtain mean value of attribute performance and importance. X-Y scatter plot was drawn with importance on Y-axis and performance on X-axis and the mean values were plotted as lines in the importance- performance grid to obtain the quadrants to assess criticality of attributes.

The service agility index (A) is computed as the product of overall assessment level of ratings, based on each criteria/ enabler (R) and the overall weights (W), as given by the experts at primary, secondary and tertiary levels.

Table 1.Conceptual model for agility assessment

Sl No	Enablers	Criteria	Attributes
A1	Customer Satisfaction	Technical Quality	Freedom from Voltage sags, Swells and Dips
A2			Frequency Stability
A3			Resilient Power Grid
A4			Low and Informed Outages
A5			Adherence to Standards of Performance of SERC
A6		Service Quality	Informed Outages
A7			Adaptiveness
A8			Empathy with Customers
A9			Fair and Friendly Deal
A10			Good maintenance of Physical & Peripheral Evidences
A11			Better presentability of VISIBLE Assets
A12			Better management of Customer Touch Points
A13			Deployment of Automation & ICTs in Sevice Delivery
A14			Online Payment Mechanism
A15			Enhanced HMIs and Automatic Telephone Answering Services
A16		Perceived Quality	No outages
A17			Quality Power Supply
A18			Reliability of Supply
A19			Accurate Metering
A20			Performance Effectiveness
A21			Convenient Location of the Servicescape
A22			Consistency of Service Providing Personnel
A23			Assurance
A24			Courtesy of Service Providing Personnel
A25			Tangibles- Facility & Personnel- Neat & Tidy
A26			Responsiveness of Service Providing Personnel
A27			No waveform distortion at the point of common coupling
A28			No harmonic distortion at the point of common coupling
A29	Safety and Security	Public Safety	Reduced chances of Accidents from assets
A30			Standardisation of Procedures and Construction
A31			Mass Education aimed at giving Electrical Safety Tips to Public
A32		Employee Safety	Adherence to Construction Standards
A33			Deployment of Personal Protective Equipments
A34			Inculcating Safe Work Culture through training & re-training
A35			Network Hardening Measures
A36			Enforcing & Promoting Safe Practices
A37			Responsibility to maintain Safe condition
A38			Bi-directional Power flow issues

The multi grade fuzzy procedures has been adopted from [1;16;17;18].

$$OA_i = R_i \times W_i$$

OA_i : Service Agility Vector
Ri: Rating of the drivers
Wi: Weightage of the drivers

The scale of assessment of Service Agility Index was graded into five levels, as per principles of fuzzy determination of the index with value between 8 and 10 representing "Extremely agile", between 6 to 8 representing "Agile", between 4 and 6 representing "Fairly agile", between 2 to 4 representing "not agile" and between 0 and 2 representing "Extremely not agile".

4. Case Study

Case Company is engaged in the business of Generation, Transmission and Distribution of Electricity within the geographical boundary of a State in India. Trading of Electricity is also undertaken by this organisation to meet the shortages which often ranges between 60 – 80% of the total need for power from private power producers or through power exchanges as per short-term, medium-term and long-term power purchase agreements (PPAs); the rest of which is met from internal generation apart from the share of generation from Centrally owned Generating Stations (CGS). The legal environment of business of Case Company is set as per the provisions of Electricity Act, 2003. Amongst the various portfolios handled by the organisation viz., Generation, Transmission, Distribution, Power System Operation, Protection & Control, SCADA & Telemetry and Power Line Carrier Communication; the power distribution plays an important role as this portfolio is closely linked with serving a large customer base spread across the length and breadth of the State. Electric Power supply is closely linked with the development of the nation and electric power industry is considered as an industry of industries as electricity is an important factor of production, the consumption of which contributes to the GNP of the nation.

4.1 Characteristics of Distribution Company (DISCOM)

The technical quality of power (or simply, power quality) and the service quality of the service delivery personnel engaged at various customer touch points of the utility do have important contributions to make, for attaining customer satisfaction. When we think of customer delight, the perceived quality of service, as seen from the customers' side are also very important. Ideally, the customer expects no outage to happen in the network with characteristics of 24 x 7 supply with no interruption at all. When non-linearity of loads are increasing in various customer installations, the quality of waveform (which ought to be perfectly sinusoidal) at the point of common coupling (PCC) of the grid (the point where the customer installation physically joins the distribution grid) gets distorted giving rise to harmonic distortion and the total harmonic distortion (THD) at the point of common coupling is to be maintained at less than 5% (IEEE 519: 1992). Soft skills possessed by the service providing personnel increases the perceived value of services provided by them to the customers and creates positively higher levels of customer satisfaction, which can be termed as customer delight. The resilience of the network or the elasticity of the network also contributes to customer satisfaction.

The resilience is the property of a power grid/ network to remain unperturbed in the event of a loss of generating unit, transmission line, transformer etc. comprising the network; so that the customer segment would not feel about the loss by way of lack of power quality such as voltage sag, flicker etc. or power supply outage. In fact, flickers and voltage sags can be treated as momentary outages for an infinitesimal duration of time; which is too short to perceive, due to persistence of sensory perception.

IOP Conf. Series: Materials Science and Engineering **954** (2020) 012010 doi:10.1088/1757-899X/954/1/012010

The visibility of service delivery process in a power distribution company is very high and this over-emphasises the importance of proper management of the customer as well as public touch points carefully and adroitly. Information Technology (IT) implementation in power distribution sector enhances the scope for increasing service quality by resorting to providing advance customer information about duration of the impending outage planned, power availability etc. to balance the dynamic situation existing in power distribution market and the ambitious expectations of customers of the utility. The flexibility of grid operations can also be met by the distribution system operator (DSO) of a utility, if dynamic forecast and customer information management can be carefully intertwined and if the customers' consumption behaviour can be influenced by the utility managers to create a 'win-win situation' to the utility as well as its customer base.

In this study, responses were gathered in a 10-point scale for attributes corresponding to criteria such as technical quality, service quality, perceived quality and public & employee safety of a power distribution company and the service agility index is evaluated using multi-grade fuzzy method. The weightages are calculated from expert responses and ratings were collected from operational level managers and importance- performance analysis (IPA) was undertaken to identify critical attributes in the study. The Importance- performance (IP) plot obtained from the study is shown in Figure 1 below.

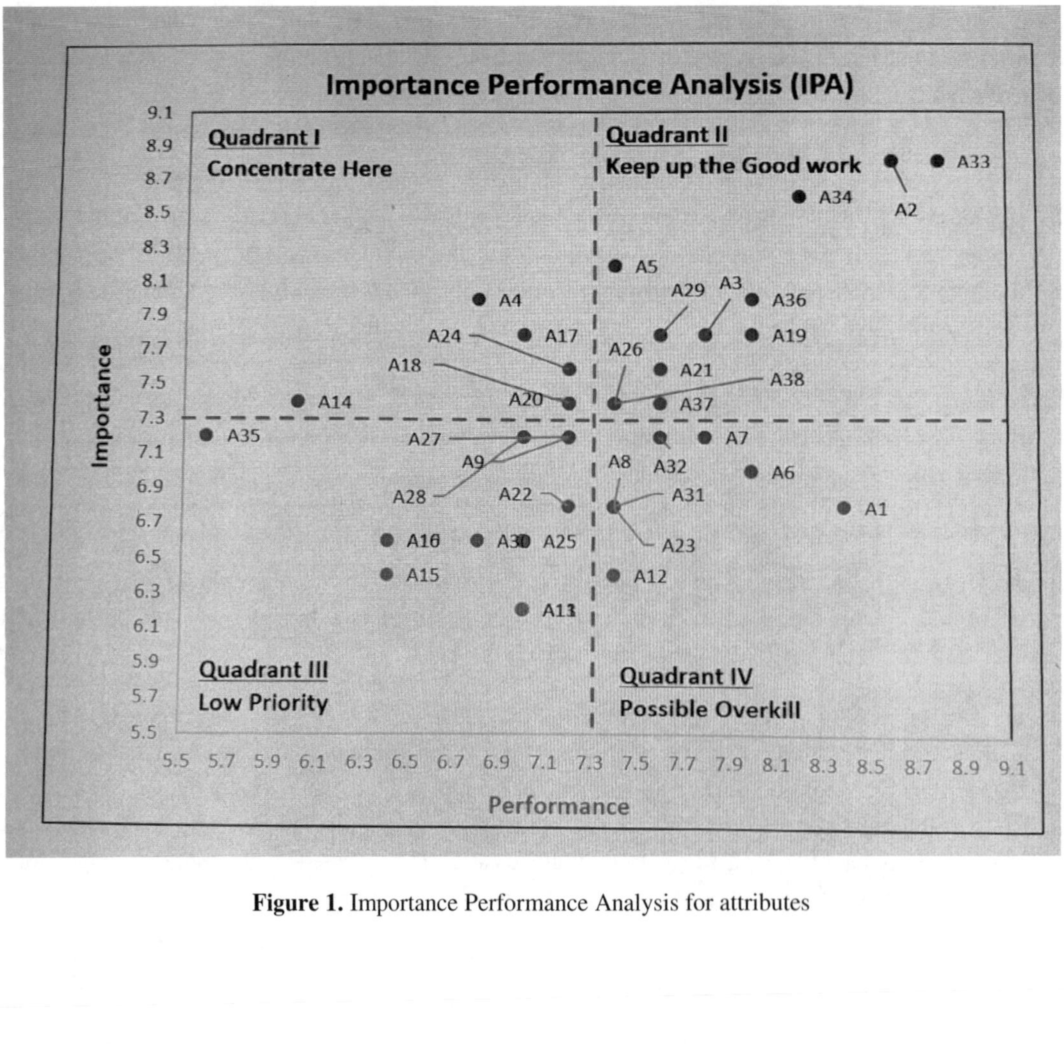

Figure 1. Importance Performance Analysis for attributes

5. Conclusion

The service agility index of Case Company was obtained, which means that Case Company is an "agile" power distribution company. As per the Ministry of Power, Government of India; the performance rating of Case Company is one among the best in the country. The performance level of service attribute A14 viz. Online Payment Mechanism is low, though it is a highly important attribute. Similarly, attributes A4, A17, A18, A20 and A24 viz. Low and Informed Outages, Quality Power Supply, Reliability of Supply, Performance Effectiveness and Courtesy of Service Providing Personnel need to have better attention of the management and their performance levels have to be improved. The management must concentrate on improvement of performance levels of these attributes in quadrant I of the IP plot. Improvement in performance levels of these service attributes could be made with greater adoption of information technology to provide timely information about outages to the customers; by investing more money in hardening the power distribution network of the utility to enhance the reliability of the network to better insulate the network from the effect of touching of tree branches and by providing soft skill training to the service providing employees of the utility aimed at improving their emotional resilience and capability to build and sustain quality & conducive interpersonal relationships.

The attributes A33, A2 and A34 corresponding to Deployment of Personal Protective Equipments (PPEs), Frequency Stability and Inculcating Safe Work Culture through training and re-training respectively have the highest levels of performance as well as importance, as elicited in the study. The other attributes which have above average levels of importance and performance are A5, A26, A38, A37, A21, A29, A3, A36 and A19; corresponding to Adherence to Standards of Performance of State Electricity Regulatory Commission (SERC), Responsiveness of Service Providing Personnel, Bi-directional Power flow issues, Responsibility to maintain safe condition, Convenient location of Service-scape, Reduced chances of accidents from Assets, Resilient Power grid, Enforcing and Promoting Safe Practices and Accurate Metering respectively.

The low priority attributes are those in quadrant III and the attributes in quadrant IV of the IP plot represent those attributes with low importance level and high performance level. For such high levels of performances, the resources spent by the management is probably a wasteful exercise, and re-allocation of resources to quadrant I attributes for improving their performance levels would be a welcome activity on the part of the management to improve the agility of the company.

In the current study, the service agility index of a typical power distribution company was evaluated and the importance and performance levels of various attributes were analysed. A power utility has other functional areas viz. Generation, Transmission etc. The agility index assessment could be done in those fields which may include operational safety Index in a generating station, where the presence of rotating machinery and energy conversion processes add to the Health, Safety & Environment (HSE) hazards, which could be assessed using multi-grade fuzzy. Similarly, in the event of an unexpected turn around in such a station, the disaster preparedness and mitigation of problems could also be analysed using multi-grade fuzzy.

References

[1]. Vinodh, S., & Chintha, S. K. (2011). Leanness assessment using multi-grade fuzzy approach. International Journal of Production Research, 49(2), 431-445.
[2]. Vaishnavi, V., Suresh, M., & Dutta, P. (2019). Modelling the readiness factors for agility in healthcare organization: a TISM approach. Benchmarking: An International Journal. 26(7), 2372-2400.
[3]. Suresh, M., & Arun Ram Nathan, R.B. (2020). Readiness for lean procurement in construction projects. Construction Innovation. https://doi.org/10.1108/CI-07-2019-0067
[4]. Suresh, M., Vaishnavi, V., & Pai, R. D. (2020). Leanness evaluation in health-care organizations using fuzzy logic approach. International Journal of Organizational Analysis. https://doi.org/10.1108/IJOA-04-2019-1752.
[5]. Vaishnavi, V., & Suresh, M. (2020). Assessment of readiness level for implementing lean six sigma in healthcare organization using fuzzy logic approach. International Journal of Lean Six Sigma. https://doi.org/10.1108/IJLSS-07-2019-0081
[6]. Vaishnavi, V., & Suresh, M. (2020). Assessing the Readiness Level of Healthcare for Implementing Agility Using Fuzzy Logic Approach. Global Journal of Flexible Systems Management, 21(2), 163-189.
[7]. Vaishnavi, V., & Suresh, M. (2020). Modelling of readiness factors for the implementation of Lean Six Sigma in healthcare organizations. International Journal of Lean Six Sigma. 11(4), 597-633.
[8]. Vignesh, V., Suresh, M., & Aramvalarthan, S. (2016). Lean in service industries: A literature review. In IOP Conference Series: Materials Science and Engineering (Vol. 149, No. 1, pp. 1-10).
[9]. Hadid, W., & Mansouri, S. A. (2014). The lean-performance relationship in services: a theoretical model. International Journal of Operations & Production Management. 34(6), 750-785.
[10].Patri, R., & Suresh, M. (2019). Agility in healthcare services: a systematic literature exploration. International Journal of Services and Operations Management, 32(3), 387-404.
[11].Patil, M., & Suresh, M. (2019). Modelling the enablers of workforce agility in IoT projects: a TISM approach. Global Journal of Flexible Systems Management, 20(2), 157-175.
[12].Menon, S. & Suresh, M. (2020). Enablers of workforce agility in engineering educational institutions, Journal of Applied Research in Higher Education, https://doi.org/10.1108/JARHE-12-2019-0304
[13].Menon, S. & Suresh, M. (2020). Organizational agility assessment for higher education institution, The Journal of Research on the Lepidoptera, 51(1), 561-573.
[14].Martilla, J., & James, J. (1977). Importance-Performance Analysis Journal of Marketing. January, 41, 77-79.
[15].Deng, W. J. (2008). Fuzzy importance-performance analysis for determining critical service attributes. International Journal of Service Industry Management.19(2), 252-270.
[16].Vinodh, S., Devadasan, S. R., Vasudeva Reddy, B., & Ravichand, K. (2010). Agility index measurement using multi-grade fuzzy approach integrated in a 20 criteria agile model. International Journal of Production Research, 48(23), 7159-7176.
[17].Ganesh, J., & Suresh, M. (2016, December). Safety practice level assessment using multigrade fuzzy approach: A case of Indian manufacturing company. In 2016 IEEE International Conference on Computational Intelligence and Computing Research (ICCIC) (pp. 1-5). IEEE.
[18]. Sridharan, V., & Suresh, M. (2016, December). Environmental sustainability assessment using multigrade fuzzy—A case of two Indian colleges. In 2016 IEEE International Conference on Computational Intelligence and Computing Research (ICCIC) (pp. 1-4). IEEE.

Friction Stir Processing – A Review

A.Arun Premnath[1] A. Tamilarasan[2] and R.Prasanth Kumar[3]

[1]Associate Professor, Centre for Composite Science and Tribology,
[2]Assistant Professor, Centre for Composite Science and Tribology,
[3]PG Student,
Sri Chandrasekharendra Saraswathi Viswa Maha Vidyalaya University
Email: arun_premnath@kanchiuniv.ac.in

Abstract. Friction stir processing (FSP) is a solid state surface modification technique which is recently used to fabricate surface composites. Friction stir processing is an extension of friction stir welding which is developed by the welding institute of UK in 1991. In FSP a rotating tool consist of a pin and a shoulder is inserted into the material which is in the form of a sheet. The main advantages of FSP includes grain refinement and the uniform distribution of reinforcement particles that improves the mechanical and wear properties of the surface composites compared with the as cast alloys. In the review paper a detailed analysis was done on the mechanical and wear properties of the composites fabricated by FSP.

Keywords: FSP. Grain refinement. Mechanical properties.

1. Introduction

The friction stir processing unlike other processing techniques like laser melt treatment, electron beam irradiation and plasma spraying does not produce interfacial reaction which cause harmful phases. FSP is basically developed to prevent from various defects that occur during casting. In FSP, non-consumable tool rotating at desirable speed is inserted forcibly (shown in figure 1.) into the work piece causes fine grained microstructure, which improves the mechanical properties and wear resistance of the composites. In this paper a thorough review was performed on the mechanical and wear properties of composites that occur during FSP.

IOP Conf. Series: Materials Science and Engineering **954** (2020) 012011 doi:10.1088/1757-899X/954/1/012011

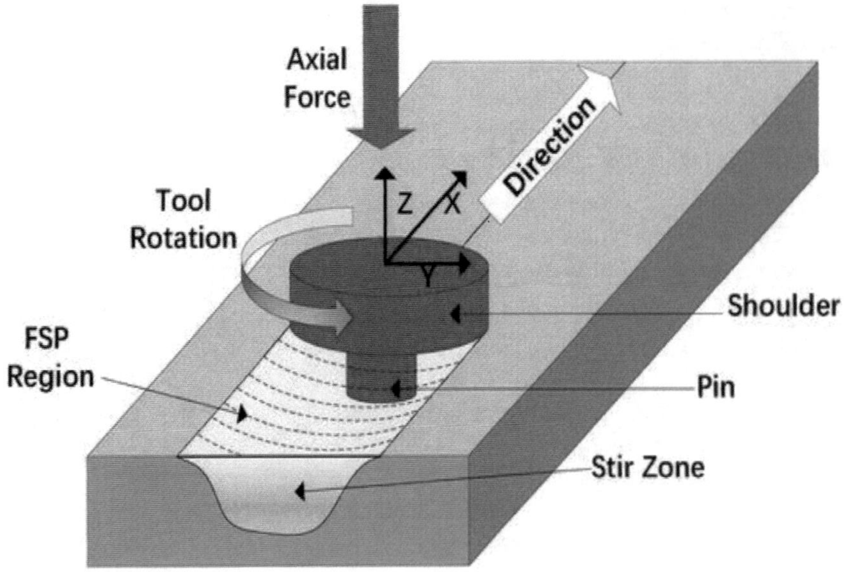

Figure 1. Schematic diagram of friction stir processing [1]

Saini et al. [1] fabricated of as cast Al-alloy surface composite via FSP. The surface composite has been developed by reinforcing MoS_2 and Zn powder in the matrix of cast Al-Si alloy and they performed observations on reinforced surfaces, microstructure, mechanical and tribological characterization. They found micro hardness value was higher in the case of composite fabricated using Zn as compare to composite fabricated using MoS_2 and Al-17Si alloy, The Ultimate Tensile Stress of FSPed modify with Zn and FSPed modify with MoS_2 has been decreased about 0.977 and 0.703 times of base metal respectively, so as that percentage elongation wasn't affected considerably within the case of composite fabricated using Zn reinforcement with in the as-cast alloy. Don-Hyun et al. [3] performed Friction stir processing of AA6061-T4 alloy with SiC particles. They found that the micro hardness of the stir zone in the midst of SiC particles about HV80 was observed due to grain refinement and scattering of SiC particles. Arun Premnath [4] investigated studies on Aluminium silicon carbide nano composites prepared by FSP. The experiments were performed on commercially available aluminium 6061 plate of size 150 × 50 × 5 mm. Initially the groove was crammed with nano-SiC particles and a pin-less tool is employed to stop escaping of the nano-SiC particles. Then, pin-tool is allowed for the processing. So as to review the effect of process parameters on microstructure and mechanical properties, the workpieces are fabricated by varying the quantity of passes, rotational speed and transverse feed. The microstructure examination shows that nano SiC particles were distributed evenly throughout the matrix for 3 pass specimen, it is observed a better values of tensile strength, micro hardness and wear loss. Mazaheri et al. [5] disquisition FSP to manufacture A356/Al_2O_3 surface composites by blending A356 samples with Al_2O_3 powder particles. A substantial increase within the average micro hardness was obtained by this method. Alidokht et al. [6] investigated the hardness of A356 plates reinforced with SiC and MoS_2. Their results showed that the FSPed specimens, which were reinforced using the 2 varieties of powders, recorded higher micro-hardness values compared to the given sample. Mahmoud et al. [7] worked on different ratios of SiC and Al_2O_3 powders as reinforcements for 5-mm-thick Al-1050-H24 plates. In the FSPed stir zone the powder particles were uniformly scattered. The results pan out that the common micro-hardness values obtained employing a 100% SiC powder increased to approximately 60 HV. They found that whenever the relative ratio of the Al_2O_3 powder particles increases the micro hardness values decreased. Darras et al. [8] used CNC vertical milling machine to conduct friction stir processing of commercial magnesium alloy with H-13 alloy steel tool is employed as pin. FSP is performed with a one pass at various combinations of rotational (1200–2000 rpm) and translational (20–30 in./min) speeds and that they processed heating and cooling rates. The material incorporates a fine microstructure with a mean grain size of about 6μm and better hardness. Mohammad

Hasan Shojaeefard et al. [9] follow out research on fabrication of particle-reinforced aluminium matrix composites. The distribution of reinforcing particle on the pin profile in terms of microstructural, mechanical, and wear properties of A356 matrix composites reinforced by SiC, TiC, ZrO_2, and B_4C particles are scrutinized. Owing to the good binding between TiC and metal matrix the composites reinforced with TIC particle shows better hardness. Eftekharinia et al. [10] found that there is a remarkable increase in the hardness by increasing the passes number because of uniform distribution of reinforcement particles and finer grains size at the stir zone. It is concluded that by increasing the rotational speed from 1000 to 1500 rpm, the temperature in the stirring zone increased to a critical value which causes grain refinement there by decreasing the wear loss of the composites.

Elangovan et al. [11] comprehended the results of tool pin profile and tool shoulder diameter on FSP zone formation in 6061 AA and opted different tool pin profiles like straight cylindrical, tapered cylindrical and threaded out of that cylindrical, triangular and square with three different shoulder diameters are acquainted with process the joints. During this regard they found that the square pin profiled tool and irrespective of shoulder diameter produced mechanically sound and metallurgically defect free welds compared to other tool pin profiles. Ma et al. [12] investigated studies on applied five-pass FSP with 50% overlap on cast Al–Si–Mg A356 alloy and identified that overlapping FSP didn't exert a major effect on the dimensions and distribution of the Si particles. The Si particles broken by FSP were uniformly scattered within the entire processed zones created by multiple-pass FSP. Wanchuck Woo et al. [13] measured the effect of the stirring pin and pressing tool shoulder on the microstructural softening occurring during FSP and ensuring natural aging behaviour of 6061-T6 AA and the microstructural softening has occurred mainly due the frictional heating from the tool shoulder during FSP. The reduction of the longitudinal residual strain outlined within the bead area was correlated to the microstructural softening. As a whole, the softening within the DRX and thermo-mechanically affected zones is expounded to the dissolution of fine needle-shape precipitates β on account of frictional heating, which reduced the micro hardness from the 110 to 70 HV. Softening has also occurred within the HAZ which was similarly related to the dissolution of β and so the expansion of coarse precipitate phase, which reduced the initial hardness to 60. Karthikeyan, et al. [14] conducted experiments on the surfaces of cast A319 alloy plates for single stir FSP, tend to find effect on the grain size, porosity level and improving the mechanical properties. Establishing most of the combinations of traverse feed rates and tool rotational speeds, the friction stir processed material displays higher strength and micro-hardness compared with the base material. The ductility of the FSP material is additionally greater than that of the base material. For an exact rotational speed ductility increases were around 1.5–5 times that of the cast alloy. The increases in ductility might be due to grain refinement within the nugget zone. The specimens processed with a tool rotational speed of 1200 rpm were free from defects and with certain combinations of parameters the processed alloy displayed an approximate raise of 50% in tensile strength and 20% in micro-hardness compared to the as-cast alloy.

Manisha Dixit et al. [15] successfully deployed to disperse and immerse nitinol particles in Al 1100 matrix by FSP. Directional residual stresses were introduced with in the prepared material and micro hardness results are compiled the cold-rolled and annealed composites show significant improvement in micro hardness value, as compared with the average value of 26VPN found for Al 1100. The hardness values for the annealed and cold-rolled composite samples were similar, with the cold-rolled samples showing slightly higher values. The strength values of the prepared composite are much higher as compared there with of the Al 1100 matrix. The anticipated yield stress values for the prepared composite were found to be 15–25% below the actually measured values, they predicted that tensile strength in the transverse direction will be expected to be less within the the longitudinal direction. The value of Young's modulus modelled was found to be 109 GPa, which predicts the presence of improved elastic properties within the prepared composite. Yuvaraja, et al. [16] found that the number of passes increases the hardness due to the refinement of grains and uniformity in particle distribution that occurs during FSP of Al5083/B4C composites. S.Soleymani et al. [16] fabricated the composites using FSP and analysed the tribological behaviour. Hybrid composites (AA5083/ (SiC + MoS_2)) exhibited higher wear resistance compared to AA5083/100% SiC or AA5083/100% MoS_2. MoS_2 particles assisted in the formation and retention of a solid lubricating film which enhanced the wear resistance of hybrid composites. Summary of investigation on Mechanical and wear properties of composite fabricated by FSP is shown in Table 1.

Table 1. Summary of investigation on Mechanical and Wear properties of composite fabricated by FSP

Investigator name	Material investigated	Characteristic studied	Prominent results
saini et al. [2]	Al-17Si alloy and Zn/MoS_2 powders.	studied on sliding wear using pin-on-disk Tribometer and UTS.	Ultimate tensile strengths for MoS_2 which is greater than Zn which are reinforced with Al-Si alloys Wear resistance is higher in MoS_2 when compared with Zn are Reinforced with Al-17Si alloy.
Arun premnath [4]	Al-6061alloy and SiC nano-particles.	Optimization of Process parameters Was done by using Taguchi orthogonal Array coupled with Durability approach	Achieved combination of optimum process parameters such as 3passes, 1500 rpm rotational speed and 70 mm/min transverse feed. Noticed better results in mechanical properties and wear loss
Mazaheri et al.[5]	A356/Al_2O_3 surface-Nano composites using FSP.	sample tests conducted to analyse the characteristics of composite.	Microstructural studies shows Al_2O_3 particles were uniformly-Distributed in Al- matrix. Improved micro- hardness and elastic-modulus were noticed when Al_2O_3 particles used as reinforcement for Al matrix.
Alidokht et al. [6]	A356 alloy and SiC/MoS_2 particles	micro structure, hardness and Tribological behaviour Was studied.	SiC and MoS_2 particles were uniformly distributed on the top of stir zone without any defects after one pass FSP.

Table 1. Summary of investigation on Mechanical and Wear properties of composite fabricated by FSP

Investigator name	Material investigated	Characteristic studied	Prominent results
Mahmoud et al. [7]	Al-1050-H24 base-material and SiC/Al_2O_3 particles as reinforcements.	Micro structure and hardness distribution.	Reinforcement particles Sic/Al_2O_3 were uniformly distributed equally on top of nugget zone using FSP. Hardness of composite get increased due to adding of SiC and decreased due to increased proportions of Al_2O_3 particles.
Woo et al.[13]	6061-T6 Al alloy	effects of stirring pin and pressing tool shoulder on micro-structural softening.	Tool shoulder creates the the friction heating which causes reduction in the longitudinal residual strain profiles in nugget zone which generates microstructural softening
Karthikeyan et al.[14]	cast A319 grade Aluminium alloy.	micro structure and mechanical properties.	Mechanical properties were improved at 1200- rpm rotational speed and ductility improved to 1.5 to 5 times for A319 cast alloy.
Solemani et al.[16]	$Al5083/(SiC+MoS_2)$	tribological behaviour	For SiC sample composite hardness increased to 102HB and also having high friction coefficient, MoS_2 has lowest friction coefficient.

2. Conclusions
The major conclusions obtained from comprehensive review of FSP in terms of mechanical properties and wear resistance is summarized below.

1. The mechanical properties and wear resistance of composites fabricated by FSP shows higher values when compared with other processing techniques
2. FSP has unique advantages compared with traditional composites in terms of surface composite fabrication which can be explored in the near future.
3. The major drawback that limits industrial application of FSP is the wear of stir tools. Modification in the tool design was required for effective usage.

3. References

[1] Li, K., Liu, X., & Zhao, Y. (2019). Research status and prospect of friction stir processing technology. Coatings, 9(2), 129.

[2] Saini N, Pandey C, Thapliyal S, Dwivedi DK (2018) Mechanical Properties and Wear Behavior of Zn and MoS2Reinforced Surface Composite Al- Si Alloys Using Friction Stir Processing. https://doi.org/10.1007/s12633-017-9710-2

[3] Don-Hyun C, Yong-Il K, Dae-Up K, Seung-Boo JUNG (2012) Effect of SiC particles on microstructure and mechanical property of friction stir processed AA6061-T4. DOI: 10.1016/S1003-6326(12)61773-7

[4] Arun Premnath. A (2018) Optimization of the Process Parameters on the Mechanical and Wear Properties of Al-SiC Nano-Composites Fabricated by Friction Stir Processing Using Desirability Approach. https://doi.org/10.1007/s12633-019-00178-6

[5] Mazaheri Y , Karimzadeh F, Enayati M (2011) A novel technique for development of A356/Al2O3 surface nanocomposite by friction stir processing. doi:10.1016/j.jmatprotec.2011.04.015

[6] Alidokht S, Abdollah-zadeh A, Soleymani S, Assadi H (2011) Microstructure and tribological performance of an aluminium alloy based hybrid composite produced by friction stir processing. doi:10.1016/j.matdes.2011.01.021

[7] Mahmoud I, Takahashi M, Shibayanagi T, Ikeuchi K (2010) Wear characteristics of surface-hybrid-MMCs layer fabricated on aluminum plate by friction stir processing. doi:10.1016/j.wear.2010.01.005

[8] B.M. Darras, M.K. Khraisheh, F.K. Abu-Farha, M.A. Omar (2007) Friction stir processing of commercial AZ31 magnesium alloy. doi:10.1016/j.jmatprotec.2007.03.045

[9] Mohammad Hasan Shojaeefard & Mostafa Akbari & Parviz Asadi & Abolfazl Khalkhali (2016) The effect of reinforcement type on the microstructure, mechanical properties, and wear resistance of A356 matrix composites produced by FSP. DOI 10.1007/s00170-016-9853-0

[10] Eftekharinia H, Amadeh AA, Khodabandeh A, Paidar M (2016) Microstructure and wear behavior of AA6061/SiC surface composite fabricated via friction stir processing with different pins and passes. DOI 10.1007/s12598-016-0691-x

[11] Elangovan and Balasubramanian (2008) Influences of tool pin profile and tool shoulder diameter on the formation of friction stir processing zone in AA6061 aluminium alloy. doi:10.1016/j.matdes.2007.01.030

[12] Z.Y. Ma a, S.R. Sharma b, R.S. Mishra b (2006) Effect of multiple-pass friction stir processing on microstructure and tensile properties of a cast aluminum–silicon alloy. doi:10.1016/j.scriptamat.2006.01.010

[13] WANCHUCK WOO, HAHN CHOO, DONALD W. BROWN, and ZHILI FENG (2007) Influence of the Tool Pin and Shoulder on Microstructure and Natural Aging Kinetics in a Friction-Stir-Processed 6061–T6 Aluminum Alloy. DOI: 10.1007/s11661-006-9034-0

[14] Karthikeyan, V.S. Senthilkumar, K.A. Padmanabhan (2009) On the role of process variables in the friction stir processing of cast aluminum A319 alloy. doi:10.1016/j.matdes.2009.08.001

[15] Manisha Dixit, Joseph W. Newkirk and Rajiv S. Mishra (2006) Properties of friction stir-processed Al 1100–NiTi composite. doi:10.1016/j.scriptamat.2006.11.006

[16] Narayana Yuvaraja, Sivanandam Aravindanb, Vipina (2015) Fabrication of Al5083/B4C surface composite by friction stir processing and its tribological characterization. http://dx.doi.org/10.1016/j.jmrt.2015.02.006

[17] S.Soleymani, A.Abdollah-zadeh, S.A.Alidokht (2011) Microstructural and tribological properties of Al5083 based surface hybrid composite produced by friction stir processing. doi:10.1016/j.wear.2012.01.009.

Optimal design and experimental investigation of the material-nanostability and deformation behaviour of Al-5.78Zn-1.45Cu-2.49 Mg, Al-5.6Zn-2.5Mg-1.6Cu and Al-Mg-0.6Si alloys under cyclic-loading for ultra-precision structural applications

Shubham Sharma[1*], Jujhar Singh[1*], Vivek Aggarwal[1], Shalab Sharma[2], Abhinav Sharma[1], Ram Narayan[1]
[1]Department of Mechanical Engg., IKG Punjab Technical University, Jalandhar-Kapurthala Road, Kapurthala, 144603, Punjab, India
[2]Deptt. of Mechanical Engg., DAV University, National Highway, Jalandhar - Pathankot Rd, Jalandhar, Punjab 144001, India

[*]Corresponding author. Tel.: +91-7009239473

[*]Correspondence E-Mail ID: shubhamsharmacsirclri@gmail.com
shubham543sharma@gmail.com
jujharsingh2085@gmail.com

Abstract: This paper presents the design and construction of a new simple test facility for the investigation and testing of the deformation behaviour of material properties, namely plastic monolithic, inelastic, elastic and spring joints under cyclic loadings. Test samples were prepared in various machining directions (orthogonal and longitudinal) of Al-5.78Zn-1.45Cu-2.49 Mg, Al-5.6Zn-2.5Mg-1.6Cu and Al-Mg-0.6Si aluminium alloys. As a pulling force for sample deformation, an electromagnet unit was used. The test load range was between 0.2N to 200N, increased by a programmable factor of 1.1-1.4 for a train time of 900s. Two capacitive sensors and interferometers sensors mounted at different sample locations were used to test the sample deformation with and without loading. In a temperature-stable housing that offers isothermal conditions, the overall measuring machine is integrated. The calculated parameters include plastic and inelastic deformation.

Keywords: Al-5.78Zn-1.45Cu-2.49 Mg alloy; Al-5.6Zn-2.5Mg-1.6Cu alloy; Al-Mg-0.6Si alloy; Spring joints; Plastic deformation; cyclic loadings; precision structure

IOP Conf. Series: Materials Science and Engineering **954** (2020) 012012 doi:10.1088/1757-899X/954/1/012012

1. Introduction

For various ultra-precision applications, such as inertial, laser and optical applications, material stability is a critical characteristic [1-5]. A significant stability requirement is considered to be stability under short time loadings (shock, impulse, and vibration), and under medium to long time loadings without load relaxation of internal stress or imbalance [6-8]. Inertial applications, namely precision weighing scales had been considered in research. The steel mono-blocks (mechanical elements) are the principal components of these scales. In order to translate the block-deformation to the output display unit in terms of weight, these elements undergo cyclic-loading and bind to an electromagnetic coil. After the block (each cycle) is released from the load, with a quick response, it should be returned to its original dimensions. In this weighing scale, which is under consideration in this report, these facts are the drawbacks of the used steel mono-blocks. Accordingly, under high weighing frequency and high-stability, the weight measurement process requires ultra-precision. Therefore, the replacement of these elements with an alternative material which validates the above-mentioned characteristics is an essential requirement for such elements.

Many common materials are currently used for high-stability applications, such as: steels containing chromium and nitrogen (carbon steels), steels containing margin (non-carbon), bronze (mostly alloys containing beryllium) and titanium. To achieve the necessary strength, these materials are solution-treated, aged quenched. Steel, on the other hand, has a high degree of unwanted magnetic permeability. The following attractive properties are provided by some aluminum-based alloys: they are cheap, simple to produce, and have a low-modulus of elasticity that results in thicker flexures [9-12]. The key objectives of the present work are to examine the mechanical-properties of various aluminium alloys experimentally, namely Al-5.78Zn-1.45Cu-2.49 Mg, Al-5.6Zn-2.5Mg-1.6Cu and Al-Mg-0.6Si, and to choose the most suitable one to replace the high frequency steel mono blocks.

2. Experimental Setup

2.1. Specimen design

A dissimilar, 220×45×25 mm aluminium alloy samples have been design, built and produced. Two thin flexures (mid cross section of 0.15×25 mm^2) with two cutting directions, i.e. longitudinal, 'L' and orthogonal, 'O' was developed to detect ultra-precision deformation of the specimens by using a CNC milling machine with diamond tools mounted on a ball bearing spindle. The specimen was placed at a 3-point, stress free, clamping-points in the milling machine. Early deformation is avoided by free cutting of the flexures at the beginning of the test. As a consequence, residual-stresses from the process of machining are held near the surface. Through heat-treatment of the samples, stress relief was achieved. For orthogonal and longitudinal cuttings, the specimen geometry is indicated in fig.1 whereas fig.2., exhibited the classification refers to the above-cuttings directions. Longitudinal and orthogonal cuttings mean that the spring cutting direction is tangential to the fibre-diameter spring at the joint spring radius and radial direction, respectively.

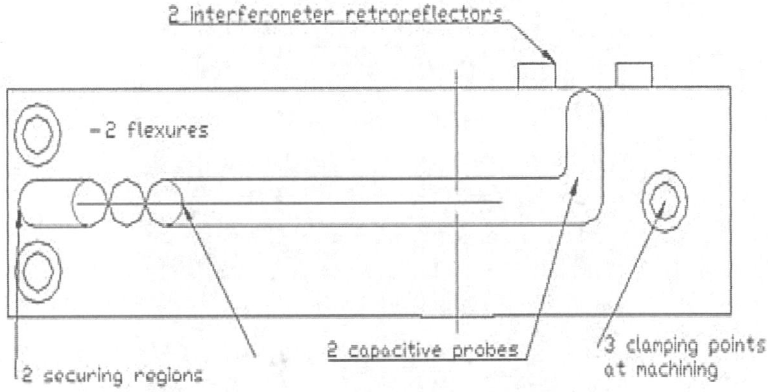

Figure 1. Specimen geometry

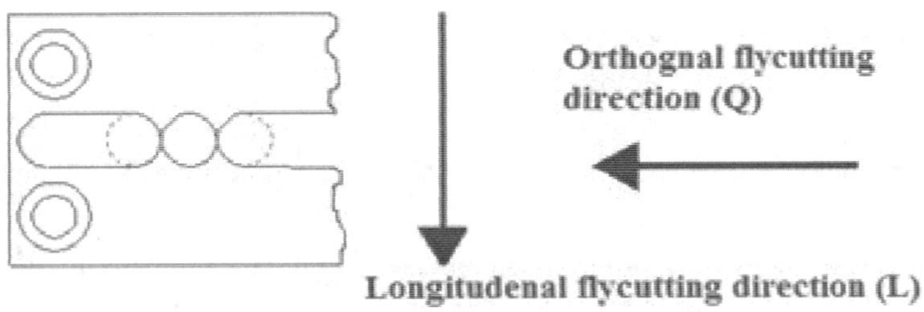

Figure 2. Specimen fly-cutting directions

2.2. Deformation measurement

Ultra-precision measurement of the deformation of the specimen is considered a significant demand in this review. For this reason, the measurement of such deformation was done using two interferometers and two capacitive sensors. The capacitive sensors are located on the opposite side of the sample groove and are glue-fixed. And on the upper surface, the interferometers are mounted. Tiny steel foils were used in order to achieve magnetic fixation with retro-reflectors. In addition, the application force hook was mounted and the interferometer initialized by retro adjustment, which can be pushed freely on the floor, only magnetically held in place. The deformation was measured at the position of the sensors and the geometry was scaled to flexure deformation.

2.3. Test machine

The test machine is designed to allow loading times to be short (1 second) to long (hours). The unit consisted of four elephant feet, an aluminium frame, i.e. a clamping frame with a quadratic opening for connecting the specimen, two cylindrical bearings for interferometer collecting, and a transmitting power. With clamping and adjustment instruments, the upper plate contains the voice-coil. Ferro fluid in the air gap in between increases the binding of the voice-coil to the magnet. In addition, special care is taken to avoid thermal-deformation during measurement by coupling the voice-coil and magnetic assembly into that of the specimen. Other sections include (not shown in the figure) a force-transducer and electronic devices. Fig.3., displays the schematic test used as machine.

Figure 3. The schematic test machine sketch

3. Experimental procedures

The mechanical properties of various cutting directions of Al-5.78Zn-1.45Cu-2.49 Mg, Al-5.6Zn-2.5Mg-1.6Cu and Al-Mg-0.6Si aluminium alloys (AA) have been analyzed herein. The experimental machine incorporates a temperature-stable housing for ultra-precision measurements. In addition, to achieve the same value as the surrounding temperature, the temperature of the stable housing was modified and regulated, with an error of 1 K for all tests. The outside vibration of 100 nm was also suppressed to 10 nm by isolators and to below 1 nm by digital data filtering. In addition, to mitigate temperature, seismic and electrical disruptions, data is collected during the night. The sample was put in the clamping frame to begin the experiment, and then cyclic loading and unloading began with 15 minutes of waiting time after releasing the unloading load to separate the inelastic recovery deformation from the plastic deformation after unloading. The loading force increased by a programmable factor of 1.1-1.4 in the next period. Usually, a test cycle takes a few minutes to allow the residual temperature compensation of capacitive and interferometer data to be processed. The test-conditions are as follows: the force load ranges from 0.2 N to 200 N; the interferometer, capacitive-sensor and temperature controller resolutions are 10 nm, 0.3 nm, and 0.001 K, respectively. The calculated parameters for aluminium alloy specimens are the inelastic and plastic deformations, and the time constant.
The abbreviations CS and IF indicate the output of the capacitive and interferometer-sensors, respectively, for each measured deformation.

4. Results and discussions

4.1. Loading and Unloading cycle (cyclic loading)

The loading-unloading period for the Al-Mg-0.6Si and Al-5.6Zn-2.5Mg-1.6Cu AA group of alloys is shown in Figure 4 (as an example for all test phases). The load cycle begins at 1130 minutes and ends at 1167.5 minutes (i.e. with 2250 s of load application time in the downward direction), as can be seen from the figure. The unloading, however, begins at 1163.33 minutes and finishes at 1248.33 minutes. In addition, it is equally important to know that a period of up to 3000 s is required for inelastic recovery to determine the plastic deformation. Furthermore, it should be noted that the inelastic deformation is much lower than the plastic deformation and therefore the loading characteristics are within a range of up to 40 MPa in their declaration of capacity.

IOP Conf. Series: Materials Science and Engineering **954** (2020) 012012 doi:10.1088/1757-899X/954/1/012012

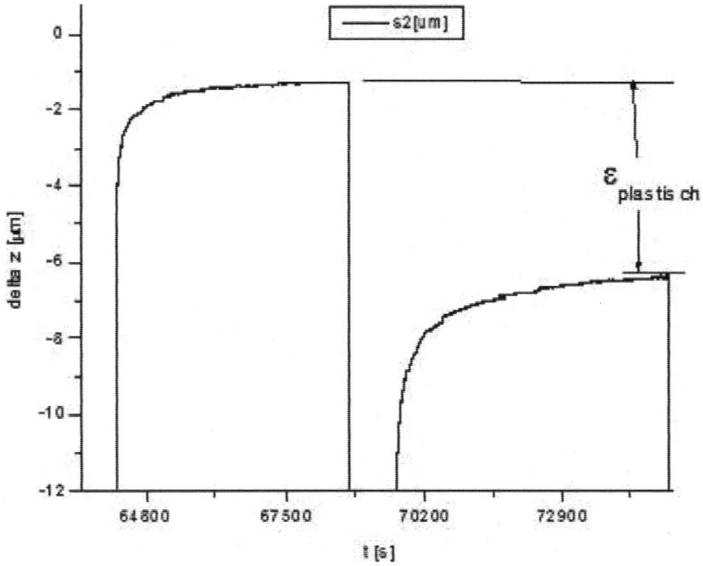

Figure 4. The loading-unloading cycle

4.2. Inelastic and plastic deformation

As a consequence of the micro-creep in the material that arises due to the motion of the atomic layers in the spring joint during the loading process, inelastic deformation occurs. The glides return to their original location after load release. Depending on the load on the joints and the pollination duration of the last few hours, this behavior will vary. In the other side, the material atoms are further away from each other without splitting as the load-rises, approaching the yield point, and there is plastic-deformation. The atomic layers are so deeply moved here that if the relief is no longer back in their original-position.

The reliance of inelastic deformation on the stress for Al-5.78Zn-1.45Cu-2.49 Mg composite with symmetrical and longitudinal cutting headings are given in Fig.5a and 5b, individually. From the figures, it can be seen that the direction of machining (orthogonal or longitudinal) negligibly affects the measure of inelastic disfigurement because of the reasonable exactness of the estimating sensors and the variation in tests properties. The base of 1 nm and the limit of 16 nm twisting were estimated under stress estimations of 10 MPa and 150 MPa, resp.

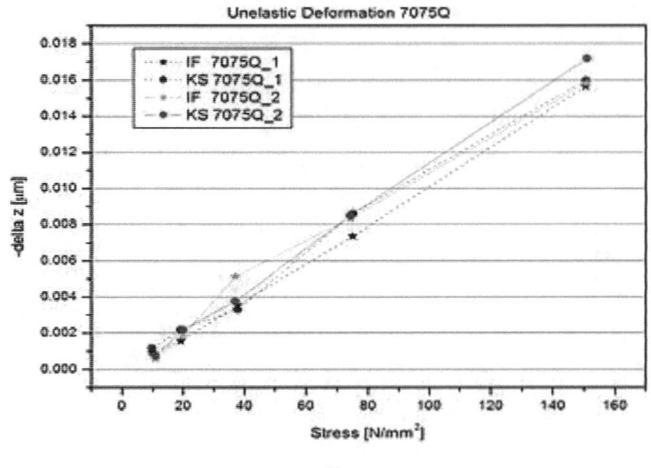

A

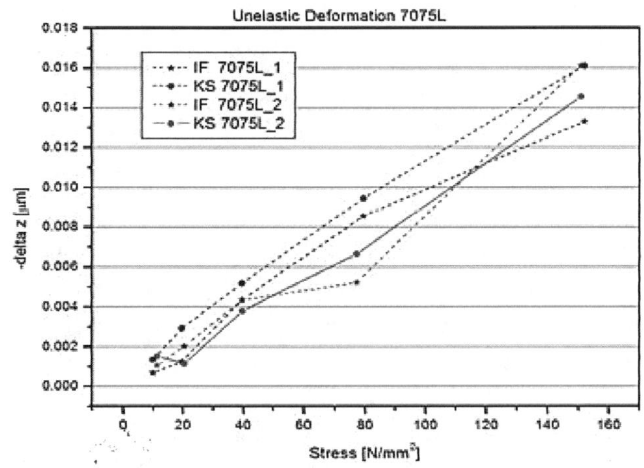

B

Figure 5. Inelastic deformation of Al-5.78Zn-1.45Cu-2.49 Mg aluminum alloy

Plastic deformation for both specimen groups was appeared in figures (6a and 6b). Aluminum alloy specimens with symmetrically machining direction has a plastic deformation of a little bandwidth for every load points. Then again, Fig.6b indicates aluminum alloy with longitudinal cutting course and could be insignificantly superior to that of symmetrically cutting, notwithstanding the enormous dispersing in the individual estimated values. Subsequently, the plastic disfigurement of the 7075-T7351-L AA is not as much as that of 7075-T7351-Q in the stress interval somewhere in the range of 40 and 110 MPa. This can be ascribed to the temperature impact and to the electrical floats of the electronic gadgets. The plastic distortion is in opposition to the tensile stress with two cutting headings. In the event that this understanding was considered here, just the capacitive sensors ought to be utilized something like 40 MPa, in light of the fact that the electrical float of the interferometer is excessively huge as appeared in Fig.6b. It is significant here to focus on the way that the plastic distortion of the whole estimation cycle adds the disfigurements comparative with the previous load. The accompanying model clarifies the plastic deformation between two stress points for both cutting directions. For AA 7075-T7351-L, the deformation mean qualities were 0.8 nm and 1.1 nm for 80 and 150 MPa stresses, individually. What's more, the outcome is plastic distortion of magnitude 0.3 nm. By a similar way the deformation of AA 7075-T7351-Q are: 1.1 nm at 80 MPa and 1.5 nm at 150 MPa, and the subsequent

plastic deformation is 0.4 nm.
Comparative with the inelastic deformation, note that the plastic deformation is lower and consequently the diagrams are in the range up to 40 MPa in their capacity explanation.

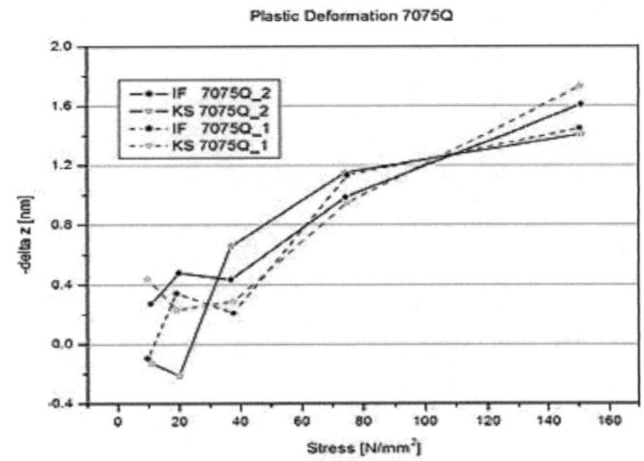

A

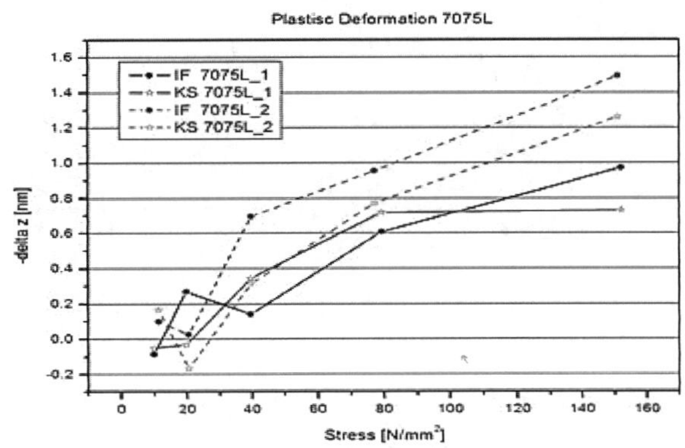

B

Figure 6. Plastic deformation of Al-5.78Zn-1.45Cu-2.49Mg aluminum alloy orthogonally machined and B) longitudinally machined

As referenced over the longitudinal cutting direction mode exhibit best outcome for estimated deformation than that of orthogonally. Consequently, just this mode utilized on the Al-Mg-0.6Si and Al-5.6Zn-2.5Mg-1.6Cu aluminum alloys.
Figure 7 indicates the inelastic deformation of Al-Mg-0.6Si aluminum alloy for the normal bending stress. The figure shows that, the inelastic deformation at the loading scope of 10 to 20 MPa can be set equivalent to zero, in light of the fact that the identified deformation for trial 1, when capacitive sensor demonstrates 0.5 nm deformation, can be deciphered as a drift. A similar explanation applies to the 0.7 nm deformation estimated by the interferometer for trail 5. Likewise, from the figure, the most extreme normal estimation of the deformation at 150 MPa was 9 nm.
Moreover, as compared to the inelastic deformation of Al-5.78Zn-1.45Cu-2.49 Mg aluminum alloy presented in figure 8, the inelastic deformation of Al-Mg-0.6Si is a bit lower. For instance, at stress value of 140 N/mm^2, the inelastic deformation as calculated by the capacitive sensor is equal to 10 nm, and 21 nm for the Al-Mg-0.6Si and Al-5.78Zn-1.45Cu-2.49 Mg respectively.

IOP Conf. Series: Materials Science and Engineering **954** (2020) 012012 doi:10.1088/1757-899X/954/1/012012

Figure 9 shows the inelastic deformation of Al-5.6Zn-2.5Mg-1.6Cu alloy as a component of the ordinary bending stress. As figure indicates at low loads up to 40 MPa, the distinction between the deformation estimated by CS and IF sensors are little. On the other hand, when stress level builds, the difference gets higher. It reflects that a sample in the range of Al-5.78Zn-1.45Cu-2.49 Mg is a further trial is worse in all tested tensile stresses. The deviation of the two estimation strategies for the both samples is inside the tolerance range. This again indicates that the principle effect on the measurement of samples is done. The distinction in thermal treatment has clearly no effect on the inelastic deformation.

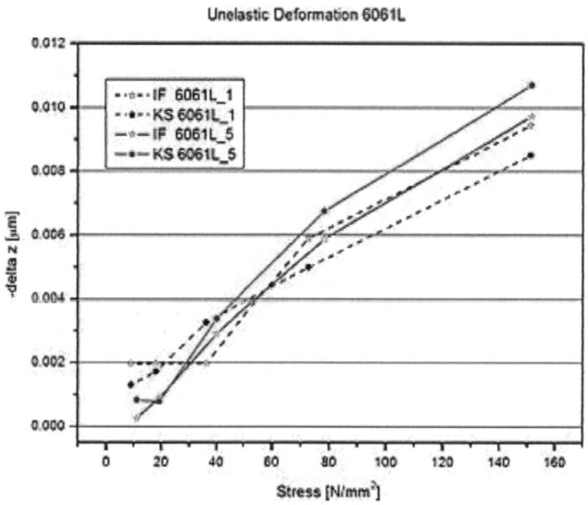

Figure 7. Inelastic deformation for Al-Mg-0.6Si aluminum alloy

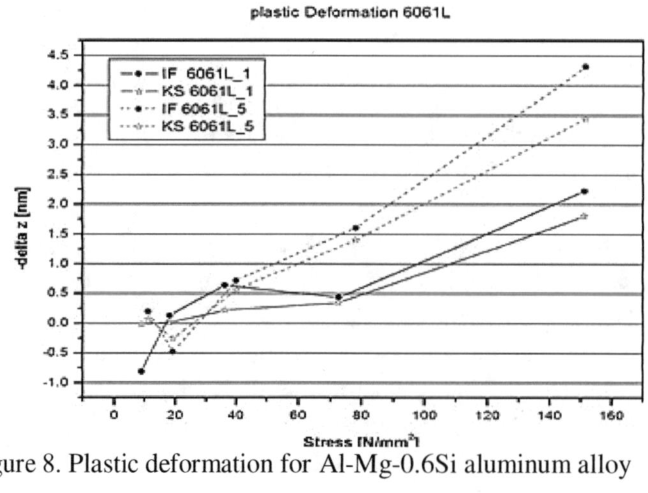

Figure 8. Plastic deformation for Al-Mg-0.6Si aluminum alloy

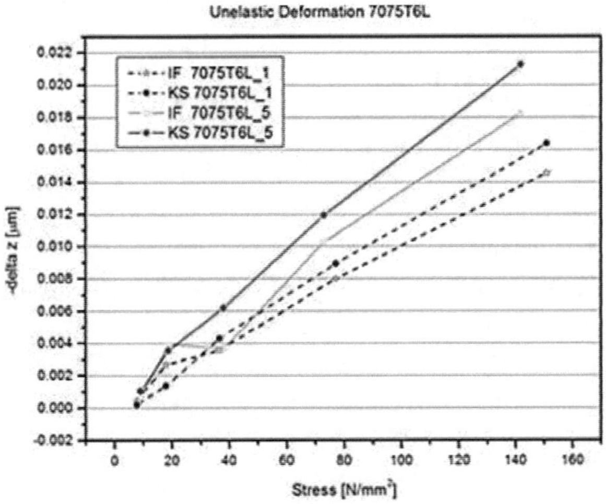

Figure 9. Inelastic deformation for Al-5.6Zn-2.5Mg-1.6Cu aluminum alloy

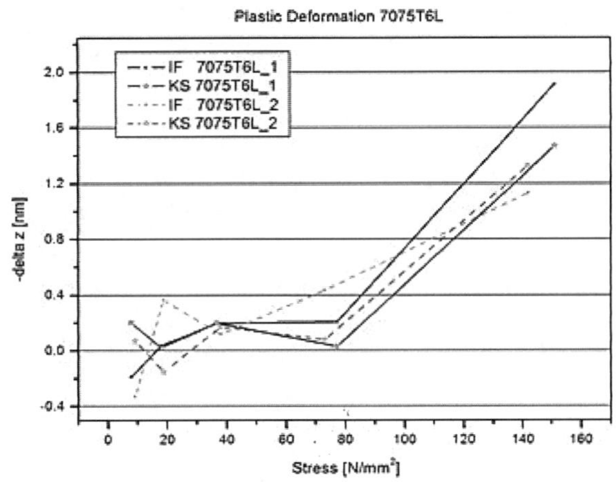

Figure 10. Plastic deformation for Al-5.6Zn-2.5Mg-1.6Cu aluminum alloy

Plastic deformation for Al-5.6Zn-2.5Mg-1.6Cu alloy specimen is presented in figure 10, from the figure it was noted that for a stress value limiting to 80 MPa, the plastic deformation is less than 0.4 nm. Then a abrupt rise in plastic deformation was noted amongst the last two points of the curve i.e. 80 and 150 MPa the corresponding increase in the average deformation of the sample is equal to 1.4-0.2 nm. The material Al-5.6Zn-2.5Mg-1.6Cu alloy compared to an even higher strength, due to the artificial aging and stretching.

5. Conclusions

This work incorporates the development of another straightforward test facility for the examination and testing of materials properties of solid spring joints for precise structures. In the test, an electromagnet was utilized to produce the pulling power important for the deformation of the unique geometry sample. The deformation and versatile recuperation of the samples in the wake of pulling power were recorded with the assistance of an interferometer, and capacitive sensor, for various aluminum alloys (Al-5.78Zn-1.45Cu-2.49 Mg, Al-5.6Zn-2.5Mg-1.6Cu and Al-Mg-0.6Si), with various machining directions (longitudinal and orthogonal).

The consequences of this work incorporate the inelastic and plastic deformation of the material, and a variety of the load of 10 MPa upto 160 MPa, at that point the naturally visible deformation after partition of the examples. The test machine carries out in a temperature stable housing. Measurement is carried

IOP Conf. Series: Materials Science and Engineering **954** (2020) 012012 doi:10.1088/1757-899X/954/1/012012

out down to a remaining strain under 10^{-6}.

a) Materials nanometer behavior is unlike from micrometer behavior.

b) The Al-Mg-0.6Si alloy material with tensile strength of 275 MPa shows 60% lower inelastic deformation than Al-5.78Zn-1.45Cu-2.49 Mg alloy with 435 MPa and Al-5.6Zn-2.5Mg-1.6Cu alloy with a tensile strength of 505 N/mm^2. Since a difference of Al-5.78Zn-1.45Cu-2.49 Mg and Al-5.6Zn-2.5Mg-1.6Cu alloys not be recognized, is near to the deduction that irrespective of inelastic deformation of the thermal post-treatment of aluminum is critical here is the alloy composition.

c) For the calculation of plastic deformation of the test materials, the deformation of 7075 L is only 20% less than in 7075Q. The Al-5.6Zn-2.5Mg-1.6Cu alloy material shows a clear difference from Al-5.78Zn-1.45Cu-2.49 Mg alloy. The lower tensile strength and lower stress are the major reasons of the toughest plastic deformation field in materials from Al-Mg-0.6Si alloy.

d) The inelastic deformation among longitudinal and orthogonal cutting direction expressed no differences to distinguish.

Declaration of conflict of Interest:

The authors declare that they have no conflict of interest with respect to the research, authorship, and/or publication of this article.

Funding:

The authors doesn't receive any research fund or grant from any organization.

Ethical approval:

This article does not contain any studies with human participants or animals performed by the author.

Acknowledgements:

The authors wish to acknowledge the Department of RIC, IK Gujral Punjab Technical University, Kapurthala, Punjab, India.

References:

[1]. Yonezawa H, Hirata Y, Sasai H 1990 Positioning table with high accuracy and high speed, *Annals of the CIRP* **39 (1)** 433–436

[2]. Sakuta S, Ogawa K, Ueda K 1993 Experimental studies on ultra-precision positioning, *International Journal of the Japan Society of Precision Engineering* **27 (3)** 235–240

[3]. Kim K D, Nam S R 1995 An improvement of positioning accuracy by use of piezoelectric voltage in piezoelectric driven micro positioning system simulation, *Mechanism and Machine Theory* **30 (6)** 819–827

[4]. Ge P, Jouaneh M Tracking control of a piezoceramic actuator *IEEE Transaction on Control Systems Technology* **4 (3)** 209–216

[5]. Choi S B, Kim H K, Lim S C, Park Y P 2001 Position tracking control of an optical pick-up device using piezoceramic actuator, *Mechatronics* **11 (6)** 691–705

[6]. Choi S B, Han S S 2007 A magnification device for precision mechanisms featuring piezoactuators and flexure hinges: Design and experimental validation, *Mechanism and machine theory* **42** 1184–1198

[7]. Ha J L, Kung Y S, Hu S C, Fung R F 2006 Optimal design of a micro-positioning Scott–Russell mechanism by Taguchi method, *Sensors and Actuators A: Physical* **125 (2)** 565–572

[8]. Van Huis M A, Chen J H, Zandbergen H W, Sluiter M H F 2006 Phase stability and structural relations of nanometer- sized, matrix-embedded precipitate phases in Al-Mg-Si alloys in the late stages of evolution *Acts material* **54 (11)** 2945-2955.

[9]. Davis J R 1998 *Aluminium and Aluminium Alloys* (USA: ASM Specialty Handbook)

[10]. Craig W 1998 D*iv. Metals Handbook*, (USA: ASM Specialty Handbook)

[11]. ASM Handbook Committee 1998 *Metals Handbook Ninth Edition, Volume 2, Properties and Selection: Nonferrous and Pure Metals (USA: ASM Handbook*)

[12]. Martin L Culpepper, Gordon Anderson 2004 Design of a low-cost nano-manipulator which utilizes a monolithic, spatial compliant mechanism, *Precision Engineering* **28 (4)** 469–482

An experimental study of wire breakage frequency on different electrodes during WEDM of Inconel-722

Vivek Aggarwal[1], Jujhar Singh[1]*, Shubham Sharma[1]*, Harish K. Garg[2], Abhinav Sharma[1], Gursharan Singh[1], Jwala Parshad[1]
[1]Department of Mechanical Engg., IKG Punjab Technical University, Jalandhar-Kapurthala Road, Kapurthala, 144603, Punjab, India
[2]Department of Mechanical Engg., DAV University, Jalandhar, 144001, Punjab, India
*Corresponding author. Tel.: +91-7009239473
*Correspondence E-Mail ID: jujharsingh2085@gmail.com
shubhamsharmacsirclri@gmail.com
shubham543sharma@gmail.com

Abstract. Inconel 722 is a nickel-based superalloy that holds high strength and impervious to oxidation and corrosion at high temperatures. Wire electrical discharge machining (WEDM) was investigated as a superior alternative to conventional machining methods for intricate and complex shapes for materials containing more strength. However, these wire electrodes also undergo breakage during WEDM of Inconel 722 which unquestionably diminishes the efficiency of machining process, and accuracy. Therefore, this work is an effort to investigate and understand the phenomenon of wire breakage in relation to the combination set of Inconel 722 and three wire electrodes which are plain brass-wire, diffused brass-wire and zinc-covered metal wire. An experimental study was carried out to check the effect of six machining parameters viz. Ton, Toff, SV, IP, WF and WT on the frequency of wire breakage. Results demonstrate that zinc-covered metal wire has low breakage recurrence according to all the parameters of the process examined. It has also determined the optimal range of process-parameters under which zinc-coated brass-wire can be used to achieve effective machining.

Keywords: WEDM, superalloys, Inconel 722, plain brass-wire, diffused brass-wire, zinc-covered brass wire

IOP Conf. Series: Materials Science and Engineering **954** (2020) 012013 doi:10.1088/1757-899X/954/1/012013

1. Introduction

Inconel 722 nickel-based superalloy have excellent corrosion resistance and oxidation properties which are temperature resistant and high strength. This superalloy has applications in medical, nuclear, food and for other industrial work [1, 2]. Inconel 722 is highly difficult to process using conventional methods, owing to its strength, resilience, low carbide presence and poor thermal conductivity [3]. WEDM is particularly close to die sinking electrical discharge machining system whereupon material abrasion occurs from an electrically conductive workpiece by inducing quick, discrete and repeated discharges of sparks in between wire device electrode as cathode and the workpiece as anode attached in a circuitry system [4-5]. Device electrode exists a wire of small diameter (0.05mm -0.3 mm) typically of brass, copper, tungsten and the workpiece is affixed upon CNC. The wire is being kept under tension to lower the chance of making imprecise sections using a mechanical positioning device. The wire is continually advanced in/take-up and supply reel to constant-diameter fresh electrode for the work. In-between the distance of 0.025mm-0.05 mm between wire electrode and the workpiece continuously, microprocessor holds it [6-7]. To flush over scrap (molten material) created from operation, a liquid dielectric medium carried through the space continuously [8]. The wire electrode plays a vital role that delivers net production of WEDM. The electrode must have sufficient high tensile strength to withstand the requisite stress, electrical conductivity to minimize energy depletion and retain higher dimensional accuracy to significantly contribute to overall performance. Over few decades, brass electrode has been commonly used in WEDM due to more tensile strength and lower cost compared to copper and an exceptional combination of high electrical conductivity [9-15]. In realistic applications, rupture of wire exhibits a steady threat to the output of the WEDM. Nonetheless, the WEDM operators can resolve the problem to make ensure the productive running of their operations by gaining knowledge of certain parameters such as work material used in the process, the phenomena and the causes of wire breakage. Factors that increase the risk of wire breakage including high thermal load originated during WEDM phase resulting in high wire strain, unjustified wire heating, inefficient material debris removal, total sparking frequency rapid raise, excess arc sparks, loss of WF and Toff, increase of pulse length, low yield and crack resistance of electrode wire and rise of wire electrode temperature due to bad flushing conditions [14-16, 18–28]. Assorted scholars have sought to gain vision into mechanism of breakage of wire during WEDM but the development is still not yet well known. Several attempts have been made to develop an innovative control device which provides electronic monitoring of rare machining circumstances and devises control methodologies that prevent wire fracture frequency and settling the output capacity of WEDM. Rajurkar and Wang [4] evolved a wire electric discharge machining sparking-frequency detector, in particularly effectively detected instant sparking frequencies throughout wire break cycle. To determine the distribution of temperature along the wire, authors used thermal model which resulted in yield of maximal temperature nearby the area of workpiece. Yan and Liao [12, 20] have tried to develop the self-fuzzy logic learning based system for monitoring and controlling the frequency of sparking. Copper wire has been used during higher cutting rate on WEDM and wire breakage frequency has been avoided. Under a steady feed rate machining condition, the pulse off time value was balanced in real time. Cabanes et al. [16] suggested knowledge-based methodology for the electronic detection of AISI D2 steel wire fracture and instability with a brass wire during WEDM. Results showed that variables like ignition delay, discharge capacity and IP significantly influenced to wire rupture. By observing wire fracture indicators, it was presumed that comprise of warning signals about increased risk of wire rupture. Cabanes et al. [29] confirmed the new wire break checking and symptomatic framework comprising of the virtual-instrumentation-framework for ascertaining the fitting extents, and the demonstrative framework for recognizing decayed cutting-systems and estimating wire breakdowns. Patil and Brahmankar [17] used a coated brass tool in order to avoid wire breakage of alumina particulate-reinforced aluminum for material CuZn50 on WEDM. They identified and set limits of machining of these composites for rupture of wire. Antar et al. [10] proposed that copper wires, Zn rich Brass and ZnCu50 wires relative to uncoated brass wires, would significantly improve WEDM operating efficiency, residual stress under the same machining conditions, and eliminate recast layer thickness. Kumar et al. [31, 32] defined outcome of (IP), (Ton), (SV), (Toff), sparking frequency, wire stream and wire electrode

surface integrity during the WEDM of pure Titanium. They found out the larger values of peak speed, pulse-time, sparking frequency and stress on wire while the lower estimations of (SV), pulse-time, (WF) persistently enhance wire wear rate and therefore, risk of wire rupture.
Assessment of literature work shows that the presented work focused mainly on studying the influence of different variables of machining on rupture of wire of variety of materials namely steels, metal matrix composites and titanium alloys during WEDM. To minimize the risk of fracture of wire while enhancing capacity of machine, authors also developed the hypothetical and existential specimens for scrutinizing and flexible command of swift alterations in machining variables, consecutively enhances machining effectiveness. Consequently, the main objective is to examine the impact of many variables on the probability of wire fracture with three dissimilar wire materials and to govern relevant variables where successful WEDM machining is feasible with minimal chances of rupture of wire.

2. Experimentation

2.1. Materials and methods

Key goal of accomplishing experimental analysis is to establish the functional set of operational variables whereupon Inconel 722 credibly efficiently machined using wire electric discharge machining processes with minimal probability of wire breakage. In this study, both tests were enacted upon WEDM machine tool (Elektra Sprint cut 734). During laboratory experiments, the plain, diffused and the Zinc-coated brass-wire performed as wire electrode materials each with a diameter of 0.25 mm. The goal of utilizing three different wire material is to identify the material of wire producing minimum wire breakage frequency. First, workpiece is affixed on a computer table to conduct experimental work with the aid from a fixture to erase the risk of poor alignment. While processing, the wire-terminal is fixed to negative polarity, while the workpiece is kept at positive polarity. Wire diameter is maintained stable throughout the methodology. Deionized water separates the wire electrode workpiece and often acted as the dielectric medium. In the current research Inconel 722 was elected as the test instrument. Selected workpiece width of 20 mm and height is 20 mm.

2.2. WEDM Process Parameters

The research aspires towards examining the result regarding various variables of WEDM on the frequency of breakage of wire when machining Inconel 722 with three different wire materials. The electrical variables appear to carry additional impact upon WEDM's machining attributes as opposed to WEDM's non-electrical cycle parameters [33]. Existing WEDM literature indicates that the main input parameters are Wire tension (WT), Wire feed rate (WF), Pulse-on time (Ton), Pulse-off time (Toff), Peak current (IP) and Spark gap voltage (SV). Thus, in this investigation, above mentioned variables were of the input process. Table 1 describes the definition of variables accompanying with measurement units and the notes used, and other parameters are enumerated in Table 2 that are kept constant throughout the experiment. These variable values were chosen based on current literature, physical capacities of the fabricator and the computer. The experiments were piloted using a one factor/time method where one input methodological variables (factor) differs by its spectrum after number of variables are elected while all others remain constant at mid-level [34]. Table 3 defines least and maximal values for the whole of six input variables amidst the one factor/time trials. Any scales of WEDM input variables were chosen, based on its capacity, and only those scales could be reached on WEDM.

2.3. Performance measure of WEDM

The variable wire breakage-frequency was held as a response variable since it shows level of effective machining in the WEDM phase without fracture of wire.

2.4. Process-Parameter's impact on Wire Breakage Frequency

IOP Conf. Series: Materials Science and Engineering **954** (2020) 012013 doi:10.1088/1757-899X/954/1/012013

Herein segment the outcome of six variables on wire breakage frequency, namely (Ton), (Toff), (IP), (SV), (WF), and (WT) was depicted in WEDM of Inconel 722. The machining period was observed employing a stopwatch with a minimum count of 0.01 s.

2.4.1. Impact of Ton upon Wire Breakage Frequency

Effect of (Ton) on wire breakage frequency using three different wire electrodes is illustrated in **Table 1** and corresponding graph is given in **Fig. 1**. That becomes obvious amidst the growth in (Ton), the frequency of wire breakage for all three different wire material types is gradually increasing. When the (Ton) factor is held to a minimum, wire breakage does not occur and the frequency factor for breakage of wire is zero. Whereas, the probability of wire rupture also decreases as time interval over which the current in-loop (Ton) flows diminishes. The explanation may be that as (Ton) decreases, discharge energy that is a variation of (Ton) and peak speed, also increases. Material extracted from material at work counts strongly on the resistance of the discharge.

Table 1. Pulse-on time effect on the wire breakage-frequency for three different wire electrodes

Pulse on Time	**Wire Breakage Frequency**		
T_{on} (µs)	**Plain Brass wire**	**Diffused Brass wire**	**Zinc-coated Brass wire**
0.2	0	0	0
0.4	0.05	0	0
0.6	0.08	0.05	0.02
0.8	0.12	0.1	0.06
1	0.16	0.11	0.09
1.2	0.24	0.19	0.12
1.4	0.28	0.21	0.18
1.6	0.32	0.29	0.26

The quantity of response decreases with the longer pulse-on length because of high discharge energy [14, 27]. This results in reduced dielectric fluid power resulting in an irregular arcing rather than simple sparking. The fracture of wire avoids Inconel 722's cutting at greater (Ton) values in this direction should be used on time [17, 23, 30, 32].

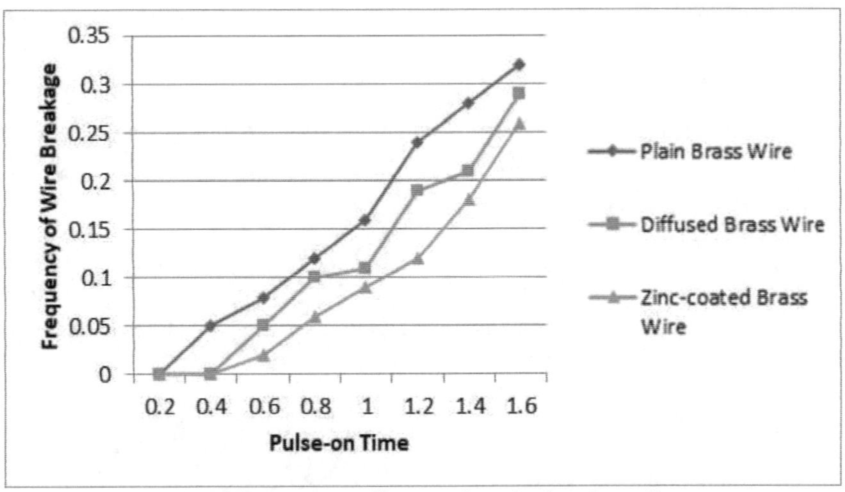

Figure 1. Pulse on time effect on the wire breakage frequency

2.4.2. Impact of (Toff) on Wire Breakage Frequency

Figure 2 demonstrates the outcome of pulse off periods upon wire breakage-frequency corresponding to values indicated in Table 2. It illustrates that wire breakage frequency was zero (minimum), while (Toff) value is held to a maximum as the (Toff) for three wire material types decreases, wire breakage frequency increases. This may due to effect that higher the (Toff) value, the lower the level of flickering resulting in a reduced molten metal (debris) within the gap. Additionally, dielectric fluid cools-down the wire at faster pulse-off speed, removing the waste in the gap, thereby avoiding detrimental arcing due to decreased dielectric power, and dramatically reducing the incidence of wire breakage. Longer off-time is also useful because of wire breakages for the machining of high-resistance materials [17, 19, 20, 30, 32]. The wire breakage frequency at different pulse-off time using three wire electrodes has been listed in Table 2 and shown graphically in Figure 2.

Table 2. Pulse off time effect on the wire breakage frequency for separate wire electrodes

Pulse off Time	**Wire Breakage Frequency**		
T_{off} (μs)	**Plain Brass wire**	**Diffused Brass wire**	**Zinc coated brass-wire**
2	0.42	0.36	0.32
8	0.31	0.27	0.23
14	0.21	0.18	0.14
20	0.17	0.13	0.10
26	0.11	0.08	0.06
32	0.10	0.07	0.05
38	0.06	0.04	0.01
44	0.04	0.02	0
50	0	0	0

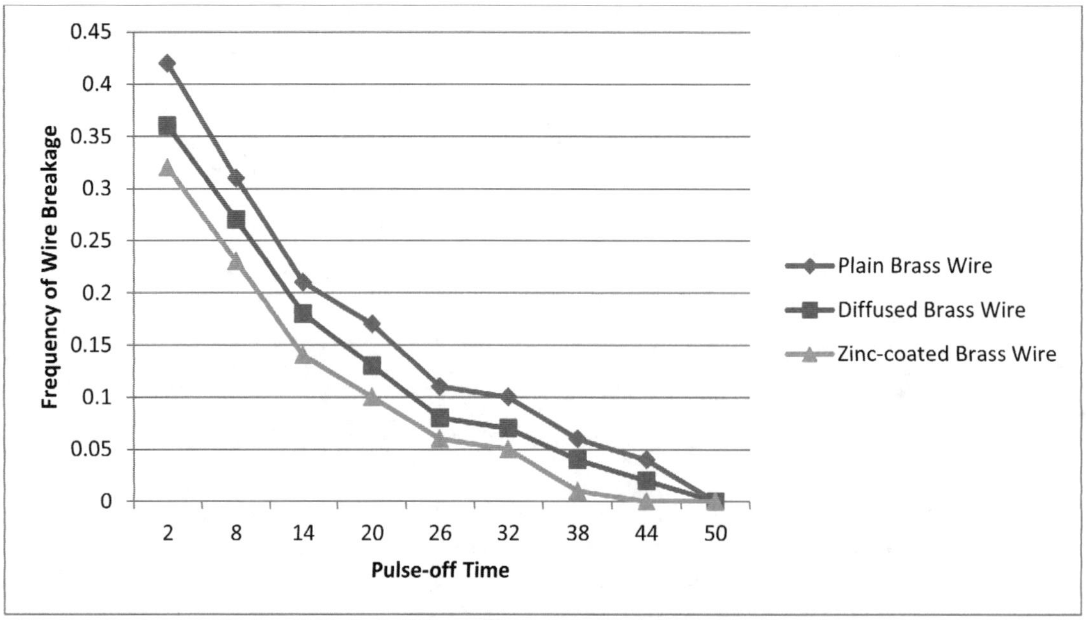

Figure 2. Pulse off time effect on the wire breakage- frequency

2.4.3. Impact of (IP) on Wire Breakage Frequency

Figure 3 demonstrates the influence of (IP) on frequency of wire rupture. It is obvious given wire breakage level lessens continually alongside (IP) value rise with three wire material types. As articulated, this is because high (IP) values produce extra discharge energy and thence, additional molten metal stockpiles in the crevice that the dielectric medium cannot properly flush. Thus, irregular arcing occurs which decreases dielectric strength and causes significant damage to the wire electrode resulting in local temperature increase. At elevated temperature the (WT) reduces allowing the wire to snap. Furtherly, attributable to confined space in the spark, some of discharge energy is consumed by wire resulting in its liquefying and thus break. [14, 32]. The values of the wire breakage frequency at different settings of peak current using three wire electrodes have been demonstrated in Table 3 and shown graphically in Figure 3.

Table 3. Peak current effect on the wire breakage frequency for three wire electrodes

Peak Current	**Wire Breakage Frequency**		
IP (ampere)	**Plain Brass Wire**	**Diffused Brass wire**	**Zinc-coated Brass wire**
50	0	0	0
70	0	0	0
90	0.06	0.04	0.03
110	0.08	0.07	0.05
130	0.13	0.10	0.07
150	0.14	0.13	0.11
170	0.19	0.15	0.12
190	0.22	0.18	0.15
210	0.24	0.20	0.17
230	0.32	0.28	0.24

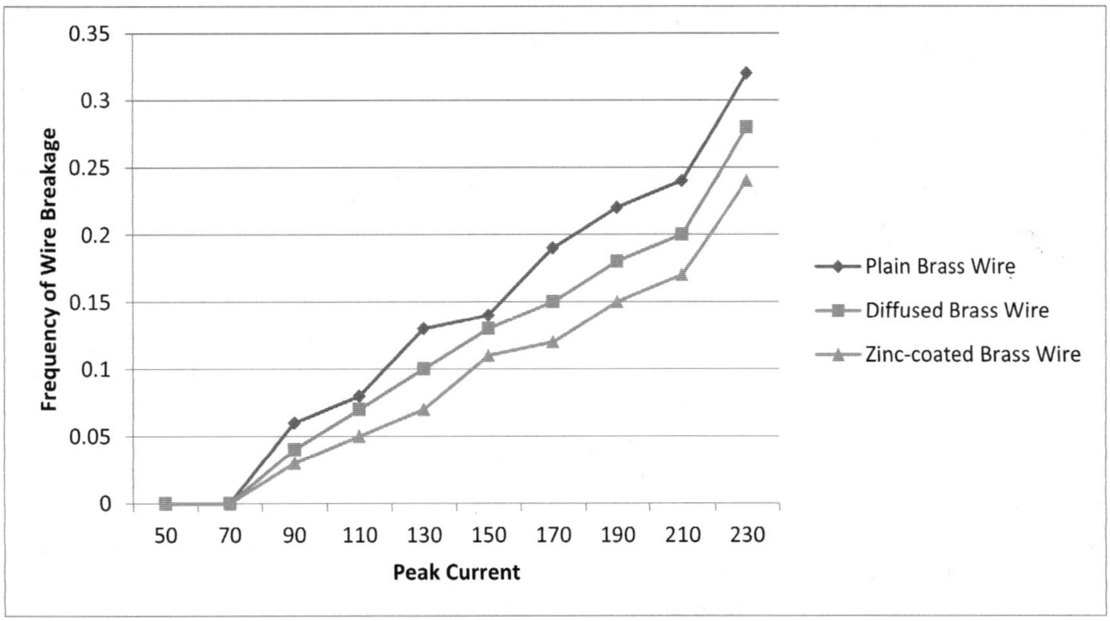

Figure 3. Impact of peak current on the wire breakage recurrence

2.4.4. Impact of (SV) upon Wire Breakage Frequency

Figure 4 illustrates the outcome concerning (SV) upon wire breakage frequency, which illustrates that wire breakage frequency rises continually alongside reduced gap voltage. If optimal value of the (SV) is retained, so for all three types of wire material no wire breakage occurs. Nonetheless, as the probability of wire breakage increases the gap voltage reduces attributable to reason that the higher (SV) levels resulting lesser spacing between wire electrode and the workpiece resulting into wire electrode being shifted from a portion of the electric discharge. Hence, the metal electrode surface temperature rises leading to a wire breakage. In comparison, the greater distance causes inadequate molten metal flushing and hence harm the small gap, sequentially directs to erratic arcs and therefore damages wire. An appropriate voltage servo difference needs to be identified to solve this problem [17, 30, 32]. The values of wire breakage frequency at different settings of spark gap voltage using three different wire electrodes have been listed in Table 4 and shown graphically in Figure 4.

Table 4. Spark gap voltage impact on the wire breakage recurrence for three separate wires

Spark Gap Voltage	Wire Breakage Frequency		
SV (volt)	Plain Brass wire	Diffused Brass wire	Zinc coated Brass wire
8	0.38	0.36	0.31
16	0.25	0.22	0.2
24	0.2	0.21	0.17
32	0.17	0.15	0.12
40	0.16	0.14	0.11
48	0.12	0.11	0.09
56	0.11	0.08	0.07
64	0.07	0.06	0.03
72	0.05	0.03	0.01
80	0.03	0	0
88	0	0	0
96	0	0	0

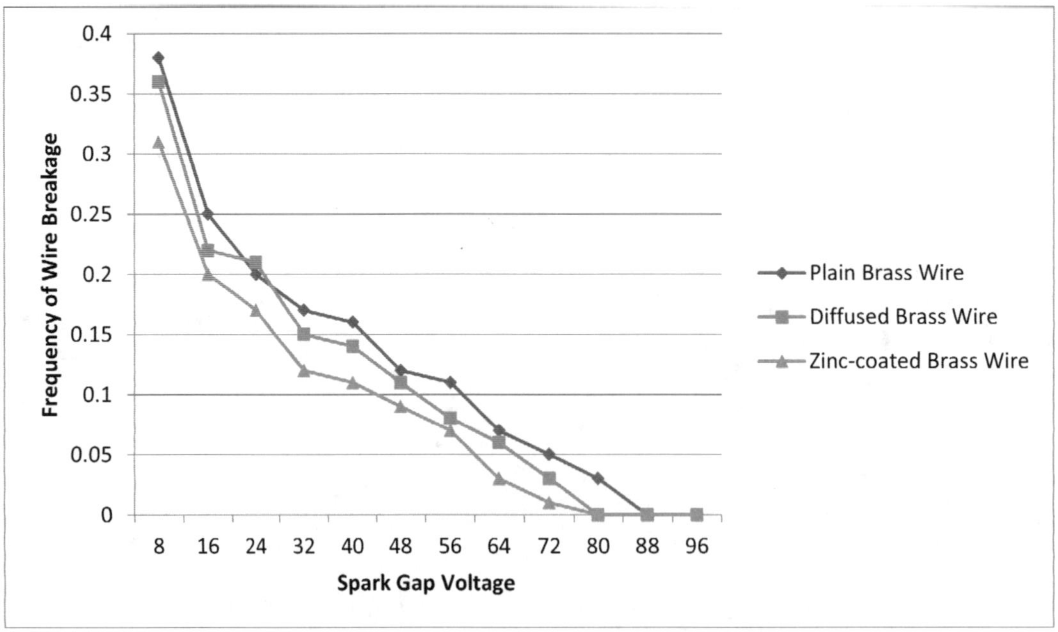

Figure 4. Spark gap voltage effect on the wire breakage recurrence

2.4.5. WF Impact on Wire Breakage Frequency

Figure 5 indicates outcome of (WF) upon wire breakage frequency. From chart this becomes obvious that wire breakage frequency for all wire material types lessens alongside increasement in volume of wire feed towards theory that lessens (WF) allows wire electrode to tolerate high discharge energy/unit time resulting into rupture of wire. Therefore, the wire electrode needs to expend more time in the gap at a decreased feed rate which reduces the amount of wire electrode craters/unit length and allows wire to fracture [14, 17, 23 and 30]. Thence, lower feed-rates would stay preferred into prevent repeated wire breakages. The wire breakage frequency at different values of wire feed rate using three different electrodes has been listed in Table 5 and shown graphically in Figure 5.

Table 5. Wire feed rate effect on the wire breakage recurrence

Wire Feed Rate WF (m/min)	Wire Breakage Frequency		
	Plain Brass wire	Diffused Brass wire	Zinc coated Brass wire
2	0.26	0.22	0.19
4	0.21	0.17	0.13
6	0.18	0.14	0.11
8	0.14	0.10	0.07
10	0.11	0.07	0.03
12	0.06	0.02	0
14	0.03	0	0

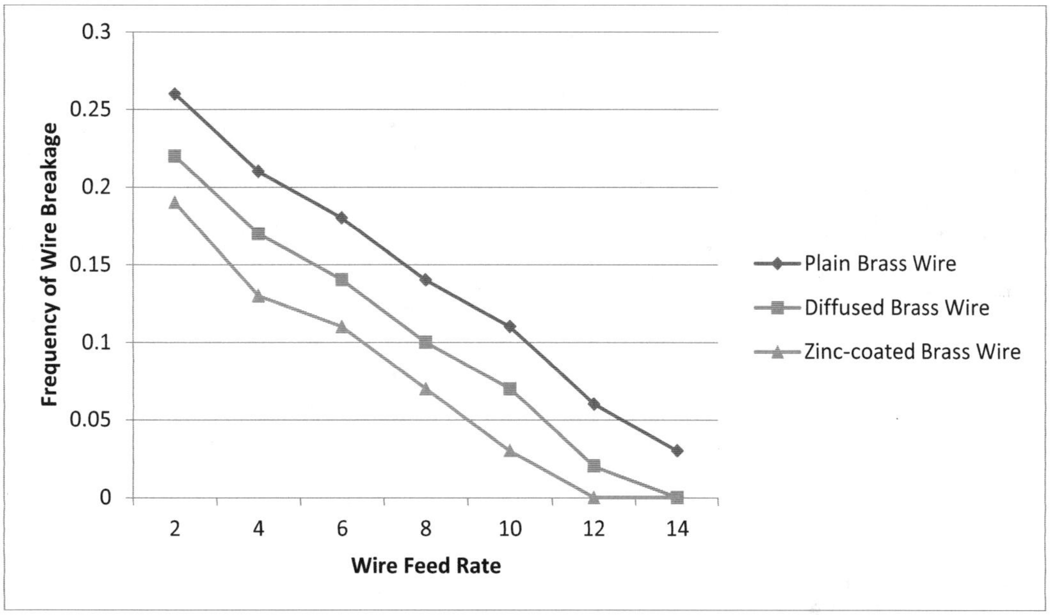

Figure 5. Impact of wire feed rate on the wire breakage recurrence

2.4.6. WT Impact on Wire Breakage recurrence

Figure 6 illustrates the outcome of (WT) on wire breakage recurrence for all wire material types. It is apparent that the level of wire rupture is marginal at low (WT) levels yet this constantly rises in little steps with a frequent rise in (WT) which could be attributed to the specifics that high (WT) holds the wire unbent as well as secure, rendering this extra susceptible to breakup due to effect of spark discharges. Additionally, once tension enforced surpassing the sturdiness of wire, this directs to the rupture of the wire and if the (WT) is held unduly massive, the rupture of the wire will occur nevertheless of distance. Thus, greater wire voltage lessens the amplitude of the wire vibration and improves the veracity of the complicated geometrical of the components originated. Thence, a relevant scale of (WT) have to be chosen to lessen the probability of wire rupture and increase the veracity of the fabricated components [22, 23, 32]. The wire strain impact on the wire breakage-recurrence for three wire electrodes has been listed in Table 6 and shown graphically in Figure 6.

Table 6. Wire tension effect on the wire breakage recurrence for three different wire electrodes

Wire Tension WT (gram)	Wire Breakage Frequency		
	Plain Brass wire	Diffused Brass wire	Zinc coated Brass wire
400	0.04	0.03	0
600	0.07	0.06	0.03
800	0.10	0.06	0.04
1000	0.14	0.09	0.07
1200	0.15	0.12	0.10
1400	0.17	0.15	0.12
1600	0.23	0.18	0.13
1800	0.30	0.25	0.20
2000	0.40	0.36	0.29
2200	0.45	0.42	0.40

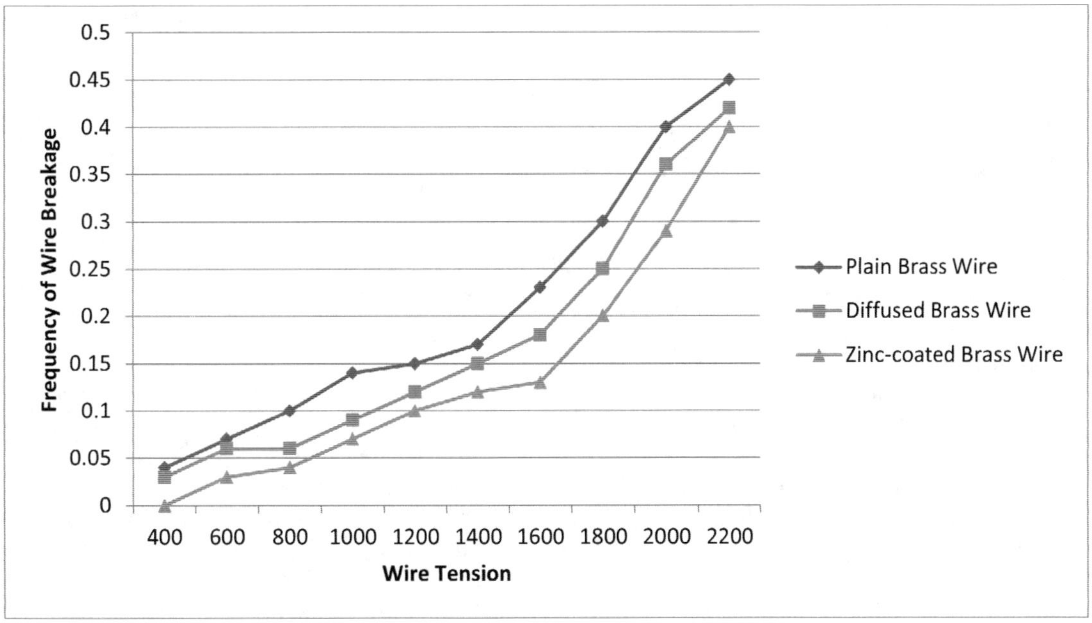

Figure 6. Wire tension effect on the wire breakage recurrence

3. Results and Discussion

Graphical analysis in Figure 1-6 demonstrates that the wire breakage point for zinc covered brass wire electrodes is the lowest relative to simple and diffused brass wire electrodes with all process parameters that may be attributed to the poor gap conductivity of the plain brass wire leading to low flushing. Adding the amount of zinc in the brass electrode is also not a solution to the issue because it leads to difficulties in drawing and producing. To solve this problem, brass is usually coated with zinc because zinc have a low liquefying point, so they disintegrate all the more rapidly and help to expand the conductivity of the gap and in this manner lead to better flushing. Furthermore, the zinc molecule 'boiling off' property assists with chilling off the wire and reduce the danger of wire breakage [9, 10]. The trial results uncovered that zinc-covered metal wire has a minimum wire-breakage recurrence for all machining parameters. In addition, detailed analysis of Figure 1-6 for zinc-covered metal wire demonstrates the range of six input-parameters for which WEDM be conducted for precise and effective cutting of Inconel 722 with no danger of breaking of the wire. The range for each process parameter (the minimum and maximum value) was calculated by closely analyzing the patterns of graphs. Where the probability of wire-breakage is least or zero was taken as the one extraordinary (low or greatest) esteem for each process parameter. Also, where an unexpected move occurs in the chart slant shows the abrupt increment in the degree of wire-breakage and consequently the other was expected as other extraordinary incentive for that process parameter. Such ranges are shown in Table 7 which indicates that while the (Ton) value was higher than 1.4 µs, the (Toff) was lower than 14 µs, the (IP) was higher than 210 A, the (SV) was lower than 16 V, the (WF) was lower than 2 m / min, and the (WT) was higher than 1600 g, the probability of wire breakages increased leading to wire electric discharge machining of Inconel 722 impossible.

Table 7. Machining Ranges of Product Parameters at WEDM Input Stage

Process Parameter	Notation	Unit	Minimum Value	Maximum Value

(Ton)	T_{on}	µs	0.4	1.4
(Toff)	T_{off}	µs	14	44
(IP)	IP	A	70	210
(SV)	SV	volt	16	80
(WF)	WF	m/min	2	12
(WT)	WT	gram	400	1600

4. Conclusions

The wire-breakage phenomenon has a negative impact on the maximum output of the WEDM process, as the total process performance is substantially diminished. It seriously hampers the nature of surface and accuracy of machined parts. In this work, wire-breakage recurrence impact of six process parameters was explored for three wire terminals viz. plain metal wire, diffused metal wire, and zinc-covered metal wire during WEDM of superalloy Inconel 722. The exploratory outcomes demonstrated that the recurrence of wire-breakage for each of the three wire electrodes materials was significantly increased by increasing (Ton), (IP) and (WT) values and a decrease in (Toff), (SV) and (WF). Furthermore, zinc coated brass wire has the least wire breakage recurrence as compared to two other wire electrodes for all machining parameters. The test investigation additionally proposed an optimal arrangement of all process parameters where proficient machining of Inconel 722 can be accomplished with no opportunity of wire breakage which can be utilized in future examinations for parametric modeling and optimization of WEDM parameters.

Declaration of conflict of Interest:

The authors declare that they have no conflict of interest with respect to the research, authorship, and/or publication of this article.

Funding:

The authors doesn't receive any research fund or grant from any organization.

Ethical approval:

This article does not contain any studies with human participants or animals performed by the author.

Acknowledgements:
The authors wish to acknowledge the Department of RIC, IK Gujral Punjab Technical University, Kapurthala, Punjab, India.

References:

[1] Newton T R, Melkote S N, Watkins T R, Trejo R M and Reister L 2009 *Materials Science and Engineering: A* **513-514** 208–215 URL 10.1016/j.msea.2009.01.061;https://dx.doi.org/10.1016/j.msea.2009.01.061

[2] Aggarwal V, Khangura S S and Garg R K 2015 *The International Journal of Advanced Manufacturing Technology* **79** 31–47 URL 10.1007/s00170-015-6797-8;https://dx.doi.org/10.1007/s00170-015-6797-8

[3] Ho K H, Newman S T, Rahimifard S and Allen R D 2004 *International Journal of Machine Tools and Manufacture* **44** 1247–1259 URL 10.1016/j.ijmachtools.2004.04.017;https://dx.doi.org/10.1016/j.ijmachtools.2004.04.017

[4] Rajurkar K P and Wang W M 1993 *Journal of Materials Processing Technology* **38** 417–430 URL 10.1016/0924-0136(93)90214-q;https://dx.doi.org/10.1016/0924-0136(93)90214-q

[5] Mandal A and Dixit A R 2014 *International Journal of Machining and Machinability of Materials* **16** 1–1 URL 10.1504/ijmmm.2014.063918;https://dx.doi.org/10.1504/ijmmm. 2014.063918

[6] Benedict G F 1987 *Non-Traditional Manufacturing Processes* (New York & Basel: Marcel Dekker, Inc)

[7] Spedding T A and Wang Z Q 1997 Parametric optimization and surface characterization of wire electrical discharge machining process URL 10.1016/s0141-6359(97)00003-2;https://dx.doi.org/10.1016/s0141-6359(97)00003-2

[8] Spedding T A and Wang Z Q 1997 *Journal of Materials Processing Technology* **69** 18–28 URL 10.1016/s0924-0136(96)00033-7;https://dx.doi.org/10.1016/s0924-0136(96)00033-7

[9] Prohaszka J, Mamalis A G and Vaxevanidis N M 1997 *Journal of Materials Processing Technology* **69** 233–237 URL 10.1016/s0924-0136(97)00024-1;https://dx.doi.org/10.1016/ s0924-0136(97)00024-1

[10] Antar M T, Soo S L, Aspinwall D K, Jones D and Perez R 2011 *Procedia Engineering* **19** 3–8 URL 10.1016/j.proeng.2011.11.071;https://dx.doi.org/10.1016/j.proeng.2011.11.071

[11] Kinoshita N, Fukui M and Gamo G 1982 *CIRP Annals* **31** 111–114 URL 10.1016/s0007-8506(07)63279-x;https://dx.doi.org/10.1016/s0007-8506(07)63279-x

[12] Yan M T and Liao Y S 1996 *The International Journal of Advanced Manufacturing Technology* **11** 267–275 URL 10.1007/bf01351284;https://dx.doi.org/10.1007/bf01351284

[13] Lee W M and Liao Y S 2003 *The International Journal of Advanced Manufacturing Technol- ogy* **22** 481–490 URL 10.1007/s00170-002-1480-2;https://dx.doi.org/10.1007/s00170-002- 1480-2

[14] Tosun N and Cogun C 2003 *Journal of Materials Processing Technology* **134** 273–278 URL 10.1016/s0924-0136(02)01045-2;https://dx.doi.org/10.1016/s0924-0136(02)01045-2

[15] Shoda K, Kaneko Y, Nishimura H, Kunieda M and Fan M X 1992 Adaptive control of WEDM with on-line detection of spark locations *Proceeding of the Tenth International Symposium for Electro-Machining* pp 410–416

[16] Cabanes I, Portillo E, Marcos M and Sánchez J A 2008 *Journal of Materials Processing Technology* **195** 101–109 URL 10.1016/j.jmatprotec.2007.04.125;https://dx.doi.org/10. 1016/j.jmatprotec.2007.04.125

[17] Patil N G and Brahmankar P K 2010 *The International Journal of Advanced Manufacturing Technology* **48** 537–555 URL 10.1007/s00170-009-2291-5;https://dx.doi.org/ 10.1007/s00170-009-2291-5

[18] Dekeyser W, Snoeys R and Jennes M 1985 *Journal of Manufacturing Systems* **4** 179–190 URL 10.1016/0278-6125(85)90024-x;https://dx.doi.org/10.1016/0278-6125(85)90024-x

[19] Rajurkar K P, Wang W M and Lindsay R P 1991 *CIRP Annals* **40** 219–222 URL 10.1016/s0007-8506(07)61972-6;https://dx.doi.org/10.1016/s0007-8506(07)61972-6

[20] Yan M T and Liao Y S 1996 *International Journal of Machine Tools and Manufacture* **36** 339–353 URL 10.1016/0890-6955(95)00050-x;https://dx.doi.org/10.1016/0890-6955(95) 00050-x

[21] Liao Y S, Chu Y Y and Yan M T 1997 *International Journal of Machine Tools and Manufacture* **37** 555–567

[22] Luo Y F 1999 *Journal of Materials Processing Technology* **94** 208–215

[23] Yan B H, Tsai H C, Huang F Y and Lee L C 2005 *International Journal of Machine Tools and Manufacture* **45** 251–259

[24] Tanimura T 1977 *Annals of the CIRP* **25** 59–59

[25] Jennes M, Snoeys R and Dekeyser W 1984 *CIRP Annals* **33** 93–98 URL 10.1016/s0007-8506(07)61387-0;https://dx.doi.org/10.1016/s0007-8506(07)61387-0

[26] Liao Y S, Huang J T and Su H C 1997 *Journal of Materials Processing Technology* **71** 487–493

[27] Tosun N and Cogun C 2003 *Proceedings of the Institution of Mechanical Engineers, Part B: Journal of Engineering Manufacture* **217** 633–642 URL 10.1243/095440503322011353;https: //dx.doi.org/10.1243/095440503322011353

[28] Saha S, Pachon M, Ghoshal A and Schulz M J 2004 *Mechanics Research Communications* **31** 451–463 URL 10.1016/j.mechrescom.2003.09.006;https://dx.doi.org/10.1016/j.mechrescom. 2003.09.006

[29] Cabanes I, Portillo E, Marcos M and Sánchez J A 2008 *Robotics and Computer-Integrated Manufacturing* **24** 287–298 URL 10.1016/j.rcim.2006.12.002;https://dx.doi.org/10.1016/j. rcim.2006.12.002

[30] Shandilya P, Jain P K and Jain N K 2012 *The International Journal of Advanced Manufacturing Technology* **61** 1199–1207 URL 10.1007/s00170-012-4095-2;https://dx.doi. org/10.1007/s00170-012-4095-2

[31] Kumar A, Kumar V and Kumar J 2013 *The International Journal of Advanced Manufacturing Technology* **68** 2645–2668

[32] Kumar A, Kumar V and Kumar J 2013 *Proceedings of the Institution of Mechanical Engineers* **227** 972–992

[33] Wang C C and Yan B H 2000 *Journal of materials processing technology* **102** 90–102

[34] Montgomery D C 2013 *Design and analysis of experiments* (New York: Wiley)

[35] Garg R K, Aggarwal V and Singh S 2014 *Applied Mechanics and Materials* **624** 124–128 URL 10.4028/www.scientific.net/amm.624.124;https://dx.doi.org/10.4028/www. scientific.net/amm.624.124

IOP Conf. Series: Materials Science and Engineering **954** (2020) 012014 doi:10.1088/1757-899X/954/1/012014

A Review on Submerged Friction Stir Welding of Light Weight Alloys

M.Sucharitha[1*], B. Ravi Sankar[2], P.Umamaheswarrao[2]

[1]Department of Mechanical Engineering, BIHER, Chennai, 60073, Tamilnadu, India.

[2] Department of Mechanical Engineering, Bapatla Engineering College, Bapatla,522101, AP, India

E-Mail: saisuchi2002@gmail.com

Abstract

Friction stir welding gained wide acceptance in many industries owing to its capability of joining lightweight materials, the subsequent development to Friction stir welding (FSW) was to submerge workpiece in liquids during welding termed as submerged friction stir welding (SFSW) resulted in improving mechanical properties. However, joint efficiency was much dependent on the process parameters. Hence, the present work aimed to consolidate the research works available in underwater friction stir welding (UFSW) primarily focusing on effect of process parameters on weld joint.

1. Introduction

Friction stir welding (FSW) is invented by The Welding Institute (TWI), U.K. in 1991 [1, 2]. It's a solid state welding process which leads welding without filler material. This welding is merely for light weight alloys where it is not possible to do the fusion welding and to regulate the welding temperatures [3]. The friction developed in between the tool and work piece thanks to rubbing, the surface on the metal get plasticized and therefore the bonding will happen. The applications found from FSW are automotives, aerospace and ship manufacturing [4-6]. From soft alloys to harder metals and for poly ethylene [7] sheets this SFSW produces good results than the friction stir welding and traditional welding. Utilization of SFSW leads to good welded grades and improves the mechanical properties. SFSW is one among the advanced technique to regulate the temperatures developed while welding and may avoid various welding defects like hot cracks, shrinkage etc. [8].

In SFSW, the rotating tool and base metal are immersed in a liquid medium, in order that when the friction is developing in between the tool and work piece the temperature will increase and thermal cycles were developed, it results the plastic deformation and may controlled by the surrounded cooling medium and maintain the temperature distribution throughout the length of the weld [9]. It produces different fine grain structures and improves the weld efficiency also mechanical properties.

2. Literature Review

H.J.Ilu et al. [10] has focused on homogeneity of 2219-T6 high strength aluminum alloy under water friction stir welded plates, divided into three layers and concluded that all the three layers had good tensile strength because of cooling effect on welding region for the soft material. Zhang et al. (2011) [11] evaluated the microstructure as the coarsening of the grains reduced compare than normal FSW because of controlling the thermal cycles by cooling effect, so that the precipitation is reduced in the weld region. When compared to normal FSW the heat distribution is narrowed at near the joint because the boiling temperature occurred at the thermal joint and the surrounding sides having less temperature. The heat flux is generated at the surface of joint, and then the mechanical properties will be improved [12]. Thermal cycles at the joint soften the surface of the material results the poor mechanical properties. The optimal welding parameters can control the temperature near the joint in under water friction stir welding [13]. As the speed increases the hardness will be increased due to the dislocation density and voids also increases with increasing the speed. To improve the tensile strength and grain structure the optimal welding parameters are required [14]. At stir zone and heat affected zone the material should not drag by pinhole because of low viscous flow of the material on the surface of the joint [15]. The size of the stir zone decreases as the tool rotational speed decreases [16].

Fratini et al. [17] concentrated on heat treatable aluminum alloys as 7075 to improve the FSW welded joints. Because of the external cooling medium the fracture area has been reduced in UFSW compared to air FSW by narrowing the thermal cycles at HAZ and TMAZ. They evaluated the microstructure as equiaxed grains have taken place by fracture determination at loci. Nelson et al. [18] observed that 7075 alloy is having more heat sensitivity compared to 2219 and low quench sensitivity. The ultimate tensile strength has lowered for reduced quench sensitivity materials in either hot or cold water medium. The mechanical properties are based on the temperature developed in FSW process and precipitation occurred on the surface of the joint. The cooling medium reduces the surrounding thermal flow and results the minimized fracture [19]. The nugget size decreases with increase in the welding speed [20].

Sabari et al. (2015) investigated and concluded that in normal FSW, the tool traverse speed vary from 25-40 mm/min and at the same conditions in UFSW the speed vary from 20-30 mm/min it results fault less joint because proper plasticization took place due to sufficient heat development and produced maximum tensile strength as well as efficiency improved one fourth of the normal FSW. The welded joint which is made at tool traverse speed of 30 mm/min has exhibited good performance, because of lower hardness [21]. When compared to normal FSW, in UFSW the size of the TMAZ and heat affected zone sizes had reduced because of cooling minimizes the size of the precipitates. So, the tensile strength was increased [22].

Kishta et al. (2014) Marine-grade 5083 aluminum alloy samples were friction stir welded in normal and submerged weld joints. The effects of different process parameters on the temperature void friction, and mechanical properties of the welded sample were investigated. Compared to air FSW, underwater FSW required higher rotational speed to produce high thermal capacity of water compared to submerging the sample in water also caused a reduction in the cooling rate. Compared to the parent material, the void area friction decreased significantly in the stirred material in the SFSW sample. The estimated void area fraction in the stir zone of the SFSW sample was almost one third of that in the parent material. The tensile strength of the underwater FSW sample was very close to that of the parent material. The elongation of the SFSW sample increased to approximately twice the elongation of the parent material [23].

Ratinasuriyan et al. (2017) performed investigation on SFSW and NFSW of 6061-T6 aluminum alloys. Results indicated that the position of the XRD peak shifts to a lower angle in SFSW compared to normal FSW. The value of SFS welds (water head of 20 mm) increasing by 12.88 % comparing to the NFSW due to the higher compressive force in the welded material during SFSW process. SFS welded sample at the water head of 20 mm exhibits higher micro hardness compared to the submerged FSW of water head 10 mm and normal FSW samples. This result is due to the increase in space lattice parameter values. Also, micro hardness value is directly proportional to the residual stress value.

The grain size of SFSW samples is smaller compared to the normal FSW samples and also the grain size is indirectly proportional to the lattice parameter and residual stress values. The significant variation is observed in displacing, lattice parameter and residual stress in SFSW as because of change in water medium leads to produce the higher cooling rate. In normal and submerged FSW, no significant variation is observed in d-spacing, lattice parameter and residual stress for a variation in rotational speed. The reason for this is that temperature variation is not much with respect to the various rotational speeds. Both the rotational speed and water head have significant influence on micro hardness since both parameters contributes for the formation of fine grain size in the welded samples [24]

Wahid et al. (2018) concentrated on 6082-T6 Aluminum alloy of a factorial-based study was successfully used to establish a relationship between the tensile strength of the welded joints and the process parameters. Among the significant parameters identified, the second-order term of rotational speed was the most influential. The optimum value of lastingness of the welded samples was 241MPa, which is like 0.79 times the lastingness of the bottom material. In comparison to welds made in FSW at these optimum conditions, a tenth increase in lastingness was realized. The low peak temperature and fewer thermal gradients thanks to water cooling limit the coarsening and dissolution of strengthening precipitates. This could be the rationale for the strength improvement within the optimal UFSW joint as compared to the traditional FSW joint. The distribution of temperature during UFSW showed similar variation with distance far away from the weld center both on AS and RS. The utmost temperature measured during UFSW was 119C on the AS closest to the weld center. Moreover, the temperature on AS was found to be higher as compared to RS, because the AS experiences a greater deformation than RS in FSW. The precipitates are fragmented into a smaller size in SZ thanks to intensive heat and severe plastic deformation. Grain coarsening and dissolution of precipitates was observed in TMAZ and HAZ. Finely populated dimples with no voids were observed, which shows defect-free joint with a traditional fracture [25].

Senthil kumar et al. (2016) conducted experimentation on the submerged FSW of AA 6063-O alloy and optimized process parameters using TOPSIS. The optimum process parameters are rotational speed of 1200 rpm, welding speed of 120 mm/min and the tapered tool pin profile. It was found that the welding speed, the rotational speed and the tool pin profile contributes 37%, 30% and 28% respectively. Due to the warmth treatment at the weld zone, normal FSW joints have finer grains compared to the bottom metal. Therefore, mechanical properties at the weld zone are improved. Also, the submerged FSW causes further improvement due to cooling. The typical grain size, hardness value at both the traditional FSW and submerged FSW joints were compared. The typical grain size reduces from 8.35 μm to 3.45 μm and also the hardness value increases from 48.85 Hv to 61.54 Hv [26].

Ghetiya et al. [27] developed a mathematical model was to predict the welded AA2014-T14 aluminum alloy. The welding speed has predominant effect on the tensile strength. The developed relationships are often wont to predict the lastingness in immersed FSW of AA2014-T14 aluminum alloy joints within the range of parameters. Detailed microstructural analysis of weld joint was administered to understand reason for strength improvement in immersed FSW. It's going to flow from to formation of fine grains in NZ and microstructure improvement at TMAZ. Intense heat and severe plastic deformation followed by water cooling were liable for fine grain formation. Fracture feature of welded joint obtained at optimal weld condition was studied. Joint was fractured at advancing side and at the interface between NZ and TMAZ. Fracture pattern shows varying size and fine dimple with some void which indicates the brittle fracture at the interface between NZ and TMAZ [28].

IOP Conf. Series: Materials Science and Engineering **954** (2020) 012014 doi:10.1088/1757-899X/954/1/012014

Wang Kauietal. (2012) concentrated that ECAP pressed 2017 Al alloy with average grain size of ~0.4 μm is underwater friction stir welded at a rotational rate of 950 r/min and a travel speed of 60 mm/min. Sound joint without obvious defect is produced by underwater FSW UFG microstructure is retained in the FSW nugget by employing external water cooling, and the size of the recrystallized grain is ~0.7 μm. Compared with the UFG based material, the underwater FSW joint exhibits softening along the cross-section, and the HAZ has the lowest hardness value owing to the coarsening of the strengthening precipitates [29].

Ammar Mofid et al. [30] has explored friction-stir and submerged friction-stir welds of AA5083/Mg AZ31 were produced in air, underwater, and under liquid nitrogen. The temperature profile, microstructure, and hardness were evaluated. The temperature profile showed a decrease of 319 K (46 C) and 326 K (53 C) from a maximum of 708 K (435 C) within the case of air-welded specimen to a maximum of 662 K (389 C) and 655 K (382 C) for welds made underwater and nitrogen. Maximum DT was realized for the coldest sample, that is, the weld initiated at 233 K (40 C) under nitrogen. The stir zone of air-welded specimen displayed massive intermetallic compound formation within the stir zone due to constitutional liquation. The stir zone of underwater welded and under nitrogen welded specimens showed a way smoother interface and fewer intermixing. Thanks to a decrease within the peak temperature, the formation of intermetallic compounds was suppressed significantly and grain growth wasn't noticeable within the dynamically recrystallized Mg alloy within the stir region. High hardness values were observed within the stir zone of the air-welded specimen suggesting the probability for the precipitation of brittle intermetallic compounds. In underwater and nitrogen weld specimens, these high hardness values weren't observed [30].

Kulwant Singh et al. [31], & Unnikrishnan et.al. [32] concluded that the tool geometry and weld parameters have taken an important place to improve the joint efficiency because the width of the stir zone will give the fine grain structure. So, that the joint will have good mechanical properties. Kouadri et al. [33] evaluated that the residual stresses developed in friction stir welding of magnesium alloy (AZ31B). The fraction volume slowly lowers than base metal at stir zone because of equiaxed grain structure. Dhanapal et al. [34] investigated on corrosion behavior of AZ61A and found that chloride ions were aggressive on Mg alloys then formed as oxides, it increases the hydrogen ions then result as the corrosion time increased. Senthil raja et al. (2015) has worked on optimization of welding parameters of AZ91D. They found that the contribution of tool rotational speed 44%, welding speed 36% and axial force 17% to improve the tensile strength of the joint [35]. Jaiganesh et al. [36] focused on AZ80A and suggested that cylindrical taper pin profile gave good joint because of sound yield strength and improved mechanical properties.

Sevvel et al. [35] investigated and found the changes being produced in the macro structural characterization and mechanical properties of the AZ31B Mg alloy by using three different tool pin profiles at the optimized process parameter values including tool rotational speed of 750 rpm, 0.5mm/min feed rate and axial force of 3KN during the FSW. Results concluded that taper cylindrical pin profiled tool exhibited improved & better mechanical properties in comparison with the opposite two pin profiles under FSW parameters of optimized values. Also the related yield strength value is almost 54.5% of its base metal and therefore the percentage of elongation being 5.45%.

Ratinasuriyan et al. [36] carried out investigations of SFSW/SFSP, concluded that in SFSW/SFSP, mainly two process parameters rotational speed, welding speed are identified more influence to grain refinement and strength improvement. In SFSW/SFSP or FSW/FSP, mainly three distinct zones are identified stirred nugget zone (NZ), thermo-mechanically affected zone (TMAZ), and heat-affected zone (HAZ) supported microstructural characterization of grains and precipitates. Maximum temperature of underwater joint is less than the traditional joint also because the time spent for the processing the fabric is low. SFSP is an improved method to FSP for the creation of ultra-fine grained bulk samples. SFSW has achieved more lastingness compared to the traditional FSW and SFSP has achieved better of elongation compared to the traditional FSP. Failure features on the traditional FSP and SFSP for grain growth and cavities coalescence during super plastic deformation [36].

3. Literature Gaps

The above literature review shows very scanty research had published on the investigations of SFSW. Submerged Friction Stir Welding required the further research to pay more attention in this field with different light weight alloys by using different fluid mediums and various optimization techniques.

4. Conclusions

In this paper work related to friction stir welding (FSW and SFSW) of aluminum alloys and magnesium alloys were reported. Scanty work has been done on SFSW of magnesium alloys as well as the researchers may pay more attention with different liquid mediums. The following conclusions are drawn from this study

- Cylindrical taper pin profile furnished good joint efficiency because of sound yield strength and improved mechanical properties.
- Rotational speed, welding speed had more influence on grain refinement and strength improvement.
- The grain size of SFSW samples is smaller compared to the normal FSW samples and also the grain size is indirectly proportional to the lattice parameter and residual stress values.

IOP Conf. Series: Materials Science and Engineering **954** (2020) 012014 doi:10.1088/1757-899X/954/1/012014

- Both the rotational speed and water head have significant influence on micro hardness since both parameters contributes for the formation of fine grain size in the welded samples.
- The cooling medium reduces the surrounding thermal flow and results the minimized fracture.

5. References

[1]. W.M. Thomas, E.D. Nicholas, J.C. Needham, M.G. Murch, P. Templesmith, C.J. Dawes, G.B. Patent Application No. 9125978.8 (December 1991).

[2]. C. Dawes, W. Thomas, TWI Bulletin 6, November/December 1995, p. 124.[3]. R.S.Mishra, Z.Y.Ma, , *Materials science and Engineering*, R **50** (2005), 1-78

[4]. Zhao J, Jiang F, Jian H G, Wen K, Jiang L, Chen X B., Journal of *Materials and Design*, 2010, **31:** 306−311.

[5]. Cavaliere P, Cabibbo M, Panella F, Squillace A. Journal of *Materials and Design*, 2009, **30**: 3622−3631.

[6]. Thomas W M, Nicholas E D. Journal of *Materials and Design*, 1997, **18:** 269−273.

[7]. Jicheng Gao, Yifu Shen, Jingqing ZhangHaisheng Xu, *Journal of Applied Polymer Science*, 2014.

[8]. Mohd.Atif Wahid, Zahid A. Khan, Arshad Noor Siddiquee, *Trans.Non ferrous met.soc. China* **28** (2018), 193-219.

[9]. Fratini L, Buffa G, Shivpuri R (2010) *Acta Materialia,* vol **58,** Issue 6:2056−2067.

[10]. Liu H, Zhang H, Yu L (2011). *ASM Int* **4.**

[11]. Liu HJ, Zhang HJ, Huang YX, Yu L (2010), *Trans Nonferrous Met Soc China* **20**:1387–1391

[12]. Zhang H, Liu H, Yu L (2012). *ASM Int JMEPEG* **21**:1182–1187

[13]. Zhang HJ, Liu HJ, Yu L (2013), *Trans Nonferrous Met Soc China* **23**:1114–1122

[14]. Zhang H, Liu H (2013). *Mater Des* 206–211

[15]. Zhang H, Liu H, Yu L (2011), *Mater Des* **32**:4402–4407

[16]. Zhang H, Liu H (2012) , *Metallogr Microstruct Anal* 269–281

[17]. Xiuli Fenga,Huijie Liub, and C. John Lippolda, *Materials Characterization*, **82**, 9 7 – 102, 2013.

[18]. Fratini L, Buffa G, Shivpuri R (2009), Int J Adv Manuf Technol 43:664–670

[19]. Nelson TW, Steel RJ, Arbegast WJ (2003), *Sci Technol Weld Joint* **8**(4):283

[20]. Fratini L, Buffa G, Shivpuri R (2010), *Acta Mater* **58**:2056–2067

[21]. Xiuli Fenga,Huijie Liub, and C. John Lippolda, *Materials Characterization*, **82**, 9 7 – 102, 2013.

[22]. S. Sree Sabari, S.Malarvizhi, V.Balasubramanian

[23]. S.Sree Sabari, S.Malarvizhi, V.balasubramanian, G.Madusudhan Reddy, (2015).

[24]. Emad Eldin Kishta and Basil Darras , (2014).

[25]. C. Rathinasuriyan and V. S. Senthil Kumar, (2017)

[26]. Mohd Atif Wahid , Zahid A Khan, Arshad Noor Siddiquee , Rohit Shandley and Nidhi Sharma, (2018).

[27]. Senthil Kumar Velukkudi Santhanam, Sankar Ramaiyan, Lokesh Rathinaraj,Rathinasuriyan Chandran, India,(2016).

[28]. N D Ghetiya , K M Patel, (2014).

[29]. Wang Kaui-she ,Wu Jia-lci, Zhou Long -hai, LIN Zhao-xia, Kong Liang , 2012 .

[30]. Mohammad Ammar Mofid , Amir Abdollah-zadeh, Farshid Malek Ghaini, and Cemil Hakan Gur., Turkey, (2012)

[31]. Kulwant singh, Gurubhinder singh, Harmeet singh, *Journal of Magnesium alloys*, 2018, 1-18.

[32]. Unnikrishnan M A, Edwin Raja Dhas, An *International Journal*, Vol.**2**, No.4, December 2015.

[33]. A.Kouadri-Henni, L.Barrallier, Riad Badji, *MATEC web of conferences*, **80**, 06003, 2016.

[34]. A. Dhanapal, S. Rajendra Boopathy, V. Balasubramanian,K. Chidambaram and A. R. Thoheer Zaman, Hindawi Publishing Corporation, *International Journal of Metals*, 2013, Article ID 317143, 2013.

[35]. R.Senthil Raja, A.Naveen Sait, *Physico chemical Mechanics of Materials*, 2015.

[36]. V. Jaiganesh, P.K. Nagarajan, P. Sevvel, J. Dhileep Kumar and S. Manivannan, *Material science forum, Trans tech publications*, Vol.**866**, pp 151-155.

37. Sevvel P, Jaiganesh V, *FME Transactions*, (2016) **44**, (1) 43-49.

38. C.Rathinasuriyan, Dr.V.S.Senthil Kumar, *International Journal of Applied Engineering Research,* Vol **10**, Number 8, (2015).

Synthesis and characterization of natural fiber reinforced laminated thermoplastic composite

S Vijayabhaskar[1], T Rajmohan[2], E Isaac Kennedy[3], T Jayanth Krishna[4]

[1,2,3,4]Department of Mechanical Engineering, Sri Chandrasekharendra Saraswathi Viswa Maha Vidyalaya Deemed to be University, Enathur, Kanchipuram-631561, India

E-mail: [1]mailvijayabhaskar@gmail.com and [2]rajmohanscsvmv@yahoo.com

Abstract. As increase in the use of eco-friendly, natural fibers are used for the reinforcement of polymer composites which has the light weight and low-cost polymer composites with better mechanical properties which offers exceptional properties over glass fiber. In this work, Basalt fiber reinforced Polypropylene (PP) laminated composite was manufactured by the compression moulding process. The variations of impact, flexural and compression strength and modules of the composite according to the number of basalt fiber layers were investigated. Scanning electron microscopy (SEM) and Energy-dispersive X-ray (EDX) conducted on the surface of the composite reveal the superior interfacial linkage between the basalt fiber and PP matrix. Hence, this offers a comprehensive discussion on synthesis and characterization of basalt fiber with polypropylene reinforced laminated thermoplastic composite.

Keywords: Polypropylene, Laminated, Thermoplastic Composite, Basalt Fiber, Scanning Electron Microscopy (SEM).

1. Introduction

The use of natural fibers for the reinforcement of the composites has received increasing attention both by the academic sector and the industries. Natural fibers have many significant advantages over synthetic fibers currently, many types of natural fibers have been investigated for use in plastics. Natural fibers are potential reinforcing materials and their usage till today has been more traditional than technical. They have been serving many useful purposes for a long time but the application of the material technology for the exploitation of natural fibers as reinforcement in polymer matrix happened in relatively recent years.[1] A promising material for usage in various applications is the basalt, which is solidified volcano lava. The basalt fibers are attractive to be used as reinforcement in composites similar to fiber glass, having better physic mechanical properties than fiber glass, but being significantly cheaper than carbon fiber. They also have a high elastic modulus, resulting in high specific strength three times that of steel.

Fiore et al. [2] observed the presence of basalt layers increases the mechanical properties to a highest extent in hybrid laminates in comparison with GFRP laminates. Salvatore et al. [3] stated that the basalt fibers provide potential improvements in interlaminar properties and flexural modulus to hybrid basalt/glass fiber reinforced polymer composite compared with the glass fiber reinforced

IOP Conf. Series: Materials Science and Engineering **954** (2020) 012015 doi:10.1088/1757-899X/954/1/012015

polymer composite. Dorigato and Pegoretti [4] found that the addition of basalt fibers in the epoxy-based carbon fiber reinforced composites enhanced the impact properties. Laminated polymer composites have many beneficial properties such as low density, good insulation, environmental resistance and good mechanical properties. Hossain et al. [5] fabricated jute epoxy laminated composite using vacuum assisted resin infiltration technique and characterized the fabricated composites by tensile and three-point bend tests.

Compression moulding is a high-pressure, high-volume method appropriate for moulding high-strength, complex fiberglass reinforcements. Advanced composite thermoplastics can also be compression moulded with chopped strand, randomly oriented fiber mat, woven fabrics or unidirectional tapes. The benefit of compression moulding is its capability to mould moderately intricate large parts. Besides, it is one of the cheapest methods when compared to other methods like injection moulding and transfer moulding. Furthermore, the material wastage is comparatively low giving it a benefit while fabricating expensive components. The process parameters include moulding time, temperature, and pressure. The present investigation deals with the fabrication of basalt fiber reinforced polypropylene laminated composites using the compression moulding method, mechanical characterization like flexural, compressive and impact properties and microstructural analysis through SEM and EDX.

2. Materials and Method

In this study, polypropylene (PP) in sheet form with a thickness of 1 mm procured from Lakshmi Corporations, Chennai, Tamil Nadu is used as the matrix material whereas the reinforcement of Basalt fiber in the mat form supplied by Go Green Products, Chennai, Tamil Nadu is used. The properties of the polypropylene and basalt fiber are shown in Table 1 and Table 2 respectively. The laminated composites are fabricated employing compression moulding process as shown in Figure 1. Two specimens of size 380 x 170 x 17 mm are prepared with different weight fractions of basalt fiber and polypropylene. Sample 1 contains 40% basalt fiber and 60% polypropylene with 8 layers of basalt fiber and 9 layers of polypropylene while Sample 2 contains 30% basalt fiber and 70% polypropylene with 7 layers of basalt fiber and 10 layers of polypropylene. Table 3 shows the process parameters of compression moulding process used in the fabrication of polypropylene laminated composites reinforced with basalt fiber. Each sample is placed in the mould and the required temperature, pressure and curing time are set after applying some amount of release agent on the upper and lower surfaces of the mould to avoid sticking of composites with the mould surface during curing.

Table 1. Properties of Basalt Fiber

Density (g/cm^2)	**Tensile strength** (GPa)	**Elastic modulus** (GPa)	**Elongation** (%)	**Maximum temperature** (°C)
2.62-3.05	3.0-3.5	79.3-93.1	3.2	650

Table 2. Properties of Polypropylene

Density (g/cm^2)	**Poisson's ratio**	**Thermal expansion** (μm/m°C)	**Thermal conductivity** (W/m-k)	**Specific heat capacity** (J/g-°C)	**Heat transfer capacity** (J/g-°C)	**Friction coefficient**
0.9	0.42	100-180	0.15	1.8	24.5	0.1

Table 3. Process parameters of Compression moulding

Specimen	**Process Parameters**		
	Temperature (°C)	**Pressure (bar)**	**Curing time (min)**
Sample 1	200	100	180
Sample 2	220	100	240

IOP Conf. Series: Materials Science and Engineering **954** (2020) 012015 doi:10.1088/1757-899X/954/1/012015

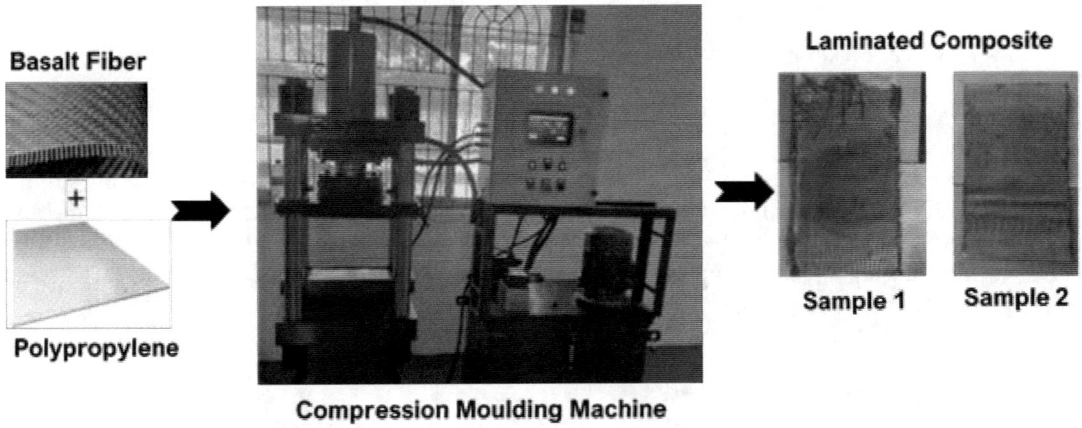

Figure 1. Compression moulding of basalt fiber reinforced PP laminated composite

3. Results and Discussion

3.1. Flexural Properties

Flexural properties denote the material flexibility and better flexural strength illustrates that the materials have high hardness and brittle properties [6]. The specimens for flexural test were prepared in accordance with ASTM D790 standard. A 3-point bending test is carried out for all two samples and the results are presented in Figure 2. The measured flexural strength determines the behaviour of the material under simultaneous compressive and tensile loading. The flexural strength increased from 68.6 MPa to 78.2 MPa as the basalt fiber proportion in the composite increased from 30% to 40% in

Test results:

No.	Specimen ID	E_H MPa	E_{Sec} MPa	$S_{0.1}$ MPa	S_1 MPa	S_2 MPa	r_M %	$r_{M\,(Corr.)}$ %	S_M MPa	r_{max} %	L mm	d mm	b mm
1	SAMPLE-1	78.2	165	-	2.00	5.16	5.0	5.1	11.0	5.0	51.2	10	13.5
2	SAMPLE-2	68.6	124	-	1.49	6.02	5.0	5.1	17.1	5.0	51.2	10	13.5

Series graph:

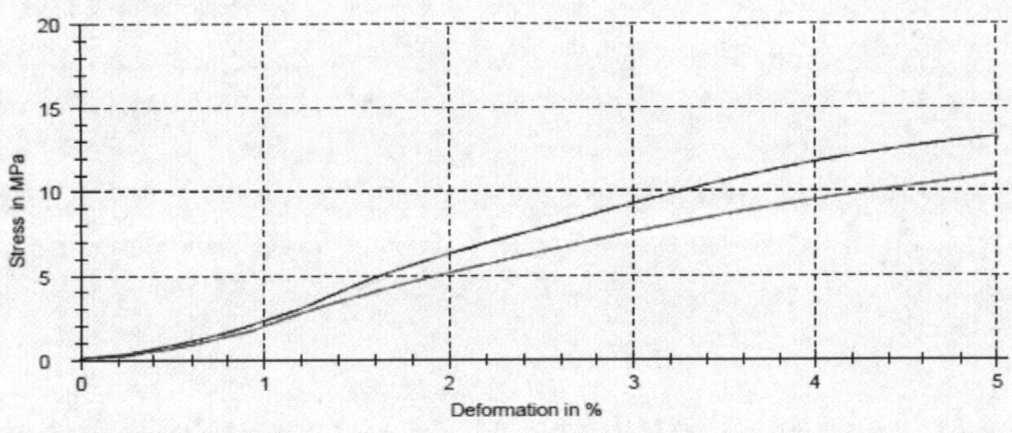

the composite. Hence the addition of fibres increased the flexural properties of the composites [7].

Figure 2. Flexural behaviour of basalt fiber reinforced PP laminated composites

3.2. Compressive properties

Compressive testing was carried out in accordance with ASTMD95 standards. The results are shown in Figure 3 and 4. The Compressive strength of a material is the maximum amount of compressive stress that it can take before failure. A maximum compressive strength of 0.22 GPa was obtained by Sample 1. The compressive properties of the composites were improved with the increase in the fiber content [8]. A substantial enhancement in compressive strength has been observed due to good interfacial bonding between the matrix and the fiber [9].

Test results:

No.	Specimen ID	E_c GPa	E_{Sec} GPa	σ_M MPa	ε_M %	ε_{cM} %	$\varepsilon_{M\,(Corr.)}$ %	L_0 mm	b mm	h mm	A_0 mm^2
1	SAMPLE-1	0.224	0.392	12.4	3.5	3.5	3.6	30.00	12	12	144.00
2	SAMPLE-1	0.0708	0.0720	11.5	5.0	5.0	5.2	30.00	12	12	144.00
3	SAMPLE-1	0.122	0.122	9.85	4.6	4.6	4.7	30.00	12	12	144.00
4	SAMPLE-1	0.0313	0.0528	12.5	5.0	5.0	5.3	30.00	12	12	144.00

Series graph:

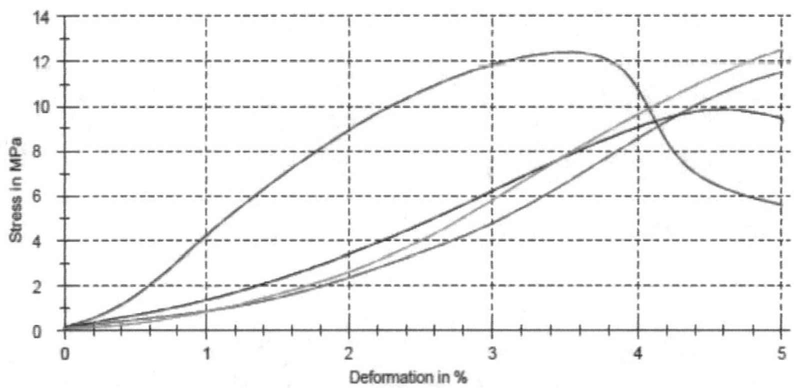

Figure 3. Compressive behaviour of 40% basalt fiber reinforced PP laminated composite

Test results:

No.	Specimen ID	E_c GPa	E_{Sec} GPa	σ_M MPa	ε_M %	ε_{cM} %	$\varepsilon_{M\,(Corr.)}$ %	L_0 mm	b mm	h mm	A_0 mm^2
1	SAMPLE-2	0.0735	0.0486	11.6	5.0	5.0	5.2	28.00	12	12	144.00
2	SAMPLE-2	0.0723	0.0836	7.15	5.0	5.0	5.2	28.00	12	12	144.00
3	SAMPLE-2	0.131	0.162	11.2	3.9	3.9	4.1	28.00	12	12	144.00
4	SAMPLE-2	0.173	0.232	9.35	3.3	3.3	3.4	28.00	12	12	144.00

Series graph:

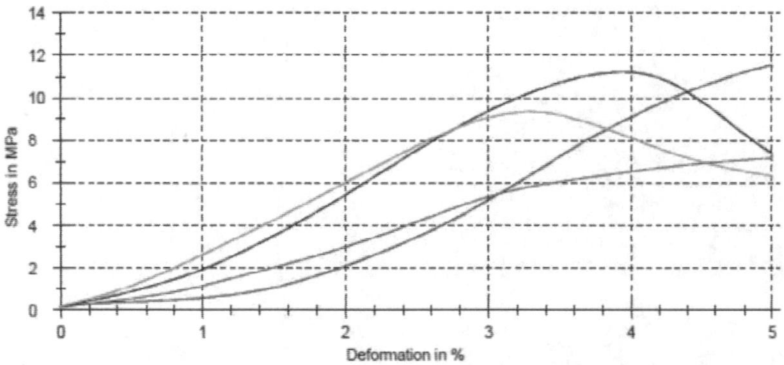

Figure 4. Compressive behaviour of 30% basalt fiber reinforced PP laminated composite

3.3 Impact properties

Impact energy is the amount of energy absorbed by a specimen during fracture and also it is the measure of toughness of the material. The specimens for the impact test are prepared in accordance with ASTM A-370 standard. The energy required for breaking the material during the test is measured and plotted in Figure 5. A maximum Impact strength of 14 J was shown by sample 1. The composite fails through the following modes namely fiber pull-out, fiber/ matrix debonding, delamination, matrix cracking during the impact [10].

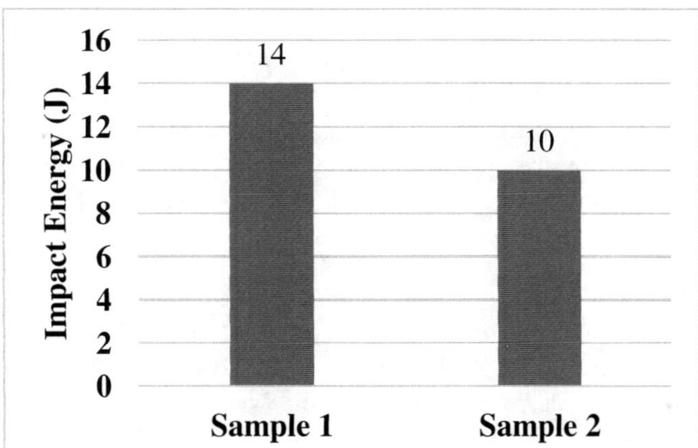

Figure 5. Impact strength of basalt fiber reinforced PP laminated composite

3. 4. Microstructure analysis

To evaluate the condition at failure, surfaces and fracture of material were observed with Scanning Election Microscopy (SEM) and Energy Dispersive X-ray (EDX). Specimen for the SEM were cut (10x10 mm in size) from the sample. Before being analyzed, the surfaces and cross- section were coated. Figure 6 shows the surface morphology of the basalt fiber reinforce PP laminated composite. The surface of the basal fiber has higher roughness and it increases the bonding area between the fiber and matrix. Hence, there is a good adhesion between the fiber/matrix leading to increase in the mechanical properties of the composites [11]. The failure due to tensile stress is incomplete and improper because of that multi-layer fabrication is used in this composite proving that the multi-layered composites has more strength compared mono layered composites [8].

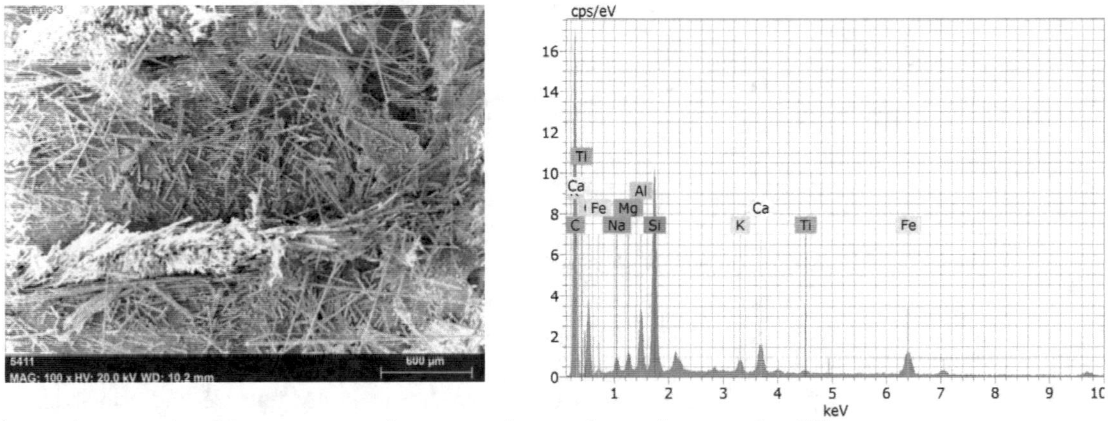

Figure 6. SEM and EDX of basalt fiber reinforced PP laminated composite

4. Conclusion

In the present investigation, basalt fiber reinforced polypropylene laminated composites were successfully fabricated through compression moulding process. The mechanical properties of the composites like flexural, compressive and impact are evaluated and also the morphological characteristics are studied using SEM with EDX analysis. The results indicated that the mechanical properties were improved with the increase in the weight proportion of the basalt fiber in the polypropylene matrix. A good bonding between the basalt fiber and the polypropylene matrix is observed through the SEM microstructure analysis.

References

[1] Sivaraja M, Kandasamy, Velmani N and Pillai M S 2010 *Bull. Mater. Sci.* **33(6)** 719–729
[2] Fiore V, Di Bella G and Valenza A 2011 *Mater. Des.* **32(4)** 2091-2099
[3] Salvatore C, Igor M D R, Sarasini F, Alessio T and Marco V 2011 *Mater. Des*. **32**, 337- 342
[4] Dorigato A and Pegoretti A 2014 *J. Compos. Mater*. **48** 1121- 1130
[5] Hossain M R, Islam M A, Van Vuurea A, and Verpoest I 2013 *Procedia Eng.* **56** 782–788
[6] El-Shekeil Y A, Sapuan S M, Abdan K and Zainudin E S 2012. *Mater. Des.* **40** 299-303
[7] Benkhelladi A, Laouici H and Bouchoucha A 2020 *Int. J. Adv. Manuf. Technol.* **108** 895-916
[8] Mohan K and Rajmohan T 2017 *J. Nat. Fibers* **14(6)** 864-874
[9] John K and Naidu S V 2004 *J. Reinf. Plast. Comp.* **23(12)** 1253-1258
[10] Eslami-Farsani R, Reza Khalili S M and Najafi M 2013 *J. Therm. Stresses* **36(7)** 684-698
[19] Sathishkumar T, Naveen J, Navaneethakrishnan P, Satheeshkumar S and Rajini N 2016 *J. Clean. Prod* **66** 610-618

IOP Conf. Series: Materials Science and Engineering **954** (2020) 012016 doi:10.1088/1757-899X/954/1/012016

Review on road marking paint machine and material

Vyankatesh Naidu[1] Vinod Bhaiswar[2]
[1]PG Student, Department of mechanical engineering, G H Raisoni College of Engineering, Nagpur.
[2]Professor, Department of Mechanical Engineering, G H Raisoni college of Engineering, Nagpur.
Email – naidu_vyankatesh.ghrcemtechcad@raisoni.net
Abstract: Road marking plays important role is maintaining traffic and guiding drivers to drive vehicles in proper way. this paper summarizes the continuous analysis and research of road marking paint machine, road marking paint applicator and road marking paint materials. The paper includes review of advance technology used in detection of road marking and recognition of road marking as well. Various types of road marking materials such as retro-reflective thermoplastic paints, cold plastic paints, water based paints, glass on bead paints and primer. Different types of road marking paints have some similar and some different properties which can be used as per requirement of work.

1. Introduction

Safety of vehicles and people can be achieved through road marking. Road marking plays vital role for road safety specially on highways. In this era of competitiveness road marking paint industry also using high end technologies. Road marking is compulsory thing to illustrate proper roads and for avoiding possible accidents of road travelers. The ministry of road transport and highways and Bureau of Indian Standards regulates standards related to road and road marking paints materials by using this standards paints are formulated by using chemicals and minerals. Each material is checked in laboratory before using it in paints to confirms it's specifications and properties according to road marking paint requirements.
This paper reveals studies of thermoplastic paints that shows features like lofty period reflectivity with outstanding ivory. Other properties are abrasion resistant, quicker shrivel time, superior durability, and magnificent adhesion. These are paints which are easily implemented with screed and spray machine and they are non hazardous as well as non harmful for atmosphere. Paints with base solvents have properties like air passing that is it has good porosity and it has bright white color. It has unique property of resisting alkali on surfaces made up of concrete and it also has property of self cleaning with hydrophobic surface. These road marking paints have property of lesser skid and excellent flow.

1.1 Road Marking Paints :-

1.1.1 Retro-Reflective Thermoplastic Road Marking Paint Thermoplastic Retro-reflective road marking paint :
One of the product of this road marking paint manufacturing machine manufactured is Shell Mark HT. Shell Mark HT has great colour balance and also don't get affected by changing weather. Shell Mark HT is thermoplastic material which is based on

IOP Conf. Series: Materials Science and Engineering **954** (2020) 012016 doi:10.1088/1757-899X/954/1/012016

petroleum resin. Mixture prepared from materials and glass beads used to make reflectivity for longer period of time. SHELLMARK HT is easy to melt and can be applied on both asphalt and concrete surface. Application of Shell Mark HT on concrete is also easy. It has excellent wear resistance to ensure long life of the marking. Life of marking is very long because it has low wearing ability.

- Characteristics :
- Better durability
- Friction Resistance
- longer period of reflectivity with magnificent brightness
- Quick shrivel & magnificent stickiness
- Fine flow & slip aversion
- Provides great aversion to changing weather &
- colour balance

1.1.2 Cold Plastic Road Marking Paint:

Shell-Mark CT is different from thermoplastic road marking paint. It is made up of two different materials. The process of making cold plastic road marking paint is road marking without solvent and only just by curing cold. Shell-Mark CT is acrylic resin based paint. The paint is made up of specifications provided by standards bureau.

- Characteristics:
- Great durability & higher reflectivity
- Great adhesion on surface
- Simple application by using screed and spray Machine

1.1.3 High Performance Waterborne Acrylic Road Marking Paint

Shell-MarkWR again one of the manufactured product from road marking manufacturing machine. Shell-Mark WR is a packed individually and shrivel quickly which makes itself all-rounder paint. Solvent is not used in Shell-Mark WR acrylic. It is solvent less road marking paint emulsion. It is also made up by using Indian Standards.

- Characteristics:
- Hard surfaces like concrete and bituminous it has splendid wear properties.
- Bitumen spill is impossible for this paint.
- Superlative and detectable in low light or no light.
- Simple application by using applying instrument like roller, brush etc.
- Better stickiness and area of highly contrasted colours.

1.1.4 Fast Drying Solvent Based Road Marking Paint:

Shell-Mark SB is made up of only one component. It is massively used for making lines. It is very solid paint. It is different from other non solvent based paint, it is solvent based paint. It is also made up on the specifications prescribed by Indian bureau.

- Characteristics :
- Sublime durability and very reflective
- Better air passage and brightness
- Resistive for alkali on concrete surfaces eminently.
- Clean by itself property which is hydrophobic surface.

1.1.5 Primer for Thermoplastic Road Marking Paint for Concrete Surface:

IOP Conf. Series: Materials Science and Engineering **954** (2020) 012016 doi:10.1088/1757-899X/954/1/012016

Primers are usually used for coating on surface before the paints. It helps to maintain adhesion between the layer of paint to surface of concrete. Shell-Prime RM is also solvent based primer. Shell-Prime RM is manufactured in such way that road marking paint and other coatings can be applied successfully.

- Characteristics :
- Simple application.
- Helps to create binding between Shell-Mark HT and concrete surfaces.

1.1.6 Water Based Gloss Finish Kerb Marking Enamel :
It is environment friendly as it is based on a acrylic resin system. It is formulated for line marking on all outdoor traffic kerbs. Synthetic Enamel is a product which is manufactured for fast shrivelling. This enamel is water based and it has glossy finish. Acrylic based enamel have no hazardous effects on environment. This enamel can used for marking any busy road.

- Characteristics :
- Weather Protective - This paint can sustain any weather like heavy rains and extremely hot sunlights which makes this product highly recommendable.
- Proper Utilized shrivelling duration - Shrivelling time is minimize with constraints so that finishing between each coatings are smooth.
- Simple application - No complex process

1.1.7 Drop on Glass Beads for Road Marking Paint:
Road marking paint can be reflective for longer period of time just by using glass bead. This helps to locate proper direction on the road in day time as well as in night also. This glass bead can withstand in any weather conditions. Glass bead can be used with all road marking materials for reflections. These are Class B Drop-on glass beads. B class glass beads are manufactured which helps to penetrate surface as much required.

- Characteristics :
- Water molecules cannot affect this glass bead.
- Smooth flowing
- Splendid adhesion on road markings.
- Sublime reflectivity with longer period of time.
- Application on roads, airports, sea ports, parking lots
- Used in parking lots, airports, highways, roads , seaports etc.

2. Review Of Papers

2.1 "Coagulation efficacy of a tanin coagulant agent compared to metal salts for paint manufacturing wastewater treatment." M. A. Aboulhassan, S. Souabi, A. Yaacoubi, M. Baudu
Research shows that the comparison between metal salts which are use for paint manufacturing and tanin coagulant's coagulation efficacy. Human health is highly affected by hazardous wastes excretes from industry. Paint manufacturing industries wastes also contain high amount of toxic water. Laboratory experiments are conducted to test PMW (Paint Manufacturing wastewater). While experimenting many elements are considered such as iron chloride, tanin based polymer and aluminium sulphate. Results reveals that TBP (Tanin based Polymer) is more efficient than other coagulant salts. In the conclusion, TBP is better than synthetic product for paint manufacturing is obtained.

2.2 "Reading the road : Road marking Classification and interpretation." Bonolo Mathibela, Paul Newman, Ingmar Posner

Author evaluate the road marking system for automated vehicle. Aim of this paper is to compare different types of condition and stages for automated vehicle to understand road marking pattern. Road marking interpretation can be easily understood by manual drivers but difficult for the automated vehicle. However road marking recognition can be done by using convolutional neural network classifier. Road marking interpretation by using automated vehicles faces some difficulties which is highlighted by using boxed junction , zigzag form, intersection form, single boundary form and special lane. Probability of getting Rusboost and conditional random field geometric functions to detect road marking is based on this classification framework. It is concluded that the work is done on larger perspective including many real world examples.

2.3 "Road Marking Recognition with computer vision system." Konstanitin V Ignatiev, Elena V Serykh, Anna V Mironuik

In this paper researchers implement robot for road marking recognition. Aim of this research is to use computer vision technology to road marking recognition. Model created by researchers have automated motion which helps it to move freely in public. Mobile robot have four wheels. Control system of mobile robot road marking is to design dashed lines on the roads, roads edges, different light conditions. GPS and other vision devices are used for locating robot. Mobile robot have car like structure. As a test, robot moves with open computer vision library and motion control according to obtained information. Robot is practiced on ground with road markings. It concludes that it is low power consuming device so it can be use single board computer raspberry pi 2.

2.4 "Influence of volatile organic compounds emissions form road marking paints on ground level ozone formation : Case study of Krakow Poland" Tomasz E Burghardt, Anton Pashkevich, Lidia Zakowska

Author stated that, ground level ozone is known in the entire world for its mail pollutants which is reason behind smog and many atmosphere pollution. When this pollutants reacts with nitrogen oxides this forms ozone which is harmful for atmosphere. Wastewater and waste gases emitting from road marking paint manufacturing plants are hazardous for environment and human health. Volatile organic compound is reacted with nitrogen oxide which affects ozone formation. Aim of this work is to test different aromatic solvent for lesser harmful effect. Results shows that, removing of the aromatic solvents is not helping to minimize the emission of volatile organic compound, on the other hand it decreases the production of ozone by half. It concluded that, waterborne paints with optimize choice of glass beads leads to selection of more sustainable option as compare to solventbased materials.

2.5 "One stage detection network with an auxillary classifier for Real Time Road marks detection." Gyan Ting Lin, Patrisia Sherryl Santoso, Che Tsung Lin, Chia Chi Tsai, Jiun In Guo

In this paper authors states that they developed road mark detector. Aim of this paper is to examine road mark detector in real world application. As a methodology for getting good processing performance 32 frame per second nvidia Titan model GPU is used. Convolutional neural network is used for classifier. Road mark detector can easily detect multiple objects in single run. Additionally it can measures accurately which makes it's performance better. Extra convolutional neural network classifier is attached in the backside of system to avoid mispositioning and escalate accuracy. In the conclusion, the detector cover eighty six percent of map. Maximum portion is covered by detector so that it is beneficial and more upgrades in area covering can be possible in future upgradation.

2.6 "Generating all roads to Rome : Road layout and randomization for improved road marking segmentation." Tom Bruls, Horia Porav, Lars Kunze, Paul Newman
Road marking is essential to solve problems of traffic. It tends to safe driving behavior from the vehicle and vehicle driver. Aim of this research is to find and implement road marking for automated vehicle. Road marking for automated vehicle requires vast networking and data. Providing such road marking networks is costly and time consuming. Data present on softwares are somewhat diffrent and not acceptable to real world. Data of virtual world may have lack of information while implementing. Therefore, author experimented with principle of randomization. Many pictures and training algorithms are generated while training. This road marking database can work more efficiently than any other manual system. Locations with more traffic may require more pair of data sets. This system can work with high efficiency in heavily traffic areas also. It concludes that, with minimum manual labelling effort this can work.

2.7 "Performance of thermoplastic road marking material." S Naidoo, W J vd M Steyn
This paper shows that, road marking paints must have some important characteristics. Aim of research is to evaluate results from different road marking paints on asphalt road surfaces. Retroreflective is main characteristics of road marking paint. Night time visibility also plays vital role while evaluating paint caliber. Day time visibility, resistive to skid and colour these three parameter which may affect decision of choosing suitable paint. Luminance and retroreflectivity of white and yellow road marking paint is checked by using different methods. Results shows that retroreflectivity can last long for one month to forty-eight months and luminance can last long up to one month to thirty months. If road marking paint is washed then there is no affect on retrorflective property but it enhances the luminance property. White road marking complied with colour specifications while there is no colour specifications for yellow road marking paint.

2.8 "Symbolic Road marking recognition using convolutional neural networks." Touqeer Ahmad, David Ilstrup, Ebrahim Emami, George Bebis
Researchers investigates the convolutional neural networks usage for road markings symbols. Road marking recognition and detection both are different from each other. Road marking recognition conventional method is by using obedient feature extraction followed by obvious training. This method of road marking recognition is not usually effective. Another method of road marking recognition is by using template matching which is also old and ineffective and inefficient. This paper concentrate only on road marking recognition not on road marking detection. Convolutional neural network have successfully runs ADAS system which makes it suitable for road marking recognition.

For experiment five different models are prepared. All five models are trained with unique convolutional neural network architecture. Each model contains different numbers of convolutional neural network connected layers and unique resolution of road markings. Data used for research is taken from publicly available data set. Resulting different recognition percentage for all five models. Most suitable data set model recognized ninety nine point zero five percentile of road markings. It concludes that symbolic road marking recognition using convolutional neural network is beneficial.

2.9 "Performance of a combined system of biotrickling filter and photocatalytic reactor in treating waste gases from a paint manufacturing plant." Peiyuan Zeng, Jianjun Li, Dongqi Liao, Xiang Tu, Melying Xu, Guoping Sun
This research suggest that biotrickling filter is combined with photocatalytic reactor. Biotrickling filter is already used in industries to treatment of waste gases. Gases emitting from paint manufacturing plant is hazardous for human health and harmful for nature too. Volatile organic compounds like ketones and esters are used for paint manufacturing process. This compounds may affect atmosphere. Removal efficiencies of biotrickling filter is ninety five percent on eightieth day. Biotrickling filter needs eighty days to accept aromatic compound. Removal of aromatic compounds from biotrickling filter was difficult. Due to some elements of waste gases it is not possible to achieve maximum efficiency of removal of dirt by using biotrickling filter. When bootlicking filter is combined with photocatalytic reactor then removal efficiency increased dramatically. It results ninety nine percent of removal efficiencies when biotrickling filter used with photocatalytic reactor. It concludes that it is beneficial to use both reactor and filter simultaneously for treatment of paint manufacturing waste gases.

3.Conclusion:- Road marking paint machine, road marking paint applicator, road marking paint material are successfully studied through various papers. Advance technologies are also mentioned in papers which can be easily implemented in future easily. Road marking paints are selected on the basis of requirement of work.

4. Reference

[1] M. A. Aboulhassan, S. Souabi, A. Yaacoubi, M. Baudu Coagulation efficacy of a tanin coagulant agent compared to metal salts for paint manufacturing wastewater treatment, 2016, 1-7.
[2] Bonolo Mathibela, Paul Newman, Ingmar Posner Reading the road : Road marking Classification and interpretation, 2015, 1-10.
[3] Konstanitin V Ignatiev, Elena V Serykh, Anna V Mironuik Road Marking Recognition with computer vision system, 2017, 165-167.
[4] Tomasz E Burghardt, Anton Pashkevich, Lidia Zakowska, Influence of volatile organic compounds emissions form road marking paints on ground level ozone formation : Case study of Krakow Poland, 2016, 714-723.
[5] Gyan Ting Lin, Patrisia Sherryl Santoso, Che Tsung Lin, Chia Chi Tsai, Jiun In Guo, One stage detection network with an auxillary classifier for Real Time Road marks detection,2018, 1379-1382.

IOP Conf. Series: Materials Science and Engineering **954** (2020) 012016 doi:10.1088/1757-899X/954/1/012016

[6] Tom Bruls, Horia Porav, Lars Kunze, Paul Newman, 2019, Generating all roads to Rome : Road layout and randomization for improved road marking segmentation,2019, 831-838.
[7] S Naidoo, W J vd M Steyn, Performance of thermoplastic road marking material, V60, 2018, 9-22.
[8] Touqeer Ahmad, David Ilstrup, Ebrahim Emami, George Bebis, Symbolic Road marking recognition using convolutional neural networks, 1428-1433.
[9] Peiyuan Zeng, Jianjun Li, Dongqi Liao, Xiang Tu, Melying Xu, Guoping Sun, Performance of a combined system of biotrickling filter and photocatalytic reactor in treating waste gases from a paint manufacturing plant, 2015, 1-9

Preparation and analysis of the thermal properties of non-toxic eco-friendly vegetable based oils

S D Sathishkumar[1] and T Rajmohan[2]

[1,2]Department of Mechanical Engineering, Sri Chandrasekharendra Saraswathi Viswa Maha Vidyalaya Deemed to be University, Enathur, Kanchipuram-631561, India

E-mail: [1]sathishsd.2@gmail.com and [2]rajmohanscsvmv@yahoo.com

Abstract. Cutting fluids play an important role in increasing the efficiency of the production system. Vegetable based cutting fluids produce less harmful effects to the environment thereby leading to the cleaner production system. In the present study, vegetable oils such as Neem and Karanja oils are blended in three different proportions which are used to improve the thermal performance of the vegetable based cutting fluids. Density, viscosity, flash point and fire point are the quality parameters used to analyze the thermal properties of vegetable oils were also studied. Triple Necked Round Bottom flask setup is used for mixing the two vegetable oils in different proportions. The experimental results reveal that the combination of 25% Neem and 75% Karanja proves to be the efficient vegetable oil for sustainable production in the modern industries.

Keywords: Cutting fluid, Karanja oil, Neem oil, Vegetable oil, Viscosity.

1. Introduction

Li et al. [1] made an investigation to study the effect of adding grapheme Nanoparticles into the vegetable-oil-based cutting fluid to enhance the machinability characteristics of TC4 alloy by using MQL technique. They have found that chosen cutting fluid meets the cleaner manufacturing demands. Gaurav et al. [2] have presented a feasibility study on jojoba vegetable oil as base oil mixed with and without MoS_2 Nanoparticles during hard turning of Ti-6Al-4V alloy with MQL technique. The results reveal that jojoba oil chosen presents as a sustainable replacement for mineral oil during MQL turning. Lawal et al. [3] have formulated oil-in-water emulsion by using palm kernel and cottonseed oils for turning operation of AISI 4340 steel with coated carbide inserts. They have found that vegetable oils have enhanced the machining characteristics when compared to conventional mineral oil-in-water emulsion cutting fluid. Gajrani et al. [4] in their present investigation have synthesized and characterized vegetable-based green cutting fluid for turning of hardened AISI H-13 steel. They have found that MQL technique using vegetable oil has improved the machining characteristics when compared to flood cooling and dry turning. Shokoohi et al. [5] have developed eco-friendly cutting fluid by using water mixed vegetable oil with anti-bacterial agent for turning operation of hardened AISI 1045 steel. They have found that the newly formulated vegetable oils are able to prevent the spread of infections caused during bacterial growth when compared to straight oil. Zhang et al. [6] have studied the effect of adding 50nm MoS_2 Nanoparticle in vegetable based oils during MQL grinding of 45 steel work piece. The results reveal that palm oil-based Nano fluids have produced the best lubrication properties when compared to conventional grinding processes. Desari and Davoodi [7] have developed vegetable oil-based Nano-lubricants by dispersing CuO and SiO_2 Nano particles in soybean and rapeseed oils in diverse concentrations for metal forming processes. They have found that 0.5-0.7 wt. % of Nano particle concentration has reduced the friction factor by 21-31%. Sen et al. [8] have developed Nano-green lubricant by reinforcing Nano-silica into palm oil for MQL milling of Inconel 690. The experimental results reveal that 1% silica deposited palm oil cutting fluid have improved the machining responses when compared to conventional machining conditions. Talib et al. [9] have formulated modified jatropha oil by mixing hexagonal boron nitride Nano particles with it. The results reveal that MJO5 base mixed with 0.5 wt. % of hBN Nano particle has provided the best physicochemical properties thereby meeting the demands of cleaner production. Silva et al. [10] have studied the effect of vegetable oil for MQL grinding process of hardened AISI 4340 steel. They have found that MQL system can enhance the machinability characteristics of grinding process thereby leading to an environmentally friendly manufacturing. Wang et al. [11] have made an experimental investigation to study the effect of water as grinding fluid during grinding process with AISI 1045 steel. They have found that small quantity lubrication system has enhanced the machining characteristics thereby leading to economic, healthy and environmental benefits. Huang et al. [12] have studied the grinding performance by using multi-walled carbon Nanotubes dispersed in vegetable oil through ultrasonic-assisted oscillator during MQL grinding of NAK80 mold steel. The results reveal that the oscillator can reduce the agglomeration of Nano particles thereby improve the machining characteristics of the MQL grinding process. Najiha et al. [13] have studied the effect of machining parameters on the wear mechanisms during end milling of aluminum alloy AA6061 by employing water based TiO_2 Nano fluids under MQL conditions. The results reveal that the volume fraction of 2.5% TiO_2 Nano particles has produced better results in terms of tool damage. Sani et al. [14] have introduced a refined modified Jatropha-based TMP ester (MJO) containing oil-miscible ionic liquids as vegetable-based cutting fluids during orthogonal cutting of AISI 1045 steel. The results reveal that the use of MQL vegetable based cutting fluids have reduced the enormous wastage of cutting fluids thereby leading to cleaner production. Susmitha et al. [15] have studied the effect of different non-edible vegetable oils on machining characteristics during drilling of mild steel work piece. The results reveal that the combination of 50% Neem and 50% Karanja proves to be the best cutting fluids in enhancing the machinability characteristics during the drilling process. Therefore, the present work deals with the preparation of non-edible vegetable oil in three different proportions and the analysis of the thermal properties of the blended vegetable oils.

2. Experimental procedure

2.1. Materials used

2.1.1. Neem oil

Neem oil comes from the seed of the tropical Neem tree, also known as Indian lilac. It has been used to treat many conditions. Figure 1 shows the photograph of the Neem oil.

Fig. 1. Neem oil

2.1.2. Karanja oil

Pongamia Pinnata is one of the few nitrogen fixing trees (NFTS) to produce seeds containing 30-40% oil. This species is commonly called Pongam, Karanja. Figure 2 shows the photograph of the Karanja oil.

Fig. 2. Karanja oil

2.2 Preparation of Bio lubricants

The bio lubricants were prepared by mixing Neem oil into the Karanja oil in three different proportions of 25% Neem-75% Karanja, 50% Neem-50% Karanja and 75% Neem-25% Karanja. The mixing was performed by using Three Necked Round Bottom Flask setup. The setup consists of flask, support base and a stirrer. The mixing was done for 3 hours. The stirrer speed was set at 300rpm. The experimental setup is shown in figure 3. The mixing process is done at Delta Laboratory, Chennai.

Fig. 3. Experimental setup

2.3 Testing the characteristics of vegetable oils

The properties of vegetable oils are tested in Delta laboratory, Chennai, India. The density of vegetable based oil samples is measured on the basis standard of ASTM D 1298. The thermal properties of the vegetable oil samples are calculated on the basis of established standards ASTM D 93 respectively. The properties of vegetable oils are shown in Table 1.

Table 1. Properties of Vegetable oil

Characteristic Test	Observation			Protocol
	Sample 1	Sample 2	Sample 3	
Density (g/cm^3)	0.907	0.914	0.916	ASTM D 1298
Kinematic viscosity @ 23°C (cSt)	30.98	34.25	39.42	ASTM D 445
Flash Point (°C)	185	165	158	ASTM D 93
Fire Point (°C)	214	201	196	ASTM D 93

Sample 1 – 25% Neem oil + 75% Karanja oil
Sample 2 - 75% Neem oil + 25% Karanja oil
Sample 3 – 50% Neem oil + 50% Karanja oil

3. Results and Discussions

3.1. Analysis of thermal properties of vegetable oil

3.1.1. Density

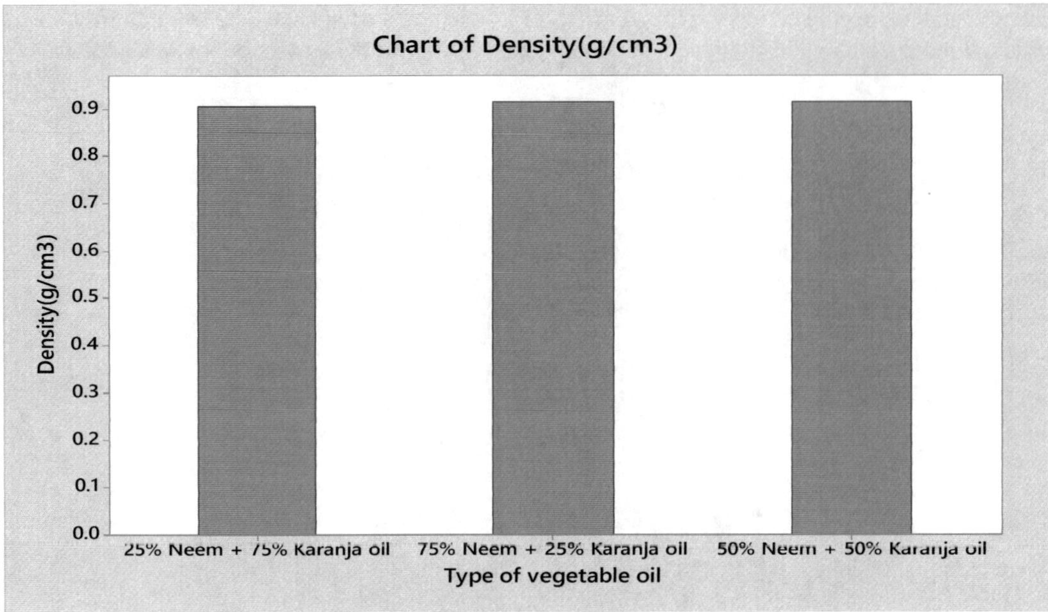

Fig. 4. Density of vegetable oils

The results of density measurements of Neem and Karanja oils mixed in different proportions are shown in figure 4. The experimental results reveal that the density of 50% Neem-50% Karanja oil sample is more when it is compared to the other two samples.

3.1.2. Viscosity

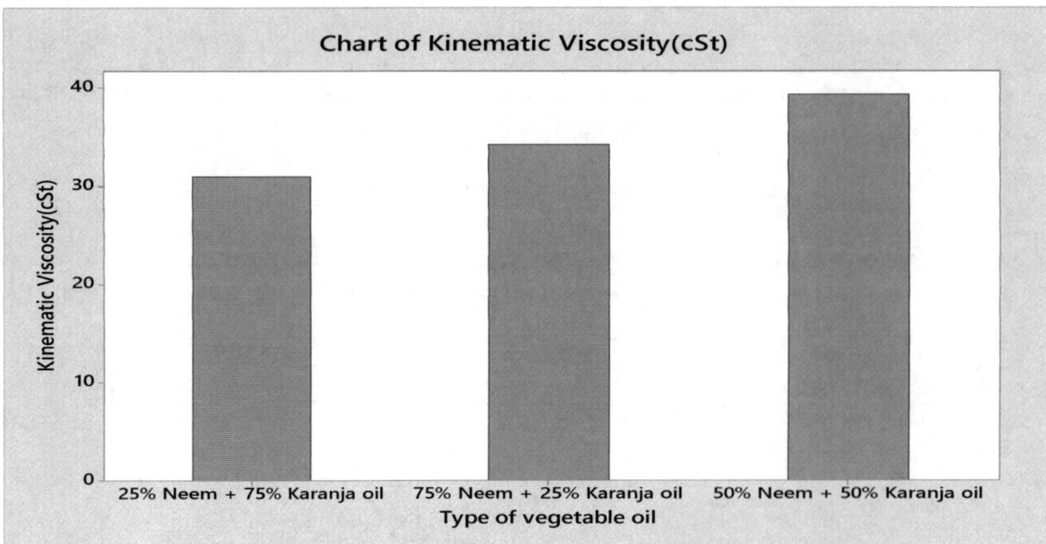

Fig. 5. Kinematic viscosity at 23°C

The results of Kinematic viscosity measurements of Neem and Karanja oils mixed in different proportions are shown in figure 5. The viscosity of the cutting fluid has its ability to reduce frictional forces between sliding surfaces in motion. It is known that the more force is required for the less viscous oil to cause motion between sliding surfaces apart. A high viscous oil requires higher energy to cause movement between sliding surfaces apart. The experimental results reveal that the amount of viscosity is optimum for 25% Neem and 75% Karanja oil which improves the machining conditions.

3.1.3 Flash point

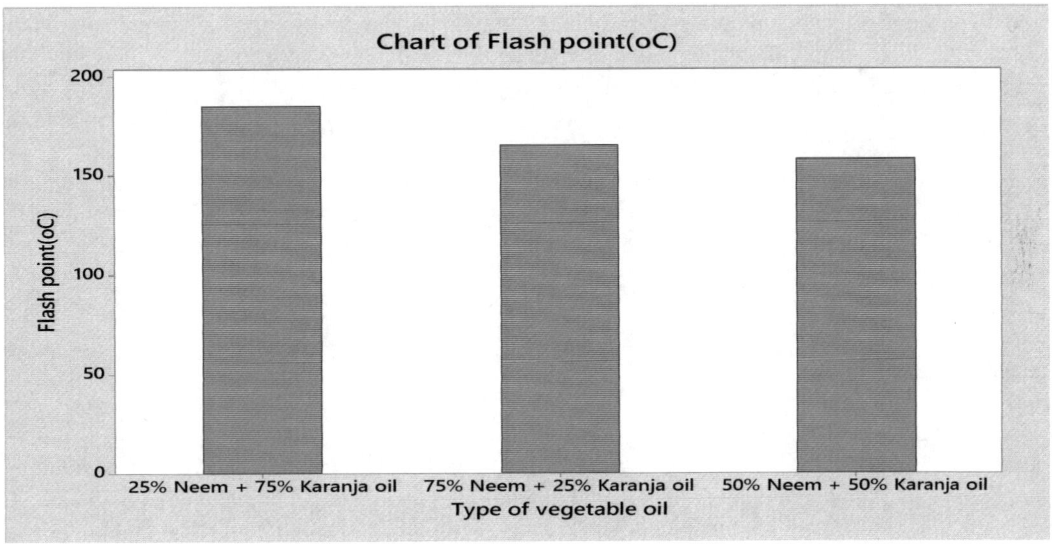

Fig. 6. Flash point of vegetable oils

Flash point is the temperature of the oil, at which its vapor ignites. The flash point of a vegetable oil should be high so that it cannot burn at elevated temperatures. The comparison of vegetable oils like Neem and Karanja mixed in different proportions in terms of flash point is shown in figure 6. The experimental values from the tables reveal that the combination of 25% Neem and 75% Karanja had its best amount of flash point of 185°C.

3.1.4 Fire point

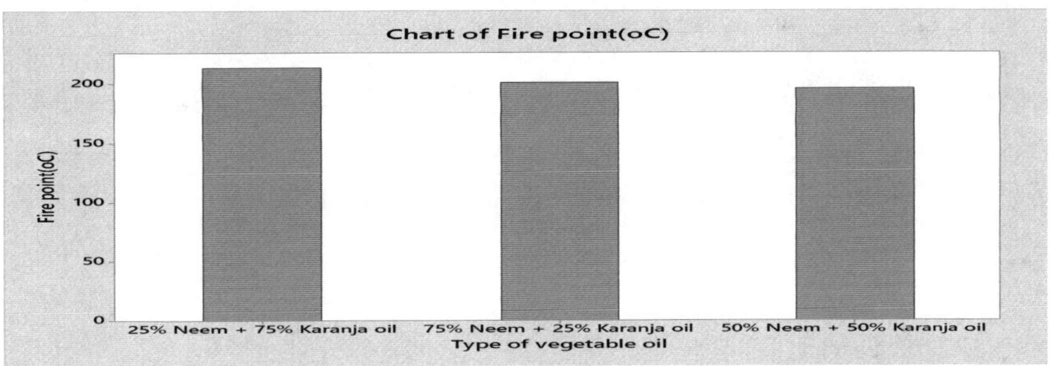

Fig. 7. Fire point of vegetable oils

The fire point is the temperature at which lubricant combustion will be sustained. The results of fire point measurements of Neem and Karanja oils mixed in different proportions are shown in figure 7. The experimental values from the tables reveal that the combination of 25% Neem and 75% Karanja had its best amount of fire point of 214°C.

3.2. Summary

In this research work, the effects of mixing Neem and Karanja oils in different proportions are analyzed. The results reveal that the 25% Neem - 75% Karanja oil sample has a significant improvement in the thermal properties when compared to that of the other samples. It indicates that the heat harvesting capability of 25% Neem-75% Karanja oil is increased.

4. Conclusion

In this research work, the effects of mixing Neem and Karanja oils at three different proportions was analyzed. The results reveal that the blend of 25% Neem and 75% Karanja had its best amount of its properties when compared to that of the other samples. It is eco-friendly, bio-degradable, non-toxic and economical.

References

[1] Li M, Yu T, Yang L, Li H, Zhang R and Wang W 2019 *Journal of Cleaner Production* **209** 1508-1522
[2] Gaurav G, Sharma A, Dangayach G S and Meena M L 2020 *Journal of Cleaner Production* 122553
[3] Lawal S A, Choudhury I A and Nuckman Y 2014 *Journal of Cleaner Production* **66** 610-618
[4] Gajrani K K, Suvin P S, Kailas S V and Sankar M R 2019 *Journal of Cleaner Production* **206** 108-123
[5] Shokoohi Y, Khosrojerdi E and Shiadhi B H R 2015 *Journal of Cleaner Production* **94** 330-339
[6] Zhang Y, Li C, Jia D, Zhang D and Zhang X 2015 *Journal of Cleaner Production* **87** 930-940
[7] Desari B Z and Davoodi B 2016 *Journal of Cleaner Production* **135** 1198-1209
[8] Sen B, Mia M, Mandal U K and Mondal S P 2020 *Journal of Cleaner Production* **258** 120755
[9] Talib N, Nasir R M and Rahim E A 2017 *Journal of Cleaner Production* **147** 360-378
[10] Silva L R, Correa E C S, Brandao J R and Avila R F D 2020 *Journal of Cleaner Production* **256** 103287
[11] Wang Z, Zhang T, Yu T and Zhao J 2020 *Journal of Cleaner Production* **253** 119896
[12] Huang W T, Liu W S and Wu D H 2016 *Journal of Cleaner Production* **137** 1553-1559
[13] Najiha M S, Rahman M M and Kadirgama K 2016 *Journal of Cleaner Production* **135** 1623-1636
[14] Sani A S A, Rahim E A, Sharif S and Sasahara H 2019 *Journal of Cleaner Production* **209** 947-964
[15] Susmitha M, Sharan P and Jyothi P N 2016 *IOP Conf. Series: Materials Science and Engineering* 149 012037

IOP Conf. Series: Materials Science and Engineering **954** (2020) 012018 doi:10.1088/1757-899X/954/1/012018

Employees stress level assessment: a case of apparel industry

M.Suresh[1] , R.Yuvaprasanth[2], R.B. Arun Ram Nathan[3] and K.Amarnath[4]
[1,2,3,4] Amrita School of Business, Coimbatore. Amrita Vishwa Vidyapeetham, India

Email: [1]drsureshcontact@gmail.com;[2]yuva.rathinasamy@gmail.com;
[3]arunramnathan@gmail.com; [4]amarnath761994@gmail.com

Abstract The assessments of employees stress level in garment industry gains vital importance to improve the efficacy of employees. Fuzzy methods are more appropriate than the conventional methods for assessment of stress level. The study demonstrates an application of a multi-grade fuzzy approach to measure employees stress level. This is the first time that the employees stress level assessment framework has been developed in the garment industry sector. Through literature review and expert opinions, sub-factors, factors and enablers of employees stress in workplace were identified and a conceptual model was developed. This study suggests that stress level assessment is essential to identify the current stress level of employees in workplace. This would help the garment industries to reduce its stress level further. In this study the stress level was found to be 4.74; which indicated the case company employees stress level is fair. Finally, the high stress sub-factors were identified to point the way for further improvements.

Keywords: *Stress level; Work system; Work culture; Employees stress; Garment Industry; Multi-Grade Fuzzy*

1. Introduction

Garment Industries play a very important role in Indian exports sector and contribute a substantial sum towards the Indian Gross Domestic Product. These industries are involved in producing wide range garments for both domestic market as well as foreign market. The processes for producing a garment start from sourcing of raw materials like cotton, wool and polyester, spinning into yarn, development of fabric, dyeing, cutting, stitching, printing, checking and packaging.

Garment industries provide employment opportunities to the sort of who are exposed to continuous and repetitive work which explains us the working environment. Garment industries mostly demand for precision and consistency in their work, it becomes important to discuss both physical and psychological behavior of those employees. Demographically garment industry employees are from low income background who were migrated from their native region. These employees earn their compensation based on the piece rate system or hourly wage. Therefore, they earn their wages based on their individual productivity. For the garment industries, productivity and on-time delivery of the products plays an important role in deciding the organizational profits. In addition to that, the work needs to be error-free to efficiently increase the productivity. Understanding the stress level of the employees becomes an important aspect in order to prevent employee attrition, injuries and also improve the efficiency of the work.

[1] has proposed a theory on role of textile industries in our Indian economy and states that garment industry is the second largest sector of industries after agriculture in terms of employability and economic contribution. This industry has undergone continuous development all around the world

and it is important for us to study the stress level of the employees working in these industries. Stress is a part of the working environment in any sector of industries, only the degree of stress varies according to the work done by the employees and analyzing this stress level of those employees is essential in these industries because psychological behavior of them plays a vital role. These industries depend upon the productivity, accuracy and quality in the work done by the employees which is explains the need to analyze the stress level of the employees.

2. Literature Review

2.1 Stress

[2] briefed about the effect of stress on employees working in emergency / rescue agencies. Stress responses are the responses generated by the human body to any demand placed on the body and they are generated by stressors. Stressors can be classified psychological or biogenic. Psychological Stressors are environmental events where interpretation of individual plays a predominant role in stimulating the particular response of the individual. Biogenic stressors are artificial stimulants such as caffeine, theophylline and nicotine which would have a direct impact on their stress response. Traumatic stress is the stress that has to be considered in case of emergency like Mining and fire department. Critical Incident Stress Debriefing (CISD) process in determining the effect of stress on the emergency workers in three different situations were Pre-incident, during incident and Post incident. Based on the analysis made by the various officials of CISD research process, traumatic stress has affected the decision-making process of the employees and in turn sufficient training must be done in order to make the employees stress free. Various methods in which the employees experience traumatic stress such as unavailability of resources, work life balance, optimal working time and this paper concludes by highlighting the various solutions that can be incorporated in order to minimize the stress experienced by the employees on all three occasions

[3] highlighted about the types of attention styles and stress factors affecting the sales and marketing managers in hotels. Analyses of different attention styles and how those attention styles have an impact on the key elements of the job like performance etc. Sample of 70 members from Hotel Sales and Marketing Association at Seattle chapter is used. Initially discusses about the varied attention styles, their impact on making mistakes and classifies the members into three categories of mistakes such as mistakes because of overloaded by external stimuli, mistakes happened as they were overloaded by internal stimuli and mistakes due to reduced attention. The second half deals with the effect of stress factors on the performance of the sales and marketing personnel of the hotel industry. Stress questionnaire was developed involving 14 stress factors and ranking them accordingly in terms of least stressful to most stressful. Responses were collected and those responses were analyzed. On those responses, it is found that the marketing personnel experience a lot of stress over the other department.

[4] conducted a research on determining the factors in the workplace that affects mental wellbeing, engagement in work and stress for working parents. Started from determining the factors responsible for suitable workplace environment in order to increase or provide optimal health conditions to the working parents. The aim of this research is to identify the connections between the objective and subjective workplace factors responsible for work fatigue, engagement in work, stress, work-family conflict among the working parents and their small children. A sample of 1562 working parents were

taken from the south Sweden region who had kids between the ages of 2 and 7 and questionnaire were distributed on the basis of differentiating the questions into three major dimensions like flexibility, attitude and benefits, then the outcomes were measured. Based on the cross-sectional analysis, flexibility in the work place and attitude towards the work had a positive impact on stress and wellbeing of the parents while the benefits in the workplace had lesser impact compared with other two. Result states that positive attitude in work as well as parenthood and workplace flexibility is responsible for good wellbeing, lesser stress and good workplace engagement among working parents.

[5] conducted a research on identifying the stress and workplace satisfaction for the employees who were servicing the people with intellectual disability. The research was done on 450 employees and questionnaire was framed based on general distress, satisfaction in work, job strain and a wide range of factors, then their outcomes were analyzed. A questionnaire consisting of more than 20 information needed such as basic characteristics, job control, role ambiguity, job feedback and a lot more were given and the results were analyzed. Path analysis results that there were three factors responsible for general distress, six factors accounted for job strain and six factors responsible for job satisfaction. This paper identified that stress on work-home conflict and also stress due to lack of staff support have direct impact on general distress and work satisfaction.

[6] described supervisor's stress based on Analytic hierarchy process (AHP) and formed an evaluation index system to analyze the stress. 16 factors were identified based on AHP which were categorized into three broad categories such as personal, organizational and environmental factors. Stress source and their related stress index were identified and analyzed based on the empirical study made on the supervisors. The result indicates that evaluation method in order to find out the hidden stressors and can be efficient used in implementing stress management in supervisors.

[7] pointed the factors affecting the stress and satisfaction of Australian nurses. This paper initially starts with the background of the study based on low employee retention in Australian hospitals. Identified 2 factors responsible for job satisfaction and stress; they were social support and empowerment. A sample of 157 registered nurses were taken, questionnaire was distributed and analysis were made, Based on analysis, Social support which was derived from colleagues, supervisors lowered stress in the work place and increased satisfaction. At the same time, the empowerment such as individual belief, value of work goal, control over work and impact increased job satisfaction and reduced the job stress.

[8] have stated the stress factors which aid in predicting the injuries of hospital personnel. Stress at work place is a major problem leading to various health problems both physically and mentally. A sample of 5111 hospital employees from 10 Finland hospitals were studied for the work stressors and injuries that happened because of stress at work place. Questionnaire was formed in order to measure the psychological distress, job control and overload of work, and then the results were analyzed. Based on the analysis it was found that psychological distress had significantly no relation with the injuries but low decision latitude, highly repetitive work which is monotonous, interpersonal problems and conflicts at workplace were the important stress related factors that had an impact towards injuries.

2.2. Health condition

2.2.1. Mental and Physical health

[9] have highlighted that the mental and physical health conditions that have been affecting the employees in American business organization in order to reduce the cost burden that happens to these health problems.

2.3. Career Development

2.3.1. Related Educational Career.

Organizational Career planning should go in line with the individual career planning because it helps the organization in employer retention**.** [10] have done a research on the important factor which contributes to career development in secretariat office in a city.

2.3.2. Demographic Characteristics

Some of the characteristics have an effect in the career development of the employees in the organization. [11] identified the factors that are affecting the career development of the employees by a comparative analysis done between two organizations. Two different factors of characteristics i.e. demographic and organizational characteristics affecting the career development of the employee as well as the organization.

2.4. Work Experience

2.4.1. Knowledge of working of an organization

Knowledge of working in an organization is essential when it comes to handling the stress. [12] proposed a research on the role of work experience that affects the future employability of the students.

2.4.2. Multi-Skilled

Multi skilled can be defined as a workplace strategy which improves productivity and reduce the indirect costs [13]. Multi-skilled worker is a person who possesses various skills which would allow them in contributing to more than one work processes.

2.5. Type A Behavior

2.5.1. Role Stress

[14] briefed about the impact of Type A behavior employee's role stress on work related outcomes. Role stress is defined as the high work tension and low job satisfaction. Role stress is measured on the basis of role ambiguity, conflict and job overload.

2.5.2. Psychological Strain

[14] had proposed a conceptual model which suggests that Psychological strain has an impact on the role stress of the employees.

2.6. Absence of Feedback

2.6.1. Perceived Feedback

[15] have proposed the impacts of individual feedback on the behavior of particular individual in an organization. Perceived Feedback is defined as the perception level of an individual who is receiving the feedback. Perception level of the employees was impacted by the nature of the source, age of the recipient, clarity of the message, consistency of the message and reliability of the source. Perception level is high when the source of the message is well known to the receiver.

2.6.2. Acceptance of feedback

[15] have proposed a research on studying the effects of individual feedback on the behavior of particular individual in an organization. The acceptance of feedback is defined as whether the individual is ready to accept the feedback given by the source.

2.7. Role Ambiguity

2.7.1. Adaptability to change and Pay

[16] have described the variables responsible for the behavior of the organization. They found that both role conflict and ambiguity are critical variables in the organizational behavior model.

2.8. Role Conflict

2.8.1. Absence of information and recognition

[17] has conducted a research on the effect of role conflict in complex organizations. This defines that role conflict exists in four forms conflict in resources and time, conflict due to expectations and demands, conflicts due to assigning different role to same person and conflict in between person.

2.9. Superior Support

2.9.1. Social Environment and Job Induced

[16] have proposed a paper on study the variables responsible for the behavior of the organization. Social environment has been explained as the opportunity to increase the friendship in the work environment. Job induced anxiety has been explained as feeling nervousness in the job.

2.10. Time Pressure

2.10.1. Incentive

[18] proposed a theory on impacts of time pressure and accountability. They developed a conceptual model to understand the impact of the time pressure and accountability. Based on that model, Incentive is referred as the less amount of pay given by the negotiators for working more hours.

2.10.2. Individual working capacity

[19] has proposed a research on degree of strain that is developed during building work which is related to individual aerobic work capacity.

2.11. Work Environment

2.11.1. Environmental Ergonomics

[20] have proposed a research on understanding the impacts of ergonomics training on the stress level of the employees. They found that that environmental ergonomics has a positive impact on the stress level of the employees.

2.12. Company Environment

2.12.1. Management Involvement

[21] have proposed a research in studying the impacts of justice perceptions of the employees to organizational commitment to leave the organization. Perception of organizational support has a positive impact on the employee retention in the company. Management involvement has been termed as the involvement of the management in every activities of the employees.

2.12.2. Organizational Legal Norms

[22] proposed a research on how ethical behavior of the organization plays a role on the organizational bases. Organizational legal norms are considered as one of the factors that is used in order to analyze the working climate of the organization. They found that organizational legal norms have a positive impact on the working climate and working behavior of the employees.

3. Research Question

The paper expects to separate into three distinct areas so that the research can be effectively and obviously comprehended in all viewpoints. In this manner the accompanying research questions were surrounded with the goal that each piece of the exploration is taken consideration independently. The study proceeds to the pathway as indicated by the stream of the exploration addresses which by finding the arrangements give the entire thought of the idea. This framework helps the managers of the Garment industry by answering the following research questions (RQ):

RQ 1: How to measure employees stress level in Garment industry?

RQ 2: What are the factors that influence employees stress level in Garment Industries?

RQ3: How to address those weak attributes to minimize the stress level?

The first question is addressed with the help of various references, literature review, research articles etc. and is verified with the help of experts. The Second question is answered with the help of literature review and Multi-Grade Fuzzy approach for stress level score. The third question is answered by using the results obtained, conclusion and the discussion.

4. About the Case Company

The Case garment manufacturing unit is located at Tirupur district in the state of Tamilnadu, India. There are approximately 230 employees working and they are a leading garment exporter. The Case Company has been in the garment sector for more than five years and they have various departments like cutting, stitching, checking, packing and inventory. Study has been carried out with the help of experts in that Case Company and various Garment units located in Tirupur, India.

5. Research Methodology

The research methodology used in this research is multi-grade fuzzy approach, primarily to analyze the stress level of employees in garment industries. This Research is limited to the Indian Garment Industries. The following driver variables (Enablers, Factors and comparing Sub factors) which influence the stress level of organization are distinguished and delineated in Table 1. With these recognized drivers, a survey with a 10-point Likert scale as per the multi-grade fuzzy methodology was developed, which was then utilized for gathering of information. The rating and weightage are collected from the experts in garment manufacturing firms. This method was used by many researchers for solving the problems in the literature ([23]; [24]; [25]; [26]; [27]; [28]; [29]; [30]; [31]).

Table 1. Enablers, Factors and Sub-factors affecting Stress level

Goal	Enablers	Factors	Sub Factors
Work stress evaluation ($\bar{I}$)	Personal Factors (I_1)	Health condition (I_{11})	Mental (I_{111})
			Physical (I_{112})
		Career development(I_{12})	Related education career(I_{121})
			Demographic Characteristics(I_{122})
		work experience (I_{13})	Knowledge of working of an organization (I_{131})
			Multi skilled (I_{132})
	Organizational Factors (I_2)	Absence of feedback (I_{21})	Perceived Feedback (I_{211})
			Acceptance of Feedback (I_{212})
		Role ambiguity (I_{22})	Adaptability to change (I_{221})
			Pay (I_{222})
		Role conflict (I_{23})	Absence of information (I_{231})
			Recognition (I_{232})
		Superior support (I_{24})	Social Environment (I_{241})
			Job Induced (I_{242})
		Time pressure (I_{25})	Inappropriate Incentive (I_{251})
			Individual working capacity (I_{252})
	Environmental Factors (I_3)	Work environment (I_{31})	Environmental Ergonomics (I_{311})
			Ambience (I_{312})
		Company environment (I_{32})	Management involvement (I_{321})
			Organizational legal norms (I_{322})

The employees stress level index of the case copmany is represented by I. It is the product of the ratings based on each drivers (R) and the weights assigned to each drivers (W) given by experts. The equation for stress index is

$$I = W \times R$$

The assessment scale has been graded into five levels since every factor involves fuzzy determination. $I = (10, 8, 6, 4, 2)$. 8–10 represents 'High Stress level', 6–8 represents 'Normal Stress level', 4–6 represents 'Fair Stress level', 2–4 represents 'low Stress level' and less than 2 represents 'extremely low Stress level. The mean normalized weightage of the experts' opinions and the attributes ratings are shown in Table 2.

Table 2. Rating and Weightage by experts

Ii	*Iij*	*Iijk*	*R1*	*R2*	*R3*	*R4*	*R5*	*Wijk*	*Wij*	*Wi*
I1	*I11*	*I111*	1	4	4	2	2	0.62069	0.3125	0.308824

		I112	0	4	4	3	3	0.37931		
	I12	*I121*	2	3	2	6	7	0.433333	0.328125	
		I122	5	4	0	3	5	0.566667		
	I13	*I131*	3	5	1	2	2	0.357143	0.359375	
		I132	1	10	2	4	3	0.642857		
I2	*I21*	*I211*	3	5	10	7	3	0.419355	0.235294	
		I212	5	3	4	4	8	0.580645		
	I22	*I221*	7	10	8	5	9	0.4	0.147059	
		I222	7	5	2	3	4	0.6		
	I23	*I231*	2	7	7	4	5	0.4	0.205882	0.308824
		I232	4	2	3	5	3	0.6		
	I24	*I241*	7	5	5	3	6	0.342857	0.176471	
		I242	9	5	10	3	8	0.657143		
	I25	*I251*	0	2	0	1	1	0.55	0.235294	
		I252	9	10	10	8	10	0.45		
I3	*I31*	*I311*	10	5	10	7	5	0.375	0.568182	
		I312	10	5	7	5	5	0.625		0.382353
	I32	*I321*	5	5	5	5	5	0.472222	0.431818	
		I322	10	5	0	0	2	0.527778		

The calculation comprises of three levels namely, primary assessment calculation, secondary assessment calculation and tertiary assessment calculation finally to obtain $\bar{I}$ the final Stress level of the organization.

The overall Stress level assessment value of the organization is $\bar{I}$ = **4.748887** which falls in the range 4–6 which represents 'Fair Stress levels'.

6. Results and Discussions

From the calculations we are able to determine that the stress level in garment industries falls in the range of 4-6 on a 10-point assessment scale in which 0 being low stress level and 10 being the high stress level. The result $\bar{I}$ = **4.748887** obtained implies that there exists a Fair stress level in the garment Industries.

Based on the experts rating, the identified sub-factors which possess a high stress level are adaptability to change, job induced, individual working capacity, environmental ergonomics and ambience.

7. Conclusion

The proposed framework identified 20 sub-factors for ten major factors contributing towards the stress level of the employees in the garment industries. They were based on the findings in literature review and experts opinion. Multi-grade fuzzy approach was used in order to analyze the weightage of those sub-factors and their implications towards the stress level in the organization. Questionnaire was developed to collect the responses from the employees in the industries. Based on the responses gathered from the employees, the overall stress level for the employees in the garment industries were found to lie within the range of 4-6 which implies that the stress level in the garment industries is "Fair stress level". The sub-factors which contribute more towards the stress level were identified from the mean value of the responses and the sub-factors which have high impact to the stress level are adaptability to change, job induced, individual working capacity, environmental ergonomics and ambience. These five sub-factors can be improved by communicating the changes to the employees before implementing, necessary periodical training for the employees, providing work based on employee's individual capability, providing incentive for working beyond the assignment, proper environmental surrounding with safety measures, proper work environment with sufficient space for production and inventory. Management should focus on these sub-factors and the improvement measures to minimise the stress level in the industries. By reducing the stress level of the employees, the overall productivity may tend to increase thus achieving organization goals.

References

[1]. Solanki, D.P (2017). A role of textile industry in Indian economy. *National Journal of Advanced Research*, *3*(3), 60-65.
[2]. Kowalski, K. M. (1995). A human component to consider in your emergency management plans: the critical incident stress factor. *Safety Science*, 20(1), 115-123.
[3]. Whitney, D. L. (1986). Attentional styles and stress factors of hotel sales/marketing managers. International Journal of Hospitality Management, 5(4), 197-200.
[4]. Eek, F., &Axmon, A. (2013). Attitude and flexibility are the most important work place factors for working parents' mental wellbeing, stress, and work engagement. *Scandinavian journal of public health*, 41(7), 692-705.
[5]. Hatton, C., Emerson, E., Rivers, M., Mason, H., Mason, L., Swarbrick, R., ... & Alborz, A. (1999). Factors associated with staff stress and work satisfaction in services for people with intellectual disability. *Journal of Intellectual Disability Research*, 43(4), 253-267.
[6]. Dong, P. (2012). AHP-Based study on the stress Evaluation System of supervisors stress. In Applied Mechanics and Materials (Vol. 174, pp. 3334-3338). *Trans Tech Publications*.
[7]. Bartram, T., Joiner, T. A., & Stanton, P. (2004). Factors affecting the job stress and job satisfaction of Australian nurses: implications for recruitment and retention. *Contemporary nurse*, 17(3), 293-304.
[8]. Salminen, S., Kivimäki, M., Elovainio, M., &Vahtera, J. (2003). Stress factors predicting injuries of hospital personnel. *American journal of industrial medicine*, 44(1), 32-36.
[9]. Goetzel, R. Z., Hawkins, K., Ozminkowski, R. J., & Wang, S. (2003). The health and productivity cost burden of the "top 10" physical and mental health conditions affecting six large US employers in 1999. *Journal of occupational and environmental medicine*, 45(1), 5-14.
[10]. Rande, S., Rahawarin, Y., & Zacharias, T. (2015). Factors Affecting the Career Development of Employees in Secretariate Office of City Samarinda. International Journal of Scientific & Technology Research, 4(3), 93-99.
[11]. Siew, J. C. (2001). Factors affecting career development: a comparative analysis of two organisations in Sarawak(Doctoral dissertation, Doctoral dissertation, Universiti Malaysia Sarawak,(UNIMAS)).
[12]. Helyer, R., & Lee, D. (2014). The role of work experience in the future employability of higher education graduates. *Higher Education Quarterly*, 68(3), 348-372.
[13]. Gomar, J. E., Haas, C. T., & Morton, D. P. (2002). Assignment and allocation optimization of partially multiskilled workforce. *Journal of construction Engineering and Management*, 128(2), 103-109.
[14]. Keenan, A., & McBain, G. D. M. (1979). Effects of Type A behaviour, intolerance of ambiguity, and locus of control on the relationship between role stress and work-related outcomes. *Journal of Occupational Psychology*, *52*(4), 277-285.

[15]. Ilgen, D. R., Fisher, C. D., & Taylor, M. S. (1979). Consequences of individual feedback on behavior in organizations. *Journal of applied psychology*, 64(4), 349.
[16]. House, R. J., & Rizzo, J. R. (1972). Role conflict and ambiguity as critical variables in a model of organizational behavior. *Organizational behavior and human performance*, 7(3), 467-505.
[17]. Rizzo, J. R., House, R. J., &Lirtzman, S. I. (1970). Role conflict and ambiguity in complex organizations. Administrative science quarterly, 150-163.
[18]. Mosterd, I., &Rutte, C. G. (2000). Effects of time pressure and accountability to constituents on negotiation. International Journal of Conflict Management, 11(3), 227-247.
[19]. Astrand, I. (1967). Degree of strain during building work as related to individual aerobic work capacity. *Ergonomics*, 10(3), 293-303.
[20]. Huang, Y. H., Robertson, M. M., & Chang, K. I. (2004). The role of environmental control on environmental satisfaction, communication, and psychological stress: effects of office ergonomics training. *Environment and Behavior*, 36(5), 617-637.
[21]. Loi, R., Hang-Yue, N., & Foley, S. (2006). Linking employees' justice perceptions to organizational commitment and intention to leave: The mediating role of perceived organizational support. *Journal of Occupational and Organizational Psychology*, 79(1), 101-120.
[22]. Victor, B., & Cullen, J. B. (1988). The organizational bases of ethical work climates. Administrative science quarterly, 101-125.
[23]. Yang, S. L., & Li, T. F. (2002). Agility evaluation of mass customization product manufacturing. Journal of Materials Processing Technology, 129(1-3), 640-644.
[24]. Vinodh, S. (2011). Assessment of sustainability using multi-grade fuzzy approach. Clean Technologies and Environmental Policy, 13(3), 509-515.
[25]. Nallusamy, S., Ganesan, M., Balakannan, K., & Shankar, C. (2015). Environmental sustainability evaluation for an automobile manufacturing industry using multi-grade fuzzy approach. International Journal of Engineering Research in Africa, 19, 123-129.
[26]. Vimal, K. E. K., Vinodh, S., &Muralidharan, R. (2015). An approach for evaluation of process sustainability using multi-grade fuzzy method. International Journal of Sustainable Engineering, 8(1), 40-54.
[27]. Vinodh, S., & Dinesh Kumar, C. (2012). Development of computerized decision support system for leanness assessment using multi grade fuzzy approach. Journal of Manufacturing Technology Management, 23(4), 503-516.
[28]. Vinodh, S., Devadasan, S. R., Vasudeva Reddy, B., &Ravichand, K. (2010). Agility index measurement using multi-grade fuzzy approach integrated in a 20 criteria agile model. International Journal of Production Research, 48(23), 7159-7176.
[29]. Ganesh, J., & Suresh, M. (2016, December). Safety practice level assessment using multigrade fuzzy approach: a case of Indian manufacturing company. In 2016 IEEE International Conference on *Computational Intelligence and Computing Research (ICCIC)* (pp. 1-5). IEEE.
[30]. Vinodh, S., & Prasanna, M. (2011). Evaluation of agility in supply chains using multi-grade fuzzy approach. International Journal of Production Research, 49(17), 5263-5276.
[31]. Sridharan, V., & Suresh, M. (2016, December). Environmental sustainability assessment using multigrade fuzzy—A case of two Indian colleges. In 2016 IEEE International Conference on Computational Intelligence and Computing Research (ICCIC) (pp. 1-4). IEEE.

IOP Conf. Series: Materials Science and Engineering **954** (2020) 012019 doi:10.1088/1757-899X/954/1/012019

Applications of Leagility in Manufacturing and Service Industries

V.Vaishnavi[1], and M.Suresh[2]
[1, 2] Amrita School of Business, Coimbatore. Amrita Vishwa Vidyapeetham, India

Email: [1]vaishnacsca@gmail.com; [2]drsureshcontact@gmail.com

Abstract Leagility is a new paradigm which combines both lean and agile concepts together. The aim of current study is to constitute a vast literature of leagility in manufacturing and service sectors categorizing the existing studies and to explore the limitations and future research directions. There are 40 research articles selected for this study and classified based on year, research methods, publisher, journal and application on leagility. Then the identification of limitations and future research directions from the articles are reviewed. The finding suggests that leagility is explored more in supply chain and it requirements to be looked into other sectors. The theoretical background is made by developing more conceptual models, frameworks and empirical testing in manufacturing, automobile, healthcare and Information Technology (IT). The validation of the models has been examined by real-time case studies. The reader can understand that leagility is applied mainly in supply chain as decoupling point and emerging concept in market. For researcher, it would have a widespread of ideas for research.

Keywords: *Leagility; Lean and Agility; Leagility in supply chain; Leagility in Service; Leagility in Manufacturing*

1. Introduction

The current business scenario has been continuously facing challenges like enhance consumer delight, modifying enterprise practices, minimizing expenses of organization and obtaining competitive advantage [1]. Customer satisfaction is the biggest challenges faced by companies to fulfill the requirement in forms of capacity, variation, quality, rate and period. An organization requires special care and attempts to confirm the existence of business in significance when customer demand varies in an unpredictable way [2]. Handling uncertainties is a critical work and various management perceptions have been anticipated to obtain this objective [3]. Quick delivery, minimization of cost, satisfying current demands and flexible system are essential to the alternative best practice of manufacturing system to improve the performance. To overcome the above challenges, firms use the most commonly applied paradigms like leanness and agility. Lean principle emphases on the removal of all kinds of non-value activities all over the organization while agile emphases on organizational behavior with consideration to market changes in a quick way. Leagility is the combination of both principles by taking advantage of lean and agile as quick reaction to the current demand of the market with effective utilization of limited resources [4]. Leagile concept first emerged in supply chain and then it spread to manufacturing sector. Leagile manufacturing benefits mainly to increase competitive advantage, improved consumer relationships, improve the quality of service, enhance sales, enhanced use of resources, improve employee spirit by dropping stock storage and have control over all other expenses and elimination of all non-value added activities.

Leagile is an emerging concept and many researchers and professionals are contributing to the exploration of a conceptual framework and empirical analysis over the past two decades. In contrast to the growth of publications, none has made studied the literature review research articles of leagility and this would be a motivation for the present study. The present study efforts to make an organized literature review of leagility from the research articles from Emerald, Taylor and Francis, Inderscience, Elsevier, Springer and Google Scholar. The following are the research questions of the current review as

RQ1: How has the leagility in different industries been looked at with regard to various parameters like year, publisher, research method and application oriented?

RQ2:What are the limitations and gaps of the existing literature?

RQ3: What are the scopes for future approach to leagility concepts in different sectors?

The rest of paper is as organized as follows: Section 2 is the brief literature on leagility. Section 3 gives the research methodology of the study. Section 4 is about the results and discussion which includes categorization, limitation and scope for future research from the articles. Section 5 is mainly focus on discussion of results and it extents the future research directions. Finally, Section 6 is incorporated the conclusion.

2. Literature review on Leagility

[5] have overlooked to describe 'leagile' by combining the lean and agile concepts. They have measured the performance of supply chains using leagility concept by considering market demand and lead time reduction, eliminating wastages and making improvement in scheduling. They also have explicated the importance of a decoupling point in a supply chain for leagility, which is the transition argument between the buffer stock and the unstable demand of products. Leagile has been defined as "the combination of the lean and agile paradigms within a total supply chain strategy by positioning the decoupling point so as to better suit the need for responding to a volatile demand downstream, yet providing level scheduling upstream from the decoupling point" [5].

Leagility framework can be separated into two divisions: one focuses on supply chains and the other one focuses on individual manufacturing units. [6] have anticipated a leagile framework which mainly operates based on both lean and agile at supply chain of manufacturing unit. The framework is about the decoupling point which divides the lean and agile process diversely. The supply chain structure is formulated based on the situation of decoupling point and decided where and when lean and agility is required. Leagility can be accomplished by aggregating the three dimensions which include efficiency, effectiveness and responsiveness [7].

3. Methodology

The systematic literature review has been made in three phases acquired from [8]. In the first phase, using key words like "Leagility", "Leagile", "lean and agile", "lean and agility" are searched to collect research articles from Emerald, Taylor and Francis, Inderscience, Elsevier, Springer and Google Scholar until July 2018. In the second phase, the research articles collected are classified based on year, author, publisher, methodology, application and the medium of language used is English. The editor notes, book reviews, preface articles and conference papers are excluded, ending up with 40 relevant articles. During the last phase, the select 40 articles are grouped under four categorizations such as year of publication, research methods, publisher and journals and application oriented. The detailed description of these categorizations is given below and identified the research gaps and future research directions.

IOP Conf. Series: Materials Science and Engineering **954** (2020) 012019 doi:10.1088/1757-899X/954/1/012019

4. Results and Discussion

4.1 Categorization of the framework

After reviewing the articles, the following categorization context is adopted in the present a complete and whole analysis of leagility.

- Categorization based on year
- Categorization based on research methods
- Categorization based on publisher and journal
- Categorization based on application oriented.

4.1.1 Categorization based on Year (1999 to 2018)

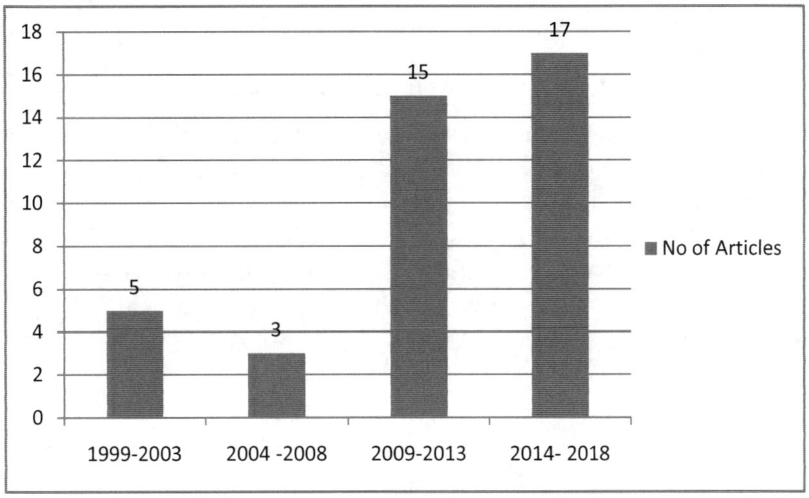

Figure 1. Distribution of articles based on year

The year wise categorization of publication indicates the review of leagility is range from 1999 to 2018. It is about the two decades but the publication is more during 2009 to 2018. The distribution of leagility literature indicates much scope in recent years with five articles published in 2018. First leagility was integrated by [5] and applied it in supply chain, but its gradual attention gained after 2009. Figure 1 shows the emerging importance of leagility over a period of 20 years.

4.1.2 Categorization based on Research Methods

The categorization of the articles with the research methods are followed as conceptual, case study, empirical, comparative analysis, theoretical contribution, assessment, report analysis and applied research. The divisions of articles with research methods are given in Figure 2.

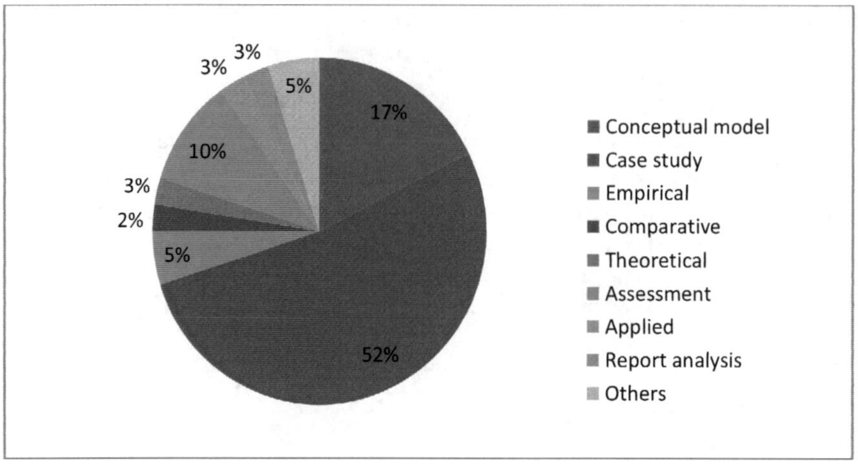

Figure 2. Categorization of paper with research methods

The literature of leagility is categorized with research method indicates that 52% of the articles on case studies, after that 17% of conceptual development and 10% of assessment. It implies that a majority of researches are done in case study from and few studies are conducted to develop a conceptual model in leagility. There is much scope for future research in developing a conceptual model for different fields and that would help in theory building. Further, the theoretical development and validity with empirical would help to generalize the leagility in various fields.

4.1 3 Categorization based on Publisher and Journal

Table 1. Literature categorization based on Publisher and Journal

Publisher	Journal	Author	No of papers per publisher
Elsevier	Int. J. Production Economics	[9]; [5]; [10]; [11]; [12]	10
	European Journal of Operational Research	[13]	
	International Journal of Operations & Production Management	[14]	
	Journal of Manufacturing Systems	[3]	
	Procedia Manufacturing	[15]	
	The Journal of Systems and Software	[16]	
Emerald Insight	Benchmarking: An International Journal	[17]; [18]; [19]	9
	International Journal of Agile Management Systems	[20]	
	International Journal of Lean Six Sigma	[1]	
	Journal of Business & Industrial Marketing	[21]	
	Journal of Enterprise Information	[22]	

	Management		
	Journal of Manufacturing Technology Management	[23]	
	Supply Chain Management: An International Journal	[24]	
Inderscience Enterprises Ltd.	Int. J. Agile Systems and Management	[25]; [26]	7
	Int. J. Industrial and Systems Engineering	[27]	
	Int. J. Information Technology and Management	[28]	
	Int. J. Process Management and Benchmarking	[2]	
	International Journal of Operational Research	[29]	
	Int. J. Services and Operations Management	[30]	
Scientific Research Publishing	Open Journal of Business and Management	[31]	1
Springer Science	Journal of Intelligent Manufacturing	[32]	1
Taylor & Francis	International Journal of Healthcare Management	[33]	8
	Production Planning & Control	[34]	
	International Journal of Logistics Research and Applications	[35]; [36]	
	International Journal of Management Science and Engineering Management	[37]	
	International Journal of Production Research	[38]; [6]; [39]	
Others	Advanced Materials Research	[4]	4
	African Journal of Business Management	[40]	
	Decision Science Letters	[41]	
	American Journal of Scientific Research	[42]	

Leagility has gained importance in publisher like Taylor & Francis, Emerald Insight and Elsevier. International Journal Production Economics having five research articles, International Journal of Production Research having 3 research articles and Benchmarking: An International Journal also have 3 research articles. The above journals have given importance and opportunities for the publication of research articles related to leagility. These three journals have repeated more than three articles on leagility. [5] have developed a concept of leagility by combining lean thinking and agile manufacturing which is published in International Journal Production Economics. The other journals who have gained importance on leagility are International Journal of Agile Systems and Management, International Journal of Logistics Research and Applications. Table 1 shows the distribution of articles based on publisher and journal.

4.1.4 Categorization based on Application Oriented

Table 2. Literature distribution based on application oriented

Application Oriented	No of papers	%
Supply Chain		
• General	16	40
• Manufacturing	3	7.5
• Fast Moving Consumer Goods (FMCG)	1	2.5
• Information Technology(IT)	1	2.5
• Personal Computer	1	2.5
• Performance	1	2.5
• Automobile	1	2.5
• Inventory	1	2.5
• Reverse supply chain	1	2.5
Healthcare	3	7.5
Food processing	2	5
House appliance industry	1	2.5
Inventory	1	2.5
Manufacturing	5	12.5
Marketing	1	2.5
Software development	1	2.5
Total	40	100

The categorization of research articles are based on the application describes the 65% of articles applied in supply chain, 12% on manufacturing, 7% on healthcare and 5% on food processing. Major portions of articles are published in supply chains which are further classified as general, manufacturing, FMCG, IT, automobile, performance, inventory and reverse supply chain. The future scope is more on other areas like manufacturing, healthcare, automobile, IT and other service industries for the development of conceptual model and it is validated by empirical test in the implementation of leagility. Table 2 shows the literature distribution based on application oriented.

4.2 Limitations

The following are the limitations derived from the research articles

- The case study findings would not be enough to the generalizability of the leagility ([3];[21]).
- Leagility is applied in various sectors and units will have different results ([36]; [34]; [11];[19]; [24];[22]).
- Secondary data analysis method is used, which has an issue associated with facts quality and accurateness [16].
- Leagility is utilized fully by healthcare in the case of cardio diabetes care case [33].

4.3 Scope for future research

The following are the future research directions derived from the articles published.

- The critical barriers are identified for future enhancement by specialized attention [17].

- The validity of empirical results is increased by examining the organized and suggested factors to diverse types of firms [18].
- Leagility is practiced in mass services and also applied professional services, service shops and mass services [19].
- Leagility is applied with several methods to identify technical gaps in supply chain for improvement activities [4].
- The suggested model of leagility is implemented to check the appropriate with real time case study [23].
- The validity and reliability of leagility is tested by applying with different sectors for supply chain [34].
- The comparative analysis of leagility with various industries on supply chains can be included leagility in diverse ways [3].
- Leagility is applied with reverse logistics generally on recycling or other environmental dimensions [36].
- ISM can be used to develop a initial model and its validation is tested by using Structural Equation Modelling (SEM) [2].
- Leagility further applied in healthcare with disease like breathing infection, HIV, cardiovascular difficulties and tuberculosis [33].
- The robustness, flexibility and appropriateness of leagility index is checked with different supply chain [15].
- The definition of leagility is tested with the performance features of supply chain. The longitudinal and action research is required on the change practice of leagility [10].
- Leagility model is further expanded with other sectors to improve the efficiency, consumer satisfaction and profit of the firm in better way.
- Leagility can be incorporated with the other techniques like neural networks and reversed mapping to solve the real-time problems [21].
- Leagility model can be utilized with other professional services like auditing, legal and financial and consultancy services [24].
- The developments of model with the variables are examined with survey. Further the performance of supply chain is tested with long term [11].
- In-depth interview can be extended further for the clarity in the application of leagility [16].

5. Futuristic Research Directions

From the present study, this the literature reviews of articles on leagility until July 2018 from reputed database is noted. Leagility is widely applied in supply chain with a decoupling point where combining leanness and agility for the reacting rapidly for consumer requirement by diminishing cost and waste [17]. This is the first study on leagility as a whole on literature review based on emerging importance in recent years. The present article elaborates the categorization based on year, journal, publication, application oriented, research method, limitation and future direction. The sample size of articles with respect to five publishers and so on. The summary and future directions of analysis are given below:

1. Leagility is mostly explored in supply chain, but not in manufacturing, service and IT ([3]; [34]). In supply chain, huge studies have emerged in conceptual model, empirical, case study, performance analysis, comparative analysis form and so on. Whereas in manufacturing, only a few studies have emerged, which include developing a model, making theoretical study, case study and so on. Validation and real time implementation would help to analyze leagility in an effective way.

2. Leagility should be applied in different fields such as automobile, home appliance, healthcare, marketing, reverse engineering, sustainability, IT, software development, textile, fast moving consumer goods, manufacturing (process layout, product design and batch processing) and so on.
3. Leagility is implemented with the change process by applying action research. This helps to increase the profit, market share, enhance competitiveness and satisfy customer demands to the dynamic market.
4. More conceptual models of leagility need to be developed and tested empirically in different sectors to expand the concepts for implementation, action and performance analysis.
5. The robustness, flexibility and appropriateness of the leagility index is tested with other supply chain for betterment of methods.
6. The comparative analysis of leagility in different sectors would help to incorporate concepts of leagility in effective ways.
7. The new definition of leagility would emerge when it is applied and tested empirically in different industries.
8. In-depth interviews are conducted to find out the opportunities and development of concept leagility in different sectors.
9. In healthcare, leagility should be applied to various departments, disease wise treatments methods, nursing, operation theatre, emergency care, pharmacy and so on. The procurement of instruments, medicines, equipment and resources like water, electricity and food can be attempted.

6. Conclusion

There has been a gradual growth of attention on the literature of Leagility in the last 10 years. By analyzing the articles from journals widely, it concluded that the concept of leagility gains importance in practitioners and researchers. The review reflects the widening of leagility by analyzing the articles which are classified based on year, research methods, application oriented, journal and publisher and the limitations and scope for future research are identified. Leagility concept is more explored in supply chain as separation of lean and agility with decoupling point. The study suggests that more articles should emerge to develop different conceptual models and to test them empirically. The theoretical background should be outlined from various aspects of findings of the study. The generalisability of concept leagility would emerge based more real-time case findings. The articles published by researcher are more than the practitioner and so this invites practitioner to contribute more research articles. Further, only one third of the selects articles are giving the limitations and future scope of research would be a drawback of the study. On the whole, leagility is explored in a tiny way and it needs to be explored in more different industries on aspects. The limitation of the current study is that the articles are collected only from Emerald, Taylor and Francis, Inderscience, Elsevier, Springer and Google Scholar. The future literature analysis can be extended by selecting articles from other publishers and by increasing the sample size.

References

[1]. Shahin, A., & Jaberi, R. (2011). Designing an integrative model of leagile production and analyzing its influence on the quality of auto parts based on Six Sigma approach with a case study in a manufacturing company. *International journal of lean six sigma*, *2*(3), 215-240.
[2]. Matawale, C. R., Datta, S., & Mahapatra, S. S. (2013). Interrelationship of capabilities/enablers for lean, agile and leagile manufacturing: an ISM approach. *International Journal of Process Management and Benchmarking*, *3*(3), 290-313.

[3]. Huang, Y. Y., & Li, S. J. (2010). How to achieve leagility: A case study of a personal computer original equipment manufacturer in Taiwan. *Journal of Manufacturing Systems*, *29*(2-3), 63-70.
[4]. Vinodh, S., & Raj, S. A. (2012). Determination of Leagile Supply Chain Index Using Fuzzy ANP Approach. In *Advanced Materials Research* (Vol. 488, pp. 1614-1618). Trans Tech Publications.
[5]. Naylor, J. B., Naim, M. M., & Berry, D. (1999). Leagility: Integrating the lean and agile manufacturing paradigms in the total supply chain. *International Journal of production economics*, *62*(1-2), 107-118.
[6]. Mason-Jones, R., Naylor, B., & Towill, D. R. (2000). Lean, agile or leagile? Matching your supply chain to the marketplace. *International Journal of Production Research*, *38*(17), 4061-4070.
[7]. Okongwu, U., Lauras, M., Dupont, L., & Humez, V. (2012). A decision support system for optimising the order fulfilment process. Production Planning & Control, 23(8), 581-598.
[8]. Sreedharan, V. R., & Raju, R. (2016). A systematic literature review of Lean Six Sigma in different industries. *International Journal of Lean Six Sigma*, *7*(4), 430-466.
[9]. Kisperska-Moron, D., & De Haan, J. (2011). Improving supply chain performance to satisfy final customers:"Leagile" experiences of a polish distributor. *International Journal of Production Economics*, *133*(1), 127-134.
[10]. Naim, M. M., & Gosling, J. (2011). On leanness, agility and leagile supply chains. *International Journal of Production Economics*, *131*(1), 342-354.
[11]. Purvis, L., Gosling, J., & Naim, M. M. (2014). The development of a lean, agile and leagile supply network taxonomy based on differing types of flexibility. International Journal of Production Economics, 151, 100-111.
[12]. Herer, Y. T., Tzur, M., & Yücesan, E. (2002). Transshipments: An emerging inventory recourse to achieve supply chain leagility. *International Journal of production economics*, *80*(3), 201-212.
[13]. Agarwal, A., Shankar, R., & Tiwari, M. K. (2006). Modeling the metrics of lean, agile and leagile supply chain: An ANP-based approach. *European Journal of Operational Research*, *173*(1), 211-225.
[14]. Krishnamurthy, R., & Yauch, C. A. (2007). Leagile manufacturing: a proposed corporate infrastructure. *International Journal of Operations & Production Management*, *27*(6), 588-604.
[15]. Banerjee, A., & Ganjeizadeh, F. (2017). Modeling a Leagility Index for Supply Chain Sustenance. *Procedia Manufacturing*, *11*, 996-1003.
[16]. Wang, X., Conboy, K., & Cawley, O. (2012). "Leagile" software development: An experience report analysis of the application of lean approaches in agile software development. *Journal of Systems and Software*, *85*(6), 1287-1299.
[17]. Matawale, C. R., Datta, S., & Mahapatra, S. S. (2016). A fuzzy embedded leagility assessment module in supply chain. *Benchmarking: An International Journal*, *23*(7), 1937-1982.
[18]. Shahin, A., & Rezaei, M. (2018). An integrated approach for prioritizing lean and agile production factors based on costs of quality with a case study in the home appliance industry. *Benchmarking: An International Journal*, *25*(2), 660-676.

[19]. Rahimnia, F., Moghadasian, M., & Castka, P. (2009). Benchmarking leagility in mass services: The case of a fast food restaurant chains in Iran. *Benchmarking: An International Journal*, 16(6), 799-816.
[20]. Mason-Jones, R., Naylor, B., & Towill, D. R. (2000). Engineering the leagile supply chain. *International Journal of Agile Management Systems*, *2*(1), 54-61.
[21]. Kalantari, T., & Khoshalhan, F. (2018). Readiness assessment of leagility supply chain based on fuzzy cognitive maps and interpretive structural modeling: a case study. *Journal of Business & Industrial Marketing*, *33*(4), 442-456.
[22]. Banerjee, A., & Mukhopadhyay, S. K. (2016). A contemporary TOC innovative thinking process in the backdrop of leagile supply chain. *Journal of Enterprise Information Management*, *29*(3), 400-431.
[23]. Rahiminezhad Galankashi, M., & Helmi, S. A. (2016). Assessment of hybrid Lean-Agile (Leagile) supply chain strategies. *Journal of Manufacturing Technology Management*, *27*(4), 470-482.
[24]. Rahimnia, F., & Moghadasian, M. (2010). Supply chain leagility in professional services: how to apply decoupling point concept in healthcare delivery system. *Supply Chain Management: An International Journal*, *15*(1), 80-91.
[25]. Matawale, C. R., Datta, S., & Mahapatra, S. S. (2015). Evaluation of leanness, agility and leagility for supply chain of automotive industries. *International Journal of Agile Systems and Management*, *8*(2), 85-115.
[26]. Raj, S. A., Jayakrishna, K., & Vimal, K. E. K. (2018). Modelling the metrics of leagile supply chain and leagility evaluation. *International Journal of Agile Systems and Management*, *11*(2), 179-202.
[27]. Banerjee, A., & Mukhopadhyay, S. K. (2015). Leagility index in supply chain-part-II: case applications, benchmarking and comparative analysis. *International Journal of Industrial and Systems Engineering*, *21*(2), 242-276.
[28]. Banerjee, A. (2015). Information technology enabled process re-engineering for supply chain leagility. *International Journal of Information Technology and Management*, *14*(1), 60-75.
[29]. Shukla, S. K., & Wan, H. D. (2010). A leagile inventory-location model: formulation and its optimisation. *International Journal of Operational Research*, *8*(2), 150-173.
[30]. Maharaja, R., Devadasan, S. R., & Sakthivel, M. (2018). Supply chain performance measurement: a future research agenda for acquiring competitiveness through the implementation of leagile manufacturing paradigm. *International Journal of Services and Operations Management*, *30*(1), 39-50.
[31]. Denavi, H. D., Mirabi, M., & Rezaei, A. (2017). Ranking of Leagility Factors Based on Job Satisfaction through a Combinatory Model of Fuzzy TOPSIS and AHP (Case Study: MRI Hospital, Shiraz, Iran). *Open Journal of Business and Management*, *6*(01), 21.
[32]. Haq, A. N., & Boddu, V. (2017). Analysis of enablers for the implementation of leagile supply chain management using an integrated fuzzy QFD approach. , *28*(1), 1-12.
[33]. Mishra, V., Samuel, C., & Sharma, S. K. (2018). Lean, agile and leagile healthcare management–A case of chronic care. *International Journal of Healthcare Management*, 1-8.
[34]. Soni, G., & Kodali, R. (2012). Evaluating reliability and validity of lean, agile and leagile supply chain constructs in Indian manufacturing industry. *Production Planning & Control*, *23*(10-11), 864-884.

[35]. Gaudenzi, B., & Christopher, M. (2016). Achieving supply chain 'Leagility'through a project management orientation. *International Journal of Logistics Research and Applications*, *19*(1), 3-18.

[36]. Banomyong, R., Veerakachen, V., & Supatn, N. (2008). Implementing leagility in reverse logistics channels. *International Journal of Logistics: Research and Applications*, *11*(1), 31-47.

[37]. Razmi, J., Hassani, A., & Babazadeh, R. (2015). A hybrid fuzzy bi-objective delivery planning model for leagile e-tailing under uncertainty. *International Journal of Management Science and Engineering Management*, *10*(1), 62-72.

[38]. Vinodh, S., & Aravindraj, S. (2013). Evaluation of leagility in supply chains using fuzzy logic approach. *International Journal of Production Research*, *51*(4), 1186-1195.

[39]. Chan, F. T., & Kumar, V. (2009). Performance optimization of a leagility inspired supply chain model: a CFGTSA algorithm based approach. *International Journal of Production Research*, *47*(3), 777-799.

[40]. Tabrizi, M. S. H. M. (2013). Modeling the metrics of lean, agile, and leagility: An AHP-based approach. *African Journal of Business Management*, *7*(13), 989-998.

[41]. Ramana, D., Rao, K., & Kumar, J. (2013). Evaluation of performance metrics of leagile supply chain through fuzzy MCDM. *Decision science letters*, *2*(3), 211-222.

[42]. Tabrizi, M. S. H. M., & Khamisabadi, J. (2013). Production, Manufacturing System Operations, and Logistics Modeling the Metrics of Lean, Agile, and Leagility: An AHP-Based Approach. *American Journal of Scientific Research*, (89), 29-40.

IOP Conf. Series: Materials Science and Engineering **954** (2020) 012020 doi:10.1088/1757-899X/954/1/012020

CuO-nano based Biodegradable Refrigeration Oil in R134a Refrigeration Systems for Energy Conservation and Green Environment

P. Vithya[1*], G. Sriram[2], S. Arumugam[3], M.D.V.Aditya Chaitanya[4], G.Surya Deepak[5], K.Sriram Kumar[6], P.Jayaram Chandra[7]

[1-5] Department of Mechanical Engineering, Sri Chandrasekharendra Saraswathi Viswa Mahavidyalaya, Kanchipuram, Tamilnadu - 631561, India.

Email: vithyagautham@gmail.com

Abstract. The present study experimentally investigated the reliability and performance of a vapour compression refrigeration system with CuO nano based biodegradable refrigeration oil. The pentaerythritol ester biodegradable refrigeration oil (PE) is blended with synthetic refrigeration oil (ISO 68 grade-POE oil) in the fraction of 50% by volume. Nanolubricant was synthesized on the basis of the concept of the nanofliuds, which was prepared by mixing CuO nanoparticles with pentaerythritol ester based biodegradable refrigeration oil blend (PE50). The concentration of nano CuO ranges in the order of 0.1% to 0.3% by weight with particle size of 40 nm to prepare CuO nano biodegradable refrigeration oil. An experimental test rig is designed and fabricated indigenously to carry out the investigations. The results indicated that CuO nano-based biodegradable refrigeration oil works normally and safely in the system. The compressor power consumption was reduced by 5.02% with the use of 0.3% CuO nano biodegradable refrigeration oil. The PE50 with 0.3% CuO nano has a considerable rise in coefficient of performance (COP) of about 45%. Hence CuO nano-based biodegradable refrigeration oil in refrigeration system considerably reduces the energy consumption and increases the COP of the refrigeration system.

1. Introduction

A lubricant is a substance introduced between two surfaces in relative motion in order to reduce the friction and wear produced by contact surfaces [1]. The main purposes of lubrication are to decrease wear and heat loss that results owing to surfaces contact in motion, and to prevent from corrosion and minimize the oxidation [2]. There has been huge attraction and interest developed in the use of lubricants from vegetable oil. Vegetable oils are biodegradable, non-toxic and have excellent lubricity properties. Earlier research findings concluded that the transesterification of vegetable oil has an increased possibility for the synthesis of lubricant with enhanced temperature and significant fluidity [3]. Janusz Nowicki et al [4] studied the synthesis of esters from oleic acid with pentaerythritol and TMP dimer using homogeneous metallic catalysts such as Sn, Ti and Zr catalysts and stated that obtained esters have superior lubrication properties in terms of low pour point of −22°C to −32°C, good viscosity index of 72.8 to 130.5mm^2/s at 40°C and improved tribological properties in terms of wear scar diameter of 0.66mm. Tirth et al [5] reviewed the present status of vegetable oil based lubricants. Vegetable oils are the primary outrider by its uses and because of their sustainable energy resources; they offer energy independence. Arianti et al [6] reported lubricants from vegetable oil provide a renewable source of ecologically associate lubricants, relating to biodegrade into harmless products. Studies show that bio lubricant derived from the various vegetable oil possesses promising lubrication characteristics. Owuna et al., (2019) re-evaluated the development of vegetable oils using polyols through various chemical modification techniques; analysed the rheological of bio lubricants and stated that TMP ester based stocks are renewable, eco-friendly and have potentials to be used for the formulation of hydraulic oils, gear oils and engine oils.

Energy demand and environmental issues of refrigerants are the major crisis of refrigeration industry. The energy consumption by refrigerators is increasing day by day due to their rapid usage. Such an increasing energy requirement will aggravate energy shortage and environment pollution [7]. The researchers proposed a number of models to predict the behaviour of a hermetically sealed compressor under different conditions. There have been many investigations on the role of refrigerant and lubricants in the working and performance of a compressor, also assumptions being different in each case. Kiranmai et al [8] investigated the refrigeration compressor work using mineral oil ISO 32, Polyolester oil ISO 22 and poly alkylene glycol oil ISO 46. The results demonstrate that the PAG oil and POE oil had the good compatibility with R134a and coefficient of performance of PAG oil is increased by 3.1% and 14.5% compared to POE oil and mineral oil respectively. Akhavan-Behabadi et al [9] assessed the influence of CuO nanoparticles on boiling and two phase phenomenon of POE oil/ R-600a mixture within a smooth horizontal tube and observed that 1.5% mass fraction of nanoparticles amplifies the heat transfer to 63% comparative to the heat transfer performance with pure POE oil/ R-600a mixture. Omer et al [10] reviewed the thermo-physical performance of nano refrigerants and nano lubricants, two phase flow boiling phenomena of nano lubricants in refrigerating systems and reported that 0.1% (w/w) of TiO_2 nanoparticles in mineral oil and HFC134a refrigerant declines the 26.1% usage of energy consumption evaluated to the R134a and polyolester oil. Vaishali et al [11] overviewed the improvement of heat transfer using nano refrigerants and nano lubricants in refrigeration systems and stated that the improvement of thermal conductivity of nano fluids augment the heat transfer rate and freezing capacity of refrigeration systems. Mohammed Ajmal et al [12] presented the impact of 0.025, 0.05 and 0.10 % mass concentrations of CuO nanoparticles on the tribological performance of Polyol ester oil through pin-on disc tribo-machine. The tribological results showed that inclusion of 0.10% CuO nanoparticles of polyol ester oil reduces the coefficient of friction from 0.18 to 0.06. The improvement of friction characteristics optimizes the energy consumption in refrigeration systems. The present study is focused on performance investigation to study the effect of addition of CuO nanoparticles (0.1% to 0.3%) on weight basis in the pentaerythritol ester based biodegradable refrigeration oil blend (PE50) on the coefficient of performance of the refrigeration system.

2. Experimental

2.1. Synthesis of Biodegradable Refrigeration oil and blend sample preparation

Rapeseed oil is used for the formulation of polyolester-based bio lubricant. A successive transesterification process was adopted for the formulation of pentaerythritol ester of rapeseed oil. Fig.1 shows the chemical reaction of successive transesterification process. The detailed procedure for the formulation of Pentaerythritol ester has adopted from Hashem et al. [13]. A synthetic grade polyol ester (POE68), a widely used lubricant in refrigeration plant was selected as a base fluid for the preparation of various lubricant blends. The pentaerythritol ester of rapeseed oil was blended with POE68 grade refrigeration oil in the fraction of 50% by volume. The blend preparation was conceded in a round bottom glass reactor and the mixture was heated up to 60°C at a stirring speed of 300 rpm for about one hour [14]. The rheological properties of different refrigeration oil samples such as kinematic viscosity, viscosity index, flash and pour point and total acid numbers was measured in accordance with ASTM standards and are listed in Table 1.

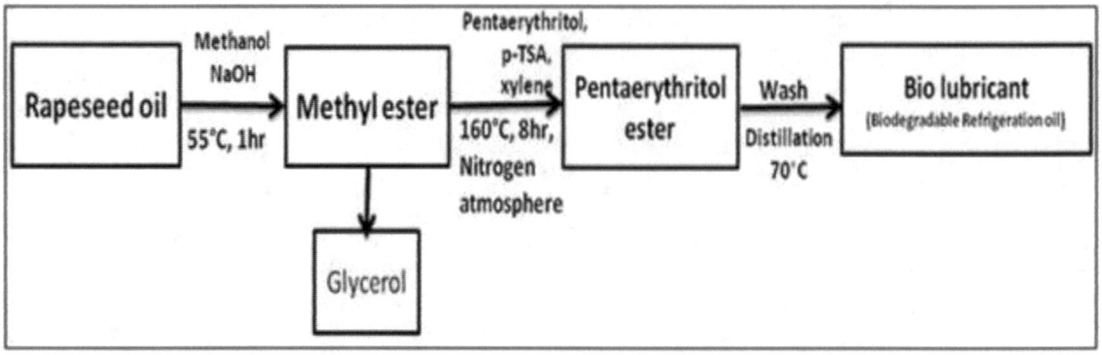

Fig.1. Chemical reaction of successive transesterification process

Table 1. Properties of CuO nano based bio degradable refrigeration oil

Test parameters	Standard	POE 68	PE	PE50	PE50 CuO 0.1	PE50 CuO 0.2	PE50 CuO 0.3
Kinematic viscosity @100°C (cSt)	ASTM D445	9.3	11	10.54	11.4	11.9	12.5
Viscosity index	ASTM D2270	114	248	244	249	252	257
Flash point(°C)	ASTM D92	270	242	220	224	226	229
Pour point(°C)	ASTM D97	-39	-25	-28	-29	-29	-29
Total Acid Number mg KOH/g	ASTM D974	0.3	1.2	0.79	0.76	0.83	0.70

2.2. Preparation of Nano-CuO based biodegradable refrigeration oil
The additive used in the synthetic and refrigeration oil was CuO in the nanoparticle form. Fig.2(a) depicts the SEM image of the CuO nanoparticles.

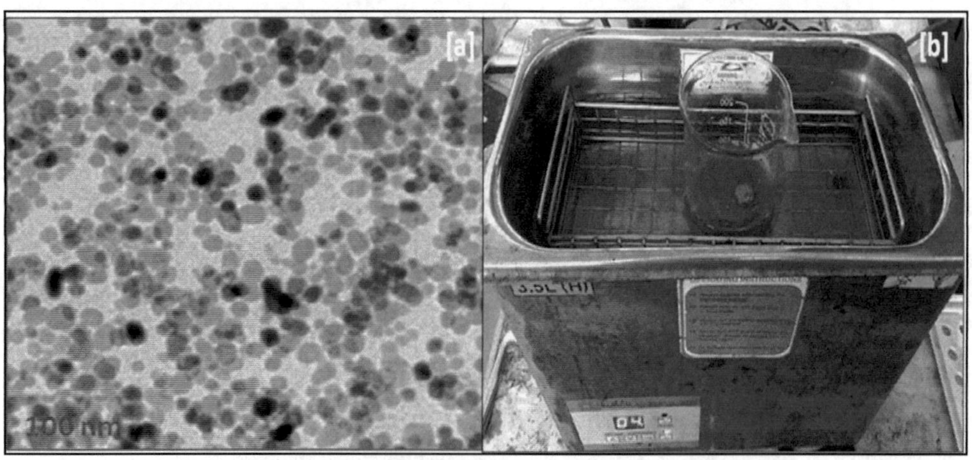

Fig.2(a) SEM image of CuO nanoparticles **(b)** Ultrasonic Agitator

IOP Conf. Series: Materials Science and Engineering **954** (2020) 012020 doi:10.1088/1757-899X/954/1/012020

The CuO nanoparticles of 40nm were procured from M/S Nano Research US. The nanoparticles of 40nm were supplied by US Nano Research Limited. The ultrasonication process was utilized to prepare the nano refrigeration oil samples of concentration of 0.1,0.2 and 0.3 (w/w) for accomplish the excellent dispersion stability of the nanoparticles without any agglomeration. Fig. 2(b) shows the photograph of the preparation of biodegradable refrigeration oil sample containing 0.3% CuO nanoparticles.

2.3. Test of Compressor performance using Nano-CuO based biodegradable refrigeration oil

The hermetically sealed reciprocating compressor (AQAW77X thermally protected) for R134a was procured from M/S Shangzhure Refrigeration Co. Ltd. The power input of the compressor was ½ HP. The condenser was air cooled and the capillary tube expansion valve was used for expansion process, the evaporator cabin dimensions are height 0.65m, length 6.54m and breadth 0.42m.

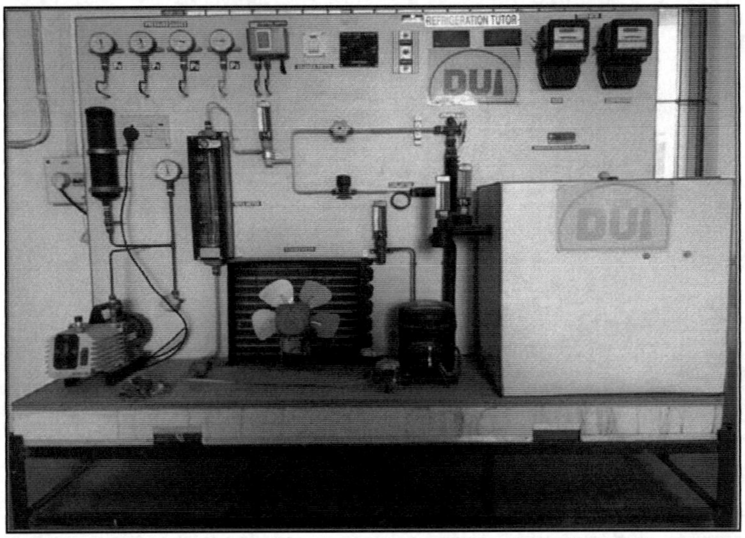

Fig.3 Photographic view of the vapour compression refrigeration system

The thermocouples used were calibrated T type for observing the temperature of the refrigerant. Fig.3 shows the photographic view of the vapour compression refrigeration system test rig. The consumed power of the whole test rig and the compressor were measured using energy meter and compressor sensor respectively. The system was evacuated first to remove the moisture content inside it for 30 minutes. Then the refrigerant R134a was charged through the charging line and ISO 68 grade oil was filled through the lubrication line of the compressor. The experiment was conducted and temperatures were observed continuously until the steady state reached. Now the test rig should be powered off for discharging the refrigerant and lubricant. The same procedure is repeated for all biodegradable and blended lubricant samples. The experiment was conducted and temperatures were observed continuously until the steady state reached. Now the test rig should be powered off for discharging the refrigerant and lubricant. The same procedure is repeated for all biodegradable and blended lubricant samples. The temperature T_1 is theinlet/outlet temperature of the compressor/evaporator, T_2 is the inlet/outlet temperature of the condenser/compressor, T_3 is the inlet/outlet temperature of the expansion/condenser, and T_4 is the inlet/outlet temperature of the evaporator/expansion valve. The performance of the refrigeration system was calculated using the standard refrigeration formulae.

3. Results and Discussion

3.1 Viscosity of Nano-CuO based biodegradable refrigeration oil

The kinematic viscosity of biodegradable refrigeration oil and nano added biodegradable refrigeration oils were measured using a Redwood viscometer. Fig.4 evaluates the result of kinematic viscosity as a temperature function. The results signify that the augment in viscosity due to the inclusion of nanoparticles. It is observed from the results that the addition of 0.3% CuO is more than that of other refrigeration oil blend samples. Fig. 4 shows that at 100°C, the increase in viscosity are 18% and 22% for a Cuo nanoparticle concentration in PE50 of 0.1% and 0.2% respectively, whereas, for 0.3% Cuo nanoparticle concentration in PE50, the increase in the viscosity was 26%. The results also indicate that the increase in viscosity index due to the addition of CuO nanoparticles up to 0.3% (w/w) in PE50. The results also indicate that the increase in viscosity due to the addition of nanoparticles at higher temperatures be attributed to a decrease in the fluid layer resistance. The tendency of viscosity deviation acquired is in accordance with the results concluded by Sabereesh et al and Kedzierski [15,8]

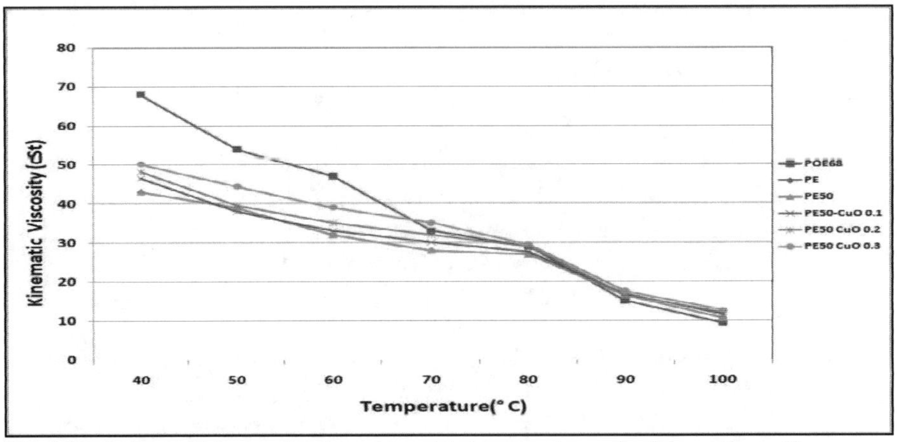

Fig.4 Variation of kinematic viscosity with temperature of refrigeration oils

3.2. Performance of Compressor using Nano-CuO based biodegradable refrigeration oil

3.2.1 Effect of Compressor shell temperature

The continuous rise of compressor outlet temperature may also cause possibility of breakdown of the lubricating oil leads to the higher friction, wear and reduction in the life span of the compressor. Fig. 5 shows the compressor shell temperature of POE68, PE and PE50 oil and nano based biodegradable refrigeration oil samples. Examination of the temperature of the compressor shell indicates that the CuO nano biodegradable refrigeration oil samples shows decrement in the value than that of synthetic lubricant which consecutively raises the performance of the system [16]. The compressor shell temperature of 0.3% CuO nano biodegradable refrigeration oil was reduced about 23%, 20% and 14% as compared to POE68, PE and PE50 respectively. The decline of friction coefficient between the compressor parts because of less frictional heating produced using nano biodegradable refrigeration oil decreases the compressor shell temperature.

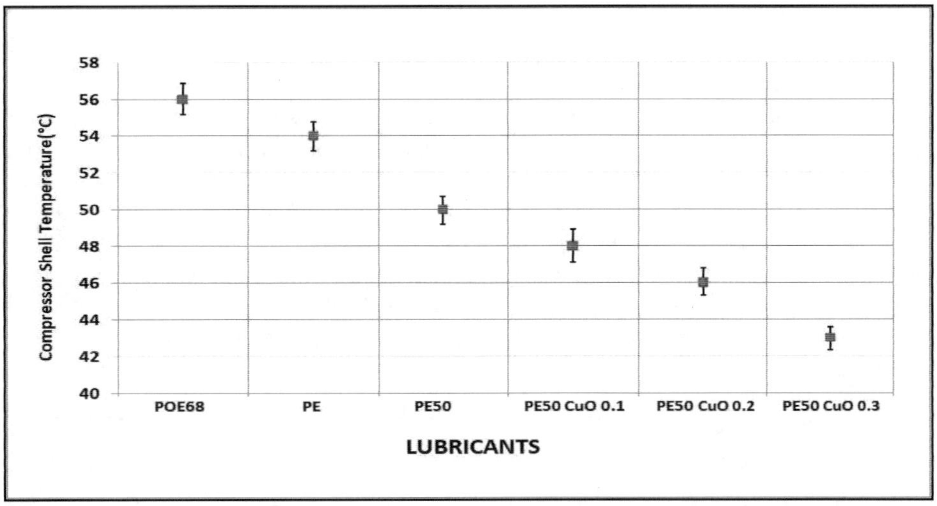

Fig.5 Effect of Cuo- nano in refrigeration oil on the surface temperature of Compressor shell

3.2.2 Power Consumption

Compressor power reduction is achieved by the refrigeration capacity increment, which in turn had a considerable effect in improving the compressor efficiency, lead to the COP enhancement. Fig. 6 shows the compressor power input using POE68, PE and PE50 oil and nano based biodegradable refrigeration oil samples. It is noted that the power consumption of 0.3% CuO nano biodegradable refrigeration oil is lower than that of all refrigeration oils.

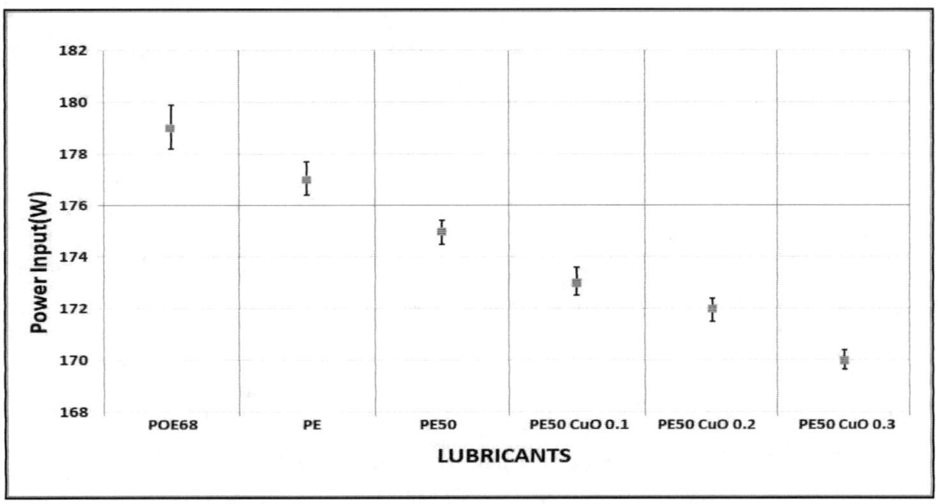

Fig.6 Effect of Cuo- nano in refrigeration oil on the Input power

The result measured from the testing of compressor shows that the power input of 0.3% CuO nano biodegradable refrigeration oil was decreased by 5.02%, 3.95% and 2.85% as compared to POE68, PE and PE50 respectively. Compressor input power was reduced due to the higher viscosity index, which in turn improves the wear characteristics [17]. Increase in refrigeration capacity of the system using nano biodegradable refrigeration oil in comparison with the synthetic refrigeration oil is achieved by the compressor power loss reduction. The heat transfer losses up to 50% take place, as the lubrication film thickness is lowered.

3.2.3 Coefficient of Performance

The coefficient of performance of the refrigeration system was determined by knowing the refrigeration effect and power input. Fig.7 shows the Coefficient of performance of POE68, PE and PE50 oil and nano based biodegradable refrigeration oil samples. The result measured from the testing of compressor shows that the coefficient of performance of 0.3% CuO nano biodegradable refrigeration oil was increased by 45%, 36% and 24% as compared to that of POE68, PE and PE50 respectively. Testing results of experimentation elucidate that the 0.3% CuO nano biodegradable refrigeration oil have a considerable rise in COP of about 45% compared with synthetic lubricant owing to the reduction of input power of the compressor and the improvement of heat transfer characteristics.

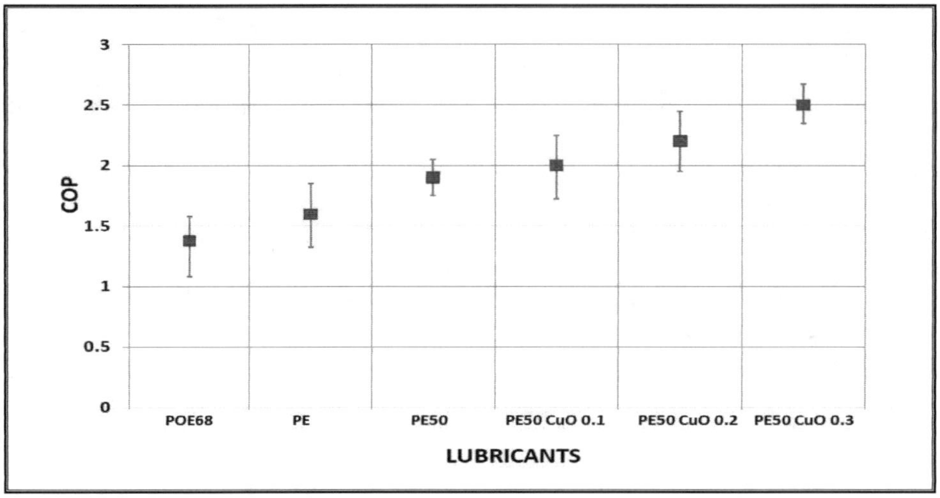

Fig.7 Effect of Cuo- nano in refrigeration oil on the COP

The biodegradable refrigeration oil (PE50) added with 0.3% CuO nano having a higher viscosity to that of POE68, which produces protective tribo- film and resulted in superior lubrication behaviour[12,18]. This is due to the reason of protective film developed by the bio lubricant provide better friction reduction between the metal surfaces.

4. Conclusions

Based on the investigations, the following conclusions are drawn:

- The viscosity of 0.3% CuO nano biodegradable refrigeration oil is higher than synthetic refrigeration oil. All the biodegradable refrigeration oil blends satisfy the ISO 68-grade requirements.
- The compressor shell temperature of 0.3% CuO nano biodegradable refrigeration oil was reduced about 23% as compared to POE68.
- The power input of 0.3% CuO nano biodegradable refrigeration oil was decreased by 5.02% as compared to POE68.
- The coefficient of performance of 0.3% CuO nano biodegradable refrigeration oil was increased by 45% as compared to that of POE68.

References

[1] Sripada P K, Sharma R, Dalai A K 2013 *Industrial Crops and Production* **50** 95-103.

[2] Nadia Salih, Jumat Salimon and Farah Nadiah Jantan 2013 *Asian Journal of Chemistry* **25** **(17)** 9751-9754.

[3] Amith Aravind, Prabhakaran Nair K, Joy M L 2018. *Inst. Mech. Eng. Part J: J. Eng. Tribol.* **08-210** 1994-1996.

[4] Janusz Nowicki, Dorota Stan´czyk, Jolanta Drabik, Jan Mosio-Mosiewski, Piotr Woszczyn ski,Marek Warzała 2016 *J. Am. Oil Chem. Soc.* **93** 973–981.

[5] Tirth M Panchal, Ankit Patel, D D Chauhan, Merlin Thomas, Jigar V 2017 *Energy Reviews* **70** 65–70.

[6] Arianti N Annisa, Widayat Widayat *MATEC Web of Conferences* **156** (06007) 1-7.

[7] Owuna F J, Dabai M U, Sokoto M A, Dangoggo S M, Bagudo B U, Birnin-Yauri U A, Hassan L G, Sada I, Abubakar A L, Jibrin M S *2020 Egyptian Journal of Petroleum 29(1)* 775-82.

[8] Kedzierskia M A 2018 *Journal of Heat Transfer* **131(4)** 1-8.

[9] Kiranmai G, John K, Samuel, Edison G, Jagan V 2005 *World Applied Sciences Journal* **32** **(9)** 2005-2010.

[10] Omer A Alawi, Nor Azwadi Che Sidik, M'hamed Beriache 2015 *Int.Communications in Heat and Mass Transfer* **68** 91–97.

[11] Vaishali P Mohod, Nishikant W Kale *2015 International Journal of Innovative and Emerging Research in Engineering* **2(1)** 191-194.

[12] Mohammed Ajmal A, SunirMajilya, Bharath Kumar H, Gobinath N 2017 *International Journal of Chem Tech Research* 10(5)93-97.

[13] Hashem A I, AbouElmagd W S I, Salem A E, El-Kasaby M, El-Nahas A M 2015 *Energy Sources Part A: Recov.Utiliz. Environ. Effects* **35(10)** 909-912.

[14] Vithya P, Sriram G, Arumugam S, Ankur Sharma, Sistila.Venkata sai sathyanarayana sastry 2018 *IOP Conf. Series: Mater. Sci. Engg.* **390** 012052 1-10.

[15] Krishna Sabareesh R, Gobinath N, Sajith V, Das S, Sobhan C B 2012 *Int. J. Refrigeration* **35** **(7)** 1989-1996.

[16] Shengshan bi, Lin shi, Li li zhang 2008 *App. Therm. Eng.* **28 (14-15)** 1834-1843.

[17] Hao P, Guoliang D, Weiting J, Haitao H, Yifeng G 2009 *Int. J. Refrigeration* **32** 1259–70.

[18] Padgurskas J, Rukuiza R, Prosyčevas I, Kreivaitis R 2013 *Tribology Int.* **60** 224-32.

IOP Conf. Series: Materials Science and Engineering **954** (2020) 012021 doi:10.1088/1757-899X/954/1/012021

Polyolester Based Lubricant from Waste Feedstock for Sustainable Environment: A Tribological Investigation

Arumugam S[1*], Baskar S[2], Sankara Subramanian V[1], Shivasankaran G[1]

[1]Department of Mechanical Engineering, SriChandrasekharendra Saraswathi Viswa Mahavidyalaya, Enathur, Kanchipuram – 631561.

[2]Department of Mechanical Engineering, S. A.Engineering College, Chennai-77.

* Corresponding Author: aru_amae@yahoo.co.in; Mobile: +91 9994699223.

Abstract: This study reported the tribological evaluation of polyolester based lubricant synthesised from waste feedstock. The transesterification process was adopted for the formulation of lubricant from waste mango seed kernel oil. The polyols used in this formulation are neopentyl glycol, trimethylolpropane and pentaerythritol. The tribological investigation was conducted using four ball tribometer as per ASTM D4172. The frictional and wear behavior of vegetable oil-based polyolester was analysed in addition to the wear morphology of investigated balls. The results indicated that the average frictional torque and mean wear diameter of pentaerythritol ester of waste mango seed oil was 0.12 Nm and 0.403 mm respectively and also found that the pentaerythritol ester showed improved tribological results as compared with other polyolesters say neopentyl glycol ester and trimethylolpropane ester of waste mango seed oil. The waste mange seed oil is found suitable candidate for the formulation of polyolester based sustainable lubricant.

1. Introduction

In general, many investigations were focused to find the alternate for petroleum-based products in the recent years [1-2]. In this connection, first and second-generation vegetable oil is found suitable for lubricant applications and it is used for many years ago specifically for lubrication purpose [3-5]. However, the formulation of lubricant from first- and second-generation vegetable oil is not cost-effective as its base stock cost is expensive than mineral oil [6]. To overcome this, base stock from waste feedstock is one such standby candidate [7]. Pierson et al experimented the use of waste mango seed oil as engine fuel [8]. The chemical modification was used to convert vegetable oils into lubricant formulations [9-10]. Here, the study reported the conversion of waste mango seed oil into polyolester based sustainable lubricant through chemical modification process. The properties and tribological behaviour of the esters formulated were assessed to find their appropriateness as lubricant candidate.

2. Experimental work

2.1 Formulation of waste mango seed oil-based polyolesters

The waste mango seed oil was used as a base candidate and polyols used in this study were pentaerythritol, trimethylolpropane and neopentyl glycol. The base oil is extracted from the kernel of waste mango seed by the procedure adopted by Dutra et al [9]. The procedure adopted for the formulation of polyolester based lubricants is as follows:

Step 1: Hundred gram of waste mango seed oil was mixed with 20 ml of methyl alcohol and 1g NaOH, a catalyst in a glass beaker and the mixture was heated to 55°C, under mechanical stirring for one hour. The resultant methyl ester of mango seed oil was separated from by product using a separating funnel. Now the methyl ester was reacted with pentaerythritol in the presence of a catalyst p-toluenesulphonic acid, and solvent xylene. The process was carried out at 150-160°C under nitrogen environment for the period of 6 h. The resultant pentaerythritol ester was methodologically purified for further investigation.

Step 2: 200 g of waste mango seed oil methyl ester was mixed with desired amount of trimethylolpropane (TMP) and it is heated to over 130°C under vacuum pressure of 0.1 mbar for 2h till the entire TMP gets melted. Then 0.9 % w/w of alkaline catalyst (sodium methoxide) was added slowly so as to avoid the spillage. The samples were drawn from the beaker at regular intervals for the completion of reaction. Then the resultant product ie., TMP ester of waste mango seed oil was cooled, vacuum filtered and distilled [13].

Step 3: Waste mango seed oil neopentyl glycol (NPG) ester were synthesized by the transesterification of 50g of mango seed oil methyl ester with 8g of NPG. The blended mixture was transferred to a flask and then, 0.3 g of calcium methoxide, a catalyst and 50 ml of isooctane were added to it. The flask mouth was attached with a stark cap so as to perform the azeotropic distillation of methanol–isooctane. The mixture temperature was maintained at 180°C at ambient pressure condition. Then the pressure was reduced slowly to 0.5 mbar by using vacuum controller. Finally, the end product of NPG ester was rotary evaporated in order to remove isooctane from the final product.

2.2 Tribological investigation using four ball tribometer

A four ball tribometer, TR-30L-IAS, supplied by Ducom Instruments, Bengaluru was used to conduct the tribological experiment. The test lubricant is charged to the ball pot where three balls of AISI E-52100 chrome alloy steel with a diameter of 12.7 mm, extra polished to grade 25 and hardness of 65 HRC are loaded. The other ball was fixed on a collet and press against the bottom balls by the supply of standard load. The experiment was carried out at lubricant chamber temperature of 75°C as per ASTM D4172. The frictional torque exerted on the three balls was measured. An optical microscope facilitated with image capturing software was used for measuring the wear scar diameter and wear scar image on the three ball surfaces and reported.

3. Results and Discussion

3.1. Physio-Chemical Properties of Polyolesters

The properties of various esters were reported in the Table 1. Table 1 was depicted the properties of the pentaerythritol ester (PEE), neopentyl glycol ester (NPGE) and trimethylolpropane ester (TMPE). The viscosity at 40°C of PEE, NPGE and TMPE were 69.1, 66.3 and 60.9 cSt respectively. The viscosity at 40°C of PEE was 4.22% and 13.46% higher than NPGE and TMPE respectively. The viscosity at 100 °C of PEE, NPGE and TMPE were 12.9, 11.4 and 10.8 cSt respectively. The viscosity at 100 °C of PEE was 13.15% and 19.44% superior to NPGE and TMPE respectively. The cloud point for PEE, NPGE and TMPE were -9 °C, −5 °C and -8 °C. The pour points of PEE, NPGE and TMPE were <−20 °C, <−31 °C and <−35 °C. The flash point of PEE, NPGE and TMPE were 311 °C, 262 °C and 249 °C. The properties of PEE showed better results as compared with NPGE and TMPE. The structure of PEE has better stable and superior when exposed at an elevated temperature, whereas NPGE and

TMPE have less stable structure as compared with the PEE. The kinematic viscosity of PEE was higher which was suitable for an oven chain application.

Table 1- Properties of polyolesters of mango seed oil.

Properties	Standard	PEE	NPGE	TMPE
Kinematic Viscosity @ 40 °C (cSt)	ASTM D445	69.1	66.3	60.9
Kinematic Viscosity @ 100 °C (cSt)	ASTM D445	12.9	11.4	10.8
Cloud point (°C)	ASTM D2500	-9	-5	-8
Pour point (°C)	ASTM D97	<-20	<-31	<-35
Flash point (°C)	ASTM D93	311	262	249

3.2. Analysis of tribometer results

Figure 1 showed the average frictional torque of various polyol esters. From Figure 1, the average frictional torque of PEE, NPGE and TMPE were 0.12 Nm, 0.14 Nm and 0.18 Nm respectively. The lower value of frictional torque and its corresponding co-efficient friction was observed for PEE as compared with the NPGE and TMPE. The higher viscosity of PEE is responsible for lower COF value. This lesser frictional behavior of PEE showed its tribological effectiveness. Figure 2 showed the mean wear scar diameter (MWSD) of PEE, NPGE and TMPE were 0.4031 mm, 0.5762 mm and 0.6651 mm respectively. The minimum mean wear scar diameter was noticed for PEE as compared with NPGE and TMPE. The presence of four carboxylate ester group in PEE could be responsible for its improved anti-wear behavior. On the counter part, the NPGE has only two carboxylate ester group. The larger pentaerythritol ester group was stable enough to protect wear. This trend is in agreement with Aziz et al [16].

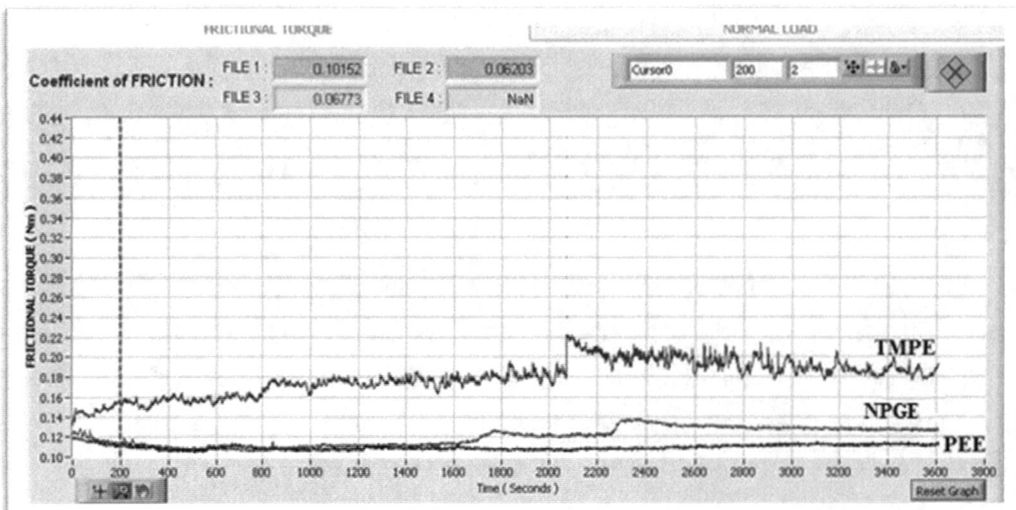

Fig. 1- Frictional torque of polyol esters of mango seed oil.

IOP Conf. Series: Materials Science and Engineering **954** (2020) 012021 doi:10.1088/1757-899X/954/1/012021

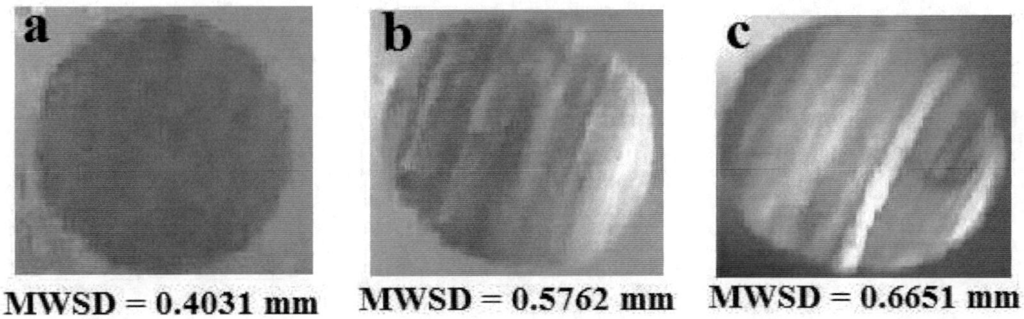

Fig 2- Wear scar image a) PEE b) NPGE c) TMPE

3.3 Surface Morphology of Ball Surfaces

The worn balls were examined by scanning electron microscope. Figure 3 depicted the surface morphology of sample ball tested under PEE, NPGE and TMPE. Fig. 3 indicated that the ball lubricated with PEE showed the smooth wear surface. The balls lubricated with NPGE and TMPE showed crack and minor voids. The micro brusqueness bonding, surface plastic deformation and damaged surfaces are the consequence of friction and wear with the use of NPGE and TMPE.

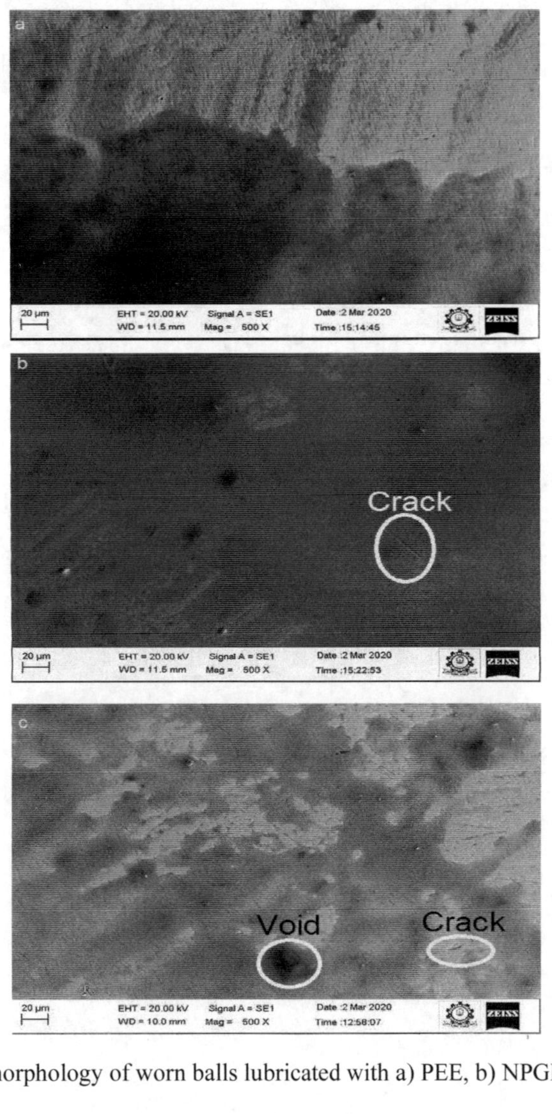

Fig. 3-Surface morphology of worn balls lubricated with a) PEE, b) NPGE, and c) TMPE.

4. Conclusions

The properties and tribological characteristics of various esters were evaluated and the findings were:

- The viscosity, cloud point, pour point and flash point of PEE was showed better values as compred with NPGE and TMPE.
- The average frictional torque of PEE was found to be 14.3% and 33.3% lower as compared to NPGE and TMPE.
- The MWSD of PEE was found to be 30.1% and 39.4% lower as compared to NPGE and TMPE.
- The worn ball surfaces lubricated with NPGE and TMPE showed the presence of crack and void. The smooth surface was observed on the worn ball surface lubricated with PEE.

5. References

[1] Zhang Z, Dube M A, McLean D D and Kates M 2003 *Bioresource Technology* **89** 1–16.

[2] Adhvaryu A and Erhan S Z 2002 *Industrial Crops & Products* **15** 247–54.

[3] Asadauskas S, Perez J M and Duda J L 1997 *Lubrication Engineering* **53** 35–40.

[4] Hwang H and Erhan S Z 2001 *Journal of the American Oil Chemists' Society* **78** 1179–84.

[5] Glancey J L, Knowlton S and Benson E R 1998 *Journal of Commerical Vehicles* **107** 266–69.

[6] Gryglewicz S, Piechocki W and Gryglewicz W 2003 *Bioresource Technology* **87** 35–39.

[7] Kaushlesh K Yadav, Neelima Garg, Anil Verma, Sanjay Kumar and Mala Trivedi 2017 Indian Journal of Agricultural Sciences **87(7)** 943-946.

[8] Jean T. Pierson, Gregory R. Monteith, Sarah J. Roberts-Thomson, Ralf G. Dietzgen, Michael J. Gidley Paul N. Shaw 2014 *Food Chemistry* **149** 253-263.

[9] Emmanuel Damilano Dutra, Thaysa Araujo de Lima, Jessica Laís de Oliveira Souza, Joanna Gabriela Vicente Silva, Katia Aparecida da Silva Aquino, Fabiana da Silva Aquino, Clecio Souza Ramos, and Romulo Simoes Cezar Menezes. 2018 *Biomass Conv. Bioref.* **8** 135–141.

[10] Wang L, He H, Xie Z, Yang A and Zhn S 2007 *Fuel Processing Technology* **88** 477–81.

[11] Arumugam S, Chengareddy P and Sriram G 2018 *Industrial Crops & Products* **123** 617-28.

[12] Gryglewicz S, Piechocki W and Gryglewicz G 2003 *Bioresource Technology* **87** 35–39.

[13] Robiah Yunus, Fakhrul-Razi, Iyuke S E and Idris A 2003 *Journal of oil Palm Research* **15** 42-49.

[14] Baskar S, Sriram G and Arumugam S 2015 *Tribology in Industry* **37** 449-54.

[15] Nagabhooshanam N, Baskar S, Ram Prabhu T and Arumugam S 2020 *Tribology International* **151** 106510.

[16] Nur Atiqah Mohamad Aziz, Robiah Yunus, Umer Rashid, Nurin Wahidah Mohamad Zulkifli. 2016. Tribology International **93** 43-49.

Investigation to Enhance the Performance of Computer Processor Cooling System

P. Gayathri Prasad[1], Prof. L.V.R.S.V. Prasad Chilamkurti[2], Santarao.K[3]

1 M.Tech Scholar, Department of Mechanical Engineering, GMRIT, Rajam, Srikakulam - 532127, Andhra Pradesh, India.
2 Professor, Principal, Department of mechanical Engineering, GMRIT, Rajam, Srikakulam - 532127, Andhra Pradesh, India.
3 Assistant Professor, Department of Mechanical Engineering, GMRIT, Rajam, Srikakulam - 532127, Andhra Pradesh, India.

E-Mail: gayathriprasad97@gmail.com, prasad.clvrsv@gmrgroup.in, santarao.k@gmrit.edu.in

Abstract. In the recent past, cooling via thermoelectric technology has arisen to be as one of the capable cooling technology which helps in lowering the energy intake in a device. In this study, the performance enhancement of cooling system for the computer processor using thermoelectric cooler is investigated. An open loop cooling system was taken into consideration, where the main parameters are focused on computer load conditions, temperature carried by nanofluid or output temperature of nanofluid at water block and heat flux. The thermoelectric cooler (TEC) made of Bismuth Telluride ($BiTe_3$) material for pn junctions is used with Delta $T_{max}$of 67^OC in this analysis to extract the heat from processor. The Aluminium Oxide (Al_2O_3) and copper oxide (CuO) ranging from 40nm - 50nm size dispersed in distilled water with various volume concentrations were used for dissipation of heat from the hot surface of the TEC. Liquid flow is varied from 0.1LPM - 0.9LPM. Results demonstrate that by maintaining the lower Delta T_{max} value, higher Coefficient of performance (COP) is achieved. There is a considerable reduction in operating temperature of processor when using the nanofluids. Both the nanofluids exhibited higher thermal conductivity on comparing with base fluids. The enhancement observed in output temperature is about 15^OC to 20^OC by using nanofluids. Higher the flowrate used higher the heat flux generated.

Keywords: Thermoelectric cooling, Thermal Conductivity, Heat Flux, Coefficient of Performance

1. Introduction

In modern world the computers are becoming a crucial technology in sectors such as business, finance, internet, education, entertainment etc., because of its speed and accuracy. The Central Processing Unit is responsible for data processing and often referred as computer processor or processor where the transistors are placed or packed inside. Advance in semiconductor technology allows one to make transistors which are small enough to pack hundreds of million of them onto a

IOP Conf. Series: Materials Science and Engineering **954** (2020) 012022 doi:10.1088/1757-899X/954/1/012022

single computer chip. While operation the chips release heat. This waste heat need to be removed or that will result in reducing the processor speed, clock speed and also lead to temporary malfunction or permanent failure. Among all the cooling technologies that are being developed, thermoelectric cooling seems to be achieving all the possibilities of reaching its demand. Kai-lun Zhang *et al.*. [1] this paper proposes a micro-scale thermosiphon thermal tube technology which utilizes a 3D chip specific microchannel structure as a vaporising segment. The maximum heat flux is improved by 50-130% for nanofluid(CuO) and critical heat flux improved by 45-90% at 1.0 wt%. M. Rafati *et al.*. [2] the main focus of this research is to utilize the enhanced properties of nanofluids(Al_2O_3, SiO_2 and TiO_2) for the cooling of computer microchips. The largest reduction of operating temperature is observed for alumina nanofluid from 49.4 to 43.9 °C for 1.0 vol% at a flowrate of 1.0 LPM. M. K. Russel *et al.*. [3] the paper investigates thermoelectric cooler (TEC) performance for the design of electronic packing. The results show that when the system operated at higher COP there is a trade-offamong the extent of ambient temperatures and off peak heat fluxes. Mohamed S. El-Genk *et al.*. [4] investigated the performance of composite spreaders(porous graphite and copper substrate) with FC-72 dielectric liquid for enhanced cooling. Dongsheng Wen *et al.*. [5] investigated the heat transfer behaviour under nucleate pool boiling conditions using nanofluids(ɤ-alumina). Mohamed S. El-Genk [6] an experimentation on composite spreader, copper spreader and porous graphite spreader upon the 10x10 mm computer chip at 0^0, 90^0, 180^0 inclination with dielectric fluids(FC-72, HFE-7100). Paisarn Naphon *et al.*. [7] an experimental study on rectangular heat fin sink with or without the thermoelectric inside a CPU by using liquid cooling. Ping-yang Wang *et al.*. [8] experiments were carried out using two unadulteratedsolutions (deionized water, R113) and two supermoist liquid solutions (deionized water+ surfactant and R113+ surfactant) for a micro-channel structure. The results showed that the critical flux is improved by 70-100%. Guohui Zhou *et al.*. [9] analysed a 1mm thick ultrathin heat pipe miniature loop with ambient air and forced air convectional cooling for a flat evaporator used in high end ultra slim laptop.

Amir F. Ali *et al.*. [10] numerical investigation for composite spreaders of 10x10 mm comprised of copper substrate and copper microporous surfaces at different thickness with dielectric liquids(FC-72, HFE-7100, PF-5060). Songkran Wiriyasart *et al.*. [11] studied experimentally by using thermoelectric air cooling and its enhancement for dual processors workstation computer. Compared to conventional cooling system the thermoelectric air module there is decrease in temperature of 5-10°C. Nizar Ahammed *et al.*. [12] experimental investigation on the performance of TEC used for cooling of electronic devices with nanoparticle Al_2O_3 is blended in water for a multiport minichannel heat exchanger. It is observed that there is 40% increment in COP for TEC at 0.2 vol% concentration. Yang Cai *et al.*. [13] analysis of this paper reveal that by increasing number of TEC's on hot side of thermoelectric system there will be a enhancement in higher heat flux. Liu D *et al.*. [14] studied the proposed micro thermosyphon cooler which is coupled with TEC used in CPU to remove the unwanted heat from the processor. M. Chandrasekar *et al.*. [15] studied experimentally about the heat transfer coefficient and characteristics of friction factor for a laminar flow at fully developed condition by using Al_2O_3 nanoparticle dispersed in water. Majid Hejazian *et al.*. [16] numerically analysed the circular tube placed horizontally which exposed to convection and saturation steam at the wall with turbulent flow of Al_2O_3/water nanofluid. Hwa Ming Nieh *et al.*. [17] investigated the performance for dissipation of heat for a radiator which cooled by air with Al_2O_3 and TiO_2 nanofluids. M. M. Elias *et al.*. [18] experimented the enhancement of thermal properties such as thermal conductivity, density, specific heat and viscosity for car radiator coolant with Al_2O_3 dispersed in EG and water.

Based on the literature reviews, the liquid cooling system has higher heat absorbing capacity. Also, the improvement in thermal performance for liquid cooling components with reduction in their geometry was a technical challenge. Research towards TEC shows that it can maintain optimum temperature for the computer processor. The main objective is to enhance the TEC cooling system for the computer processor by replacing nanofluids instead of basefluids. The main parameters focused on heat carried away by the nanofluid and heat flux. The monitored results guide to the enhance of cooling performance for processor.

2. Preparation of Nanofluid

The preparation of nanofluid involves two steps. In first step addition of nanoparticles and second step involves the stabilization of nanofluid. The volume fraction for nanofluid varying the range from 0.1%-1.0% is calculated by using the equation,

$$\emptyset = \frac{\frac{m_{NP}}{\rho_{NP}}}{\frac{m_{NP}}{\rho_{NP}} + \frac{m_{BF}}{\rho_{BF}}} \quad (1)$$

The basefluid Deionised water is prepared by adding SDS(Sodium Dodecyl Sulfonate) inorder to maintain proper dispersion of nanoparticles. The solution is checked for pH scale inorder to avoid the damage for instruments. If the pH scale lies in acidic range then the solution is diluted by adding NaOH, 1M solution. The nanparticles of Al_2O_3 and CuO were considered for experimentation in the range of 40 nm-50 nm diameter. The TEM (Transmission Electron Microscopy) of Al_2O_3 and CuO nanoaprticle is shown in figure(1) and (2). The specifications of basefluid and nanoparticles are indicated in Table(1) and (2).

Table 1. Nanoparticles parameters

Properties	Al_2O_3	CuO
Average Size of Nanoparticle (nm)	43	50
Density(Kg/m^3)	3970	6510
Color	White	Black and Brown
Morphology	Spherical	Spherical
Melting Point (^{O}C)	2055	1201
Specific Heat (J/Kg-K)	765	540
Thermal Conductivity (W/m-K)	40	18

Table 2. Base fluid parameters

Properties at 30°C	Deionized Water
Density(Kg/m^3)	997
Specific Heat (J/Kg-K)	4178
Thermal Conductivity (W/m-K)	0.613
Viscosity (Kg/m-s)	0.000798

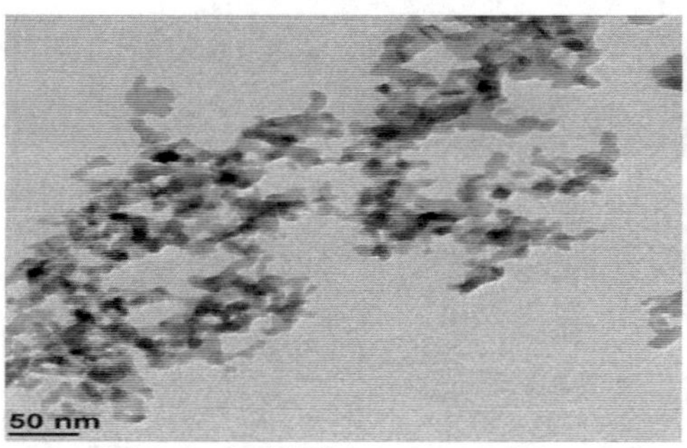

Figure 1. Al_2O_3 nanoparticle

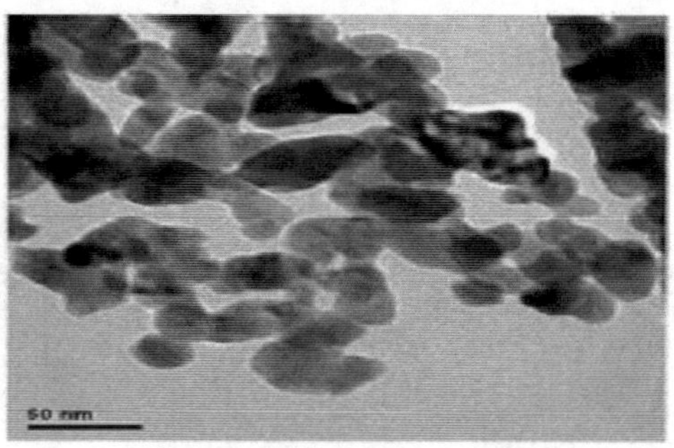

Figure 2. CuO nanoparticle

The fractions of 0.2%, 0.4%, 0.6% and 1.0% were added to the basefluid. The addition of nanoparticles is done by using homogenizer at 6000rpm for 30 minutes to avoid agglomeration. For the proper suspension of nanoparticles and its constant dispersion in the fluid is carriedout under ultrasonification at 600rpm for 6 hours. The ultasonicator used here with specifications of 400W power and frequency of 36±3 KHz.. The prepared nanofluid is kept under observation for sedimentation upto 72 hours in stable condition.

Figure 3. Ultrasonication

2.1 Thermophysical Properties of Nanofluids

The thermophysical properties of nanofluids such as Thermal Conductivity, Specific Heat, Viscosity, Density are calculated by using the correlations. [5] [11]

2.1.1 Thermal conductivity

The overall effective thermal conductivity is calculated by adding static and dynamic thermal conductivity. The static thermal conductivity is based on Maxwell Theory and dynamic thermal conductivity is based on the Brownian motion of nanoparticles.

$$k_{eff}=k_{static}+k_{Brownian} \tag{2}$$

for spherical nanoparticles static thermal conductivity is given by,

$$k_{static}=\mathrm{k_{BF}}\left[\frac{(k_{NP}+2k_{BF})-2\emptyset(k_{BF}-k_{NP})}{(k_{NP}+2k_{BF})+\emptyset(k_{BF}-k_{NP})}\right] \tag{3}$$

Brownian motion of nanoparticle is calculated by

$$k_{static}=5.10^{4}\beta\emptyset\rho_{BF}c_{p,BF}\sqrt{\frac{k_{B}T}{\rho_{NP}D_{NP}}}f(\mathrm{T},\emptyset) \tag{4}$$

where, ρ_{NP} and D_{NP} are the density and diameter of nanoparticles. k_B is Boltzmann constant($1.3807\mathrm{x}10^{-23}$ J/K) and f(T,ø) value is 1

$$\text{where,} \quad \beta = 0.0011(100\emptyset)^{-0.7272} \ \emptyset > 0.01 \tag{5}$$

2.1.2 Dynamic Viscosity

Viscosity of nanofluid is calculated by using the correlations

$$\mu_{eff}=\mu_{\mathrm{static}}+\mu_{\mathrm{Brownian}} \tag{6}$$

$$\text{where, } \mu_{\mathrm{static}}=\mu_{\mathrm{BF}}\frac{1}{(1-\emptyset)^{2.5}} \text{ and } \mu_{\mathrm{Brownian}}=\frac{k_{Brownian}}{k_{BF}}\frac{\mu_{BF}}{Pr_{BF}} \tag{7}$$

2.1.3 Specific Heat

The Specific Heat of nanofluid for different volume concentrations is calculated by following equation,

$$c_{pNF} = \frac{(1-\emptyset)(\rho c_p)_{BF} + \emptyset(\rho c_p)_{NP}}{(1-\emptyset)\rho_{BF} + \emptyset\rho_{NP}} \quad (8)$$

2.1.4 Density

The nanofluid density for volume concentration 0.1%-1.0% is calculated by below equation,

$$\rho_{NF} = (1-\emptyset)\rho_{BF} + \emptyset\rho_{NP} \quad (9)$$

where, ρ_{BF} and ρ_{NP} are the densities of basefluid and nanofluid.

2.1.5 COP of TEC

The efficiency of the cooling or heating device is calculated by using Coefficient of Performance (COP), which is defined as the ratio of the cooling power to the input power.

$$\text{COP} = \frac{Q_c}{W}$$

Where, $W = Q_h - Q_c$ and $W = SI(T_h - T_c) + I^2R$ (10)

$Qc = -[SIT_c - 0.5I^2R - k(T_h - T_c)]$ and $Qh = SIT_h + 0.5I^2R - k(T_h - T_c)$ (11)

3. Experimental Apparatus and Method

3.1 Experimental Apparatus

The experimental apparatus consists of high performance computer chip of current generation. During experiment the processor is placed with the heater based on the temperatures of computer processor during different load conditions. The heater used in this experiment has specifications of 220AC/DC input voltage with 200W power supply and the surface maximum temperature of 200°C. Thermoelectric cooler is placed above the heater in such a way that the cold surface of TEC is placed upon heater and hot side of TEC attached to the water block. TEC model used here is TEC-12706, which consists of 127 p-n junctions. The material used for p-n junctions are Bismuth-Tin(BiSn). The details of TEC-12706 is given in below table.

Table 3. TEC Parameters

HOTSIDE TEMPERATURE	25°C	50°C
Q_{max}(W)	50	57
Delta T_{max}(^{O}C)	66	75
I_{max}(Amps)	6.4	6.4
V_{max}(Volts)	14.4	16.4
Module Resistance(Ohms)	1.98	2.30

The TEC has dimensions of 40mmX40mmX3.8mm. The top surfaces of TEC are made up of ceramic material Alumina(Al_2O_3) and it has a maximum operating temperature of 138°C. The voltage and current for the TEC is controlled by TEC controller. Water block is placed above the TEC i.e, hot side of TEC with dimensions, 40mm X 40mm X 12mm as shown in figure 1. The water block is made up of aluminium material. The TEC, heater and waterblock are made in contact with each other by surfaces using the Thermoelectric paste inorder to have the maximum heat transfer between the surfaces. The thermoelectric paste used here has a thermal conductivity of 1.85W/m-K. The inlet and outlet connections of water block is connected to the pump and radiator respectively.

A Diaphragm pump of 12V, 80W is used in this current experiment with a capacity of 5LPM. A flow meter is used in between pump and inlet such that required flow is maintained. The main advantage to use diaphragm pump is it has less moving parts.

A Storage Tank Capacity of 2-3 Litres were used. The inlet pipe of pump is directly connected to the storage tank. The storage is filled with nanofluid in required amount as specified above. The Outlet of water block is connected to the radiator. The radiator used in this setup has dimensions of 60mmX82.5mmX25.7mm. The hot fluid enters into the radiator from waterblock and it is cooled by air cooling as the liquid flows inside the radiator. A CPU fan is fixed above the radiator for faster cooling purpose. A CPU fan with rated power of DC 12V and 0.1A and dimensions of 60mmX60mmX15mm is utilized.

Thermocouples are placed at radiator outlet, inlet and outlet of waterblock. Thermocouples are connected to Digital Temperature indicator. A digital temperature indicator with multichannel is used to monitor the readings.

3.2 Experiment Method:

The experiment is conducted for three different temperature that were obtained during load conditions of processor i.e, idle condition, half load and full load conditions. These different load conditions are also tested for flow rates ranging from 0.1LPM-0.9LPM. The setup is also tested for different nanofluid concentrations upto 1.0vol%.

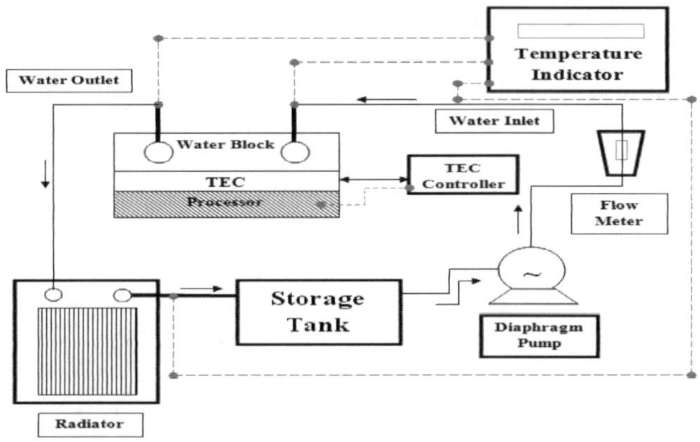

Figure 4. Block Diagram

Initially the storage tank is filled with nanofluid and checked for leakages. The connections are made according to the circuit. Now, the heater is supplied with power, so that the required temperature is attained on surface of heater by using voltage regulator. Now, the power is supplied to the TEC and it starts creating the cold and hot side surfaces(Peltier effect). The TEC controller is used to control the TEC for maintaining constant Delta T_{max} and temperature on cold side surface. Here, a constant delta T_{max} is considered inorder to reduce the power consumption and maintain higher COP of TEC. The pump is switched on and the flowrate is maintained or varied for different experiments. As the nanofluid flows inside the waterblock the maximum amount of heat is absorbed basing on the properties of nanofluid. The output temperature of nanofluid is measured and displayed on digital temperature indicator. A heat flux sensor is used to measure the heatflux at waterblock surface. Now, the hot fluid enters into the radiator and cooled. The output temperature of radiator is also recorded. The experiment is conducted for 300 seconds and values are noted.

4. Results and Discussion

4.1 Effect of computer operating load conditions:

The experiment was conducted at room temperature that is the inlet fluid is maintained at $30^{O}C$. In present study, the cold side of TEC is attached to the processor and the hot side of TEC is attached to the waterblock(heat exchanger). The setup is tested for three load conditions such as Idle, Half and Full load conditions. The processor temperature is removed by TEC when power is supplied to it. The maximum amount of heat is present on hot side of TEC since the processor is being cooled by the cold surface of TEC, this heat is removed by circulating the nanofluid inside the waterblock which is present on the hot surface of TEC. The experiment is tested for different nanofluid concentrations inorder to carryout maximum heat.

IOP Conf. Series: Materials Science and Engineering **954** (2020) 012022 doi:10.1088/1757-899X/954/1/012022

4.1.1 Output Temperature(^{O}C):

The Al_2O_3 and CuO nanofluids are tested for different load conditions. The variation in output temperature with respective both nanofluids are as depicted in the figures 5-10. The maximum amount of heat is carried out at 0.1LPM when compared to the other flowrates. This is because at lower flowrate the volume of liquid circulated is less compared to higher flowrate which means the fluid particles flow through the component is less compared to higher fluid volume circulation. Also, the vast amount of heat carried by Al_2O_3 nanofluid compared to the CuO nanofluid and basefluid. The 1.0 Vol% concentration of nanofluids show the maximum heat transfer among the other concentrations. But, for copper the heat carried away by the different volume concentrations at different flowrates is almost same (with a difference upto 1-2 ^{O}C) which is lesser when compared to Al_2O_3 nanofluid.

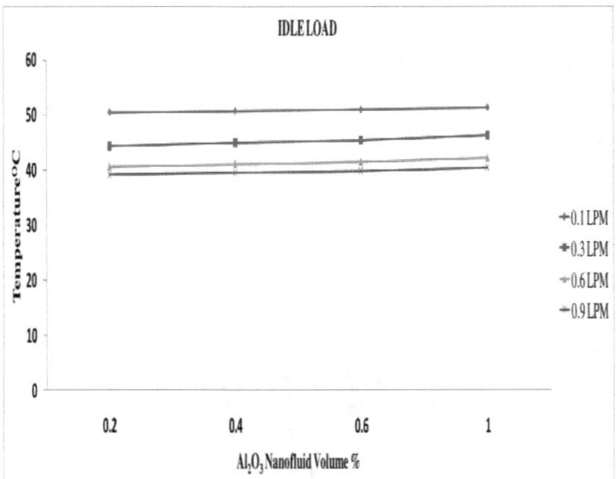

Figure 5. Al_2O_3 NF at idle load

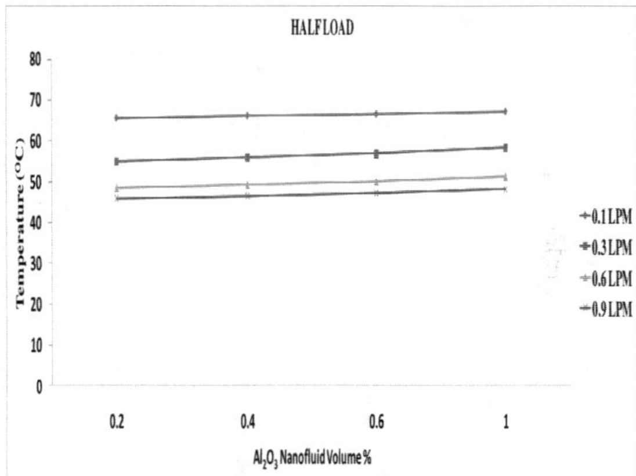

Figure 6. Al_2O_3 NF at half load

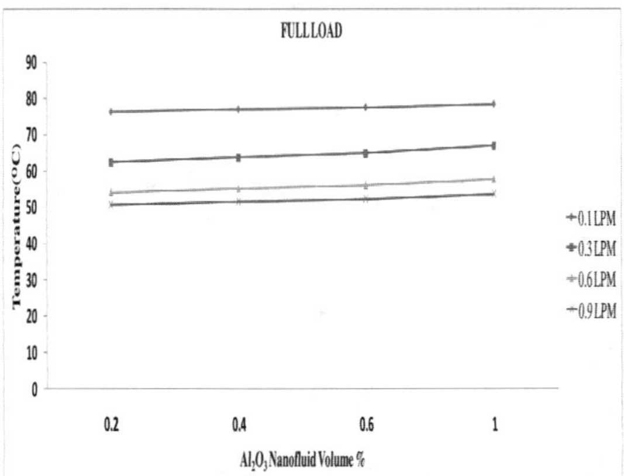

Figure 7. Al_2O_3 NF at full load

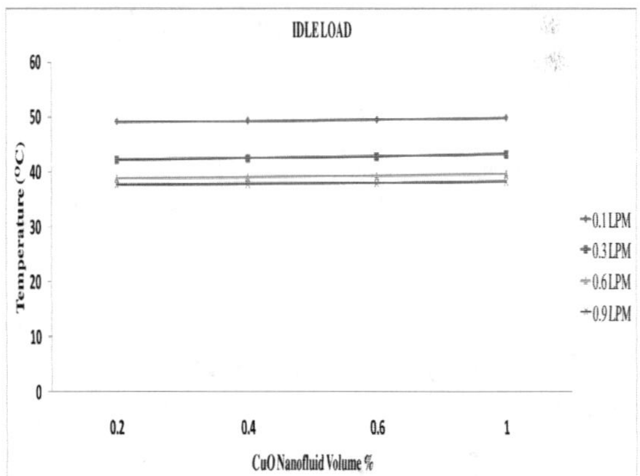

Figure 8. CuO NF at idle load

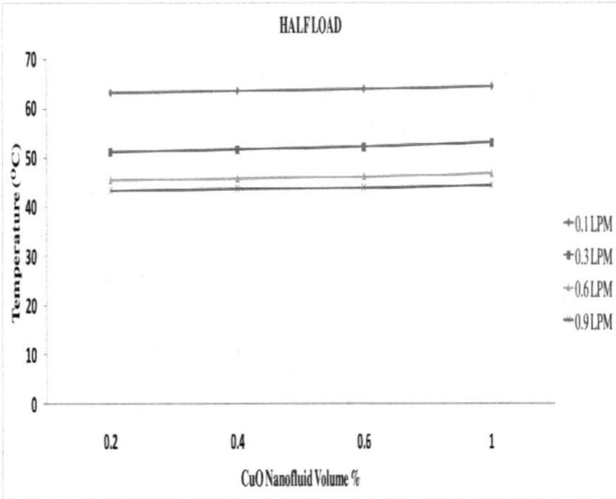

Figure 9. CuO NF at half load

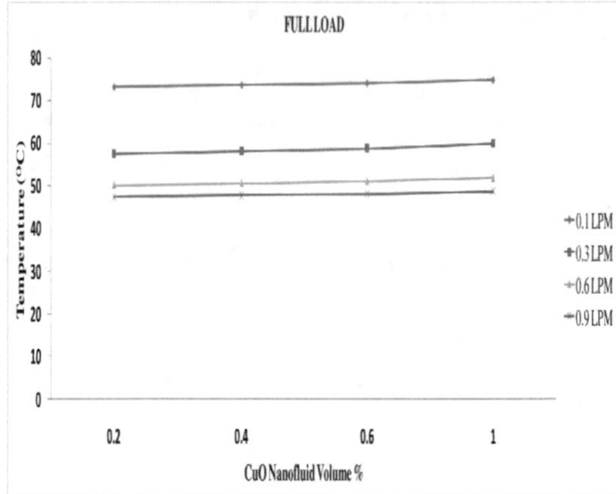

Figure 10. CuO NF at full load

4.1.2 Heat Flux(W/m^2):

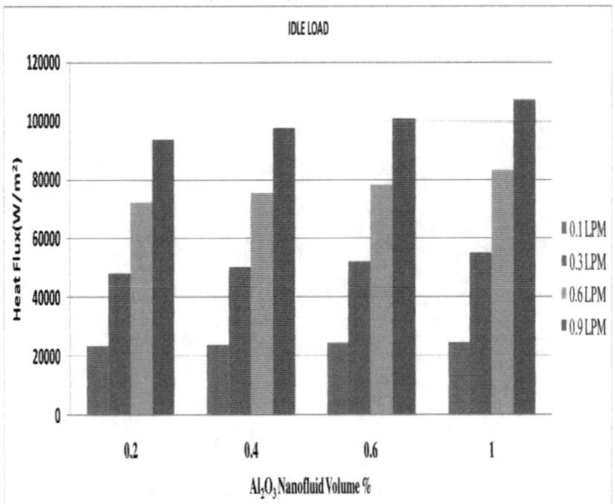

Figure 11. Al_2O_3 NF at idle load

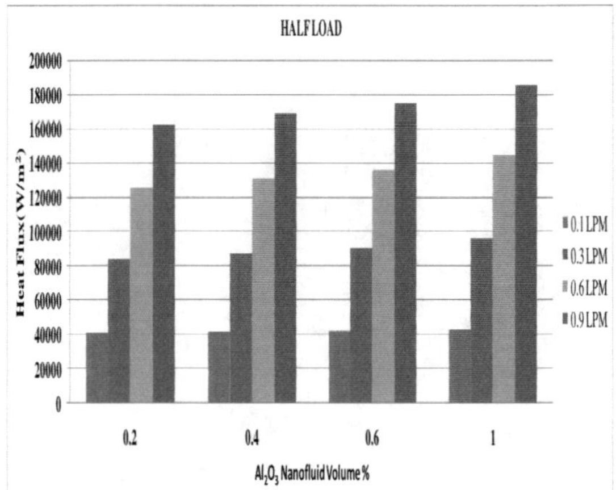

Figure 12. Al_2O_3 NF at half load

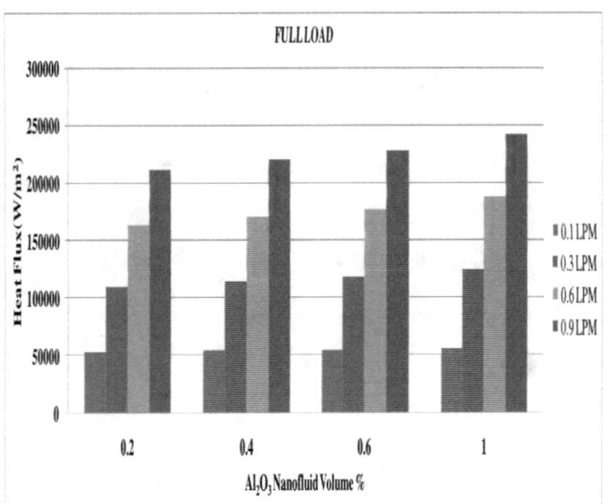

Figure 13. Al_2O_3 NF at full load

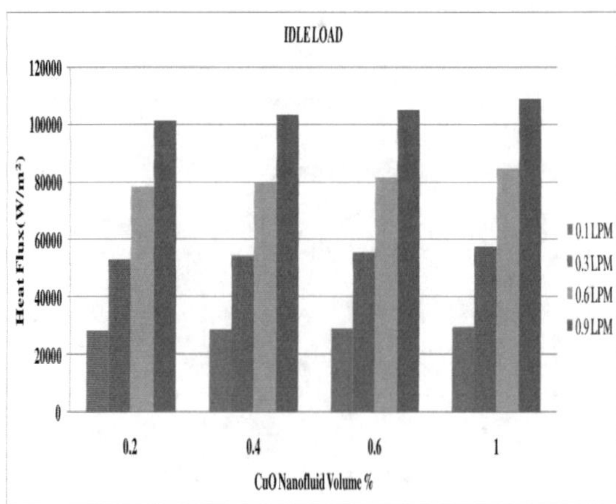

Figure 14. CuO NF at idle load

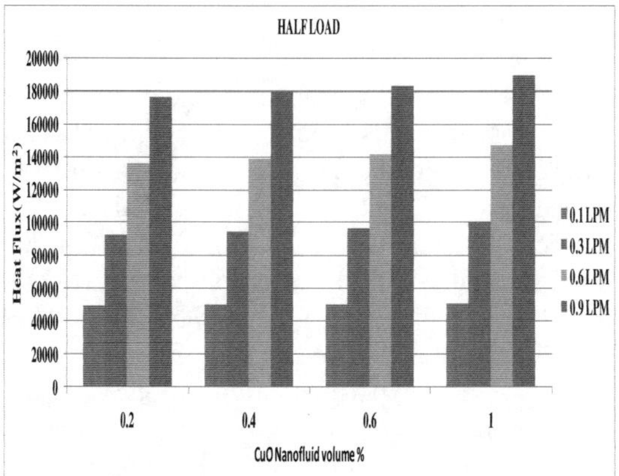

Figure 15. CuO NF at half load

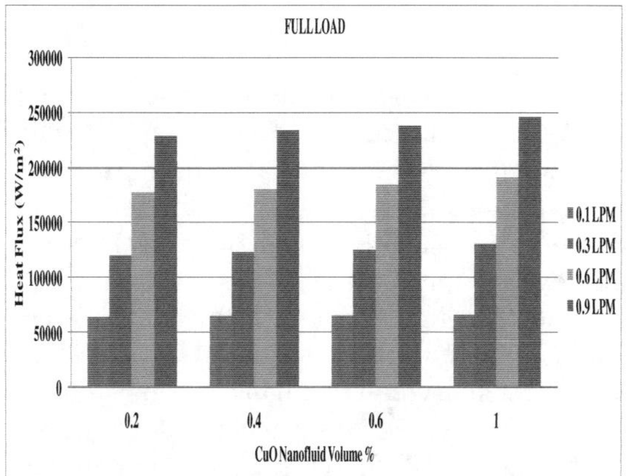

Figure 16. CuO NF at full load

At idle, half and full load condition, from experiment observed that, there is a decrease in temperature as the flow rate increases. Also, observed there is rise in heatflux as flowrate increases. At idle load condition, compared to basefluid the enhancement was observed from 42°C to 50°C and 49°C for Al_2O_3 and CuO nanofluids at 0.2Vol% for 0.1LPM respectively. At half load condition, compared to basefluid the enhancement was observed from 51°C to 65°C and 63°C for Al_2O_3 and CuO nanofluids at 0.2Vol% for 0.1LPM respectively. At full load condition, compared to basefluid the enhancement was observed from 58°C to 76°C and 73°C for Al_2O_3 and CuO nanofluids at 0.2Vol% for 0.1LPM respectively. At idle load condition, compared to basefluid the enhancement was observed from 42°C to 51°C and 49°C for Al_2O_3 and CuO nanofluids at 1.0Vol% for 0.1LPM respectively. At half load condition, compared to basefluid the enhancement was observed from 51°C to 67°C and 64°C for Al_2O_3 and CuO nanofluids at 1.0Vol% for 0.1LPM respectively. At full load condition, compared to basefluid the enhancement was observed from 58°C to 78°C and 74°C for Al_2O_3 and CuO nanofluids at 0.2Vol% for 0.1LPM respectively.

This phenomenon is observed because as the concentration of nanoparticle is increased in the base fluid the thermal properties such as thermal conductivity, specific heat etc., will increase. Hence the fluid tends to absorb more amount of heat. The higher heatflux is observed for full load condition at 0.9LPM is 242053 W/m^2 and 246664 W/m^2 for Al_2O_3 and CuO nanofluids respectively. The CuO nanofluid has higher heat flux in all load conditions and flowrates when compared to Al_2O_3 nanofluid. This is because the specific heat is higher for CuO nanofluid.

4.2 Performance of TEC:

Delta T_{max} is the temperature difference between the cold side and hot side plate. From the data sheet of the model used in this experiment TEC1-12706, for 25°C and 50°C on hot side plate of TEC it can create a temperature difference of 66°C and 75°C. Nizar Ahammed et.al [12] observed that for Al2O3 nanofluid 0.2Vol% there is 40% enhancement of COP of TEC. The maximum temperature difference can be attained by supplied maximum current and voltage. But, increase in Delta T_{max} value results in decrease in COP of TEC. In present experiment a TEC controller is used to maintain the required temperature difference by regulating the power supply to the TEC. In this problem the Delta Tmax for TEC is varied from 12°C-22°C. The COP value at idle, half and full load conditions are 0.718, 0808 and 0.844 respectively. Hence, we can conclude from experiment that for same current value at lower Delta T_{max}, maximum COP is attained for TEC.

5. Conclusion

In this work, a open loop thermoelectric liquid cooling system with nanofluid is developed for computer processor cooling. The TEC is characterized by considering its input power and coefficient of performance response. The experiment is performed for different load conditions, flowrates and two nanofluids(Al_2O_3 and CuO) with different volume concentrations. The results show that thermal conductivity of nanofluid increases with increase in particle concentration. From the data it can be concluded that higher heat flux is achieved for CuO nanofluid at 0.9LPM when compared to Al_2O_3 nanofluid. The heat carried away by the Al_2O_3 coolant is 2-5°C higher than CuO coolant. The higher coefficient of performance of 0.84 was achieved at Delta T_{max} of 22°C. The thermal effectiveness of cooling system increases with increase in volume concentration. The experimentation is done by using TEC1-12706 thermoelectric module with nanofluid for a computer processor cooling, the further analysis can be done on less power consuming thermoelectric modules and also the cooling technology in a compact model.

6. References

[1] Zhang K L, Liu Z H and Zheng B C 2016 *Energy Conversion and Management* **128** 44-56.
[2] Rafati M, Hamidi A A and Niaser M S 2012 *Applied Thermal Engineering* **45** 9-14.
[3] Russel M K, Ewing D and Ching C Y 2013 *Applied thermal engineering* **50** (1) 652-659.
[4] El-Genk M S, Saber H H and Parker J L 2007 *Applied thermal engineering* **27** (5-6) 1072-1088.
[5] Wen D and Ding Y 2005 *Journal of Nanoparticle Research* **7** (2-3) 265-274.
[6] El-Genk M S 2012 *Energy conversion and management* **53** (1) 205-218.
[7] Naphon P, and Wiriyasart S 2009 *International Communications in Heat and Mass Transfer* **36** (2) 166-171.
[8] Wang P Y, Li S F and Liu Z H 2017 *International Journal of Heat and Mass Transfer* **115** 479-487.
[9] Zhou G, Li J and Jia Z 2019 *Applied Energy* **239** 859-875.
[10] Ali A F and El-Genk M S 2012 *Energy Conversion and Management* **53** (1) 259-267.
[11] Wiriyasart S, Hommalee C and Naphon P 2019 *Case Studies in Thermal Engineering* **14** 100445.
[12] Ahammed N, Asirvatham L G and Wongwises S 2016 *Experimental Thermal and Fluid Science* **74** 81-90.
[13] Cai Y, Wang Y, Liu D, and Zhao F Y 2019 *Applied Thermal Engineering* **148** 238-255.
[14] Liu D, Zhao F Y, Yang H X, and Tang G F 2015 *Energy* **83** 29-36.
[15] Chandrasekar M, Suresh S, and Bose A C 2010 *Experimental Thermal and Fluid Science* **34** (2) 122-130.
[16] Hejazian M, Moraveji M K, and Beheshti A 2014 *International Communications in Heat and Mass Transfer* **52** 152-158.
[17] Nieh H M, Teng T P and Yu C C 2014 *International Journal of Thermal Sciences* **77** 252-261.
[18] Elias M M, Mahbubul I M, Saidur R, Sohel M R, Shahrul I M, Khaleduzzaman S S and Sadeghipour S 2014 *International Communications in Heat and Mass Transfer* **54** 48-53.
[19] Peyghambarzadeh S M, Hashemabadi S H, Hoseini S M and Jamnani M S 2011 *International communications in heat and mass transfer* **38** (9) 1283-1290.
[20] Sundar L S, Farooky M H, Sarada S N and Singh M K 2013 *International Communications in Heat and Mass Transfer* **41** 41-46.
[21] Yang X F and Liu Z H 2012 *International journal of heat and mass transfer* **55** (25-26) 7375-7384.
[22] Goudarzi K and Jamali H 2017 *Applied Thermal Engineering* **118** 510-517.

[23] Peyghambarzadeh S M, Hashemabadi S H, Jamnani M S and Hoseini S M 2011 *Applied thermal engineering* **31** (10) 1833-1838.
[24] Huminic G and Huminic A 2013 *International communications in Heat and Mass transfer* **44** 52-57.
[25] Vajjha R S, Das D K and Namburu P K 2010 *International Journal of Heat and fluid flow* **31** (4) 613-621.
[26] Naraki M, Peyghambarzadeh S M, Hashemabadi S H and Vermahmoudi Y 2013 *International Journal of Thermal Sciences* **66** 82-90.
[27] Leong K Y, Saidur R, Kazi S N and Mamun A H 2010 *Applied Thermal Engineering* **30** (17-18) 2685-2692.
[28] Tijani A S and bin Sudirman A S 2018 *International Journal of Heat and Mass Transfer* **118** 48-57.
[29] Narrein K, Sivasankaran S and Ganesan P 2015 *Numerical Heat Transfer, Part A: Applications* **68** (11) 1266-1279.
[30] Kole M and Dey T K 2010 *Experimental Thermal and Fluid Science* **34** (6) 677-683.

IOP Conf. Series: Materials Science and Engineering **954** (2020) 012023 doi:10.1088/1757-899X/954/1/012023

Investigation of mechanical properties on nano cuprous oxide coated/uncoated spur gear

K.Kumar[1*] and S.Arul[2]
[1]Department of Mechanical Engineering, Sri Chandrasekharendra Saraswathi Viswa Mahavidyalaya, Enathur, Kanchipuram – 631561, Tamilnadu, India.
[2] Department of mechanical Engineering, Annamacharya Institute of Technology and Science, Tirupati, Andhra Pradesh, India
[1*]Corresponding Author: kummarr@gmail.com

Abstract. Metals and their alloys have been widely used in all aspects of science and engineering. However the science of nanotechnology is driving newer demands and requirements for better performance of existing materials and also at a higher level of precision. This is naturally presenting complicated demands on the surface of these metals with a need for surface modification. In the present scenario it is noteworthy that nano scaled particle coatings enhances the mechanical properties in the metals rather than the macro or micro sized particles. Hence an attempt was made to characterize the mechanical properties of cuprous oxide nano particles coated spur gear (Mild Steel). The mechanical properties such as tensile, hardness and corrosion (salt spray) were experimentally investigated at ambient condition. A deep compression had been made between the coated and uncoated gear. The results reveal that the coated gear shows better performance compared with uncoated gears.

1. Introduction

The possibility of lightening mechanical components is becoming more and more important for the future, particularly in the automotive and aerospace fields, where the availability of more power with low gas polluting emission is fundamental. However, very often the fatigue behaviour, as well as the wear resistance and load bearing capacity of light alloys are poorer than the ones of steels. From this point of view, thin hard coatings deposited by means of PVD (physical vapor deposition) technique enables enhancement of hardness, wear, corrosion resistance and mechanical properties of mechanical components and are, for the time being, widely used in an increasing number of applications [1, 2]. A few decades ago, the cutting and machining tool field was the first to be interested in the improvements of the tribological properties achieved with these coatings, and from then on, their development has interested many other industries such as aerospace, automotive and biomedical ones.

Research studies have also demonstrated that the mechanical and corrosion properties of coated components can be increased by some coatings [3–10]. In fact, the residual compressive stresses induced at bulk material surface by some deposition processes prove beneficial to the fatigue resistance, provided that the coating is uncracked and free of defects. Therefore, coated light alloys with similar or even better performances than the ones of traditional construction steels would be really important for structural and competition applications (11). In the present research work a spur gear mad of mild steel is considered. Further the Cuprous oxide nano particles were synthesized through bio synthesis precipitation method and it is coated on the gear with high temperature and normal temperature. The coated gears were subjected to mechanical and corrosion examination with certain parameters respectively.

2. Experimental work

2.1. Extraction of cuprous oxide nano particles

The Cuprous Oxide (Cu_2O) nanoparticles is synthesized with green synthesis method The synthesis process uses copper as precursor and ammonium chloride with lactobacillus bulgaricus as reducing agent in presence of datura metal leaf extract as bio-surfactant in an aqueous medium. Initially, 1000 ml of Lactobacillus bulgaricus is added with 100 gm of Ammonium chloride followed by continuous rapid stirring to prepare a solution. Next 250 gm of melted copper is mixed with the prepared solution. Thus upon mixing the solution changes its colour to yellow which indicates the reaction between the copper and the prepared solution came to an end. Now the copper can be separated from the solution and allowed to settle overnight and the supernatant solution was then discarded cautiously. Secondly melt the copper which is obtained from the above step and mix it with a solution made of Datura Metal Leaf liquid. Wash it with de-ionized water and ethanol for three times to take out the excessive starch bound with the Nano-particles. Ocher color precipitates Obtained are dried at room temperature and is shown in Figure 1. After drying, Nano-particles were stored in glass vial.

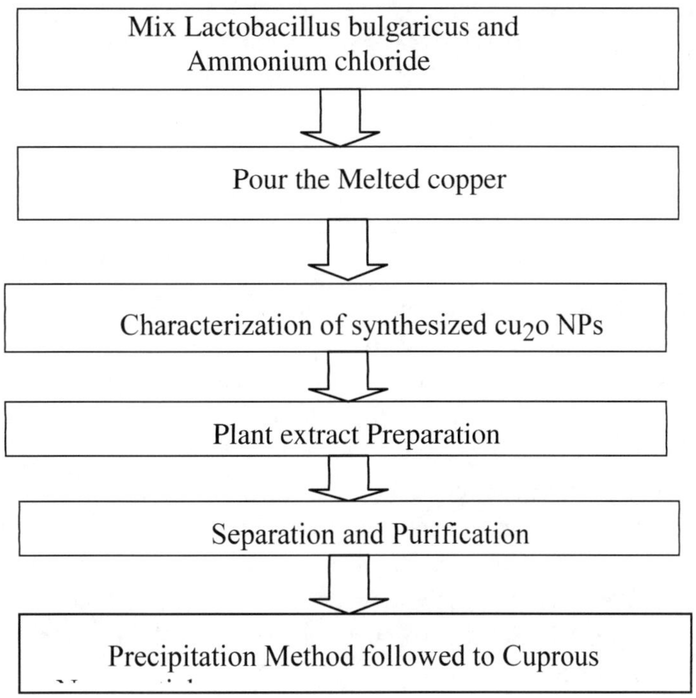

Figure 1. Steps for Cu_2O nano particle extraction

2.2 Coating of cuprous oxide nano particles on spur gear

In the present research work the green synthesized cuprous oxide nanoparticles had been coated on the spur gear substrate made of mild steel. The thin film coating is achieved by means of physical vapor deposition process namely sputtering. The sputtering method of thin film deposition involves introducing a controlled gas, usually chemically inert argon, into a vacuum chamber, and electrically energizing a cathode (Cu2O nanoparticles) to establish a self-sustaining plasma. The exposed surface of the cathode, called the target, is a slab of the material to be coated onto the substrates. The gas atoms lose electrons inside the plasma to become positively charged ions, which are then accelerated into the target and strike with sufficient kinetic energy to dislodge atoms or molecules of the target material. It can be thought of as a sort of atomic scale bead blasting. This sputtered material now constitutes a vapour stream, which traverses the chamber and hits the substrate, sticking to it as a coating or "thin film".

2.2. Characterization of coated spur gear

2.2.1. Tensile behaviour

Tensile strength and percentage elongation at break measurements are among the most important indications of the strength of the material and are the most widely specified properties of the materials. The tensile property data are more useful in preferential selection of a particular type of plastics from a large group of plastic materials. Tensile properties such as tensile strength and percentage elongation at break were tested using microprocessor controlled universal testing machine (UTM), with an accuracy of ± 2 MPa. The specimens were prepared as per ASTM standard. A minimum of five samples were tested at room temperature for each composition and an average value was taken.

2.2.2 Tensile strength

The ultimate tensile strength (UTS) is the maximum stress a material can withstand while breaking. The tensile strength of a coated spur gear depends on the bonding between the coating and the substrate. The function of the substrate is to transfer the stresses to the load bearing coatings.

Tensile strength = Load at break / original cross-sectional area

= P / b x d

Where, P is the load applied in N, b is the width in mm and d is the thickness in mm

2.2.3 Elongation at break

Ductility is a qualitative, subjective property of a material and the measurement of ductility is of interest to indicate to the designer, in a general way, the ability of the metal to flow plastically before fracture. The conservative measures of ductility that are obtained from the tensile test are the engineering strain at fracture (typically called elongation) and the reduction of region at break. In this examination, the elongation of the specimen at break has been computed by the accompanying expression (Budinski1998).

Elongation at break = L_f x 100 /L_o

Where L_f is the final gauge length after failure and L_o is the initial gauge length before loading (here the gauge length has been maintained as 50 mm).

IOP Conf. Series: Materials Science and Engineering **954** (2020) 012023 doi:10.1088/1757-899X/954/1/012023

2.2.4. Vickers Microhardness Test (VMH)

Three cylindrical specimens (5mm in diameter × 2 mm in length) were prepared from coated/uncoated gears. Each specimen was polymerized for 20 seconds. Vickers microhardness test (Shimadzu HMV; Shimadzu Corporation, Tokyo, Japan), was performed at the top (depth=0 mm) and bottom (depth=2 mm) surfaces of each specimen (three indentations for each specimen), using a 50-g load for 15 seconds.

3. Result and discussion

3.1 Tensile behaviour of coated and uncoated gear

3.1.1 Tensile properties of Non-coated and coated gear tooth

The variation in the tensile strength of the cuprous oxide coated and uncoated spur gear is displayed in Table 1. The maximum tensile strength obtained was 555.16 MPa. Theoretically, the maximum coating strength of the substrate is gained by incorporating a high weight fraction of coating material. Thus there is an efficient stress-transfer mechanism at the coating-substrate interface occurs. Strong interactions allow more efficient load transfer and, hence, better mechanical performance. In comparison to the other metals, mild steel has the unusual ability to accept hard coatings. Besides that it might be due to the homogenous dispersion of nano cuprous oxide coating on metal substrate matrix the interfacial adhesion between Cu2O and mild steel increased which results in the enhancement in the tensile properties of resultant material.

Table 1. Tensile Test Results

Sl.No.	Property	Coated	Non-Coated
1	Tensile Strength (MPa)	555.16	510.17
2	Yield stress (MPa)	389.55	371.77
3	Elongation(%) in Gauge length of 50mm	39.54	35.86
4	Reduction (%) area	55.09	61.25

3.1.2 Elongation at break

Elongation at break (E_b) of the mild steel (Spur gear) substrate coated by nano cuprous oxide is shown in table. 1 The addition of Cu_2O coating increases the tensile elongation at break from 35.86% to 39.54%. This is due to the interference of coating in the mobility or deformability of the substrate. This interference is created through the physical interactions and immobilization of the mild steel substrate by the presence of mechanical restraints thereby reducing the elongation at break [12].

3.2. Chemical composition

In the present research after the coating process had been finished the specimens are subjected to chemical composition test to evaluate the materials present in the specimen. This test is assisted with Optical Emission Spectroscopy (OES). Optical Emission Spectroscopy (OES) is found to be the most accurate and reliable technique for producing qualitative and quantitative analysis of materials composition. It works by exciting atoms with energy that comes from a spark formed between the sample and electrode. The material composition of coated and uncoated specimens where depicted in table 2.

Table 2. Chemical composition

Element	% weight	
	Without coating	**With coating**
Carbon	0.42	0.40
Manganese	0.63	0.63
Silicon	0.22	0.20
Sulphar	0.004	0.009
Phosphorous	0.008	0.010
Chromium	0.026	0.051
Nickel	0.021	0.014
Cobalt	0.004	0.004
Copper	0.034	0.002
Niobium	0.003	0.004
Titanium	0.004	0.003
Vanadium	0.009	0.008
Tungsten	0.01	0.02
Lead	0.007	0.007
Bismuth	0.016	0.019
Nitrogen	0.081	0.074
Molybdenum	-	0.005

3.3. Vickers micro hardness

The variation in the hardness of coated and uncoated spur gear is evaluated using Vickers micro hardness tester and it was displayed in table 3. From the table it is inferred that there is an increase in hardness when compared with uncoated spur gear. This improvement in hardness was attributed to the formation of adhesive properties on the substrate. The microhardness of the Cu_2O coated samples under were carefully observed and studied.. It is evident from the study that the microstructure evolved in coating depends on the processing parameters and hence such metallurgical parameter influence the grain size which is paramount to the buildup of surface hardness [13].

IOP Conf. Series: Materials Science and Engineering **954** (2020) 012023 doi:10.1088/1757-899X/954/1/012023

Table 3. Micro Vickers Hardness Test

Sample No	Non coated sample	Cu_2o Nano Coated
1	823	848
2	810	844
3	806	842

3.4 Corrosion Test

Corrosion test provides controlled corrosive environment and it has been used to produce relative corrosion information for specimens of metals. Salt spray Test is used to predict the corrosion. The Test Parameters are listed in the Table 4 and the results where depicted in table 5.

Table 4. Salt spray Test Parameters

Sl. No.	Parameter	Value
1	Concentration of Sodium chloride	5.4% NaCl
2	Chamber Temperature	34.3^0C -$34.7\ ^0$C
3	pH of Salt solution	7.1
4	Air Pressure	15 psi
5	Collection of Solution per Hour	1.4 ml
6	Method of cleaning of specimen before loading and after completion of Testing	a) Specimen cleaned gently prior to loading b) Specimen washed gently in clean running water to remove salt deposits from their surfaces and then dried immediately.
7	Required exposure period	24 Hrs
8	Exposure period	24 Hrs

Table 5. Corrosion Test Results

Sl. No.	Time	Observation
1	At 24 Hrs (Non Coated)	Red rust Observed
2	At 24 Hrs (Coated)	No Corrosion

IOP Conf. Series: Materials Science and Engineering **954** (2020) 012023 doi:10.1088/1757-899X/954/1/012023

4. Conclusions

In the present work, green synthesized cuprous oxide nanoparticles where coated on spur gear and subjected to mechanical and corrosion test. It is noted that the tensile strength increases in coated gear when compared with the uncoated gear. The elongation at break increases with the coating on spur gear. The Vickers hardness test show there is a gradual decrease in hardness compared to coated and uncoated gears.

The corrosion resistance of the coated spur gear is high compared with uncoated gear. It is seen from the specimen that for uncoated upon 24 hrs of salt spray a red rust is observed whereas in coated gear no corrosion is seen. It is concluded that cuprous oxide coated gears shows better properties compared with uncoated gears.

5. References

[1] Merlo AM 2003 *Surf Coat Tech* ***174*** 21–26
[2] Vetter J, Barbezat G, Crummenauer J and Avissar J 2005 *Surf Coat Tech* ***200*** 1962–1968
[3] Stewart S and Ahmed R 2002 *Wear* ***235*** 1132–1144
[4] Kim KR, Suh CM, Murakami RI and Chung CW 2003 *Surf Coat Tech* **171** 15–23
[5] Inoue K, Lyu S, Deng G and Kato M 1996 *Proc VDI Beric* ***1320*** 357–369
[6] Su YL, Yao SH, Wei CS, Kao WH and Wu CT 1999 *Mater Sci and Tech* ***15*** 73–77
[7] Baragetti S, La Vecchia GM and Terranova A 2003 *Int J Fatig* ***25*** 1229–1238
[8] Baragetti S, La Vecchia GM and Terranova A 2005 *Int J Fatig* ***27*** (10–12) 1541–1550
[9] Baragetti S 2007 *Int J Fatig* ***29*** 1893–1903
[10] Puchi-Cabrera ES, Staia MH, Lesage J, Gil L, Villalobos-Gutie ́rrez C, La Barbera-Sosa J, Ochoa-Pe ́rez EA and Le Bourhis E 2008 *Int J Fatig* ***30*** 1220–1230
[11] Calheiros FC, Kawano Y, Stansbury JW and Braga RR 2006 *Dent Mater* ***22*** 799–803
[12] Jiang ZY, Zhang Z and Friedrich K 2007 *Comp Sci and Tech* **67** 168-176
[13] Fayomi O S I and Popoola A P 2012 *Int J electrochem sci* ***7*** 6555- 6570

IOP Conf. Series: Materials Science and Engineering **954** (2020) 012024 doi:10.1088/1757-899X/954/1/012024

An Effect of EDM on AL6061-5% SiC as stir cast MMC

S.Bharani kumar [1&3],Dr.S.Arul[2]

1 Resarch scholar, SCSVMV University, Kanchipuram, Tamil Nadu, India.
2 Professor, Annamacharya Institute of Technology and Sciences, Tirupati, Andhra Pradesh, India
3Assistant professor, Rajalakshmi Institute of Technology, Kuthambakkam,Tamil Nadu, India.
[1*]Corresponding Author: sanbharani@gmail.com

Abstract. Usage of advanced machining method in the formation of composites made of aluminum metal matrix (Al-MMC) has produced substantial ability because the manufacture of intricate die shapes in these tough material is of elevated precision and surface finish Impossible.EDM has proven to be one of the many unorthodox processing techniques that are successful in shaping These materials which are difficult to process. The main aim of this investigation work is to find the outcome of Pulse ON-time (Pon) ,Current (Cu)and flushing pressure (Fp) on tool wear rate (TWR), metal removal rate (MRR) taper (T), radial overcut (ROC), and surface roughness (SR) on machining Al6061 with 5% SiC reinforcement.[1,2] On this Investigation we are deputed Electronica PSR 35 and dielectric fluid was used as kerosene with the help of flushing machine to pump out the dielectric fluid. Copper is used as the electrode and it also used as a tool of diameter 3mm to pierce the work piece for the required length. Three machining parameter at each having three stages was used to perform the experiment on the orthogonal array of L27 taken into account. The experiments are conducted for three following trails in random order. The maximizing taken into account for optimization and ANOVA statistical tool is used and to find out the surface morphology SEM is taken on different parts of the work piece.[3]

1. Introduction

Expeditious advances in materials expertise are currently being encountered concurrently with various shifts in performance and economic demands on new resources. A compound material is a comprehensive mixture of more than two unlike materials with a familiar line. Since composites are typically use of their properties like structural, however the term can be limited to simply these materials that include reinforcement supported by a combination of material. So merged materials usually have a stiffer and stronger broken fibre or unit phase than the permanent matrix stage. Metal Matrix Composites are other resources that are centered on the metal matrix, typically aluminium, which is reinforced by aluminium oxide, silicon carbide or boron carbide ceramic fibres, particles or whiskers[4]. Any diamond-plated grinding wheel or polycrystalline diamond cutting machines with carbide can be machined with them. With traditional machining, given the heavy tool wear and high tooling costs, the non-contact material removal process provides an acceptable alternative. Along with

IOP Conf. Series: Materials Science and Engineering **954** (2020) 012024 doi:10.1088/1757-899X/954/1/012024

the countless numbers of unorthodox processing methods for the composite materials, EDM has proved successful.This present study some of the findings of the ongoing new examination of 5 per cent EDM-related SiC reinforced Al6061-MMC. The intention of this effort is to scrutinize and report on the responses viz the impact of Cu, Pon and Fp on. MRR, TWR, T, ROC and SR for Al–5 per cent SiC electrical discharge machining

2.Experimental Procedure

The input parameters of this experiments are defined as three parameters and the stages are set as given in Table 1.and the output variables and their reaction are TWR,MRR,ROC,T & SR; symbol

	Current	Pulse on time	Flushing Pressure
Symbol	Cu	Pon	Fp
Units	A	μs	Lb in^{-2}
Stage 1	5	250	5
Stage 2	10	500	10
Stage 3	15	750	15

Table 1: Machining inputs and stages

i) Material Removal Rate is taken as the ratio of the weight dissimilarity of the workpiece to the machining time before and after machining,

$$MRR = \frac{Mrb-Mra}{t}$$

Mra = Workpiece weights before machining
Mrb = Workpiece weights after machining
t = Time of Machining

ii) Tool Wear Rate is is taken as the ratio of the ratio of the weight difference of the workpiece to the machining time before and after machining,

$$TWR = \frac{Twb-Twa}{t}$$

Twa = Tool weights before machining
Twb = Tool weights after machining
t = Machining time

iii) From the expression T can be computed:

$$\theta = \tan^{-1}\frac{db-dt}{2h}$$

db = Machined hole diameters at the bottom of the material
dt= Machined hole diameters at the top of the material
h = height of the material removed

iv) The ROC is expressed as half the diameter variation of the Holes created to the diameter of the tool

$$ROC = \tan^{-1}\frac{db-dti}{2h}$$

db = Machined hole diameters at the bottom of the material
dti= diameter of the tool

For a setting of three factor-three levels, sum of experiment to be performed shall be 27. An L27 OA, in this investigation was selected. With a few additional factors to be applied Studying with the same form of material, the decision was made to use Setup L27, which in turn reduces the number of subsequent experiments. Also the comparison That would be simpler outcomes[5,6]

IOP Conf. Series: Materials Science and Engineering **954** (2020) 012024 doi:10.1088/1757-899X/954/1/012024

Al-MMC was drilled with 5 per cent SiC using a 3 mm diameter copper tool. For the workpiece positive polarization was connected, and for the tool negative polarization was connected. Kerosene was used as dielectric fluid with the help of flushing The workpiece and tool material particularization are given below for know the material characteristics, work piece particulars for Al6061 are Al- 92.5%, SiC- 7.1% and Mg – 0.4 and the strengthening material of SiC is 5% and this percentage by volume and particle size is 20 micrometre. Tool materials particles for copper is copper 70% and Zinc 30%[7,8]investigational details

Trail no	Machine Settings		
	Current	Pulse on Time	Flushing Pressure
	Cu (A)	Pon (μs)	Fp (Lb in-2)
1	5	250	5
2	5	250	10
3	5	250	15
4	5	500	5
5	5	500	10
6	5	500	15
7	5	750	5
8	5	750	10
9	5	750	15
10	10	250	5
11	10	250	10
12	10	250	15
13	10	500	5
14	10	500	10
15	10	500	15
16	10	750	5
17	10	750	10
18	10	750	15
19	15	250	5
20	15	250	10
21	15	250	15
22	15	500	5
23	15	500	10
24	15	500	15
25	15	750	5
26	15	750	10
27	15	750	15

Table 2: Experimental outline

3.Results and discussion

3.1Effect of current(Cu) and pulse ON-time(Pon)

EDM's mach inability based on the electric supply of the material at work. Despite low power supply Reinforced SiC conductivity and high thermal resistance Particulate matter, which eventually reduces electrical conductivity The results obtained indicate that of the work material That Al–5 per cent SiCP can be efficiently machined with EDM.They output shows that superior concurrence with the explanation of a lot of researchers.

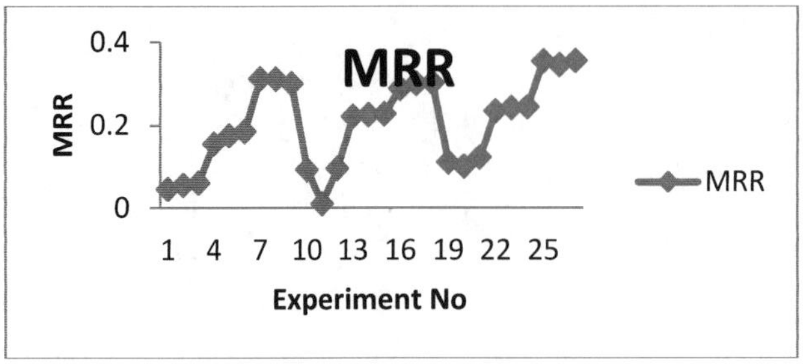

Fig 1: Variation of MRR

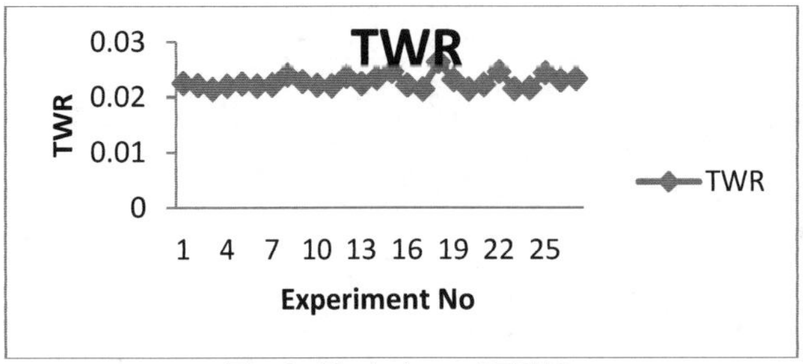

Fig 2: Variation of TWR

With current and pulse ON time, the material removal rate (Fig 1) began to increase.. It is also apparent that with current and pulse ON, the surface roughness is increased. The current flows high results in increased thermal output load to copper tool electrode followed by loading superior quantity of stuff detached from mutually copper tool and thus guide also to high Metal Removal Rate and Tool Wear Rate(Fig 2) [9].So this will leads to high yield volume and surface roughness would be rougher from earlier stages and in addition of greater pulse on duration results in large removal of stuff in larger hollow size and tends to higher end of roughness of the surface.

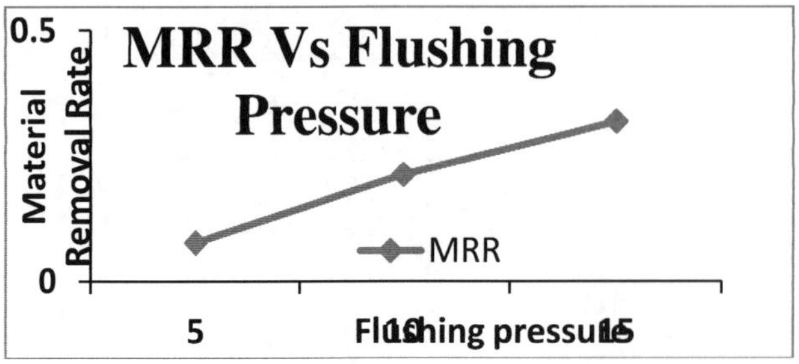

Fig 3: MRR Vs Flushing Pressure

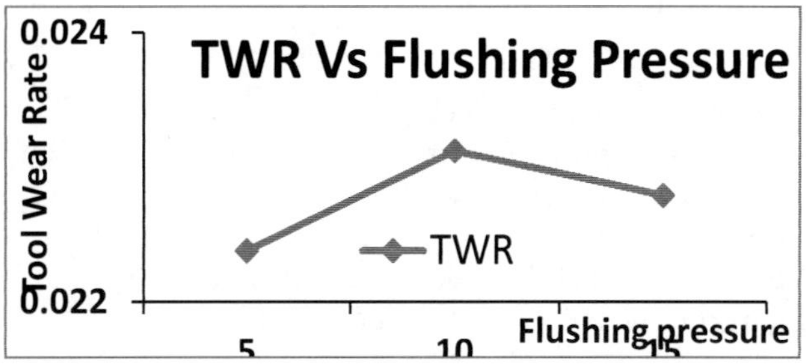

Fig 4:TWR Vs Flushing Pressure

At the beginning, due to the entanglement of SiC particles in the spark gap, the material removal rate was noted to be faster and slightly lower. This experience was seen in this practice to be higher at superior machine settings. Later, surface sparking was induce a TWR has also been found to rise with current and pulse on-time increases The high wear rate is due to the presence of the copper tool with a low melting alloy, zinc. (Fig 3 & Fig 4) However, with the copper tool, the machining stability is satisfactorily accomplished, as its high erosion rate makes zinc vapours in the plasma tube, which can decrease arc resistance and help speed up ionisation, so the decrease in dimensional stability was realistic. [10]

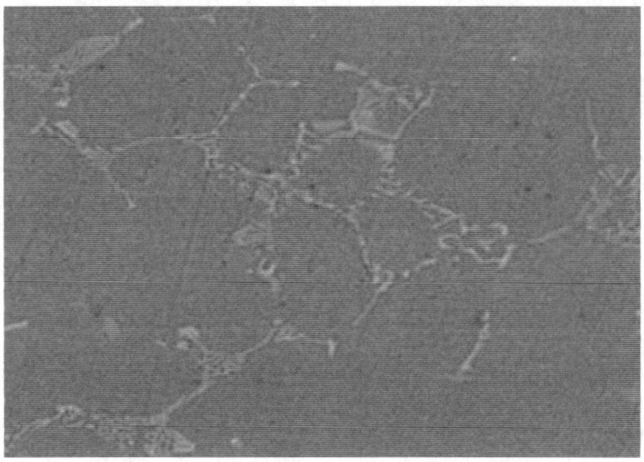

Fig 5: SEM image for 5% SiC- Al6061

In the observation of SEM (Fig 5) Photographs the cracks on the recast layer are formed and they are hardly ever go through beyond the layer of recast was observed damage on the surface may be taken into account at the end of the layer thickness of the recast region. It is noted that the SiC particle are smallest amount there on the recast area and that most of the particles might contain been racked up and distant in machining [11].Also, the bottom alloy surrounding the reinforcement changed into soften for the duration of EDM. It is more obtrusive at better scores and consequently the SiC debris may want to have been deposited below the recast level. In addition, random void are determined which might also additionally were because of missing of becoming a member of of the molten droplets or which might were due to the trapped chatter during resolidification

3.2Effect of flushing pressure

When the dielectric fluid is compelled into the spark hole at low velocity, the gathered debris cause short-circuiting to come to be much less pronounced . This will help enhance the output of work and hence increase Metal Removal Rate. High amount of flushing impedes the development of ionized link along the slit, resulting in a larger detonation setback and reduced release force and thus low level Metal Removal Rate. In addition, the tool 's cooling pace goes up with increased flushing pressure and consequently reduced tool wear is observed.

3.3 Experimental Research

Table 3 shows A description of ANOVA findings and the parameter levels viz. To be set to maximise the individual factors viz. Cu, Pon, and Fp. TWR, ROC, MRR and to SF.

Source of Variance	SS (sum of square)	d.f (degrees of freedom)	M.S (mean square)	F Value	Percentage Contribution
Current	0.922	2	0.462	18.472	72.72
Pulse on Time	0.256	2	0.131	4.637	18.92
Flushing Pressure	0.061	2	0.025	1.507	4.64
Error	0.053	2	0.024	1	3.72
Total	1.292	8			100

Table 3 – ANOVA Results

4. CONCLUSION

This experimental investigation evaluate the possibility of machining Al6061–5% SiCP Metal Matrix Composites

- For greater current and pulsed ON-time settings, the Metal Removal Rate was found to be greater at the higher rate of tapering, radial overcut and surface finish.
- Tool Wear Rate was also create to be high, superior than the Metal Removal Rate, for bigger current setting.
- The dimensional at elevated current flow and promptly ratings Precision is impacted.
- The flushing pressure used as Dielectric fluid have a significant result on both the Metal Removal Rate and Tool Wear Rate.

References

[1]. D.C. Montgomery, Design and Analysis of Experiments, fourth ed.,Wiley, New York, 1997.

[2]. M. Ramulu, M. Taya, EDM machinability of SiCW/Al composites,J. Mater. Sci. 24 (1989) 1103–1108.

[3]. S. Bharanikumar, S. Arul,Examination on surface roughness in EDM of aluminium 6061 reinforced with 5% SiC using design experiments, Applied Mechanics and Materials Vols. 813-814 (2015) pp 526-530

[4]. Jong Hyuk Jung, Won Tae Kwon, J. Mech. Sci Tech. 24 (5) (2010) 1083-1090.

[5]. T. Muthuramalingam, B. Mohan, Ind. J. Eng. Mat. Sci. 20 (2013) 471-475.

[6]. Vikas, Apurba Kumar Roy, Kaushik Kumar, Proc. Mat. Sci. 6 (2014) 383-390.

[7]. S. Bharanikumar, S. Arul ,influence of silicon carbide particle addition in the aluminium (al6061) composite on edm process parameters, Int. J. Chem. Sci.: 14(4), 2016, 3157-3166

[8]. V. Senthilkumar and B. U. Omprakash, Effect of Titanium Carbide Particle Addition in the Aluminium Composite on EDM Process Parameters, J. Manuf. Proc., 13(1), 60-66 (2011).

[9]. P. Pecas and E. Henriques, Electrical Discharge Machining using Simple and Powder- Mixed Dielectric: The Effect of the Electrode Area in the Surface Roughness and Topography, J. Mater. Proc. Tech., 200(1), 250-258 (2008).

[10]. T.V.Christy, 2010. A Comparative Study on the Microstructures and Mechanical Properties of Al 6061 Alloy and the MMC Al 6061/TiB2/12P, Journal of Minerals & Materials Characterization & Engineering, 9 (1) pp.57-65.

[11]. S. Bharanikumar, S. Arul , A Comparative study on MRR, TWR and Surface Roughness properties of machined surface by electric discharge machining (EDM) in Metal Matrix composite (MMC) of Al6061-SIC, ciencia e tecnica, Vol. 34 (n. 1, 2019).

IOP Conf. Series: Materials Science and Engineering **954** (2020) 012025 doi:10.1088/1757-899X/954/1/012025

Analysis of Thermal Barrier Coating's Behaviors on Alloys – A Review

[1]K. Muthukumaran, .[2]J. Francis Xavier, [3]R. Ramadoss, [4]P. Balu, [5*]C. Ezilarasan

[1]Assistant Professor, Department of Mechanical Engineering, Mailam Engineering College, Tindivanam, Tamil Nadu, India

[2]Professor, School of Mechanical Engineering, VIT Bhopal University, Bhopal, Madhya Pradesh, India-466114.

[3]Professor, Department of Mechanical Engineering, Easwari Engineering College, Chennai, India-600089.

[4]Associate Professor, Department of Mechanical Engineering, VSB Engineering College, Tamilnadu, India.

[*5]Professor, Department of Mechatronics, Chennai Institute of Technology, Chennai, Tamil Nadu, India-600122.

*Corresponding Author: ezhilshriram.c@gmail.com

Abstract: Thermal barrier coatings (TBC) are made from alloys such as super alloys, titanium alloys and advanced steel materials. These alloys are widely used to make parts of aircraft, and these parts are used for in corrosive environments and high temperature operations. TBCs are therefore miserably needed to improve fatigue life, creep strength and corrosion resistance of these alloys. Thermal barrier coating consists of a bond coat and a top coat. The behaviour of the coating, such as oxidation resistance, hardness, microstructure and thermal oxide formation, depends on the coating technique and the bond and top coating materials. Therefore, an attempt has been made in this paper to review recent research on the Thermal Barrier Coating Behavior. Most researchers used NiCoCrAlY as a bond coat and YSZ as a top coat. HVOF and plasma spray techniques were found to be effective than other techniques such as Electron beam physical vapor deposition in this review.

1.Introduction

TBC materials are applied to the surface of the turbine to withstand the very high temperature gas effect. Many TBC techniques are available, such as physical vapor deposition electron (PVDE), plasma spray type (PST), high-speed oxygen fuel (HVOF), electro spray vapor deposition (ESVD) etc., TBC has layers such as metallic bond coat, thermal oxide and ceramic coat [1]. The TBC is required to withstand thermo-mechanically loaded working conditions. TBC should also withstand heat stress induced by heating and cooling and the coating should stabilize its phase and structure under the load on the coating. TBC consists of layers such as substrate, bond coat, TGO and top coat. YSZ, Mullite, Alumina, Ceria, rare earth zirconates and Meta glass composites are all used as top coat. TBC can be produced on the substrate of materials using Electron Beam Physical Vapor Deposition (EBPVD), Air Plasma Spray (APS), High-Speed Oxygen Fuel (HVOF), Electrostatic Spray Assisted

Vapor Deposition (ESAVD) and Direct Vapor Deposition. TBC is widely used to coat aircraft materials, steam engines and heat exchangers. It is understood that the importance of the TBC and the application of techniques are identified. An attempt is therefore made to review the performance of various TBC materials and to apply oxidation resistance techniques, coat life, formation phases and structural change.

2. Effect of TBCs on Mechanical Properties

Hitesh Vasudev et al. [2] carried out an analysis of the effects of TBC coated cast iron. Alloy 718 and NiCrAly were used as a source of coating. Micro hardness, high temperature oxidation resistance and erosion resistance have been reported to have been improved by TBC iron coating compared to the uncoated iron. The HVOF method was used to coat Inconel 718 and the rate of oxidation of the coated alloy was reported. Nickel chromium was used as a source of coating. Uniform thermally grown oxide (TGO) was observed after a test of up to 1000C, but cracks in the coating layer and non-uniform TGO were observed after an oxidation test of 1000C and 50 hours. The SEM images of the substrate after the oxidation test at 1100C for a time period of 8 hours, 24 hours and 50 hours are shown in Figure.1 [3].

Inconel 600 alloy coated by Ravi Shankar et al. [4] using Ni-50 % Cr and Alumina. The HVOF method was used to coat the surface of the substrate and Alumina was coated on the top using a plasma spray technique. They investigated the thermal life cycle of the coated Inconel alloy and reported an improvement in the thermal cycle performance at 650 ° C for 100 cycles, as well as an improvement in the adhesion bond strength. Zhe Lu et al. [5] made TBC on the nickel alloy substrate using the EBPVD technique and the bond coat was produced using the APS and HVOF techniques. The thermal cycle life, thermal fatigue and thermal shock tests were analyzed. The hardness rate was increased on the top coat after thermal fatigue test at 1429 cycles and the adhesive strength decreased when the HVOF bond coat was implemented.

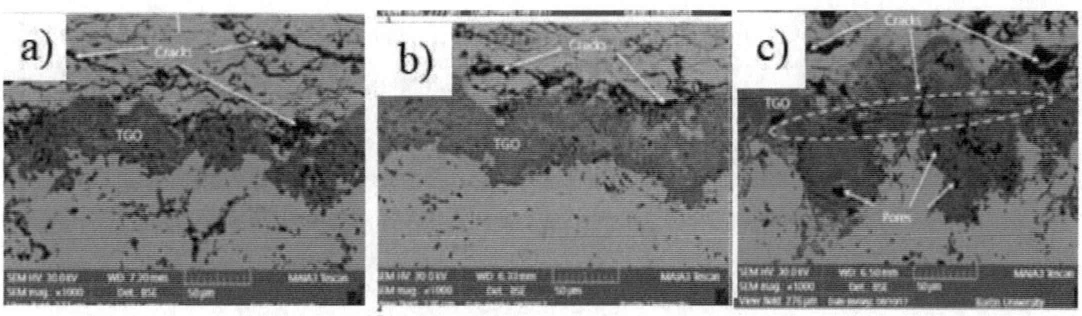

Figure1. (a-b) SEM Images of the structure after the test at a) 1100C for 8hours, b) 1100 for 24 hours c) 1100C for 50 hours

Omid Khanali et al. [6] discussed the effect of the composite coating on Inconel 718. Materials such asYSZ-Nano size and aluminum oxide have been used as composite coatings. They noted that the sintering temperature influences the size of the YSZ particles and that the best nano-composite coating of YSZ and Al was obtained by sintering at 1250 ° C for four hours. Two samples of Inconel 738 coated TBCs were made with respect to coating materials such as NiCrAly / YSZ and NiCrAly / CSZ and their coating performance was investigated for hardness test and hot corrosion test. The method of plasma spraying was used for coating. Al2O3 was used as a top coat for both samples. The resistance of hot corrosion to NiCrAly / CSZ and nanostructured Al2O3 / CSZ coating was improved when compared to YSZ [7].

Eh Hovsepian et al. [8] used the Magnetron sputtering technique for depositing CrN and NbN material on P92 steel. The P92 steel is used to make a steam turbine. The resistance of oxidation to the coated steel has been improved. Hardness and young modulus have been improved in the low Nb coating of the mixture of CrN and NbN, although the creep rate for the coated sample has been reduced. It was concluded that the Magnetron sputtering method did not have a good impact on the coated sample. Khushdeep Goyal et al. [9] used CNT reinforced coating materials Cr3C2–20NiCr and normal coating materials Cr3C2–20NiCr to coat steel substrates SA213-T-22. The HVOF technique has been used to coat the steel. They noted that the porosity, roughness and corrosion resistance decreased when CNT mixed Cr3C2–20NiCr was used as a coating material.

Fatoba et al. [10] used a laser system to coat titanium and cobalt on the Ti-6Al-4V substrate. They studied the impact of the laser system parameters on corrosion resistance and the mechanical properties of the coated alloy. The laser power of 750W, scanning rate of 1.2m / min, was observed as the best solution for alloy coating with 60 % titanium and 40 % cobalt. The refined structure was seen at at a higher rate of scanning speed; however, the structure became coarse at a lower rate of scanning for the same laser power. Tensile strength of this alloy has also been improved if the cobalt content is more than aluminum in the coating mixture. Boissonnet et al. [11] analyzed the performance of coated aeronautical engine parts, such as combustion chambers, deflectors and injectors. HVOF was used for bond coating of Haynes-188 & AM1 with NiCoCrAlY and Inconel 600 with NiCrAlY and YSZ ceramic coating was produced using a plasma method. They concluded that sintering temperature played a role in slowing down the coarse structure, inducing cracking.

Kadir Mert Doleker et al. [12] have deposited Nichrome as a coat bond and YSZ as a top bond on Inconel 718 alloy. The coting was made using a plasma technique and it has been observed that this has been shown to be the best environment up to 1000 ° C. Vibha et al. [13] stated that after loading, the TBC must maintain the same microstructure and permanent thermal insulation. They carried out a test of fatigue on the coated Inconel 718. Coating materials such as NiCoCrAlY as bond coat and

8wt %Y2o3ZrO2 as top bond were used to coat the substrate of Inconel 718 and it was concluded that the coated alloy performed well without damage up to less than 16 cycles. Delamination and damage were observed when this alloy was subjected to more than 16 cycles. Elbert Contreras et al. [14] deposited a single layer and several layers of coating on a steel substrate using the Sputtering method. They used (Ti, Al) (N) and TaN as coating material. They concluded that the growth of the columnar was seen in more layers of coating. In the case of more layer coatings, the size of the grain and the surface roughness were reduced. After heat treatment, Pasquale Cavaliere et al. [15] sprayed Inconel 625 powder on Inconel 718. They studied the impact of the cold spray parameters on the structure and mechanical properties of the coated alloy. The best solutions for reducing porosity and increasing adhesion are low gas temperature, high pressure and medium distance between the spray gun and the steel surface. In addition, the repaired parts were deposited and the fatigue test was analyzed with regard to the change in cyclic loadings.

Adam Khan et al. 16] analyzed the impact of coatings such as NiCr – Cr2O3 and (A40 T) in the salt-H2O condition. The plasma spray technique was used for the IN617 coating and the IN617 coated A40 T coating proved to be best for corrosion resistance than the IN617 coated NiCr – Cr2O3 coating. Masayuki Araia et al.[17] performed a finite element analysis to investigate the behavior of the IN738 coated alloy in terms of deformation and crack formation. CoNiCrAly was used as bond coat and 8wt % Y2O3-ZrO2 as top coat. They reported that the crack initiated at the concave place of the top coat and the strain rate was low at crack initiation at high temperature.

Shailesh Dhomne et al. [18] coated the spark ignition engine and stated that the TBC gave a good heat loss insulation. Coating materials such as YSZ, Mullite, Al2O3, Alsi, NiCrAl, Mg-PSZ, Y-PSZ, CaZrO3, MgZrO3 were used. Obrtlik Karel et al. [19] analyzed fatigue life of the coated IN713 and CoNiCrAlY and ZrO2 with stabilized Y2O3 were used as bond coat and top coat respectively. The fatigue life response of the coated alloy was compared to the uncoated alloy. Cracks were mostly initiated at the top and bond coat interfaces, and these cracks are homogeneous. The crack formation is shown in Figure.2.

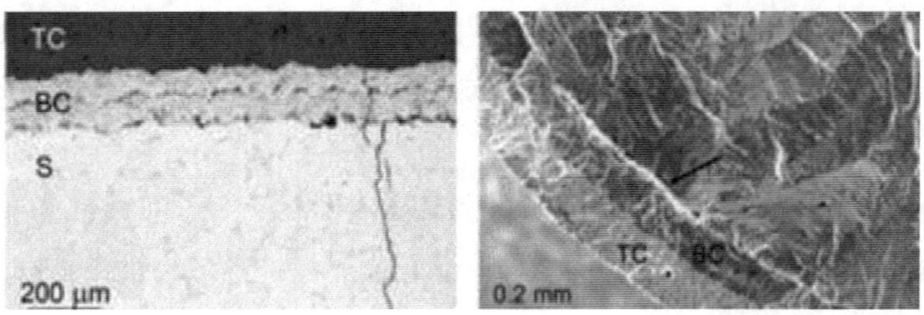

Figure 2. (a-b) Crack Initiation a) Growth of the crack from top to bond coat b) Surface of the fracture

Gupta et al. [20] stated that the porosity and delamination of the defective coating had a significant impact on the mechanical properties, oxidation and corrosion resistance. Therefore, the TBCs should be free from defects. They therefore attempted to analyze the performance of the IN71 coated with NiCoCrAlY as bond coat and ZrO2.7 % Y2O3 as top coat. Thermal spray gun has been used for coating. They predicted thermal conductivity and the Youngs modulus of top coats. They made analysis of changes in the coat structure, porosity and coat life. It was noted that the optimized coat showed low thermal conductivity and was associated with low porous and crack levels. The lowest current and high spray distance were identified as the best coating structure parameter.

Kromer et al. [21] coated the AM1 alloy with no bond coat and made only the top coat with YSZ. The coating was made using the Plasma spray method. They analyzed the thermal mechanical fatigue, the creep test and the structure of the coat. Coting was performed with AM1 aging and without AM1 aging to compare the performance of the test. Increased Creep elongation and low yield point were observed during the non-aging AM1 alloy test. Surface treatment was also performed on the coating and there were no identical defects. Joao et al. [22] analyzed the performance of the coated Nimonic C263 alloy and MCrAlY was used as a bond coat and 8 wt. percent YSZ as a top coat. The coating was made using a plasma spray method. Figure.3 (a-c) shows the microstructure observed at the lowest life, middle life and good life of the coated alloy after the fatigue test. The microstructure captured at the lowest lifespan consists of a minimum of micro surface roughness and a minimum of larger radii peaks. The microstructure captured at mid-life performance consists of a greater number of swelling micro roughnesses. The microstructure captured at higher life performance consists of a surface that is curved and rounded outwards.

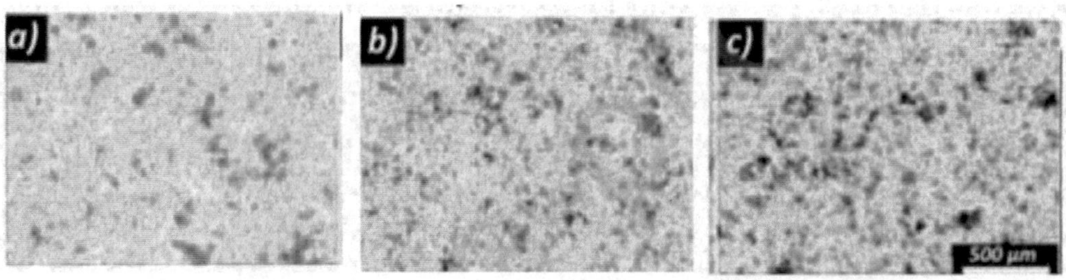

Figure 3. (a-b) Micro structure of Sample obtained after Fatigue Test. a) Low performance b) Middle performance C) Good performance

Components such as gas and steam turbines are heavily affected by heat, and the surfaces of these components must be coated to resist corrosion and improve performance. The resistance of hot corrosion depends on the type of coating that is either liquid or solid-state. Cr content in super alloys has superior resistance to hot corrosion, but the Cr content would not withstand enough resistance in the Sulfur container situation. Working temperature and the presence of phase at that temperature would also have an impact on the resistance of hot corrosion [23].

Hitesh Vasudev et al. [24] coated gray cast iron using Inconel 718 powder and aluminum oxides as a composite coat by HVOF technique. The impact of the composite coat on hardness and oxidation resistance was found to vary the weight percentage of aluminum (10 %, 20 %, 30 %) with Inconel powder. It was noted that Wt-30% of Al2O3 with Inconel 718 powder improved resistance to oxidation and increased hardness. Furthermore, the presence of phases in the composite coat, namely chromium oxide, titanium oxide, nickel chromium oxide and chromium niobium oxide, has been reported to improve oxidation resistance.

Pasupuleti Kirti Teja et al. [25] used NiCrAly as a bond coat and 8 % YSZ as a top coat on INL718 alloy. The coating was made with a plasma spray. They observed that the sintering temperature had a significant influence on the oxidation resistance, the failure mechanism of the shock test. The failure mechanism was induced by the presence of thermal stress and the TGO layer in the bond and top coat interfaces. Jinku Yu et al. [26] carried out an analysis of the performance of coated copper and nickel-based coating materials used as coating materials. Jet electrode deposition has been used for coating. The best resistance of oxidation at 900 ° C in copper coating with nickel and nickel with iron was reported as nickel-iron-tungsten coating material.

Zhang Na et al. [27] investigated the effect of micro arc oxidation and solution gel method on the coating of AZ91D magnesium alloy. .It was concluded that the combination of MAC and TiO2 solution gel produced good resistance to oxidation compared to MAC single coating. Nikhil Rajendra Kadam et al. [28] evaluated the indentation on the coated surface using a finite element analysis. They considered INL718 alloy to be a substrate, NiCoCrAly to be a bond coat and 7YSZ to be a top coat. They concluded that the flat shaped indenter exhibited compressive and von-mises stress than the spherical and conical shapes of the indenter.

Dutta et al.[29] coated iron-nickel-chromium alloy 800 by spray technique. Aluminum was used to coat the substrate of this alloy. Analyzes such as the microscopic, hardness test and scratch test of the coated alloy were performed and the Al / rich oxygen, iron / aluminum and aluminum oxides were observed in the intermediate layers of the coating. Kadir Mert Doleker[30] used the HVOLF method to coat the 316L steel substrate. Iron and aluminum powders have been used as coating material and coated steel analysis, such as oxidation testing, XRD and SEM analysis, has been carried out. The laser coat melting showed good performance in resistance to oxidation than the iron & aluminum coat.

3. Conclusions

- Thermal barrier coating effects of substrates such as Inconel alloys, magnesium alloys and steels have been reviewed.
- The effect of various TBC bond and top coat materials was discussed.

- Microstructure, micro hardness, fatigue life, creep life, oxidation resistance, crack formation in the coating and coating defects were discussed with respect to the coating of different coating materials.
- Resistance to oxidation at different temperatures has been reviewed with respect to different coating materials.
- Impact of different coating techniques and methods have been reviewed and HVOF and plasma spray techniques have been found to be effective than other techniques.
- The TBC investigation of nickel-based alloys, titanium alloys, various grades of steel, magnesium alloys and other alloys used in the hostile environment is inadequate.
- It was found that most researchers used NiCoCrAlY as a bond coat and YSZ as a top coat.
- The impact of the various TBCs on porosity, TGO formation, phase changes, alteration of the microscope, fatigue life and creep life was found to be inadequate with respect to coatings by different coating materials and different substrate materials.

4. References

1. Morumpalle Sai Sahith, Giridhra G, R Suresh Kumar, Development and analysis of thermal barrier coatings on gas turbine blades-A Review, ICAMA 2016, Materials Today: Proceedings,5 pp.2746-2751, (2018).
2. Hitesh Vasudev, Lalit Thakur, Amit Bansal, Harmeet Singh, Sunny Zafar, High temperature oxidation and erosion behavior ofHVOFsprayedbilayerAlloy-718/NiCrAlY, coating,362pp. 366-380,(2019).
3. Abdullah Selim Parlakyigit, Dervis Ozkan, Mecit Oge, Yasin Ozgurluk, Kadir Mert Doleker, Turgut Gulmez, Abdullah Cahit Karaoglanli, Formation and growth behavior of TGO layer in TBCs with HVOF sprayed NiCr bond coat, Emerging Materials Research, https//doi.org/10.1680/jemmr.18.00085,1-9 (2018).
4. A. Ravi Shankar, K. Thyagarajan, Chandrakanth Upadhyay, C. Mallika & U. Kamachi Mudali, Plasma-sprayed alumina coating on Inconel 600 for fast reactor neutron detector application, Surface Engineering, https://doi.org/10.1080/02670844.2017.1404695, (2017) 1-8.
5. Zhe Lu, Guanlin Lyu, Abhilash Gulhane, Hyeon-Myeong Park, Jun Seong Kim, Yeon-Gil Jung and Jing Zhang, Experimental and Modeling Studies of Bond Coat Species Effecton Microstructure Evolution in EB-PVD Thermal Barrier Coatings in Cyclic Thermal Environments, Coatings,9.626,(2019).
6. Omid Khanali, Sina Ariaee, Masoud Rajabi1 and Saeid Baghshahi, An investigation on the properties of YSZ/Al2O3 nanocomposite coatings on Inconel by electrophoretic deposition, Journal of composite materials, pp.1-9, (2017).

7. A. Keyvani & M. Bahamirian, Hot corrosion and mechanical properties of nanostructured Al2O3/CSZ composite TBC, Surface engineering, pp.1-12 (2017).
8. P.Eh Hovsepian, A.P.Ehiasarian, Y.P.Purandare, P.Mayr, K.G.Abstoss, M.Mosquera Feijoo, W.Schulz, A.Kranzmann, M.I.Lasanta, J.P.Trujillo, Novel HIPIMS deposited nanostructured CrN/NbN coatings for environmental protection of steam turbine components, journal of Alloys and Compounds, 746, pp.583-593, (2018).
9. Khushdeep Goyal and Rakesh Goyal, Improving hot corrosion resistance of Cr3C2–20NiCr coatings with CNT reinforcements, Surface engineering, pp.1-12,(2019).
10. O. S. Fatoba, O. S. Adesina, A. P. I. Popoola, Evaluation of microstructure, microhardness, and electrochemical properties of laser-deposited Ti-Co coatings on Ti-6Al-4V Alloy, International Journal of Advanced Manufacturing Technology, **97**, pp.2341–2350 (2018)
11. G. Boissonnet, G. Bonnet, A. Pasquet, N. Bourhila, F. Pedraza. Evolution of thermal insulation of plasma-sprayed thermal barrier coating systems with exposure to high temperature. Journal of theEuropeanCeramicSociety,39,6,pp.2111-2121 (2019).
12. Kadir Mert Doleker, Yasin Ozgurluk, Abdullah Selim Parlakyigit, Dervis Ozkan, Turgut Gulmez, Abdullah Cahit Karaoglanli, Oxidation Behavior of NiCr/YSZ Thermal Barrier Coatings (TBCs),Open Chem., 16, pp,876–88 (2018).
13. Vibha, Vijay Kumar Dwivedi, and Dipak Kumar, Nano-Microstructural Characterization of Inconel-718 with Thermal- Barrier Coatings after thermal shock tests, International Journal of Innovative Technology and Exploring Engineering, 8,5 pp.36-41, (2019).
14. Elbert Contreras, Gilberto Bejarano, Maryory Gómez, Synthesis and microstructural characterization of nanoscale multilayer TiAlN/TaN coatings deposited by DC magnetron sputtering, International Journal of Advanced Manufacturing Technology, (2018).
15. Pasquale Cavaliere, Angelo Perrone and Alessio Silvello, Fatigue behaviour of Inconel 625 cold spray coatings, Surface Engineering, (2017).
16. M. Adam Khan, S. Sundararajan and S. Natarajan, Design and statistical analysis of plasma coatings on superalloy for gas turbine application, Materials at High Temperature, (2016)
17. Masayuki Araia, Hiroaki Katorib, Kiyohiro Itoa,Numerical simulation of inelastic deformation and crack propagation in TBC-multilayered Ni-based superalloy subjected to thermo-mechanical loadings, Surface & Coatings Technology, 399,pp.126159, (2020)
18. Shailesh Dhomne, Ashish M. Mahalle, Thermal barrier coating materials for SI engine, journal of materials research and technology, vol. 8,1pp.1532-1537, (2019).
19. Obrtlik Karel, Hutarova Simona, Celko Ladislav, Julis Martin, Podrabsky Tomas, and Sulak, Effect of thermal barrier coating on low cycle fatigue behavior of cast Inconel 713LC at 900°C, Advanced Materials Research, 848-853,(2014)

20. M.Gupta, P. Nylen, J. Wigren, A Modelling Approach to Designing Microstructures in Thermal Barrier Coatings, Journal of Ceramic Science and Technology, 04, 02, pp.85-92. (2013)
21. R. Kromer, F. Mauget, L. Despres, S. Costil, J. Cormier, Thermo-mechanical fatigue evaluation of a thermal barrier coating bond coatless system, Materials Science and Engineering A, 756 pp.130-141, (2019)
22. Joao P. Martins, Ying Chen, G. Brewster, R. McIntyre, Ping Xiao, Investigation of the bond coat interface topography effect on lifetime, microstructure and mechanical properties of air-plasma sprayed thermal barrier coatings, (2020).
23. Gaurav Prashar, Hitesh Vasudev, Hot corrosion behavior of super alloys, Materials Today: Proceedings, https://doi.org/10.1016/j.matpr.2020.02.226, (2020).
24. Hitesh Vasudev, Lalit Thakur, Harmeet Singh, Amit Bansal, An investigation on oxidation behavior of High Velocity Oxy-Fuel sprayed Inconel718-Al2O3 composite coatings, Surface & Coatings Technology, https://doi.org/10.1016/j.surfcoat.2020.12770, (2020).
25. Pasupuleti Kirti Teja, Parvati Ramaswamy, S.V.S. Narayana Murthy, Thermal Shock and Oxidation Stability Tests to Grade Plasma Sprayed Functionally Gradient Thermal Barrier Coatings, 1 (2019).
26. Jinku Yu, Qinyang Li, Xican Zhao, Qi Qiao, Sen Zhai, and Jia Zhao, Correlation Between Microstructure and High-Temperature Oxidation Resistance of Jet-Electrodeposited Ni-Based Alloy Coatings. Journal of materials engineering performance, https://doi,.org/10.1007/s11665-020-04811-z, (2020).
27. Zhang Na, Yu Shengxue, Xing Qian, Chen Xiaolei, Zhang Mingxian, and Shen Dejiu, Corrosion Performance of Composite MAO/TiO2 Sol–Gel Coatings on Magnesium Alloy AZ91D, https://doi.org/10.1007/s11665-018-3653-7, (2018).
28. Nikhil Rajendra Kadam, Karthikeyan, G, Dhananjay Kulkarni,Finite element analysis on nano-indentation behavior in columnar 7YSZ EB-PVD thermal barrier coatings,J urnal Tribologi, 22,1-17,(2019).
29. R. S. Dutta, K. Singh, B. Vishwanadh and G. K. Dey, Aluminide formation on Alloy 800 by plasma spraying and heat treatment, DOI: 10.1080/10426914.2017.1317794, (2017).
30. Kadir Mert Doleker, The Examination of Microstructure and Thermal Oxidation Behavior of Laser-Remelted High-Velocity Oxygen Liquid Fuel Fe/Al Coating, Journal of materials engineering performance, 29 (3220–3232, (2020),

IOP Conf. Series: Materials Science and Engineering **954** (2020) 012026 doi:10.1088/1757-899X/954/1/012026

Influence of nanoparticles on the characters of polymeric composites

R. Vinayagamoorthy[1*], K. S. Subrahmanyam[2], K. Murali Krishna Murthy[3], K. Arun Prajwar[4], P. Gopinath[5], M. Sai Lahari[6] and K. Pruthvi Rangan[7]

[1-7] Department of Mechanical Engineering, Sri Chandrasekharendra Saraswathi Viswa Mahavidyalaya, Kancheepuram, Tamilnadu - 631561, India.

Email: vin802002@gmail.com

Abstract. The need of structural materials is inevitable for machineries, equipment's and tools. These materials are mostly made up of metals and polymeric composites and with other materials in rare cases. Polymeric composites nowadays replace almost all materials due to its improved characters and reduced manufacturing costs. The characters of these polymeric composites are improved by reinforcing multiple fibers in it and thus enhances its suitability for different applications. Among the characters of a composites, weight reduction is a vital property that decides the usability of the material in high end applications like aircraft, robots etc. Weight of composites could be enhanced without losing its characters by reinforcing nanoparticles in different forms. Several nanomaterials have been utilized by researchers in various forms and they are proven to be the best in enhancing the characters of the composites. The present research gives a clear picture on the different challenges during the synthesis, fabrication and characterization of nanoparticles reinforced plastic composites.

1. Introduction

Polymeric composites are produced by taking a polymer as the matrix and a fiber as the reinforcement. The matrix gives the structure whereas the reinforcements add the properties to the component [1, 2]. At the earlier stage, man-made fibers have been utilized but later, green fibers taken from nature are used as reinforcing fibers [3-5]. Before using a natural fiber, it must be subjected to several pre-treatment techniques in order to remove the unusable matters [6-10]. Polymeric composites are mainly intended for boats and radial fan blades and other components in automotive, aircraft and marine industries [11-14]. Due to the increasing need of polymer-based materials, there is an urge to develop materials without sacrificing the characters. This made the scientists to use nanoparticles in polymeric materials [15-17]. Nanoparticles are made from chemical compounds especially from oxides like zirconium oxide (ZrO_2), hafnium oxide (HfO_2), zinc oxide (ZnO), ferrous oxides (Fe_2O_3 and Fe_3O_4), titanium oxide (TiO_2), aluminium oxide (Al_2O_3) and silicon dioxide (SiO_2). Some polymers adaptable to these nanomaterials are polymethylmethacrylate (PMMA), polyacrylic acid (PAA), polyethylene (PE), polystyrene (PS), acrylic acid etc. Nanomaterials are used in different ways in a polymeric composite. Particles and fillers of nanomaterials are added in the polymer before composite fabrication. Nanoparticles in the form of fibers like carbon nanotubes (CNT) are directly used as reinforcement as a replacement to commercial fibers during composite fabrication [18, 19]. The other forms of nanomaterials are shown in Figure 1. The influence of nanoparticle in a composite depends upon the size and shape of it and also its characters but majority of research have proved that the addition of nanomaterials enhances the characters of the composite [20, 21]. The present review gives a clear view on the various nanomaterials and their influence on the mechanical characters of the composite materials.

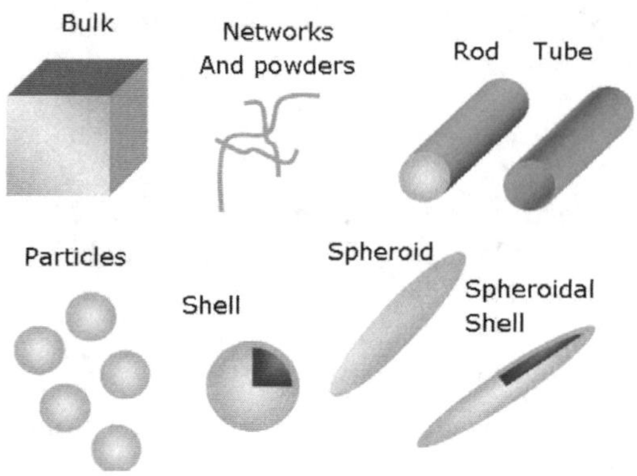

Figure 1. Different shapes of nanomaterials

2. Carbon based nanoparticles in composites

CNT is found to be one of the vital compounds among the nanomaterials. CNT when mixed as a filler in acrylonitrile butadiene styrene (ABS) matrix is proved to be a good insulator of sound and hence are more apt for sound proofing devices. The composition of CNT in the material has a predominant influence and it is found that a minimum of 15 % of CNT is to be retained in order to achieve optimum sound proofing [22, 23]. CNT is also found to be more suitable for epoxy matrix and the mixing is to be done through ultrasonic probe sonicator. A research on CNT reinforced composites used short fibers of CNT as reinforcement in PMMA matrix. It has been identified that, a fiber length of 10 nm at a composition of 10 % elevated the Youngs modulus and tensile strength as compared to that of the virgin polymeric composite [24, 25]. CNT particles in polymeric composites also elevates the interlaminar shear strength (ILSS) and flexural strength of the composites to a notable extent. In addition, the presence of CNT particles enhances the stability of the composites to retain its mechanical strengths and storage modulus until 200^0C. [26, 27]. Graphene is another allotrope of carbon which has almost similar characters like CNT. A research has been made by using graphene nanoparticles and glass as reinforcements in epoxy matrix and reported that, the load bearing ability of the material enhances as a result of graphene addition. The optimum value of graphene is observed as 0.1 % [28]. Like this, many results have been evolved by using carbon nanoparticles. Some of them with their influences on composite characters are presented in Table 1.

Table 1 Influence of carbon-based nanoparticles on composite characters

S. No.	Nanoparticles used	Matrix used	Findings	Reference
1.	CNT	Epoxy	0.6 % of CNT inclusion improved the stiffness of the composites by 60 %.	29
2.	Carbon and silica particles	Epoxy	0.25 % of carbon and 0.25 % of silica in epoxy produced optimum tensile and fatigue performance.	30
3.	Expanded Graphite (EG)	Poly tetrafluoro	Addition of nano-EG enhanced the hardness of the matrix whereas the stress	31

		ethylene	relaxation time is extended.	
4.	CNT	Ultra-high-weight polythylene	CNT improves the composite stiffness, impact strength and damping capacity. The developed composites are more suitable for aerospace and naval industries.	32
5.	CNT	Epoxy	Addition of CNT alleviates the tensile strength and impact strength of the composites. The natural frequency of the composite increases with a hike in CNT content.	33
6.	Carbon nano fibre (CNF)	Epoxy	Improvement in the tensile modulus and hardness in the CNF reinforced composites as compared to that of the neat resin.	34
7.	CNF	Epoxy	Inclusion of 16 % CNF in the matrix produced optimum electromagnetic interference shielding effectiveness. CNF is better than CNT, EG and carbon black.	35
8.	CNT	Epoxy	Mechanical properties such as flexural strength, interlaminar shear strength and tensile strength goes up with an elevation in CNT content until 0.3 %. After 0.4 % of CNT content, the properties decline due to agglomeration and poor wetting of fibers.	36

3. Natural and nanomaterial reinforced hybrid composites

Nanoparticles may also be produced from natural fibers like jute, hemp, sisal etc and this could be achieved by using ball milling technique. A research has been made to synthesize jute nanoparticles in size ranging from 10 nm to 30 nm using ball milling and reinforced in to the epoxy matrix to form composites. Thermal properties of the composites have been tested and revealed that, the nanoparticles enhance the crystallization and thermal degradation temperature. In addition, the nanoparticles stabilize the resin molecules and eliminates the occurrence of cracking as a result the storage and loss modulus of the composites and the interfacial adhesion are enhanced [37, 38]. CNT is used as one of the reinforcements in a hybrid composite along with natural fibers such as banana, jute and flax. Investigations on the influence of CNT composition and fiber stacking sequence showed that, a hike in the CNT composition elevates the mechanical strengths of the composite. On the other hand, the stacking sequence also plays a vital role in determining the characters. Among the tested samples, it is observed that a 1 % CNT composition and using the jute layers at the extremities gives optimum mechanical properties to the composite material [18]. In a research, calcium carbonate nanoparticles are deposited on to the bamboo fibers by an ironic solution reaction at various temperatures. The deposited bamboo fibers are then reinforced in polypropylene matrix in the form of structural composites. It is revealed that, a highest absorbance of calcium carbonate is achieved at a temperature of 25^0C. Also, it is suggested that the precipitating method is a best method to densify and hydrophobize the natural fiber alleviating the cell wall defects [39, 40].

4. Ceramic nanomaterial-based composites

Ceramic materials are majorly used in the form of nanoparticles. In a research, clay and silica are organically synthesized using 3-aminopropyltriethoxysilane by sonication process before using it as a reinforcement in the composite material. Composites have been prepared with epoxy as the matrix and glass as the supporting reinforcement. Wear characters have been studied and reported that, a load and sliding speed showed a lesser debris formation. Also, it is noticed that the presence of nanoparticles alleviates the wear rate of the composites as compared to that of the virgin composite [41, 42]. Aluminium nitride (AlN) and aluminium oxide are a kind of naturally occurring ceramics and have several usages in the ceramic industry. A study has been made to investigate the behaviour of AlN and Al_2O_3 nanoparticles in PEEK matrix composites and revealed that, the microhardness of the composites goes up with an elevation in the fiber loading. In comparison, AlN composites showed an enhanced microhardness than the Al_2O_3 composites [43, 44]. A study on the influence of silica and Si-C_3H_7 treated silica has been made by using polyimide matrix and concluded that, the elevation in the nanoparticle size alleviates the mechanical properties of the composite. On the other hand, when the composite is reinforced with treated silica, the mechanical strengths of the composite goes up to certain extent. It is also identified that, the grafting ratio has a direct impact on the characters such as Youngs modulus, shear modulus and poisson's ratio [45, 46]. The ratio between degree of polymerization (P) and the degree of polymerization of the grafted bush (N), P/N ratio has also had a predominant influence on the mechanical characters of the composite and this has been identified in a research on nanosilica reinforced on to the polystyrene matrix [47, 48]. Several studies have been to assess the influence of ceramic nanomaterials. Some of them are presented in Table 2.

Table 2 Influence of ceramic-based nanoparticles on composite characters

S. No.	Nanoparticles used	Matrix used	Findings	Reference
1.	Aluminium oxide	Epoxy	Addition of nanoparticle alleviates the tensile strength and impact strength of the composites. The natural frequency of the composite increases with a hike in nanoparticle content.	33
2.	Silica	Polyester	Residual strength and stiffness of the composites goes up when the silica nanocoating is given to glass fibers.	49
3.	Titania and hydroxyapatite	Poly-lactic-co-glycolic acid	Well dispersed nanoparticles in the matrix enhanced the tensile strength, tensile modulus and compressive modulus of the composites	50
4.	Barium sodium niobite	Polystyrene	Improved adhesion between the filler and polymer leading to elevate tensile properties of the composite.	51
5.	Barium titanate, strontium titanate	Poly ε-caprolactone	Improvement in the mechanical properties as a result of nanoparticles. In addition, nanoparticles elevated the permittivity and reduced the losses in composites.	52
6.	Barium titanate	Epoxy	Dielectric constant and dissipation factor of the composites elevates with the inclusion of nanoparticles.	53
7.	Barium titanate	Polyvinyl alcohol	Dielectric constant goes up with a hike in the ceramic concentration whereas dielectric loss tangent goes down with	54

			the increase in ceramic concentration.	
8.	Aluminium oxide and silica	Polymethyl methacrylate	Composites with aluminium oxide nanoparticles showed an increased impact strength and fracture toughness as compared to silica reinforced composites.	55

5. Synthetic nanomaterial-based composites and applications

Processing of nanomaterials are usually done through artificial synthesis of chemicals. Among the several synthetic materials, TiO_2 has been utilised as the nanoparticle with a size of 300 nm in epoxy and polyamide matrices during composite preparation. A detailed analysis on the wear performance has been made and it is revealed that, the stress concentration of the fibers in the composite is alleviated and thus there seems to be an enhancement in the wear resistance [56 57]. A mixture of ZrO_2 and SiO_2 are found to be apt for dental applications when used in combination with matrices based on monomeric di-methacrylate. A comparative study between micro-SiO_2 and a mixture of nano ZrO_2 and SiO_2 particles has been made and showed that, the hardness and reduced modulus are heterogeneous in microfiber composites whereas they are homogeneous in nanofiber composites. Hence, the wear behaviour of nanoparticle composites ae superior to the microfiber composites [58, 59]. Magnetic iron oxide (Fe_3O_4) prepared in the size of 25 nm by co-precipitation of $FeCl_2$ and $FeCl_3$ using sodium hydroxide solution has been utilised as the nanoparticles in epoxy matrix for composite fabrication. Penetration tests have been conducted on the fabricated materials and it has been found that, addition of 5 % nanoparticles improved the resistance to penetration by 9 % in comparison to the composite without nanoparticles [60, 61]. In another research, polymethyl phenyl siloxane is filled with Co_3O_4 nanoparticles and then reinforced on to poly butyl methacrylate and reported that, after grafting the hydrophilicity of nanoparticles has been alleviated and the hydrophobicity has been elevated. Moreover, it has been concluded that the endothermic characters of the nanoparticle reinforced composites are far better than the virgin composites [62-64]. A research used calcium carbonate nanoparticles as reinforcement in ethylene-propylene-diene rubber during composite formulation and reported that, nanoparticles alleviated the glass transition temperature of the composite to a notable extent in comparison to the microparticles [65, 66]. Like this there several other research works which are primarily made to enhance the characters of the composite in any way and makes it suitable for variety of applications [67, 68]. Nanoparticle based polymer composites finds several applications. The major phenomenon of weight reduction enriches the usage of nanocomposites in aircraft industry. The various possible ways that a nanocomposite could be used in an aircraft is presented in Figure 2.

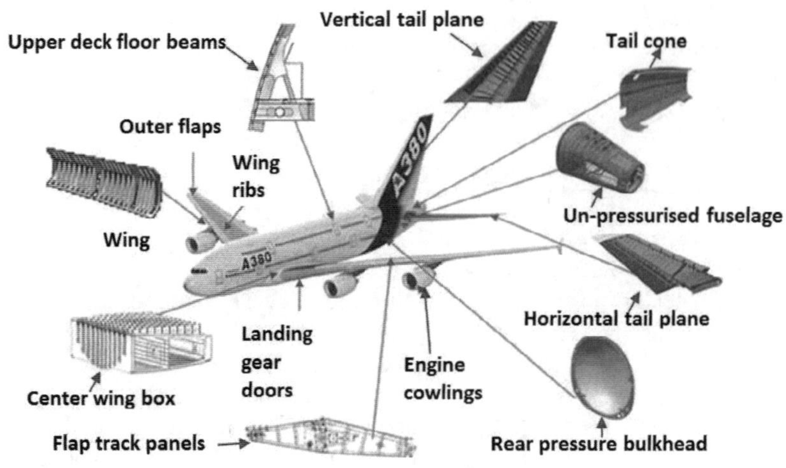

Figure 2 Applications of nanoparticle-based polymer composites

6. Conclusions

This paper has made a comprehensive survey on various research pertaining to the nanoparticle reinforced plastic composites. Many nanoparticles have been utilized by the researchers and majority of it have produced a positive sign in enhancing the mechanical characters of the composite. Very few researches have produced negative sign but at the same time they are capable of elevating some other physical or chemical characters of the composites. The processing techniques used for a nanomaterial is different for each material and thus the selection synthesis technique is vital as it directly affects the characters of the composite material. A major research work has been done by using carbon in any form as a nanomaterial and all these resulted in elevating the strengths and other physical characters of the composite. Natural fibers on the other hand have also produced superior results in comparison to that of the synthetic material in nano form and thus, natural nano material is apt for polymeric composites. When polymeric composites are used in high temperature applications ceramic nanomaterials are proven to be the best among all nanomaterials.

References

[1] Mahmoud Nassar M A, Ramanathan Arunachalam and Khalid Alzebdeh I 2017 *Int. J. Adv. Manuf. Technol.* **88** 2985.
[2] Vinayagamoorthy R 2017 *J. Reinf. Plast. Compos.* **36** 1577.
[3] Vinayagamoorthy R and Rajeswari N 2014 *J. Reinf. Plast. Compos.* **33** 81.
[4] Vinayagamoorthy R 2019 *Polym. Polym. Compos.* doi: 10.1177/0967391119883163.
[5] Michael Fuqua A, Shanshan Huo and Chad Ulven A 2012 *Polym. Rev.* **52** 259.
[6] Vinayagamoorthy R, Sivanarasimha,S, Vijay Padmanabhan, Vedula Ganesh and Karthikeyan S 2015 *Int. J. Appl. Eng. Res.* **10** 663.
[7] Vinayagamoorthy R 2019 *J. Nat. Fiber.* doi: 10.1080/15440478.2019.1598916.
[8] Sunil Kumar Ramamoorthy, Mikael Skrifvars and Anders Persson 2014 *Polym. Rev.* **55** 107.
[9] Vinayagamoorthy R, Sivanarasimha S, Vinay Kumar K R and Vijay Padmanabhan 2015 *Appl. Mech. Mater.* **813** 14.
[10] Vinayagamoorthy R and Venkatakoteswararao G 2019 *J. Thermoplast. Compos. Mater.* doi: 10.1177/0892705719828783.
[11] Rajani Samal K and Giti Sudha Giri 1992 *J. Macromol. Sci. Part C: Polym. Rev.* **32** 55.
[12] Vinayagamoorthy R, Karthikeyan S, Prem Bhargav R S and Rajivalochan T V 2015 *Appl. Mech. Mater.* **813** 101.
[13] Vinayagamoorthy R and Madhavan S 2011 *Int. J. Bus. Rev. Manuf. Manage.* **1** 5.

[14] Vinayagamoorthy R, Mothilal T and Madhavan S 2009 *Int. J. Des. Manuf. Technol.* **3** 63.
[15] Thomas Hanemann and Dorothee Vinga Szabo 2010 *Mater.* **3** 3468.
[16] Vinayagamoorthy R, Saswath Koundinya, S L, Mani Teja G L S N and Kovuru Adithya 2016 *Indian J. Sci. Technol.* **9** 1.
[17] Vinayagamoorthy R 2018 *J. Thermoplast. Compos. Mater.* doi: 10.1177/0892705718815529.
[18] Mohan K and Rajmohan T 2017 *J. Nat. Fiber.* **6** 864.
[19] Vinayagamoorthy R, Tumuluri Nikhil Koteshwar, Vajhala Venu Madhav, Twarakavi Karthik Sai and Vikas Konda 2019 *Mater. Today: Proc.* **16** 277.
[20] Vinayagamoorthy R, Vikas Konda, Parikshit Tonge, Tumuluri Nikhil Koteshwar and Premkumar M 2019 *Mater. Today: Proc.* **16** 567.
[21] Zhanhu Guo, Ta Kim Y, Kenny Lei, Tony Pereira, Jonathan Sugar G and Thomas Hahn H 2008 *Compos. Sci. Technol.* **68** 164.
[22] Vinayagamoorthy R, Ankur Shama, Vignesh Iyer and Navneeth G 2019 *Adv. Manuf. Processes., Lect. Notes Mech. Eng.* **48** 521.
[23] Jae-Chul Lee, Young-Sun Hong, Ri-Guang Nan, Moon-Kyu Jang, Caroline Lee S, Sung-Hoon Ahn, and Yeon-Jun Kang 2008 *J. Mech. Sci. Technol.* **22** 1468.
[24] Rajmohan T, Vinayagamoorthy R and Mohan K 2018 *J. Thermoplast. Compos. Mater.* doi: 10.1177/0892705718796541.
[25] Behrouz Arash, Harold Park S and Timon Rabczuk 2015 *Compos. Struct.* **134** 981.
[26] Vinayagamoorthy R, Rajeswari N and Balasubramanian K 2014 *Jordan J. Mech. Ind. Eng.* **8** 385.
[27] Ming-Chuen Yip, Yi-Chieh Lin and Chung-Lin Wu 2011 *Polym. Polym. Compos.* **19** 131.
[28] Tuncer C and Canyuri OE 2019 *Acta Physica Polonica A.* **135** 752.
[29] Abdullah Algarni, Numan Salah, Mostefa Bourchak, Asim Jilani, Ahmed Alshahrie and Mahmoud Nadim Nahas 2016 *J. Compos. Mater.* **51** 2675.
[30] Megahed M, Megahed AA and Agwa MA 2018 *J. Indust. Tex.* **49** 181.
[31] Yanhong Yana, Zhining Jiaa and Yulin Yang 2011 *Procedia Environ. Sci.* **10** 929.
[32] Maksim Kireitseu, David Hui and Geoffrey Tomlinson 2008 *Compos. Part B: Eng.* **39** 128.
[33] Mohammed SM, Gamil M, Mohammed SS, Mahmoud TS and El-Kady MEY 2017 *Nanosci. Nanotechnol.* **11** 1.
[34] Lee H, Mall S, Nalladega V, Sathish S, Roy A and Lafdi K 2006 *Polm. Polm. Compos.* **14** 549.
[35] Moumita Ray, Jinu Jacob George, Ajay Chakraborty and Anil Bhowmick K 2010 *Polm. Polm. Compos.* **18** 59.
[36] Salam MBA, Hosur MV, Jahan N, Rahman MM and Jeelani S 2013 *Polm. Polm. Compos.* **21** 495.
[37] Vinayagamoorthy R 2018 *J. Reinf. Plast. Compos.* **37** 49.
[38] Padal KTB, Srikiran S and Surya Nagendra P *All India Manuf. Technol. Des. Res. Conf.* IIT Guwahati, 12 Dec 2014.
[39] Vinayagamoorthy R 2017 *Mater. Manuf. Processes.* **32** 645.
[40] Gao Jie, Wang Ge, Cheng Hai-tao and Sheldon Q Shi *55th Int. Conven. Soc. Wood Sci. Technol.* Beijing, 27-31 Aug 2012.
[41] Vinayagamoorthy R 2019 *J. Nat. Fib.* **16** 163.
[42] Lingaraju D, Ramji K, Pramila Devi M and Rajya Lakshmi U 2011 *Bull. Mater. Sci.* **34** 705.
[43] Vinayagamoorthy R, Rajeswari N, Vijayshankar S and Balasubramanian K 2014 *Int. Rev. Mech. Eng.* **8** 952.
[44] Goyal RK, Tiari AN and Negi YS 2008 *Mater. Sci. Eng. A* **491** 230.
[45] Kannan P, Balasubramanian K and Vinayagamoorthy R 2015 *Int. Rev. Mech. Eng.* **9** 536.
[46] Hui Zhang and Jiachu Xu 2016 *J. Chem. Pharm. Res.* **8** 39.
[47] Vinayagamoorthy R, Rajeswari N and Karthikeyan S 2015 *Appl. Mech. Mater.* **766-767** 812.
[48] Marissa Giovino, Julia Pribyl, Brian Benicewicz, Ronald Bucinell and Linda Schadler 2019 *Nanocompos.* doi: 10.1080/20550324.2018.1560988.

[49] Yongchao Zhang and Changwen Mi 2019 *J. Appl. Polym. Sci.* **48652** 1.
[50] Huinan Liu and Thomas Webster J 2010 *Int. J. Nanomedicine* **5** 299.
[51] Abraham R, Thomas SP, Kuryan S, sac J, Varughese KT and Thomas S 2009 *eXPRESS Polym. Lett.* **3** 177.
[52] Amrit Bagchi, Sai Rama Krishna Meka, Badari Narayana Rao and Kaushik Chatterjee 2014 *Nanotechnol.* **25** 485101.
[53] Pratap A, Joshi NJ, Rakshit PB, Grewal GS and Shrinet V 2013 *Int. J. Modern Phy: Conf. Ser.* **22** 1.
[54] Beena P and Jayanna HS 2019 *Polym. Polym. Compos.* **27** 619.
[55] Ahmed Alhareb O, Hazizan Md. Akil and Zainal Arafin Ahmad 2016 *Polym. Polym. Compos.* **24** 71.
[56] Venkatakoteswara Rao G, Vinayagamoorthy R, Abinesh K, Sudharsan M, Ponmeganathan S and Deepak Kumar LS 2020 *Adv. Mater. Manuf. Eng., Lect. Notes Mech. Eng* 533.
[57] Li Chang and Klaus Friedrich 2010 *Tribol. Int.* **43** 2355.
[58] Vinayagamoorthy R, Rajeswari N, Vijayshankar S, Vivekanandan M, Sri Rama Murthy Bellala and Venkata Subramaniam KR 2014 *Procedia Mater Sci.* **5** 2075.
[59] Devaprakasam D, Hatton PV, Möbus G and Inkson BJ 2007 *J. Phy: Conf. Ser.* **126** 012057.
[60] Vinayagamoorthy R, Shubham Kumar, Suraj Kumar, Mote Sai Sharan, Afzal GMD and Rajamurugan TV 2020 *Adv. Mater. Manuf. Eng., Lect. Notes Mech. Eng.* 43.
[61] Hany Fouda, Khalid Ghith and Lin Guo 2017 *MATEC Web Conf.* **88** 01002.
[62] Vinayagamoorthy R, Manoj I V, Narendra Kumar G, Sai Chand I, Sai Charan Kumar G V and Suneel Kumar K 2018 *J. Mech. Sci. Technol.* **32** 2011.
[63] Vinayagamoorthy R and Rajmohan T 2018 *J. Reinf. Plast. Compos.* **37** 1037.
[64] Sun JT, Huang YD, Cao HL and Gong GF 2004 *Polym. Polym. Compos.* **12** 739.
[65] Vinayagamoorthy R, Manoj IV, Narendra Kumar G, Sai Chand I, Sai Charan Kumar GV and Suneel Kumar K 2018 *IOP Conf. Ser: Mater. Sci. Eng.* **390** 012029.
[66] Yabin Zhou, Shifeng Wang, Yong Zhang and Yinxi Zhang 2006 *Polym. Polym. Compos.* **14** 527.
[67] Ambiga K, Annadurai R and Vinayagamoorthy R 2016 *Indian J. Sci. Technol.* **9** 1.
[68] Vinayagamoorthy R, Kanakadandi Guru Subramanyam, Tuduru Nikhil Kumar and Yerasi Harshavardhan Reddy 2016 *Indian J. Sci. Technol.* **9** 1.

IOP Conf. Series: Materials Science and Engineering **954** (2020) 012027 doi:10.1088/1757-899X/954/1/012027

Effect of change of cross section and Frequency Analysis of Anti-Roll Bar Using FEA

T. Srinivasan[1], CH.V.L.N. Ram Gopal[1], P. Sasidhar[1], P. Vineeth Kumar[1] and ShaikSohail[1]

[1]Dept. of Mechanical Engineering, Koneru Lakshmaiah Educational Foundation, Vaddeswaram, Guntur District, Andhra Pradesh, India.

Corresponding Author: seenusreeja@gmail.com

Abstract. Anti-roll bars are the suspension system component in automobiles that are used to control the body roll effect. They have a strong impact on the handling functions of the engine. The design effects of the anti-roll bar are normal in different vehicles and the study is mandatory for each design change. Finite Element Analysis (FEA) is used for an accurate study of the anti-roll bar style. The finite element analysis is carried out in ANSYS. The paper comprises of pre-processing, evaluating, post-processing, and using the work bench to interpret the FEA results. Be that as it may, because of a high number of rehashed plans looks at, deciphering time and cost issues related with the utilization of general FEA bundle projects may create critical inconveniences in utilizing these bundle projects to perform against move bar structure examination. Vibration characteristics such as typical set vibration levels and frequency are varied within the automotive environment. Even if an automobile is driven, the random excitations from road inputs and intermittent vibrations from engine firing are felt by its components and structures. It can sometimes cause problems with reliability if the construction of the components is often not completely evaluated.

1. Introduction

Anti-roll bar is also known as a sway bar it's almost like a steel tube or rod structure that links the left and right suspension components that prevent body roll and the swaying of the vehicle that occurs during the vehicle's turning or due to the topography of the road. Anti-roll bar is an integral component of vehicle suspension system. The Figure 1 shows the torsional stiffness of the anti-roll bar gives the ability to decrease body roll which is called Rolling Stiffness. Anti-Roll Bar (ARB) is responsible in the Suspension System for comfortable driving, skidding behaviour. It can be installed in the front frame or Rear frame or both sides, irrespective of the suspensions are rigid axle type or independent type [1]. The body roll is described as the angle by which the body of the vehicle rotates around its longitudinal axis. The motion is not sophisticated for riders.

IOP Conf. Series: Materials Science and Engineering **954** (2020) 012027 doi:10.1088/1757-899X/954/1/012027

A stabilizer bar used in vehicle suspension is optimized to reduce the tension concentration for certain systematic limitations at the corner turns. The advantage of the anti-roll bar is that, by restricting the increase in camber angle due to body roll, it improves traction. Rider safety, handling and road retention are the three parameters a vehicle suspension device as to provide consensus solutions and is presented in Figure 1. The car and its passengers are safeguarded from shocks and road surface vibrations [2].

Figure 1. Effect of anti-roll bar on vehicle body roll

First factor for this is the reason that, even in secure cornering, so much roll disturbs the driver and gives an impression of possibility for roll over. And the driver is unable to push. The second reason for it is impact at the camber angles of the tyres the angle between central plane the wheel's symmetry and the vertical plane in the touch patch core. The camber angle has the function of connecting the wheel load and the tire's point of contact with the ground surface. The Figure 2 shows when vehicle is turning the effect of camber angle. This balance is disrupted as camber angle is modified due to body roll, and the tire touch patch often becomes smaller.

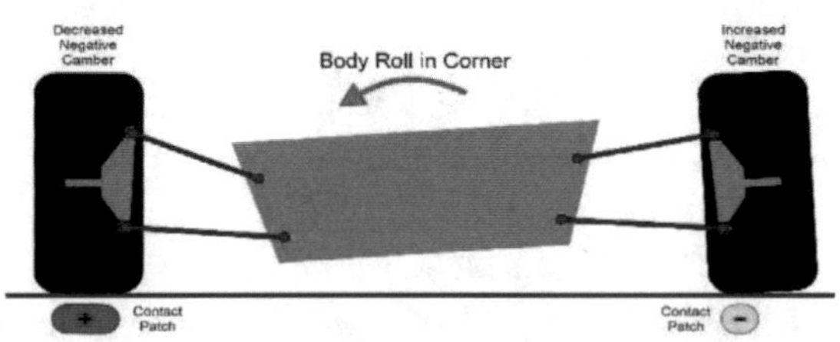

Figure 2. Camber angle on body roll in corner

2. Finite Element Modelling and Boundary Conditions

The Figure 3 shows three-dimensional computer aided model of the antiroll bar for rear suspension was developed using Solid works software and transferred by STEP to ANSYS Workbench 16.0.

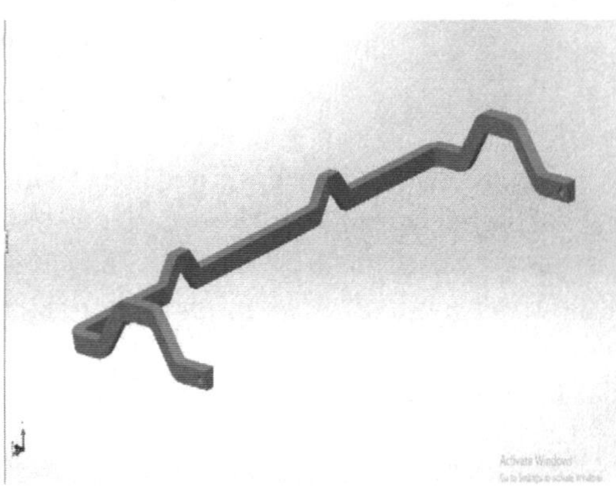

Figure 3. CAD Model of Trapezoidal Section

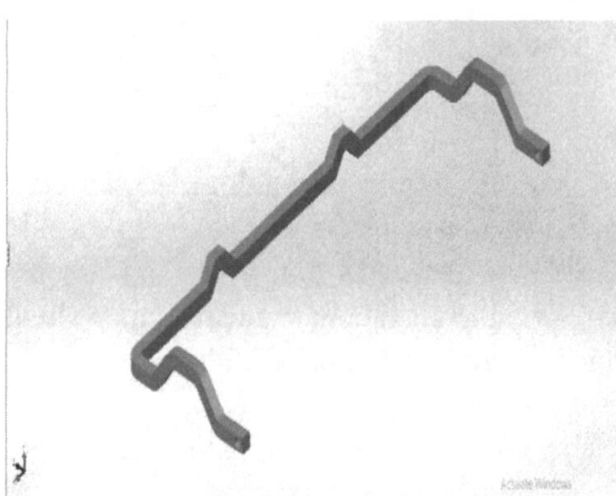

Figure 4. CAD Model of Square Section

Simulation is purely static means eliminated dynamic conditions which is affected during rolling. About the displacement constraints which is exist at two locations, first at the bar ends and second at bushing locations. At the bar ends considering pin joints displacement in x and z i.e. UX, UZ degrees of freedom and rotation in y and z directions i.e. ROTY and ROTZ degrees of freedom were constrained. The centre of the antiroll bar should be free to rotate and constrained on axial movement. The loading Force , applied at the bar ends, upward direction at one end and in downward direction at another end [1]. The Figures 3 and 4 show the CAD and finite element model of anti-roll bar. The individual anti roll bar was considered for the purpose of analysis and the effects of natural frequency and magnification factor were defined for different bars.

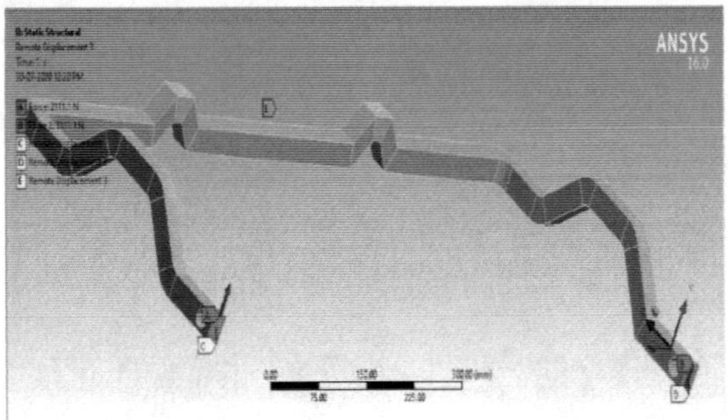

Figure 5. Loading and boundary conditions of Anti Roll Bar

3. Static Analysis

From static analysis, the results for square section are maximum value of von-mises stress is 659.79 MPa. The maximum principal stress is 647.18 MPa and the anti-roll bar deformation up to 73.193 mm. The results for trapezoidal section are maximum value of von-mises stress is 1151.3 MPa. The maximum principal stress is 1190 MPa and the anti-roll bar deformation up to 92.572 mm. For applying maximum for load of 2111.1 N. The Figures 5 and 6 are the corresponding values of Square section whereas the Figures 7 and 8 are the corresponding values of Trapezoidal section respectively.

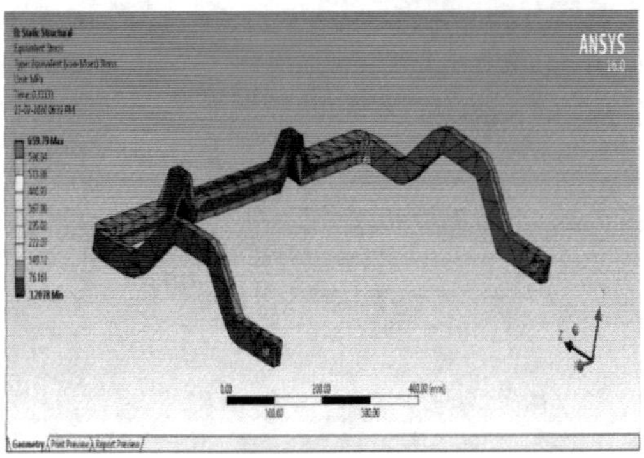

Figure 6. Square section: Von - mises stress

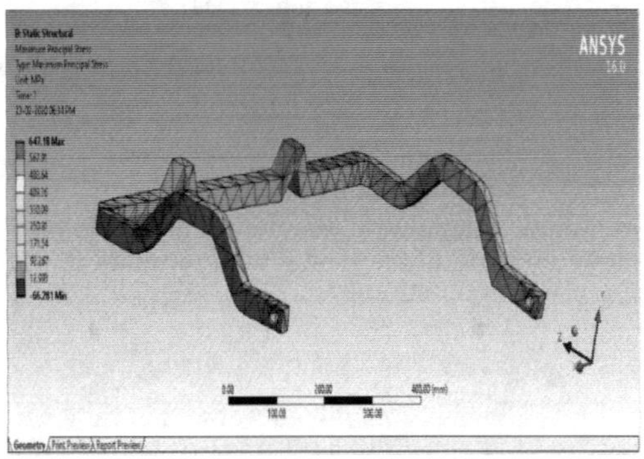

Figure 7. Square section: Maximum principal stress

4. Modal Analysis

The first six natural frequencies and the related mode shape of a bar can be identified during modal analysis. Any type of load or boundary conditions are not applied [7]. This research gives major importance for car riding comfort. The vibration and the disturbance under those limits. The natural frequency of both square and trapezoidal are in range 10-250 Hz. The Figures 10 and 11 represent the first and third modal shapes of Square, whereas Figures 12 and 13 represent the first and third modal shapes of Trapezoidal respectively.

IOP Conf. Series: Materials Science and Engineering **954** (2020) 012027 doi:10.1088/1757-899X/954/1/012027

The natural frequency and characterization of the modals of Square section and Trapezoidal section for Anti roll bar are tabulated in Tables 1 and 2 respectively.

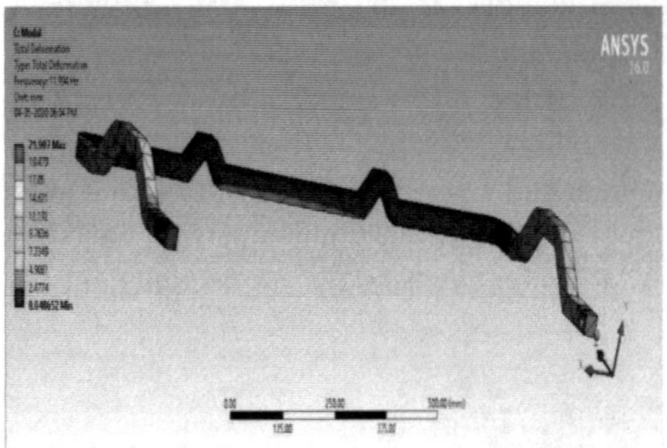

Figure 10. Anti-roll bar of Square: Modal shape of First

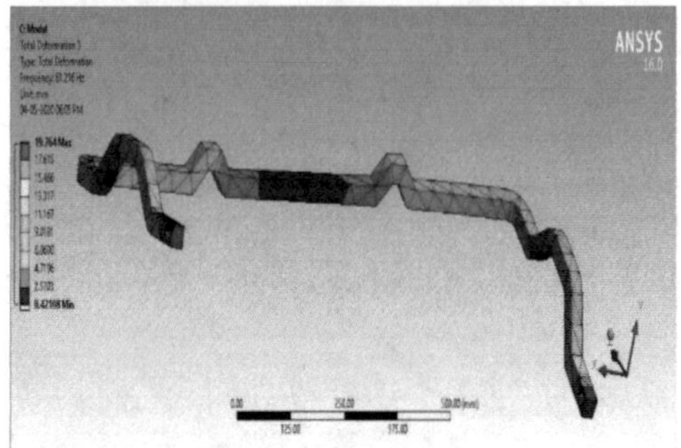

Figure 11. Anti-roll bar of Square: Modal shape of Third

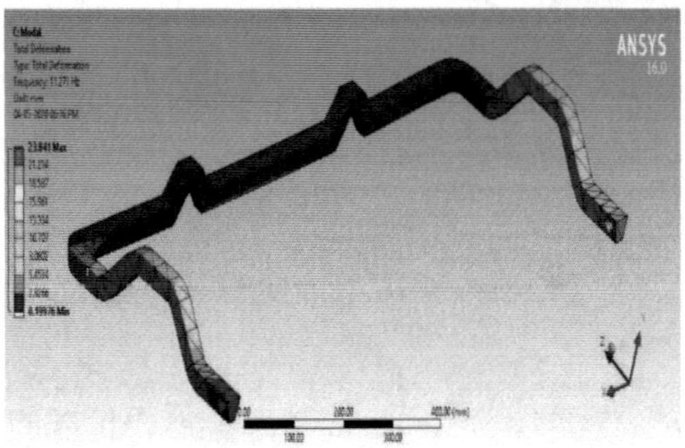

Figure 12. Anti-roll bar of Trapezoidal: Modal shape of First

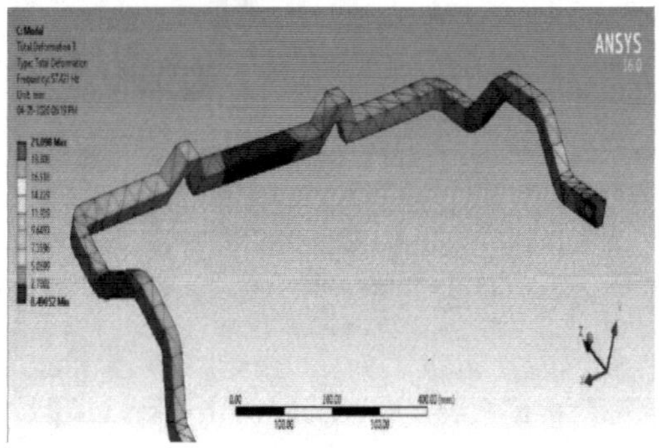

Figure 13. Anti-roll bar of Trapezoidal: Modal shape of Third

Table1. Natural Frequency and characterization of modal of Square section Anti roll bar

Frequency (Hz)	Characterization of modal
11.994	Vertical bending along y direction at free ends
14.869	Twisting at free ends with respect to x axis
61.216	Twisting at free ends
87.238	Bending at fixed ends
201.02	Twisting at fixed ends
245.06	Bending and Twisting at fixed ends

Table-2. Natural Frequency and characterization of modal of Trapezoidal section Anti roll bar

Frequency (Hz)	Characterization of modal
11.994	Vertical bending along y direction at free ends
13.901	Twisting at free ends with respect to x axis
57.421	Twisting and bending at free ends
82.144	Bending at fixed ends
189.93	Twisting at fixed ends
231.01	Bending and twisting at fixed ends

IOP Conf. Series: Materials Science and Engineering **954** (2020) 012027 doi:10.1088/1757-899X/954/1/012027

5. Random Vibration Analysis

Random vibration model is used to analyse the responses of systems under spontaneous loading. It refers to the non-periodic and non- deterministic properties. Due to this kind of vibration is statistical or probabilistic methods. The power spectral density uses load as input factor as random vibration analysis. Tables 3 and 4 show that the response of power spectral density for Square section and Trapezoidal section respectively. It is used to measure the frequency information and energy excitation. The random vibration analysis can be performed in high range frequency [7]. The standard way to describe the frequency of random vibration damage is in terms of its power spectral density [8]. The representation of random signals called zero mean Gaussian process. This major aspect of the random process provides a method for simple reproduce of random signals through vibration test system. They are three predominantly factors complicated vibration of various disruption for the frame: the first factor is caused by the rough path, the second factor is the harmonic vibration when the produced by engine while operating, the third factor is caused by the bar [9].

Table 3. The Response of power spectral density for Square

Frequency (Hz)	Deformation (mm)	Displacement (mm)	Stress (MPa)	Strain
11.994	21.907	40.01306061	3448.9	3.51E-07
14.869	13.93	13.05029928	260.62	1.48E-09
61.216	19.764	6.380941192	28.038	1.53E-09
87.238	16.925	3.28361064	37098	1.51E-06
201.02	14.119	0.991673271	4.3699	2.12E-10
245.06	12.695	0.657647209	285.93	2.05E-08

Table 4. The Response of power spectral density for Trapezoidal

Frequency (Hz)	Deformation (mm)	Displacement (mm)	Stress (MPa)	Strain
11.994	23.841	3.14E+05	49652	7.72E-09
13.901	15.899	8427.1	1889.5	3.11E-09
57.421	21.098	3893.2	40.285	4.87E-10
82.144	18.299	2614.3	53221	5.78E-07
189.93	15.125	1.04E-02	11.853	3.21E-11
231.01	13.384	2.4066	213.96	7.32E-09

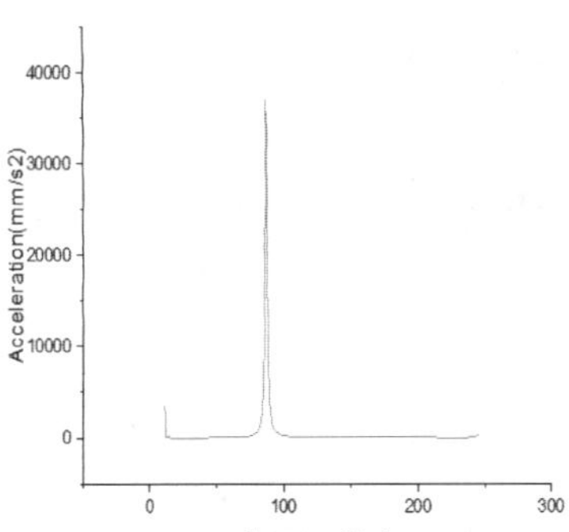

Figure 14. The results of Acceleration response spectrum analysis for Square

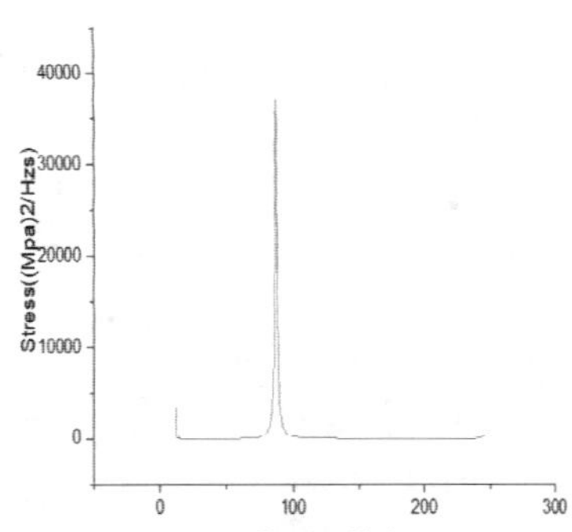

Figure 15. The results of stress response spectrum analysis for Square

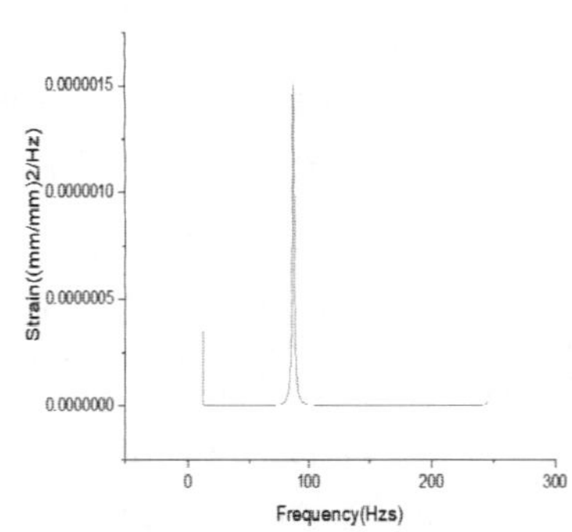

Figure 16 . The results of Strain response spectrum analysis for Square

IOP Conf. Series: Materials Science and Engineering **954** (2020) 012027 doi:10.1088/1757-899X/954/1/012027

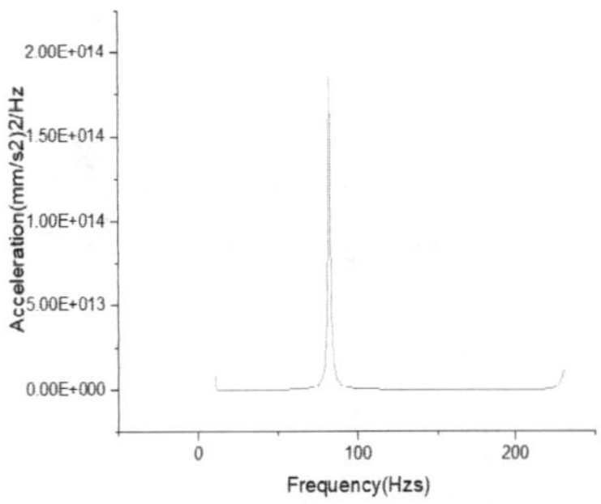

Figure 17. The results of Acceleration response spectrum analysis for Trapezoidal

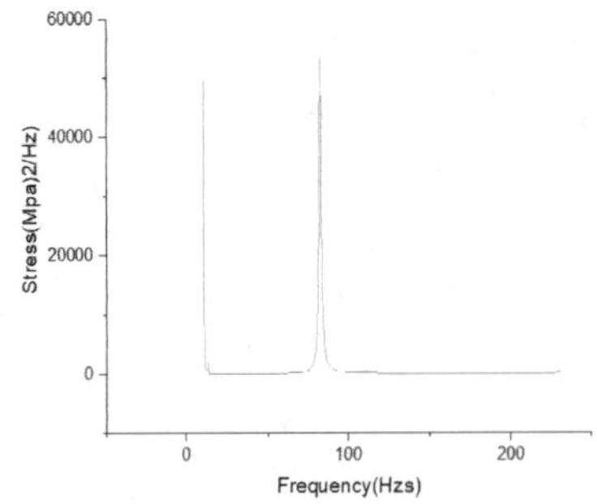

Figure 18. The results of Stress response spectrum analysis for Trapezoidal

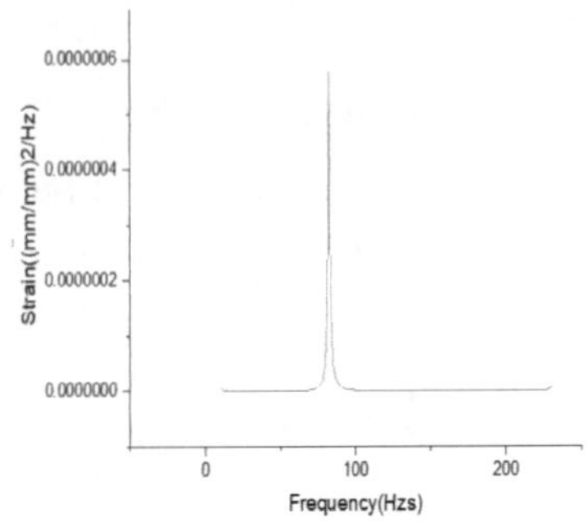

Figure 19. The results of Strain response spectrum analysis for Trapezoidal

The predicted results of power spectral density of acceleration, stress, strain graphs are shown Figures 14-19. For square section graphs are almost are similar but initial line is more strain parameter. For trapezoidal section graphs are not similar stress curve is more and strain curve is straight. These due to the variation of set natural frequency in different cross section which are predefined in ANSYS software [10 -12].

6. Conclusions

The anti roll bars are turn able vehicle components that direct effect on the performance of car. By changing the bar factors, properties can be enhanced. To have required roll stiffness is achieved with lowering bar by changing bar properties. The factor of size and endurance limit are totally depending on the diameter/ depth of bar. In the application ANSYS workbench, FEA static analysis for square are predicted maximum value principal stress is 647.18 MPa and von misses 659.79 MPa. Trapezoidal are predicted maximum value principal stress is 1190 MPa and von misses 1151.3 MPa. The minimum deformation is obtained for square section 73.193 mm. The frequency values of square and trapezoidal sections range from 10-250 Hz. For different sections deformation based on natural frequency are fluctuation. The main purpose is comfort ride the natural frequency frame should be 60 Hz.

7. References

[1] Amol Bhanage K and Padmanabhan, 2015 Static and Fatigue Simulation of Automotive Anti Roll Bar before DBTT, *Int J Appl Engg Res* ***10*** (71).

[2] Kumar G D, Kommuri N, Kota H, Prasanth V and Hameed S. 2018 Vibrational analysis of a human body on an automobile *J Adv Res Dynam and Cont Sys* ***10*** (6) 39-45.

[3] Durali M and Kassaiezadeh A, 2019 Design and Software Base Modeling of Anti-Roll System, *SAE Tech Paper* 2002-01-2217.

[4] Rama Sastry D.V.A., Ramanaa K.V., Mohan Raob N. 2019 Analysis and prediction of performance of mr damper at different currents and control strategies for quarter suspension system of a roadway vehicle', *Inter J Vehic Struct Sys* ***11*** (1) 27-32.

[5] Trinath Sai Kumar M., Md R., Narayana M.V., Murali P., Kammuluri B 2018 Design and analysis of crane hook with different cross sections *Inter J Engg Tech(UAE)* ***7*** 262-263.

[6] Bala Subramanyam P.N.V., Rao B.N., Prakash R.L., Rajavardhan D.S.K., Sankarudu B.S.K., Ganesh P 2019 Modal analysis on go-kart chassis *Inter J Engg Adv Tech* ***8*** (4)1701-1705.

[7] Kekan A.H., Kumar B.R 2019 Natural frequency response of rotor shaft to crack depth and crack location *Inter J Mech Produ Engg Res Devel* ***9*** (6) 499-510.
[8] Deepak Kumar G., Kommuri N., Kota H., Prasanth V., Hameed S 2018 Analysis of a human body on an automobile *J Adv Res Dynam Cont Sys* ***10*** (6) 39-45.
[9] Ravi Prakash Babu K., Rao P.K.V., Ali M.A., Raghu Kumar B 2019 Crack simulations in shaft-hub-blade system using modal analysis *Int J Rec Tech Engg* ***8*** (1) 2646-2650.
[10]Dama K.K., Surya S., Babu C., Pavan, Kiran R 2018 Modal analysis of an automotive subsystem using cae techniques *Int J Mech Eng Tech* ***9*** (3) 1157-1162.
[11]Rao P.K.V., RaghuKumar B., Veera Sai Chandh Y., Teja A 2019 Finite element analysis of CNT reinforced aluminium composite subjected to mechanical loading *Mater Today: Proceed* ***16*** 308-313.
[12]Venkatesulu M., Sambaiah C.H 2018 Modelling and analysis of static and dynamic of connecting lugs and mechanical bracket using abaqus *Int J Mech Prod Engg Res Devel* ***8*** (2)1249-1260.

IOP Conf. Series: Materials Science and Engineering **954** (2020) 012028 doi:10.1088/1757-899X/954/1/012028

Impact of mechanical properties on sheet metal forming processes by using single-point incremental shaping method

T.Srinivasan[1], P.Ramu[2], G.Suresh[3], Y.S.Govardhan[4] and S.Srinivasan[3]

[1] Dept. of Mechanical Engineering, Koneru Lakshmaiah Educational Foundation, A.P., Vaddeswaram, Guntur, India

[2] Dept. of Mechanical Engineering, Sri Sai Ram Institute of Technology, TN, Chennai, India.

[3] Dept. of Mechanical Engineering, Rajalakshmi Institute of Technology, TN, Chennai, India.

[4] School of Mechanical Engineering, Vellore Institute of Technology, TN, Chennai, India

[1]Corresponding Author: seenusreeja@gmail.com,

Abstract

The usage of single tool using Computer numerical control (CNC) machines in processes. These flexible processes satisfying the need of the manufacturer through the Single point incremental forming (SPIF). Modern and die less sheet metal forming technique applied in this technique mainly for small batch and custom made items of different shapes and dimensions. It offers many advantages compared with traditional sheet processes, especially in terms of flexibility and material formability. The basic idea is to build the knowledge about the advanced manufacturing process and do preliminary investigation on the applicability of single point incremental forming process. In obtaining the mechanical properties, two types of tests are being conducted. They are tensile test and composition test. These test conducted according to ASTM E8M and ASTM B209 respectively. The present work is to be carried out on Aluminium alloy which have very important industrial application with a light weight to strength ratio. In this work the stress, strain, formability, wrinkles, deformation will be studied.

1 Introduction

Forming processes are known to be the basic manufacturing processes that allow use of acceptable stresses such as compression, friction, shear or combined stresses, which in turn induces plastic deformation to the materials creates the necessary shapes. Metals are the main material used in these processes because they have very strong properties of formability and malleability. Because of the huge need for different goods demanded by the public metallic materials plays the lead role as the raw materials in the processes of formation. Nevertheless, due to a wide demand for plastic based goods, other compounds like plastic also nowadays used as the raw materials in most of the fabrication industries, where as during the forming operation of metallic materials, no substance is extracted during the entire process of formation of materials instead materials are deformed and displaced. Forging, extrusion, rolling, rotary swaging, explosive formation, electromagnetic formation and sheet metal working are some of the following examples of forming techniques. Single point incremental forming (SPIF) is a new fabrication methodology which allows complicate profileusing the Computer Numerical Control (CNC) movement of a hemispherical forming tool. Due to size factor, the development of thin sheet metal forming processes is limited in the manufacturing processes since it requires the massive state of art technology as it needs the miniature size of CNC machines. The initial grain size during the forming processes also plays a major role in during formability processes [1].

Aluminum alloys combination are highly usage in aero and automobile industrial appllications owing to their density low and good tensile properties, lower corrosive, wear and less thermal co-efficient of expansions as matched to regular metals alloys combination [2]. Normally the unique contrast during the operation is happen between the two hardened metals during the deforming operation by considering the spring back effect and thickness of the metals. Similarlly the outcome of the compared models are checked with two consecutive specimens. Atlast the final formed part has to come across the cumulative of all parameters of the forming operation [3]. Most of the leading fabrication industries having wide knowledge in the forming procesess are limited their service in making the minature size products with help of the forming technology, since it needs the massive machine investment and special type of state of art technology as well as the formation of the grains after the fabrication processess must be elaborately studied into microscale [4].

The sheet metal is one of the leading metal forming process considered as the economically accepted one in most of the industries, small intricate shapes and size products can be easily formed with help of the forming processes. Moreover during the forming metals are often coming across the thin sheets and subjected with the tension, compression and shear forces. Most of the everyday usage products are completely manufactured with help of this forming technology. Similarly depends upon the usage of the raw materials the metals are produced with various thickness, it varies from 0.1 mm to 6 mm, beyond the size of the 6 mm materials are considered as the structural steels. Importance to analysis the different reasons for affecting the friction surface. There are so many scientific journals discussed and reported about friction in different way [5].The strength of the aluminium alloys based on the strain hardening effect of alloying elements such as silicon,iron and magnesium [6-7]. Increasing the weight of an automobile can decrease fuel efficiency and maximise its impact on the environment. Our main need is to reduce the vehicle weight therefore to arrange alternate aluminium alloys [8]. The micro level analysed for the evenscattering of the material and the mechanical studies such as Tensile strength, hardness and impact strength of the composition were evaluated for varying alloying materials with varying percentage [9-12].

2 Materialsand Tool

A 6061 T6 square aluminium foil (130mm×130mm×1mm) was hold on the specially design fixture frame and was formed by SPIF. The raw material or semi - finished or finished piece which is formed by executing various processes. The workpiece is commericially obtainable aluminium alloy sheet of grade 6061 T6 series soft-Temper Foils with 1mm thickness and 2.7 gcm^3, young's modulus of 70 GPa and Poisson's ratio of 0.3. The Figure.1 shows worksheet and single point tool of varies size and shapes. A hemi spherical end tool is used for making various shape elements, flat shape single point tool is usedfor fabricating of flat shape components. The shape and diameter of a tool has a highly stimulus on forming characteristics and process.

(a) (b) (c)

Figure 1. (a) Work piece, (b) single Point tool and (c) Tool with different shape [13]

3. Methodology

3.1 Mechanical Test

The widely considered parameters such as r-value, n-value, K-value force and their consecutive percentage elongations are the prime and main technical interest in tensile studies during forming processes. The tensile specimens are been cut from the rolling direction, from the sheet with different orientations, namely transverse (90), rolling (0) and diagonal (45) are prepared according to the ASTM standard specifications like ASTM-E8M-9. The above set of parameters for all three sheet orientations is tested in this work, and this has been used in this work. Using the relationships, the mean value of parameters such as n-value, r-value, K-value and planar anisotropy can be determined. The figure 2 shows the specimens before subjecting with the tensile test.

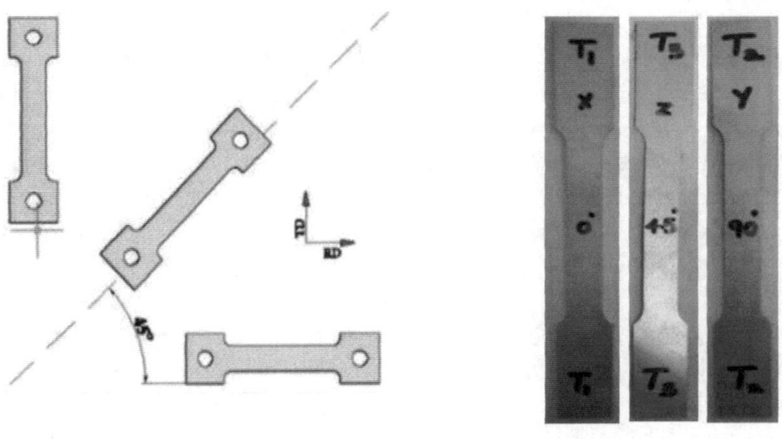

Figure 2. Before Tensile Test Specimen

Where,

T1 (x-axis) 0° = Rolling Direction,
T2 (y-axis) 90° = Transverse Direction,
T3 (z-axis) 45° = Diagonal Direction.

3.2 Single Incremental Forming

Single incremental shaping process is based on the technique of creating the desired portion and shape by attempting to move a punch, following a predetermined tool direction as shown in Figure 3. A comprehensive investigational and numerical analysis of sheet incremental formation technique by using a universal purpose CNC machine, fitted with a special instrument and forming the sheet with distinct die geometries, is shown in the following set up. The key challenge associated with this method is to manufacture parts with adequate precision and consistency in the measurements. The forming tool feed rate is 10 mm/min and oil coolant was supplied as a coolant to minimize the friction.

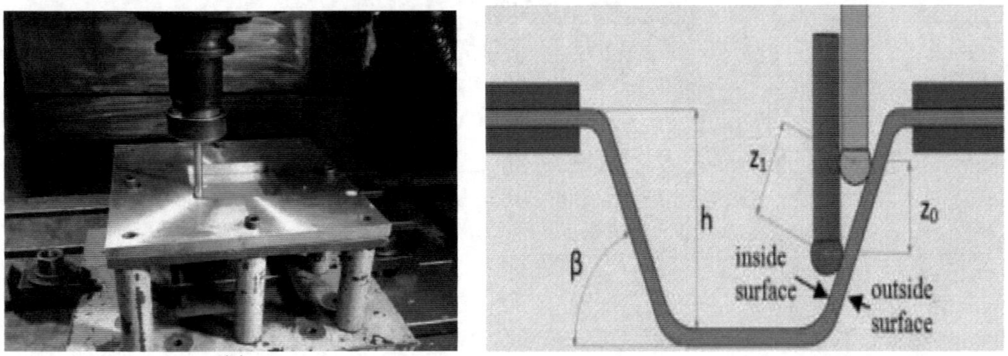

(a)

(b)

Figure 3. (a) Experimental setup and (b) Input parameters of the part processed [13]

4 Result and discussion

4.1 Mechanical Properties

Where, n is that the strain hardening exponent and K is that the strength coefficient. A log-log plot of true stress and true strain to maximum load will end in are line. The linear slope of this is often n and K is that the true stress at $\varepsilon = 1.0$ the strain hardening exponent may have values from n = 0 to n = 1. If n = 0, perfectly plastic solid and n = 1, elastic solid. Most of the metal have n values between 0.1 to 0.5. The strain hardening exponent (n-value), plastic strain ratio (r-value) as well as strength coefficient (K-value) and finally nr-value formability properties are described in table 1. The void parameters for the experiment are enlisted in the response of major strain resulting from fracture and forming limits for the experiment. Figure.4 shows the after tensile test specimen.The trend of the force is regular which have single peaks and valleys as shown in the figure 5.

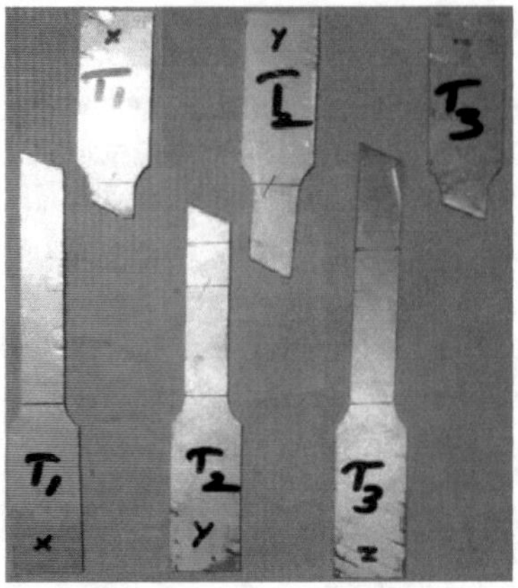

Figure 4. After Tensile Test Specimen

Average plasticisation strain ratio,

$R_{av} = (R_0+R_{90}+2R_{45})/4$ Average strain hardening exponent, (1)

$n_{av} = (n_0+n_{90}+2n_{45})/4$ Average strength coefficient, (2)

$K_{av} = (K_0+K_{90}+2K_{45})/4$ The average strength coefficient (3)

In a similar manner, the average intensity coefficient was also estimated. The metal sheet's compressive strength, as well as the common stresses like tensile stress, ultimate tensile stress, strain rate, percentage of elongation, strain hardening exponent component, plastic strain ratio of the metals and atlas the strength coefficient are stated. The forces obtained from the experimentsare plotted in the form of graphs with respect to stroke.

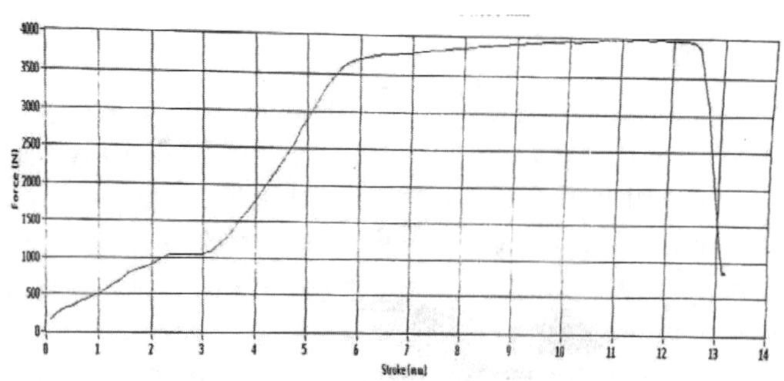

Figure 5. Force vs Stroke response graph

Table 1. Tensile properties of AA6061 sheets rolled at room temperature.

Orientation	Yield Stress (MPa)	True Tensile Stress (MPa)	% Elongation	Strain Hardening Exponent, n	Strength Co-efficient, K	Plastic Strain Ratio, R
0°	286.49	309.27	8.06	0.508552	422.4	0.489
45°	241.35	270.00	6.64	0.516908	393.96	0.578
90°	257.20	278.07	7.42	0.566855	677.43	0.672
Average	261.68	285.78	7.37	0.39807	373.45	0.43475

4.2 Forming With a Tumbler

The initial trial was carried out on single point incremental process to optimize the force and surface finish of the fabrication of aluminium metal forming. The final part of the profile are shown in Figure 6. First phase of forming, the shear and compression force were governing the process of Tumbler shape profile of Al6061 T6 alloy. The second phase of plastic deformation, the uniaxial strength is extremly influence sheet forming. The major strain, stress & displacement within the limit. The speindle speed, depth and forming angle is used for SPIF in 500 rpm, 40mm and 60° repectivley. Figure 7 shows the Crack and Poor surface finish, Tool Failure during the SPIF process.

Figure 6. Final Product

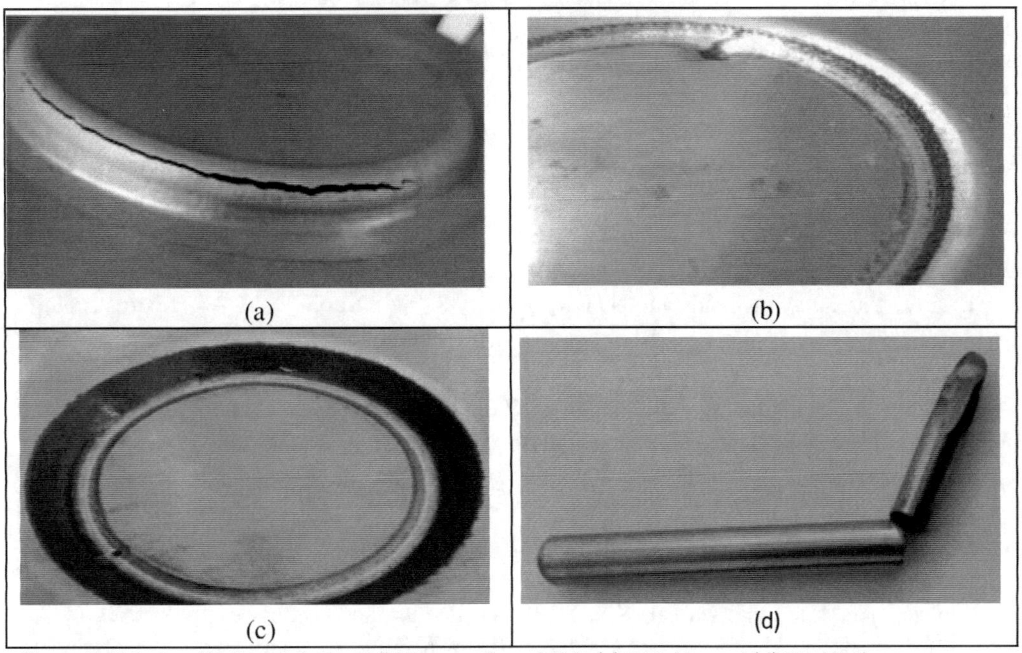

Figure 7. Mode of Failure (a) Crack (b)Poor surface finish (c) Wrinkle and (d) Tool Failure

Conclusion

Thus, the advanced manufacturing process study and preliminary investigation on the applicability of the incremental fabrication process for a single point was carried out. From the results of tensile test, it is confessed and clear that the three operating parameters that influences the formability of single point incremental forming was derived and studied for the analysis. The outcomes of the modelling study which evolves empirical relations in the view to find out the formability parameters such as n-value, K-value and R-value have been found out. From the above work, it is also observed that the parameters or factors which are responsible for the failures such as crack, wrinkles, roughness and tool failure were found from the iterations. Based on the iterations,the suitable parameters for the forming was evaluated and the wrinkle-free parts have been formed out successfully.

Reference:

[1] R. Ben Hmida, S. Thibaud, A.Gilbin, F. Richard, 2013, Influence of the initial grain size in single point incremental forming process for thin sheets metal and micro parts: Experimental investigations, Materials & Design, Vol.45, pp: 155-165.

[2] Sathyaraj S., Venkatesan K., Srikanth J.(2018), 'Fabrication of aluminium 6061- SiC-Al2O3 mmc and hmmc by strir casting technique and comparing the mechanical properties', International Journal of Mechanical and Production Engineering Research and Development,Vol.8(1), PP. 635-642

[3] GabrielCenteno, IsabelBagudanch, A.J.Martínez-Donaire, M.L.García-Romeu, C.Vallellano, 2014, Critical analysis of necking and fracture limit strains and forming forces in single-point incremental forming, Materials & Design, Vol.63, pp: 20-29.

[4] Sivakandhan C., Babu Loganathan G., Murali G., Suresh Prabhu P., Marichamy S., Sai Krishnan G., Pradhan R. (2019), 'Material characterization and unconventional machining on synthesized Niobium metal matrix', Materials Research Express, 7(1), PP:1-22

[5] Seshacharyulu K., Bandhavi C., Naik B.B., Rao S.S., Singh S.K.(2018), 'Understanding Friction in sheet metal forming-A review', Materials Today: Proceedings ,5(9), PP. 18238-18244.

[6]Chakravarthi, KVA; Koundinya, NTBN; Murty, SVSN; Rao, BN, 2018, Microstructure, properties and hot workability of M300 grade maraging steel, DEFENCE TECHNOLOGY, 14, 51-58.

[7]T. Sampath Kumar, A.VinothJebaraj, E.Shankar, N.Tamiloli, K.Sivakumar, 2019, Metallurgical and mechanical characterization of TiCN/TiAlN and TiAlN/TiCN bilayer nitride coatings, Surfaces and Interfaces,Vol.15, pp: 256-264.

[8] Nagabhyrava R., Kota S., Geda A., Gudi S.H. (2017),'Investigation of mechanical properties of AL 7075/SIC/GR hybrid metal matrix composites',International Journal of Mechanical Engineering and Technology,8(5),PP.265-269.

[9] Venkatesulu M., Rama Kotaiah K. (2019), 'Production and mechanical properties of AL 6063/B4C composites', Journal of Mechanical Engineering Research and Developments, 42(1), PP.46-49.

[10]Rao T.B., Goutham Karthik M.S.V.S., Raviteja Ch.K.S., RajaSekhar Reddy G., Sravan Kumar Reddy N., Shanmukha Prasad V.(2018), 'Microstructural and mechanical properties of AL6061/GR composites processed though stir casting', International Journal of Mechanical Engineering and Technology ,9(4), PP. 28-34.

[11]Rao T.B., Venkata Sai Prem K., Yashika N., Chevvakula S., Nalli V., Prasad V.S.(2018), Microstructural and mechanical properties characterization of Al6061/nano TICP composites fabricated through stir casting, International Journal of Mechanical Engineering and Technology ,9(5), PP. 193-199.
[12]Rao, TB. 2018, An Experimental Investigation on Mechanical and Wear Properties of Al7075/SiCp Composites: Effect of SiC Content and Particle Size, Journal of Tribology-Transactions of the Asme, 140(3).
[13] M. Rimašauskas, K. Juzėnas, R. Rimašauskienė, E. Pupelis, 2014, The research of single point incremental forming process for composite mould production, MECHANIKA, Vol. 20(4), pp: 414–419.

IOP Conf. Series: Materials Science and Engineering **954** (2020) 012029 doi:10.1088/1757-899X/954/1/012029

Performance and Emission Characteristics of C.I Engine using Jatropha and Pongamia Mixed Biodiesel

S Arunprasad[1], Hari Jagadeesh Iyer[1], K Omkaar[1], M Jayaganth[1] and K Palani[2]

[1] Department of Production Engineering, Sri Sairam Engineering College, Chennai, 600 044, Tamil Nadu, India

[2] Department of Mechanical Engineering, Sri Chandrasekharendra Saraswathi Viswa Mahavidyalaya (Deemed to be University), Kanchipuram, 631561, Tamilnadu, India

E-Mail: sarunprasad75@gmail.com

Abstract. In the present study, experiments were conducted on compression ignition engine and the performance and emission characteristics of a diesel engine using jatropha and pongamia mixed biodiesel were observed. The mixed biodiesel was mixed with diesel in different combination of MB25 (25% mixed biodiesel and 75% diesel), MB50 (50% mixed biodiesel and 50% diesel), MB75 (75% mixed biodiesel and 25% diesel), and MB100 (100% mixed biodiesel) and the results were compared with diesel. From the results, it was found that the brake thermal efficiency is increased when increasing diesel amount to mixed biodiesel and it was lower for MB100. The reduction in brake specific fuel consumption has been found for MB25, MB50, MB75 when compared to MB100. The reduction of HC (hydrocarbon), CO (carbon monoxide), smoke emissions, and increase of NOx (nitrogen oxides) for different loads were observed and compared with diesel.

1. Introduction

Owing to the enormous increase in the number of automobiles and the increasing electricity demand has resulted in greater demand for petroleum products. The increasing demand for petroleum products has led to the oil crisis in recent times. Therefore, there is an urge for finding out a replacement and attention has been focused on developing renewable or alternative fuels to replace petroleum based fuels for vehicle usage. Senthil Kumar et al. [1] conducted experiments with jatropha oil and methanol on a dual-fuel engine. He reported that dual fuel engines can use a wide range of fuels and yet operate with low smoke emissions and high thermal efficiency. HC and CO emissions were reported to be higher in the dual fuel mode. Pramanik [2] showed that 50% pure Jatropha curcas oil can be substituted for diesel to use in an IC engine without any major operational difficulties with acceptable BTE and BSEC.

The performance of jatropha oil blends in a diesel engine was investigated by Forson et al. [3]. The test showed that jatropha oil could be conveniently used as a diesel substitute in a diesel engine. The test further showed increases in BTE, brake power, and reduction of BSFC for jatropha oil and its blends with diesel. Biodiesel from Jatropha curcas and Pongamia pinnata were used as fuel in a direct injection diesel engine by Kumar et al. [4] and reported that 20% blend gives comparable performance and less emission. Hebbal et al. [5] carried out experimental investigations on a single-cylinder DI diesel engine to examine the suitability of deccan hemp oil as alternative diesel fuel. The BSFC was reported to be higher and the BTE was lower slightly than that of diesel. From the emission analysis, it was observed that the smoke, CO and HC emissions at full load for neat deccan hemp oil are higher than that of diesel.

The authors also validated the performance of deccan hemp oil, as the results are in good comparison with the results of jatropha and pongamia oils. The experiments were conducted to examine properties, performance, and emission characteristics of different blends (B10, B20, and B40) of pongamia, jatropha and neem in comparison to diesel. They reported that B20 had a closer performance to diesel with less emission of HC, CO and smoke. Venkateswara Rao et al. [6] concluded that the pongamia methyl ester gives an improved performance compared to jatropha and neem methyl esters. It is attempted single cylinder four stroke diesel engine on the performance, combustion, and emission characteristics of different methyl ester of biofuels including thevetia peruviana seed oil, jatropha oil, pongamia oil, mahua oil and neem oil. It is concluded that methyl ester of thevetia peruviana seed oil (METPSO) has comparable engine performance with a lesser amount of emission compared to other blends. Brake thermal efficiency increases with increasing brake power for all fuels. This is due to a decrease in heat loss and raise in power developed with an increase in load. It is observed that with the brake thermal efficiency curves of five blends of biodiesel closely follow that of diesel and the maximum deviation is found to be 9.39% for neem oil at the maximum load. It is only 2% for METPSO. This is due to the higher energy content and lower density of the METPSO blend compared to other bio-diesel blends. The experiment was conducted by Balusamy et al. [7].

Bhupesh Sahu et al. [8] conducted an experiment in CI engine using Jatropha oil and evaluated the performance and emission characteristics of a diesel engine using different blends of methyl ester of Jatropha with mineral diesel. From the results, it is concluded that the decrease in brake thermal efficiency and an increase in brake specific fuel consumption was achieved. There was a reduction of CO, HC, NOx, Smoke opacity, and an increase of CO_2 and NOx emission. Suryawanshi et al. [9] conducted experiments on various blends using pongamia oil methyl ester with diesel on CI engine. There is significant progress in engine performance with a reduction in HC, CO and Smoke emissions. NOx emissions are found to be slightly high. Abed et al. [10] conducted experiments on a single cylinder diesel engine using Jatropha, palm, algae, and waste cooking oils. Emissions are measured and compared with diesel. CO, HC, CO_2 and smoke emissions are lower for biodiesel mixtures B10 and B20 (Jatropha, algae, and palm) compared to diesel fuel. NOx emissions from all biodiesel mixtures B10 and B20 increases than diesel fuel for all biodiesel blend B10 and B20. Arunprasad et al. [11] made an attempt to study the influence of injection pressure (IP) and injection timing (IT) on the performance and emission characteristics of diesel engines by using mixed biodiesel (Thevetia peruviana, Jatropha, Pongamia, and Azadirachta indica). The injection pressure is varied from 200 to 230 bar and the injection timing is varied from 23 to 29° bTDC at an increment of 10 bar and 2° bTDC, respectively, and the results were compared with diesel. From this study, the results showed that the brake thermal efficiency (BTE) was increased by 2.4% with an increase in injection pressure and 1.5% with an increase in the injection timing for the maximum load, but lesser than diesel. Furthermore, a reduction of 5.08% of brake specific fuel consumption (BSFC) has been noticed for the rise in IP and IT with loads but higher than diesel. The reduction was 34.17%, 53.85%, and 29.7% and 29.17%, 53.85%, and 21.95% of hydrocarbons (HC), carbon monoxide (CO), and smoke emissions, respectively, at 230 bar injection pressure and 27° bTDC injection timing. Also, a significant increase in nitrogen oxides (NOx) and carbon dioxide (CO_2) emissions at the maximum load was observed by increasing the injection pressure and injection timing. Arunprasad et al. [12] experimented with neem oil biodiesel blends and diesel value was compared with these results and presented. The result clearly shows that blend B60, compression ratio 18:1, and 6 Kg load was the optimum parameters.

IOP Conf. Series: Materials Science and Engineering **954** (2020) 012029 doi:10.1088/1757-899X/954/1/012029

2. Experimental Test Setup

A 3.5 kW, 1500 rpm, Kirloskar diesel engine [13] is used in this investigation as shown in Figure 1. The detailed specification is given in Table 1. Two separate fuel tanks with a fuel switching system are used, one for diesel (D100) and the other for biodiesel (MB100). Fuel consumption is measured using an optical sensor. A differential pressure transducer is used to measure the air flow rate. The engine is coupled with an eddy current dynamometer to control engine torque through a computer. Engine speed and load are controlled by varying excitation current to eddy current dynamometer using a dynamometer controller. A piezoelectric pressure transducer is installed in the engine cylinder head to measure combustion pressure. A high precision crank angle encoder is used to give signals for the top dead center and crank angle. An AVL exhaust gas analyzer and AVL smoke meter are used to measure emission parameters and smoke intensity respectively. Thermocouples (chrommel alumel) are used to measure exhaust temperature, coolant temperature, and inlet air temperature. The properties of fuel are shown in Table 2.

Table 1 Engine Specifications

Engine Specifications	
Make	Kirloskar –TV1
Power and Speed	3.5 kW and 1500 rpm
Type of engine	Single cylinder, DI and 4 Stroke
Compression ratio	16.5:1
Bore and Stroke	80 mm and 110 mm
Method of loading	Eddy current dynamometer
Method of starting	Manual cranking
Method of cooling	Water
Type of ignition	Compression ignition
Inlet valve opening	4.5^0 before TDC
Inlet valve closing	35.5^0 after BDC
Exhaust valve opening	35.5^0 before BDC
Exhaust valve closing	4.5^0 after TDC
Fuel injection timing	23^0 before TDC
Nozzle opening pressure	210 bar
Lube oil	SAE40

Figure 1 Experimental setup

Table 2 Properties of fuel

Property	Diesel	Mixed Vegetable oil	Mixed Biodiesel (MB100)	ASTM code
Calorific value, kJ/kg	43200	39854	38720	D4809
Specific gravity	0.823	0.892	0.831	D445
Kinematic viscosity (at 40°C) cSt	2.5	38	4.16	D2217
Cetane number	50	48	52	D4737
Flash point, °C	56	148	88	D92
Fire point, °C	64	158	96	D92
Cloud point, °C	-8	8	10	D97
Pour point, °C	-20	6	9	D97
Ash content, %	0.001	0.003	0.003	D976
Kinematic viscosity (at 40°C) cSt	2.5	38	4.16	D2217
Cetane number	50	48	52	D4737
Flash point, °C	56	148	88	D92

3. Results and Discussion

3.1 Performance Characteristics

The variation of brake thermal efficiency with brake power for various blends of mixed biodiesel is shown in Figure 2. The BTE increases with all loads for all blends of biodiesel [14]. It is noticed from the graph that B25 has given maximum BTE when compared to other blends and minimum for MB100.The BTE values for MB25, MB50, MB75 and MB100 are 29.2 %, 28.0 %, 27.8 % and 27.6 % respectively, whereas 30.2% for diesel at full load. This is due to the lower calorific value and low viscosity of the fuel blends.

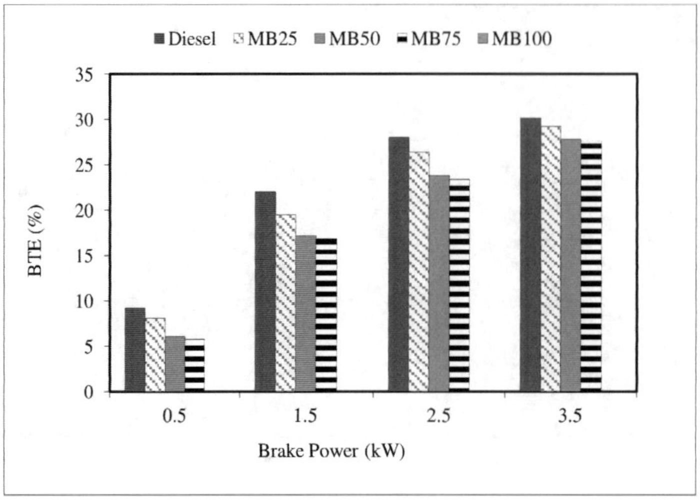

Figure 2 BTE Vs Brake Power

The variation of brake specific fuel consumption with brake power for various blends biodiesel is shown in Figure 3. The BSFC of mixed biodiesel is higher than the value when compared to diesel. For all blends, the BSFC decreases with an increase in brake power. At the higher load, the BSFC for MB25 and MB100 is found to be higher by 3.33 % and 13.33 % respectively when compared to diesel. It is noticed that BSFC values for blends such as MB25, MB50, MB75, and MB100 are 0.31 kg/kWh, 0.32 kg/kWh, 0.32 kg/kWh and 0.34 kg/kWh respectively, whereas 0.3 kg/kWh for diesel This is due to more viscosity and specific gravity and lower heating value of MB25 compared to diesel.

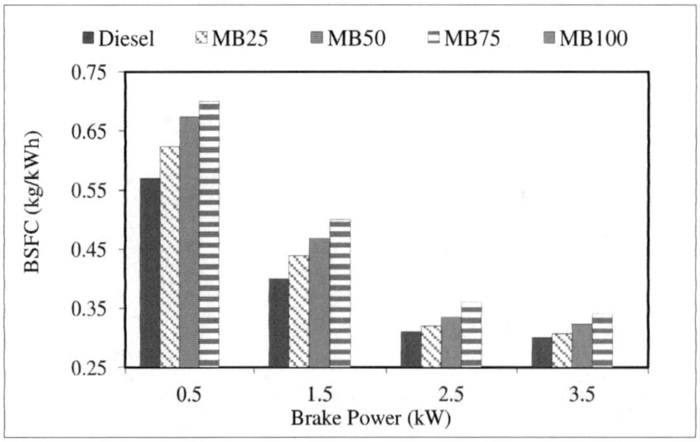

Figure 3 BSFC Vs Brake Power

It is concluded that the exhaust gas temperature increases gradually for various blends of mixed biodiesel as the load increases. It is shown in figure 4. In comparison with diesel, MB100 has the extreme EGT at the higher load. The maximum deviation for MB100 is 17.72 % and MB25 is 6.33 % compared to diesel fuel at the higher load condition. It is due to the mitigation of friction power at higher load conditions.

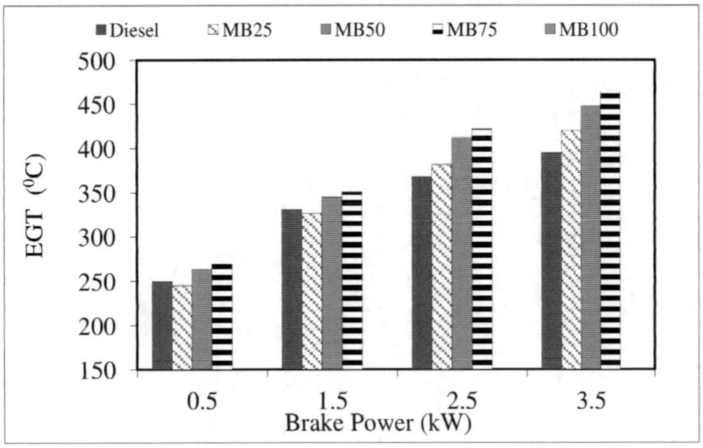

Figure 4 EGT Vs Brake Power

3.2 Emission Characteristics

The CO emissions for various blends with brake power is shown in figure 5. It is noted from the figure that the CO emissions are decreased with an increase in load. The growth of CO emission is mainly due to the shortfall of oxygen in the fuel-rich mixture during halfway stages of combustion, which leads to incomplete combustion of the fuel. CO emissions are lower for all mixed biodiesel blends for all loads compared to diesel. The maximum variation is for MB100 (33.33%) at higher load and for MB25 it is only 8.33 %. This is due to the higher combustion temperature in the engine cylinder accelerates the oxidization rate. The HC emissions [15] for different blends with load is shown in the figure. It is shown in Figure 6. HC emission decreases when the load increases for different blends of mixed biodiesel. In MB100, HC emission is lesser than diesel by 13.48 % at higher load and for MB25 the variation is 4.35 % when compared to diesel. This is due to the higher content of oxygen which leads to more complete burning than diesel fuel.

The variation of NOx emission with brake power for various blends of mixed biodiesel is shown in figure 7. It is noted that NOx contents are more for all blends of mixed biodiesel than that of diesel. Due to lesser combustion chamber temperature NOx value is lower at low load and increases when the load increases. The NOx emission variation for MB100 and MB25 are 20 % and 6.36 % respectively when compared to diesel. The NOx emissions for all the blends are noticed as 620 ppm and 625 ppm for MB50 and MB75 respectively while 550 ppm in the case of diesel. It is due to the higher adiabatic flame temperature during combustion and also due to an increase in the local temperature and the concentration of oxygen within the fuel spray envelope.

The variation of smoke emissions with brake power is shown in figure 8. The smoke emissions of mixed biodiesel for all blends are lower than that of diesel. At full load, the smoke emission of MB100 is found to be lesser by 9.75% compared to that of diesel due to oxygen present in the mixed biodiesel and for MB25, the smoke emission is lower by 4.88% when compared to diesel. The various values of smoke emissions recorded for the different blends such as MB50 and MB75 are 7.32 %, 8.54 %. It is due to the presence of oxygen in the mixed biodiesel and its blends that contributes to improved combustion at maximum load conditions.

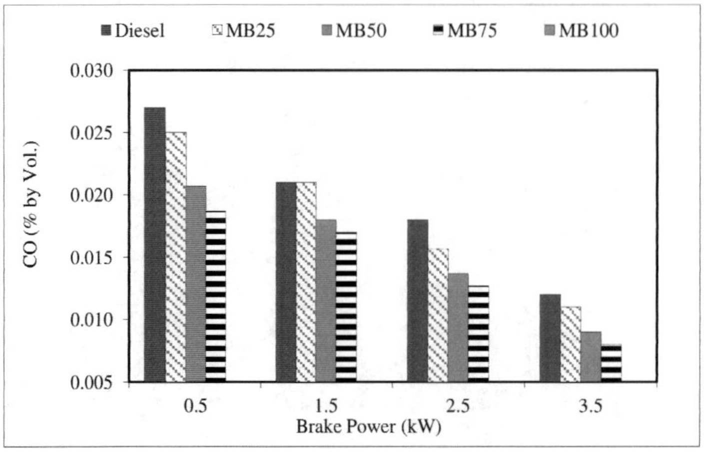

Figure 5 CO Vs Brake Power

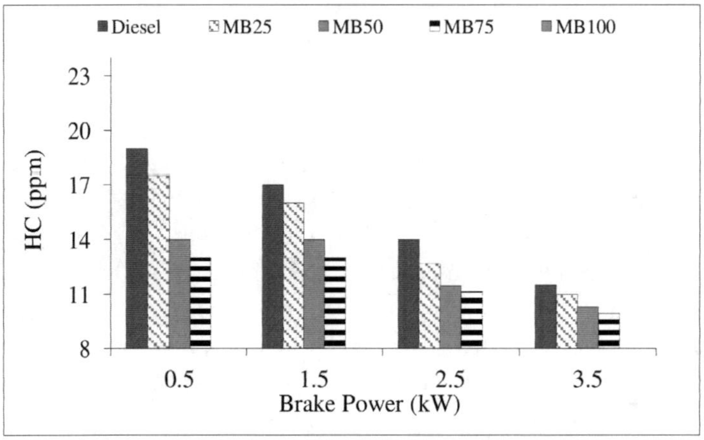

Figure 6 HC Vs Brake Power

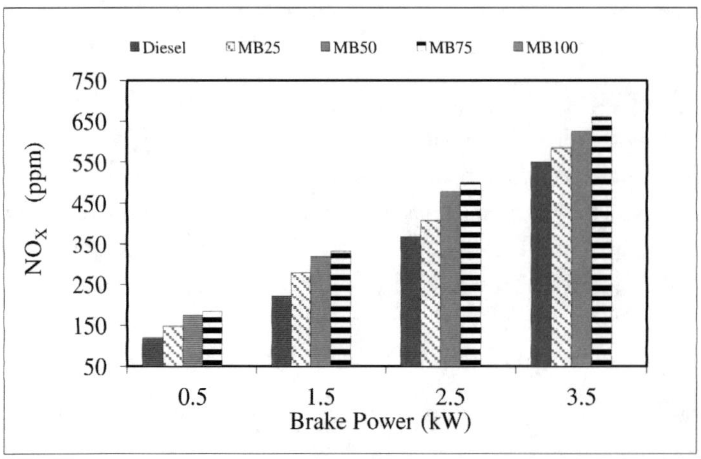

Figure 7 NO_x Vs Brake Power

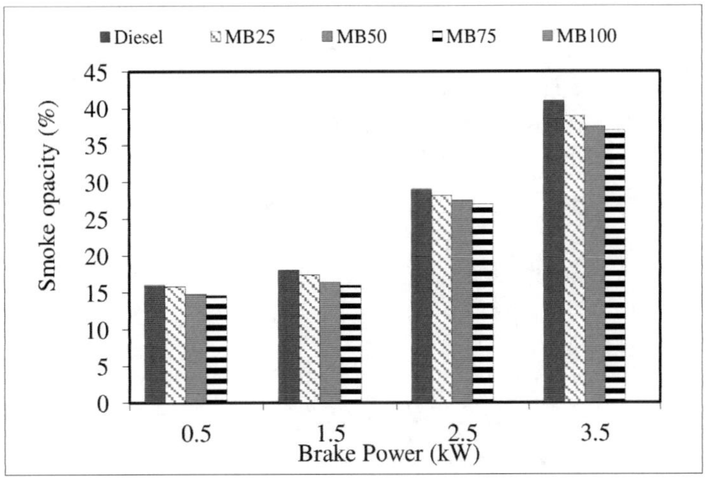

Figure 8 Smoke Opacity Vs Brake Power

4 Conclusions

Experiments were conducted with different blends of mixed biodiesel in CI engine for various loads and the results as follows

- Mixed of biodiesel MB25 has given higher BSFC than diesel by 3.33 % and lower BTE than that of diesel at maximum load.
- The EGT of mixed biodiesel blends is comparable with diesel. It is found that 6.33% and 17.72 % is higher for MB25 and MB100 than that of diesel.
- The presence of HC, CO in the exhaust gas of MB25 is lower than diesel by 4.35 %, 8.33 % at maximum load.
- NO_x content is higher for MB100 by 20 % and MB25 by 6.36 % when compared to diesel.
- The smoke emission is reduced by 9.75 % at full load compared to that of diesel

5 References

[1] Senthil Kumar M, Ramesh A and Nagalingam B 2001 *Soc.of Auto.Eng.* 2001-01-0153
[2] Pramanik K. 2003 *Renewable Energy.* **28** 239-248
[3] Forson F.K., Oduro E.K. and Hammond-Donkoh E. 2004 *Renewable Energy.* **29** 1135-1145
[4] Kumar R, Sharma M, Ray S S, Sarpal A S, Gupta A. A, Tuli D K , Sarin R, Verma R P and Raje N R. 2004 *SAE* 2004-28-0087
[5] Hebbal O D, Vijayakumar Reddy, K and Rajagopal K 2006 *Fuel* **85** 2187-2194
[6] Venkateswara Rao T, Prabahar Rao G and Hema Chandra Reddy K 2008 *Jourdan Journal of Mechanical and Industrial Eng.* **2** 117-122
[7] Balusamy T and Marappan R 2009 *International Energy Journal.* **10** 93-100
[8] Bhupesh Sahu, Ajay Singh Paikra, Dilbag S Mondloe and Himanshu Agrawal 2018 *International Research Journal of Eng. and Tech.* **5** 682-688
[9] Suryawanshi JG & Deshpande NV 2004 *Society of Automotive Engineers.* 280018
[10] Abed KA Gad, MS El Morsi, AK Sayed, MM and Abu Elyazeed, MM 2019 *Egyptian Journal of Petroleum.* **28** 183–188
[11] Arunprasad S, Balusamy T 2018 *International Journal of Green Energy.* **15** 376–384

[12] Arunprasad S, Balusamy T 2014 *Adv. Materials Research.* **984-985** 867-872
[13] Sivalakshmi S. Balusamy T 2012 *Society of Automotive Engineers.* 2012-01-0870
[14] Nurun Nabi, 2010, *Applied Thermal Engineering*, **30** 839–844
[15] Zafer Utlu and Mevlüt Sureyya Kocak 2008 *Renewable Energy*, **33** 1936-1941.

Development of Portable Sugarcane Harvester

Elanchezhian C[1], Vijayaramnath B[1], Nethaji TP[2*], Purusothaman S[2], Singaaram K[2], and Yogeswaran M[2]

[1]Professor, Department of Mechanical Engineering, Sri Sairam Engineering College, Chennai, Tamilnadu – 600 044, India.

[2]UG Student, Department of Mechanical Engineering, Sri Sairam Engineering College, Chennai, Tamilnadu – 600 044, India.

*Email: ksriram2706@gmail.com

Abstract. The present study is focused on development of sugarcane harvester with less manufacturing cost. Now-a-days, the fast production of agriculture product is needed even though the labour demand increased. In this work, the portable sugarcane harvesting machine is designed instead of manual machine which can cut the two sides from the upper and lower portion of sugarcane. The height of the upper blade is adjustable which is completely with electrical motor. This machine will reduce the harvesting time and labour fatigue efforts with increased productivity to all class farmers. Comparing with manual harvesting only 10% of labour is required for entire operation. Also, it reduces the labour cost with faster process. The machine is used by maximum number of farmers and overcome the labour shortage problem with increased productivity. This work is designed for two important sustainable development goal which are responsible and consumption production industry innovation and infrastructure.

Keywords: *Harvesting machine; Sugarcane; Portable; Productivity*

1. Introduction

1.1 Harvesting machine

The modern combine harvester is a versatile machine designed to efficiently harvest a variety of grain crops. The name derives from its combining three separate harvesting operations such as reaping, threshing, and winnowing into a single process.

1.1.1. Etymology of Harvest

Harvest is a noun which came from the Old English word (coined before the Angles moved from Angeln to Great Britain) meaning "autumn" (the season), "harvest-time", or "August". (It continues to mean "autumn" in British dialect, and "season of gathering crops" generally.) "The harvest" came to also mean the activity of reaping, gathering, and storing grain and other grown products during the autumn, and also the grain and other grown products themselves. "Harvest" was also "To harvest" means to reap, gather, and store the harvest (or) the crop. People who harvest and equipment that harvests are harvesters; while they do it, they are harvesting [1].

1.1.2. Harvesting process

Harvesting is the process of gathering a ripe crop from the fields. Reaping is the cutting of grain or pulse for harvest, typically using a scythe, sickle, or reaper. On smaller farms with minimal mechanization, harvesting is the most labour-intensive activity of the growing season. On large mechanized farms, harvesting utilizes the most expensive and sophisticated farm machinery, such as the combine harvester. Process automation has increased the efficiency of both the seeding and harvesting process [2]. Specialized harvesting equipment utilizing conveyor belts to mimic gentle gripping and mass transport replaces the manual task of removing each seedling by hand. The term "harvesting" in general usage may include immediate post harvest handling, including cleaning, sorting, packing, and cooling. The completion of harvesting marks the end of the growing season, or the growing cycle for a particular crop, and the social importance of this event makes it the focus of seasonal celebrations such as harvest festivals, found in many religions.

Harvesting commonly refers to grain and produce, but also has other uses: fishing and logging are also referred to as harvesting. The term harvest is also used in reference to harvesting grapes for wine. Within the context of irrigation, water harvesting refers to the collection and run-off of rainwater for agricultural or domestic uses. Instead of harvest, the term exploit is also used, as in exploiting fisheries or water resources. Energy harvesting is the process of capturing and storing energy (such as solar power, thermal energy, wind energy, salinity gradients, and kinetic energy) that would otherwise go unexploited [3-4]. Body harvesting, or cadaver harvesting, is the process of collecting and preparing cadavers for anatomical study. In a similar sense, organ harvesting is the removal of tissues or organs from a donor for purposes of transplanting. In a non-agricultural sense, the word "harvesting" is an economic principle which is known as an exit event or liquidity event. For example, if a person or business was to cash out of an ownership position in a company or eliminate their investment in a product, it is known as a harvest strategy.

1.2 History of harvesting machine

The Bell machine was pushed by horses. A few Bell machines were available in the United States. In 1835, in the United States, Hiram Moore built and patented the first combine harvester, which was capable of reaping, threshing and winnowing cereal grain. Early versions were pulled by horse, mule or ox teams.

1.3 History of sugarcane harvester

A sugarcane harvester is a large piece of agricultural machinery used to harvest and partially process sugarcane. The machine, originally developed in the 1920s, remains similar in function and design to the combine harvester. Essentially a storage vessel on a truck with a mechanical extension, the machine cuts the stalks at the base, strips the leaves off, and then cuts the cane into segments [5]. These are then deposited into either the on-board container, or a separate vehicle travelling alongside. Waste material is then ejected back onto the field, where it acts as fertilizer.

IOP Conf. Series: Materials Science and Engineering **954** (2020) 012030 doi:10.1088/1757-899X/954/1/012030

1.4 Inventor of sugarcane harvester

The inventor of sugarcane harvester is Mr.Wurtele invented sugarcane harvester in 1938.He is attempting to repair the machine, Mix, Louisiana at the time he found the sugarcane harvester. The wurtele machine has to be handled only in hand. So, it needs a lot of effort from the operator. The machine cannot be handled for more time due to its weight and cutting vibration.

2. Experimental work

In worldwide production of sugar main countries are Brazil, India, china produced up to 78% of global sugar. The chassis frame of the main base of vehicle on which body is mounted with wheels and machinery is presented in Figure 1. Sheet metal is cut by a cutting machine as per the chassis dimension and it is weld on chassis frame for carrying load on the vehicle. Two cutters are fixed on the chassis frame at upper and lower portion [6]. Cutter is controlled by the sensor for protection. The machine is completely electrical with the help of motor and the height of the upper blade is adjustable [7]. The design methodology of harvesting machine is represented in Figure 2.

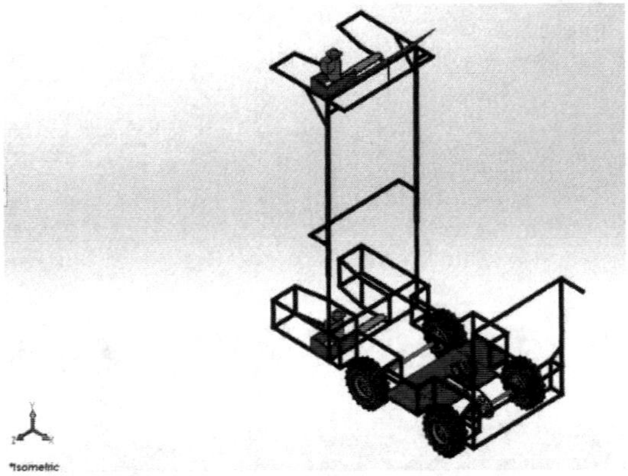

Figure 1. Sugarcane harvesting machine

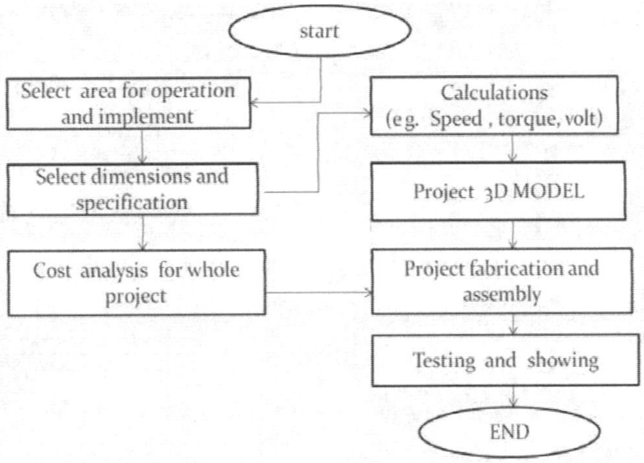

Figure 2. Design Methodology

IOP Conf. Series: Materials Science and Engineering **954** (2020) 012030 doi:10.1088/1757-899X/954/1/012030

3. Model development and ANSYS analysis

This is conceptual CAD model of laboratory setup for cutting unit for estimated optimum speed and force coming on cutting disc for smooth and good quality cutting of sugarcane [8] is represented in Figure 3.

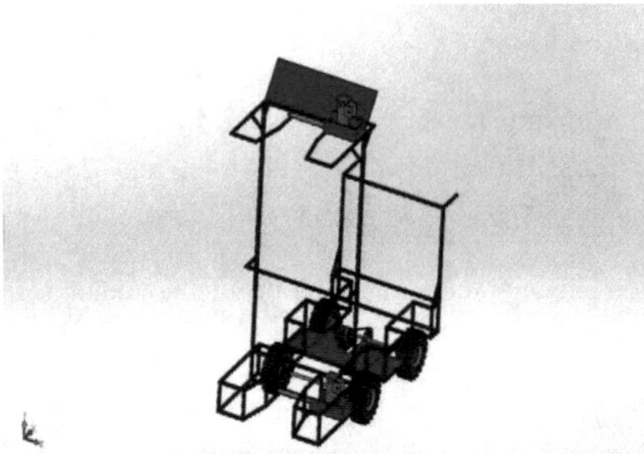

Figure 3. View of Harvester

The four different views of sugarcane harvester are presented in Figures 4 to 7 respectively using the CAD design.

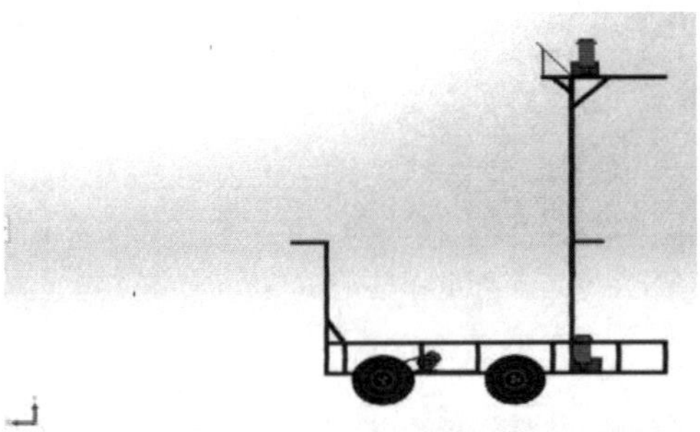

Figure 4 Back view

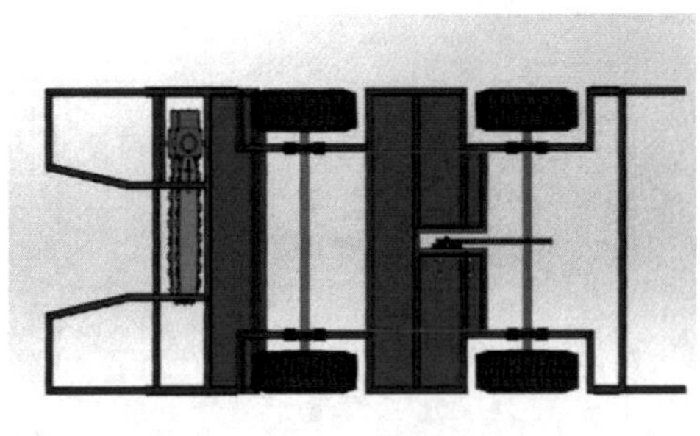

Figure 5 Bottom view

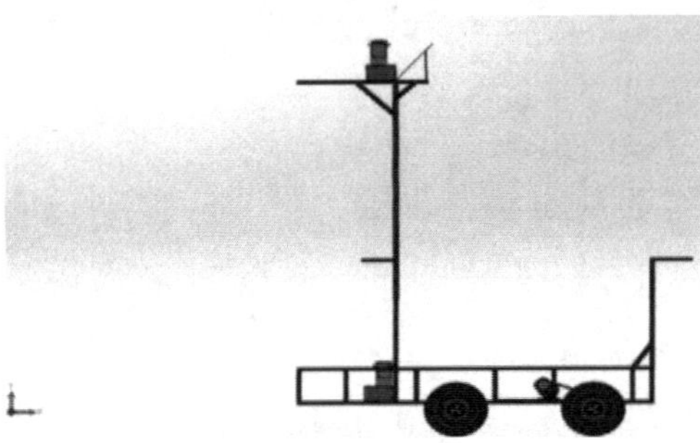

Figure 6 Front view

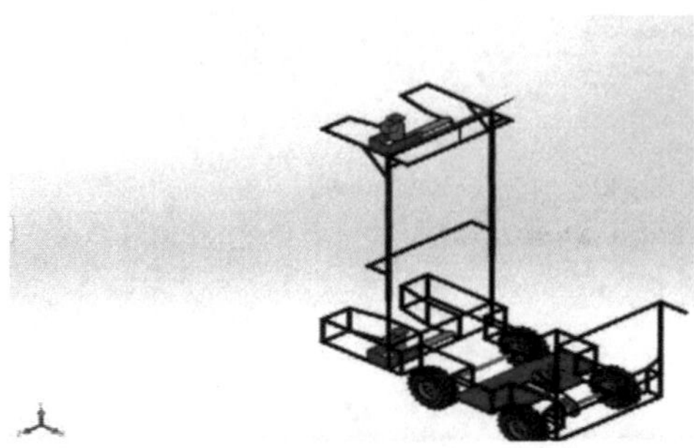

Figure 7 Isometric view

4. Results and Discussion

A detailed static analysis was done on the component using ANSYS software [9-10]. This was mainly done to identify the main points of the component where the majority of the load acts. This, in turn helps us identify the points of constraint necessary for effective sugarcane harvester optimization.

4.1 Analysis of Different Parameters

The analysis results of all the simulations are represented in Figures 8 and 9 respectively for the two different approaches, such as (i) component load is applied on Front and (ii) Fixed support applied on back. The mass properties of the portable harvesting machine are represented in Figure 10. Our vehicle is completely analysed by using ANSYS software. The vehicle is completely safe.

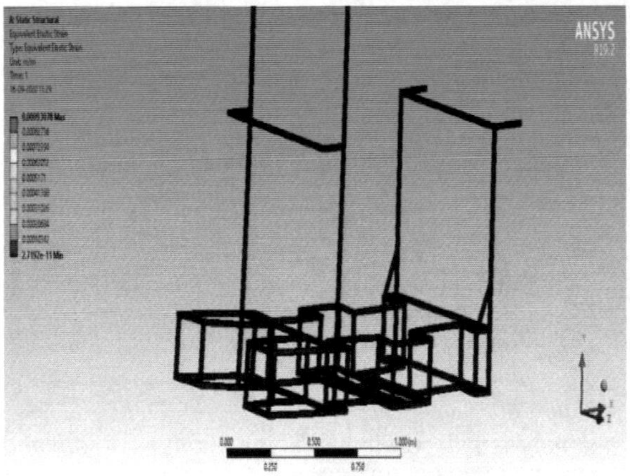

(a) Equivalent Elastic Strain

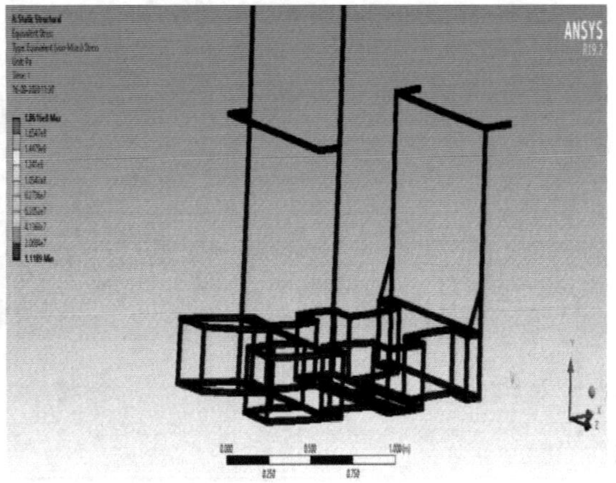

(b) Equivalent Stress

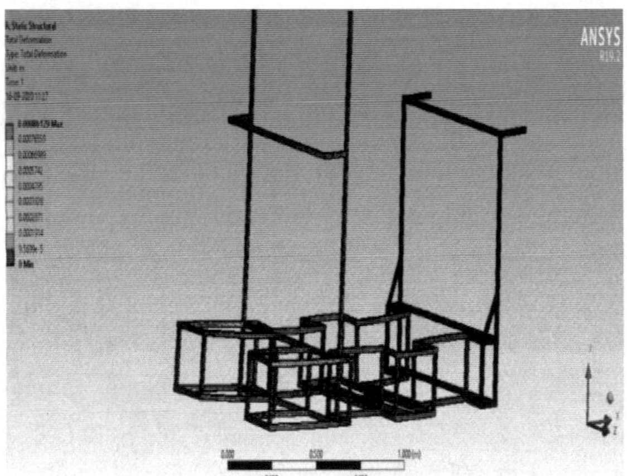

(c) Total Deformation

Figure 8. Load is applied on Front and Fixed support applied on back

IOP Conf. Series: Materials Science and Engineering **954** (2020) 012030 doi:10.1088/1757-899X/954/1/012030

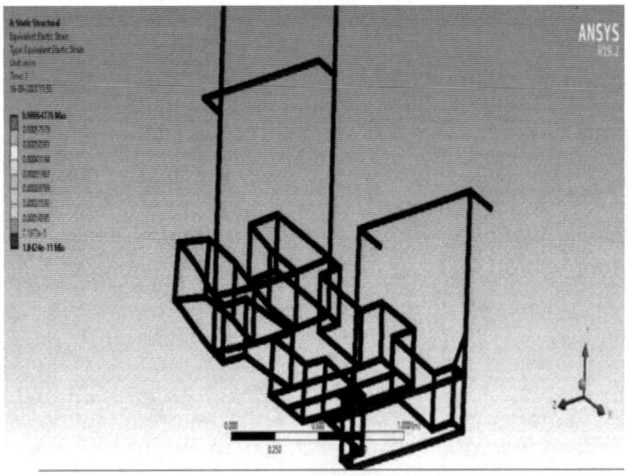

(a) Equivalent Elastic Strain

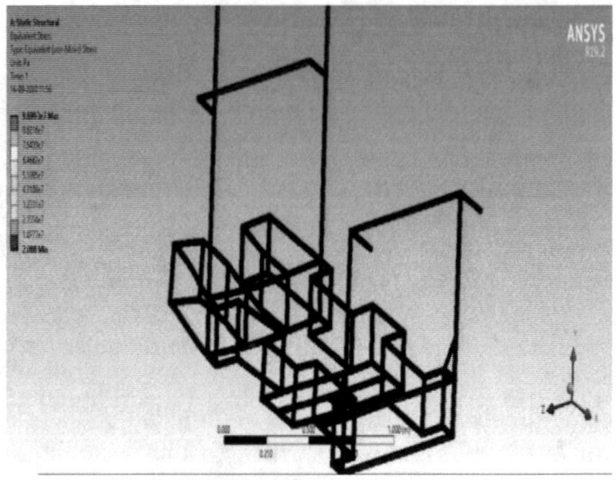

(b) Equivalent Stress

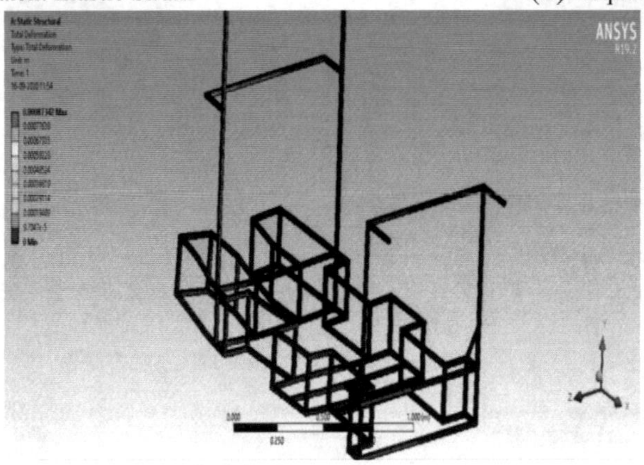

(c) Total Deformation

Figure 9. Fixed support is applied on front and load is applied on back

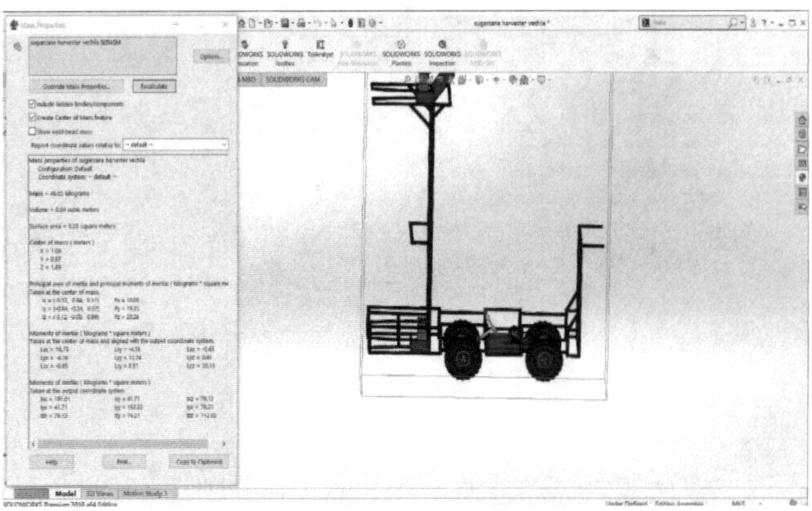

Figure 10. Mass properties of Harvester

5. Conclusions

Comparing with manual harvesting only 10% of labour is required for entire operation. So, it reduces the labor cost and process become faster and easy. The machine is used by maximum number of farmers definitely and farmer can overcome the labour shortage problem. The productivity is also increased. There is a view that Industrialization is harmful to farmers, but the same Industrialization could be moulded to promote farming. The farming scale has reduced exponentially and one of the reasons is that it requires great human skill and effort and even then the cost yield is less. In extended vision, the reduction of the human efforts and the natural investment proves significance in the mere future. The attempt of making small scale sugarcane harvester might have few drawbacks but few modifications could be made. The following modifications could be made in the machine:

- Solar panel could be installed so that the machine is environment friendly.
- Complete automation could be brought. Installation of sensors improves the efficiency.
- Installation of pallets or trolleys so that the cut canes be carried to the tractors or even to the mills.
- Packing mechanism for the cut canes could be designed and developed effectively.

6. References

[1] Shelke GD, Borikar SS, Awathale MP, Khante AP and Zode MP 2015 *Int J Innov Res Sci and Tech* ***1***(11) 122-127

[2] Gupta C and Oduori M 1992 *Trans* ***35***(6) 1747-1752

[3] Masute RJ, Chaudhari SS, Khedkar SS and Deshmukh BD 2014 *Int J Res Engg and Appl Sci,* ***2*** (1) 52-55

[4] Li S, Shen Z, Ma F, Gao J and Yu X 2013 *Int J Engg-Trans C: Aspects,* ***26*** (9) 975 -984

[5] Lin J, Yan W and Lin J 2012 *Res J Appl Sci, Engg and Tech* ***4*** (14) 2183-2185

[6] Siddaling S and Ravaikiran BS 2015 *Inter J Engg Res and Gen Sci* ***3***(4)

[7] Samaila S, Al-Sharief HM and Abdulkadir SA 2012 *AU J Tech* ***16***(1), 59-62

[8] Li SP, Meng YM, Ma FL, Tan HH and Chen WX 2002 *J Mater Proc Tech,**129*** (1) 418-422

[9] Mahadevan K and Balaveera Reddy K 2013 *Design Data Hand Book*, 3rd Edition

[10] http://en.wikipedia.org/wiki/Sugarcane harvester

Experimental investigation of radiator heat transfer efficiency using Water + Ethylene Glycol based Al_2O_3, SiC nanofluids

S.SatyaPrasad[1],V.Rambabu[2], S.Ravibabu[3]

1 M.Tech Scholar, Department of Mechanical Engineering, GMRIT, Rajam, Srikakulam - 532127, Andhra Pradesh, India.
2 Professor, Hod, Department of mechanical Engineering, GMRIT, Rajam, Srikakulam - 532127, Andhra Pradesh, India.
3 Assistant Professor, Department of Power Engineering, GMRIT, Rajam, Srikakulam - 532127, Andhra Pradesh, India.

Satyaprasad442@gmail.com,vrb.gmr@gmail.com, ravibabu.s@gmrit.edu.in

Abstract. The experimental work to boost forced heat transfer performance in an Al_2O_3, SiC-Water+ ethylene glycol nanofluid radiator was discussed in this paper. In this analysis , different volume fractions of nanofluids in the range of 0.01 to 0.1 percent were prepared and sonicated for 2 hours using an ultrasonic sonicator to achieve stable suspension with the incorporation of Al_2O_3, Sic nanoparticles into the water. The maxium-performable heat transfer was found to be 50 percent higher than water at a fraction of 0.1 per cent. The effective thermal conductivity of the nanofluid is increased with increased particle concentration, which helps to boost radiator heating efficiency. The rate of heat flow ranges from 3 to 12 lpm. It is noted that the efficiency of heat transfer has increased with an increase in flow rate.

Keywords: Forced convective heat transfer, Radiator, Heat exchanger, Nano fluids, Al_2O_3 SiC – Water + ethylene glycol.Introduction.

1. Introduction

Radiators are thermal exchangers which are used to transfer thermal energy from one medium to another for cooling and heating purposes. The majority of radiators are designed to work in automobiles, houses, and appliances. The radiator is also an environmentally friendly source of heat, and this can be either for the purpose of heating the environment or for the purpose of cooling the fluid or coolant supplied, as is the case for cooling the vehicle's engines. Such liquids are colloidal nanoparticle concentrations shaped in a working fluids. Nanofluids can be viewed as single-phase fluids in studies such as computational fluid dynamics. Near to all recent research papers are using a two-phase premise. The classical theory of single-phase fluids, where the physical properties of nanofluids are taken as a function of the properties and concentrations of both constituents, can be applied. Using a two-component model, an alternative technique simulates nanofluids. Thermodynamic and mechanical heat transfer measures the rate of heat transfer, the proportionality between the heat flux and the thermodynamic mass transfer coefficient for the heat flow. The heat flux is a quantitative definition of the Victoria heat flow over a surface. Over several years, researchers have concentrated on overcoming the restricted heat transfer capacities of traditional means of heat transfer, such as liquid, motor oil, and ethylene glycol (EG). Nanofluid is a particle sized nano metre, nothing but S.M.Peyghambarzadeh et al [1] measured the capacity of heat EG / Al_2O_3 nanofluids and has been measured experimentally and a 40 percent increase in temperature capacitance has been observed compared to the base 1 percent of fluids by volume. ChilamkurtiLvrsv Prasad et al [2] found lubricant friction coupled with Al_2O_3 nanopowder and noted a 46.35 percent increase in the heat flow rate and a 16.85 percent decrease in the coefficient of friction at room temperature (35 °C) with 0.5 percent volume fraction. Kole and Dey et al [3] Sonication has been used to prepare ZnO-ethylene glycol nanofluids. In thermal conductivity, there was a maximum gain of 40 percent for 3.75 percent vol as opposed to base fluid with clusters of ZnO in ethylene glycol. Dattatraya Subhedar and Bharat Ramani et al [4] observed a linear rise in the thermal conuctivity of nanofluid as the volume concentration rises, with an 8.5 percent increase in the volume fraction of Al_2O_3 nanoparticles of 20 nm size in water / MEG base fluid. It is observed in heat capacity. K.B Anoop et al [5] stated that, two volume fraction were used with a mean particle size varying from 45 nm to 150 nm. The coefficient of heat transfer of particles of 45 nm was greater than that of particles of 150 nm. The coefficient of heat transfer indicate greater enhancement than in the developed region. Neguen et al. [6] Using Al_2O_3 water nanofluid, instead of water base fluid, they studied the overall heat transfer behaviour of a traditional cooling system. Compared to the 6.8 percent base fluid, the heat transfer coefficient increased by more than 40 percent, they revealed. Yu et al [7] claimed that using various types of nanofluids it was possible to achieve around 15-40 percent of the heat transfer enhancement. With all these superior features, the size and weight of an automotive car radiator can be decreased without compromising its heat transfer effectiveness. Pak and Cho et al. [8] performed an investigation of convective turbulent nanofluid (Al_2O_3-water) heat transfer characteristics of 1-3 percent vol. with increased volume concentration and Reynolds number, the Nusselt number for nanofluids increases. Lai et al. [9] estimated the flow behaviour of nanofluids (Al_2O_3-water; 20 nm) in a millimetre-stainless steel test tube under constant wall heat flux and low Reynolds ($Re < 270$) was examined. Nusselt 's overall nanofluid enhancement of 8 per cent at 1 vol. Percentage reported. Guo et al. [10] determined convective co-efficient increase in heat transfer was achieved by 60 % compared to (60/40) ethylene glycol / water base fluid with 2 vol percent nanofluid under circular convective tube fluid flow. Sundar et al. [11] made an observation with turbulent Fe_3O_4 / water nanofluid flow at 0-0.6 vol. in a single circular tube. Percent concentrations were reported and the coefficient of convective heat transfer enhancement factor increased by 30.96 percent and 10.01 percent compared to water at 22,000 Reynolds. Devireddy Sandhya et al. [12] appealed that, the development of vehicle radiators was experimentally tested with 40 percent ethylene glycol and 60 percent water based TiO_2 nanofluids at volume concentrations of 0.1 percent, 0.3 percent and 0.5 percent. Their findings showed that, at small concentrations, the heat transfer increased by up to 37 percent compared to water. Ravikanth et al. [13] claimed that, in the flat tubes of the radiator, the heat transfer properties of ethylene glycol and water mixtures based on Al_2O_3 and CuO nanofluids were numerically analysed. Their findings showed that thermal energy efficiency improved with increasing the volumetric concentration of nanofluids in particles. R.Saidur et al.[14] claimed that by adding 2 percent copper nanoparticles to ethylene glycol, around 3.8 percent of the heat transfer increase could be achieved. Bhogare et al.[15] conducted experimental tests on the thermal efficiency of 0-1 Vol percent Al_2O_3 / water-EG nanofluids in a car radiator have been conducted. The analyses were carried with the air side Reynolds number ranging from 83491-91290. 0.54 percent and 3.8 percent at 1 Vol percent and 83491 air side Reynolds number were reported as the highest improvement in efficiency and heat transfer.

2. Preparation of Nanofluid

A first step in evaluating nanofluid to heat transfer studies is the readying of suspension nanoparticles. Throughout this research, the nanofluid Al_2O_3 SiC -water+ glycol is prepared using a two-step process. One is process of one stage and the other is the process of two stages. Al_2O_3 nanoparticles with an average size of 30 to 50 nm and SiC nanoparticles with an average size of 10 to 200 nm were developed by Nanolabs. In the absence of a dispersant or surfactant, Since the nanoparticles in question had a rough nature when spread in water, we were agglomerating or precipitating. Addition, adding any agent can change the properties of the fluid. Lastly, in two

IOP Conf. Series: Materials Science and Engineering **954** (2020) 012031 doi:10.1088/1757-899X/954/1/012031

measures, 10 litres of distilled water and 4 litres of ethylene glycol mixture are taken together as the base fluid.

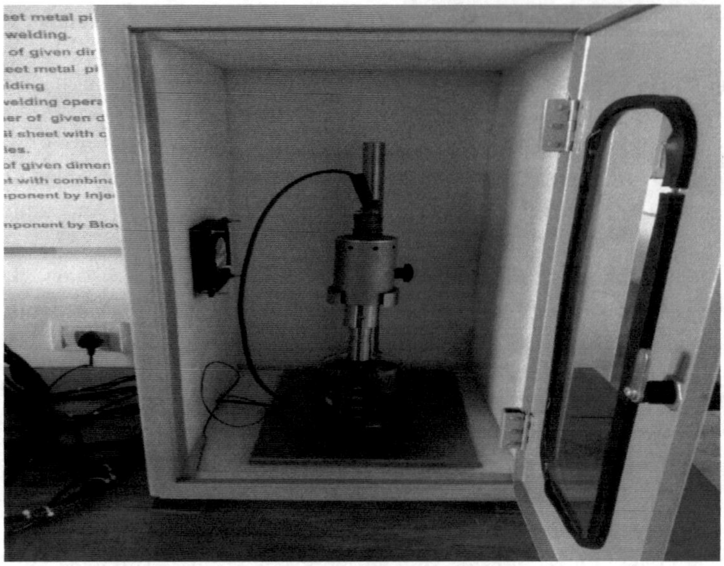

Figure 1: ultra sonicator Pr-1000

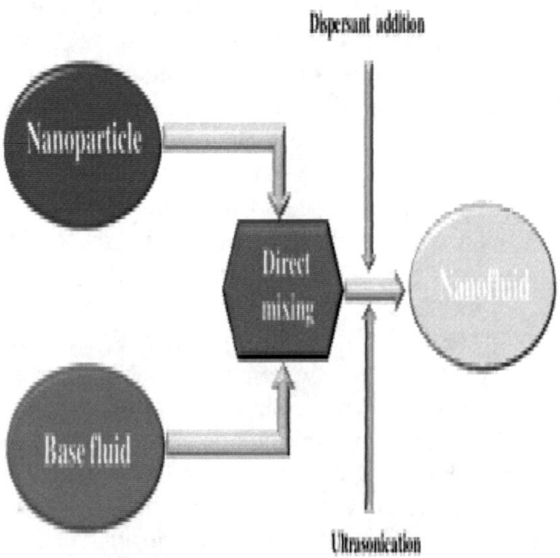

Figure 2:Schematic diagram of nanofluid

3.Properties of Nanofluids

Before examining the radiative efficiency of heat transfer of nano fluids, proper understanding of the effect of nano fluid is required. The concentration of nanoparticles in the tube can be called uniform, assuming the nano particles in the fluid are well distributed. While this statement may not be valid in actual systems due to certain physical processes such as particle movement, this can be a useful method for evaluating a nano fluid's physical properties.

The mixing principle measures the density of nano fluids:

$$\rho = \o\, \rho_p + (1-\o)\rho bf$$

In accordance with the thermal equilibrium model, the total heat ability of nanofluids can be calculated:

$$c = (\o \rho_p c_p + (1-\o)\rho_{bf}\, c_{bf}) / \rho_{nf}$$

Table 1. Nanoparticular specifications

Name of chemical	Aluminium oxide	Silicon Caride
Absolute Purity	99%	99%
Outlook	White	Black
Smell	Alcoholic	No Smell
Mean Amount of particles	30-50nm	20-100nm
Density	3.5-3.9g/cm^3	3.16g/cm^3

Table 2. The physical thermal properties of water, water+EG(3:1)

Base fluid	Density (kg/m^3)	Thermal conductivity (W/mK)	Specific heat (J/KgK)	Thermal Expansion coefficient (K^{-1})	Dynamic viscosity
Water	998	0.61	4178	0.31×10^{-3}	0.808×10^{-3}
Water+EG	1025	0.52	3530	0.40×10^{-3}	0.96×10^{-3}

Table 3. Nanoparticulate thermo-physical properties

S No	Properties	Al_2O_3	SiC
1	Mass density (Kg /m^3)	3160	3580
2	Relevant heat (J/Kg k)	773	1340
3	Thermal Conductivity(W/m K)	40	350

Note that these transport properties are temperature functions. Consequently, the properties were determined using the mean liquid temperature between the inlet and outlet. Using small quantities of aluminium oxide and nanoparticles from Silicon Carbide cane more or less all of the basic fluid's physical properties

4 Experimental setup

Schematic diagram and configuration of experimental test system. The experimental setup includes Pipelines, Centrifugal pump, Automobile radiator, Storage tank, Forced draft fan, temperature indicator, magnetic stirrer, Heater, Thermostat and Thermocouples. The operating fluid is filled up to 33 per cent of the storage tank in all tests. The car radiator used in our research is automobile, consists of 48 capsules cross-sectional. Table 4 defines the specifications for vertical formed tubes made of aluminium and for automotive radiators.

Table 4. Requirements of the Automobile Radiator

Length of Tube	38cm
Width of Tube	1.1cm
Thickness of Tube	0.5mm
The radius of a tube semicircle	0.00381 m
Number of total of Tubes	48
Maximum surface area	0.5731 m

Mean diameter of hydraulics	0.0051009 m

Nanofluid is heated by a 3 KW electric heater which has a thermostat-based temperature control system for conducting experiments at various temperatures. A total of 7 thermocouples to calculate the temperature around the radiator (K Type) was set. Two thermocouples control the temperature of inlet and outlet, and the remaining 5 thermocouples are used to denote the radiator tubes outside the wall temperature. More focus is placed on the transfer coefficient of fluid side heat in the present research work, which can be determined by the temperatures in the inner wall. After all, the inside pipe temperature profile is believed to be equal to the outside surface wall temperature, as the thickness of the surface of the radiator tube is insignificantly thin. All experiments were carried out by variing the inlet of the radiator Temperature and flow rates vary from 50-70 ° C to 3-12 l / min respectively

Figure 3: Schematic of Experimental Setup.

5. Heat Transfer Coefficient Calculations

The following procedure was carried out to determine the coefficient of heat transfer and the corresponding Nusselt number:

$Q = hA\Delta T = hA(T_b - T_w)$

You can measure the heat transfer coefficient as follows:

$Q = mCp\Delta T = mCp(T_{in} - T_{out})$

As for Q equality, in the formulas above

$$Nu_{Exp} = \frac{h_{exp}D}{k} = \frac{\frac{mc_p(T_{in}-T_{out})}{A(T_b-T_w)}D}{K}$$

For the entire radiator, Nu is the total rate of Nusselts, m is the flow rate which is the combination of flow velocity and water flow of volume, Cp is the basic capacity of fluid heat. A is the lateral region of the Tin and Tout radiator tubes. Inlet and outlet temperature Tb is the bulk value of the fluid flowing through the average inlet and outlet temperature of the radiator, and Tw is the tube's wall temperature, which is the mean value of two surface thermocouples. In this equation, K is the heat conductivity of the fluid, and dhy is the canal 's hydraulic diameter. It also should be remembered that all physical properties were measured at temperature of the fluid bulk.

6. Results and Discussions

6.1 Base Fluid

The tests were performed by the base fluid (Water+EG) to verify the experimental test rig before conducting the experiments with nanocoolants. Figure 3 provides experimental results at constant reservoir of radiator 85°C temperature. An increase in the number of nussels was observed based on the results, with an increase in the number of Reynolds. The conclusions of the experiments are compared with the theoretical data from the Dittus-Boelter[16] eq(10) equation and the Gnielinsky[17] eq(11) correlation for the validation of the experimental configuration.

$$Nu = 0.026 Re^{0.8} Pr^{0.3}$$

Nu=(f/8(Re-1000)Pr)/(1+12.7 (f/8)^(0.5) (〖Pr〗^(2/3)-1))

Where f= $(\ln Re-1.69)^{-2}$

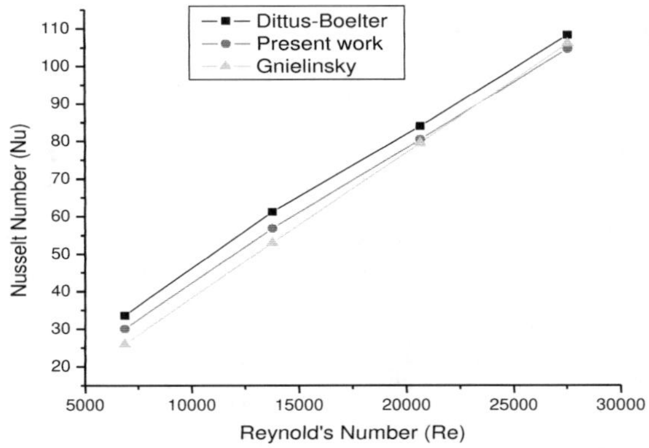

Figure 4: Nusselt Number Variation with Reynold Number Variation

On the basis of the above plots, the experimental data was found to agree with conceptual information gathered from correlation of Dittus-Boelter with a maximum error of 5 percent. Then we need to verify the precision of the experimental test rig and the results calculated.

6.2 Nano Fluid

Using the nanofluids prepared at different volumetric concentrations (0.1% to 0.3%), the nanofluid was circulated through the radiator tubes at various flow rates (3-12lit/min) and temperatures (60°C) for each concentration. The heat transfer convection mode takes place between the nanofluids and the radiator walls as the nanofluid moves through the radiator and then, through the forced convection, the heat absorbed by the radiator walls is rejected into the atmosphere using a draught fan.

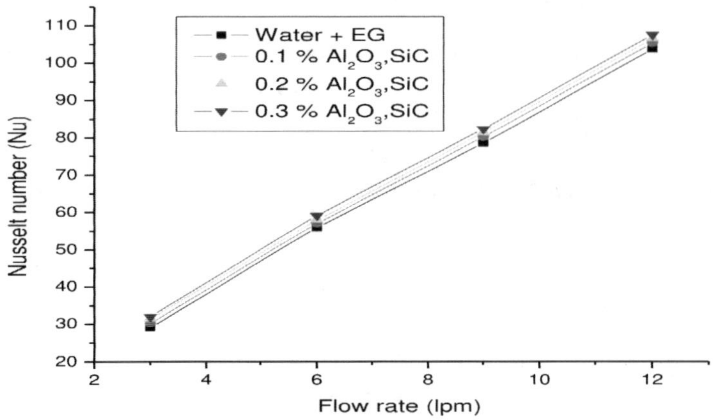

Figure 5: Nusselt number variance with different flow rate(3-12lpm) at 60°C inlet temperature.

At different volume concentrations of nano-particles of Al2o3, SiC, the experimental values of the coefficient of forced convective heat transfer. The Nusslet number will give 30.05 at the flow rate of 3lpm, just as the flow rate of 6lpm the Nusslet number will give 56.84 similarly at 9lpm the Nusslet number value 80.47 and finally at 12 lpm the Nusslet number gives the value 104.6.

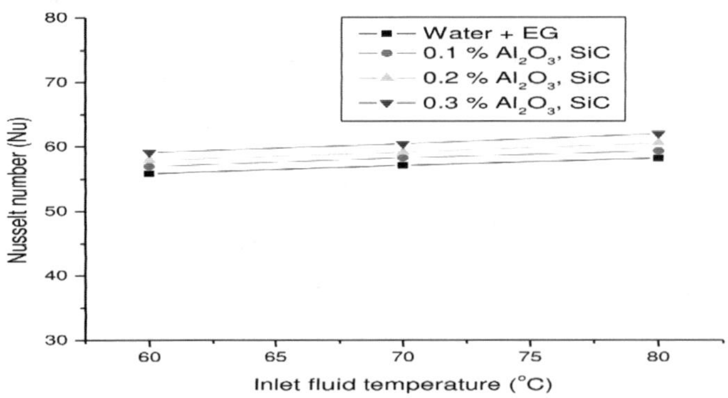

Figure 6: Nusselt number variance with inlet temperatures for different nanoparticles volume fractions.

To understand the thermal efficiency of an automotive radiator, tests were carried out at varying temperatures from 60 ° C to 90 ° C. Figure 6 shows that difference in the nussel number with the temperatures of the inlet radiator. Improvement in the number of nussels is observed with rising working fluid temperatures, based on the findings. The same pattern is noted for all nanofluid concentrations. At 0.3 vol percent concentration, 21 percent improvement in Nu with various working fluid temperatures between 60 °C to 90 °C with a flow rate of 12L/ min was recorded.

7. Conclusion

The involvement of alumina SiC nanoparticles in water+ glycol will increase heat transfer efficiency from radiators. The amount of thermal energy improvement depends on the amount of nano particles added to water + glycol. Finally, as opposed to pure Water + Etylene Glycol, the rise in thermal energy was found at a level of 0.1vol per cent around 50 per cent.

IOP Conf. Series: Materials Science and Engineering **954** (2020) 012031 doi:10.1088/1757-899X/954/1/012031

Through the growing base fluid flow rate (3-12lpm), the heat transfer coefficient for both liquid + ethyl alcohol and nanofluids is greatly increased. The improvement of efficient heat capacity does not appear to be responsible for the production of the great heat transfer variety of the other mechanical properties.

8.References

1. Subhedar, D. G., Ramani, B. M., & Gupta, A. (2018). *Case studies in thermal engineering*, *11*, 26-34.

2. Peyghambarzadeh, S. M., Hashemabadi, S. H., Hoseini, S. M., & Jamnani, M. S. (2011). *International communications in heat and mass transfer*, *38*(9), 1283-1290.

3. Ali, H. M., Ali, H., Liaquat, H., Maqsood, H. T. B., & Nadir, M. A. (2015). *Energy*, *84*, 317-324.

4. S.SURESH, C., PRASAD, C., & SANTA RAO, K. Performance Assessment of an Automobile Radiator using AL2O3/Water + EG Based Nanocoolant.

5 Subhedar, D. G., Ramani, B. M., & Gupta, A. (2018). *Case studies in thermal engineering*, *11*, 26-34.

6 Heris, S. Z., Shokrgozar, M., Poorpharhang, S., Shanbedi, M., & Noie, S. H. (2014). *Journal of dispersion science and technology*, *35*(5), 677-684.

7 Oliveira, G. A., Contreras, E. M. C., & Bandarra Filho, E. P. (2017).. *Applied Thermal Engineering*, *111*, 1450-1456.

8 Elsaid, A. M. (2019).. *International Communications in Heat and Mass Transfer*, *108*, 104263.

9 Saidur, R., Leong, K. Y., & Mohammed, H. A. (2011). A. *Renewable* and sustainable energy reviews, *15*(3), 1646-1668.

10 Kole, M., & Dey, T. K. (2012). *Applied Thermal Engineering*, *37*, 112-119.

11 Devireddy, S., Mekala, C. S. R., & Veeredhi, V. R. (2016). *International communications in heat and mass transfer*, *78*, 121-126.

12 Bhogare, R. A., & Kothawale, B. S. (2014). *IOSR Journal of Mechanical and Civil Engineering*, *11*(3), 23-30.

13 Vajjha, R. S., Das, D. K., & Namburu, P. K. (2010). *International Journal of Heat and fluid flow*, *31*(4), 613-621.

14 Mojarrad, M. S., Keshavarz, A., Ziabasharhagh, M., & Raznahan, M. M. (2014). *Experimental Thermal and Fluid Science*, *53*, 111-118.

15 Li, X., Zou, C., & Qi, A. (2016). *International Communications in Heat and Mass Transfer*, *77*, 159-164.

16 Sundar L S, Farooky M H, Sarada S N and Singh M K 2013 *International Communications in Heat and Mass Transfer*, 41, 41-46.

17 Yang X F and Liu Z H 2012 *International journal of heat and mass transfer* 55(25-26) 7375- 7384.

18 Huminic G and Huminic A 2013 International communications in Heat and Mass transfer 44 52-57.

19 Tijani A S and bin Sudirman A S 2018 *International Journal of Heat and Mass Transfer* 118 48-57.

20 Narrein K, Sivasankaran S and Ganesan P 2015 *Numerical Heat Transfer, Part A: Applications* 68(11) 1266-1279.

Performance and Emission Characteristics of Four Stroke Diesel Engine Using Sesame Biodiesel Blends with Addition of TiO_2 Nano-Particles

R Satya Santosh Uday[1], K Simhadri[2]

[1] M. Tech scholar, Department of Mechanical Engineering, GMR Institute of technology, Rajam, srikakulam-532127, Andhra Pradesh, India.
[2] Asst. professor, Department of Mechanical Engineering, GMR Institute of technology, Rajam, srikakulam-532127, Andhra pradesh, India.

E-mail:ssudayrejeti@gmail.com, simhadrikambala@gmail.com

Abstract. Rising demand for fuel in various applications will now pose a major risk in terms of global emissions levels. Taking this issue into consideration takes significant attention. Hence the production of alternating fuels is required. Bio-diesel is environmentally friendly fuel consisting of fatty acid alkyl esters. This experimental study includes a comparative analysis to determine the effect of nano additives and their performance characteristics. The theme of the research is the combination of sesame oil biodiesel with titanium dioxide (TiO_2), a metal-based nanoparticle with diesel to investigate its effects on the performance of diesel engines. Different fuel blend ratios with TiO_2 nanoparticles, ethanol and biodiesel have been developed. Test fuels were diesel (D100) and biodiesel blends like B10, B20 and B30. Thermo-physical characteristics of all test fuels have been calculated and the engine output parameters, such as torque, power, fuel consumption, etc., were examined. The impact has also been realized on pollutants such as CO, CO_2, HC, NO_X and smoke ability.

Keywords: Emission and combustion, Diesel engine, Performance, Sesame oil methyl ester.

1. Introduction

The key alternative source of energy is bio-fuels, since they are sustainable, organic and environmentally friendly. It can be obtained from various organic products, like mahua, jatropha, coconut, and vegetables, etc. Which is safe, non-combustible, harmless, environmentally friendly biofuel and bio-degradable without harm to the environment in the soil. Supplier emissions are also cleaner and lower from diesel. [1]. Biodiesel is a source of renewable energy which can help reduce greenhouse gas emissions. It does little towards support increasing of earth's environmental temperature, since plant feedstock has extracted carbon from the environment [11]. Bio-fuels are mostly used as fuel for transport. Over the past decade, global bio-fuel production has steadily grown from 16 billion liters in 2000 to around 110 billion liters. Biodiesel fuels can be used without engine changes in standard diesel cars.With better lubrication properties and significantly lower emissions than those of today low-sulphur diesel fuels, biodiesel is an environmentally sustainable source of renewable energy [2]. In addition, biodiesel reduces fuel system wear and enhances the life of fuel injection equipment that relies on low-pressure lubricating fuel in high-pressure systems. Biodiesel generates more total combustion, thereby increasing the performance of engine power and partly offsetting the higher energy density of petro-diesel.The growing industrialization of petroleum products worldwide tends to a sharp increase in requirement. Terrestrial fuels are considered fossil fuels. These stored fuels have, and are irreplaceable, a limited quantity of fuel. With our current known reserves and rising consumption rates, it is thought that they will not last long [2]. In some areas of the world, these small supplies of petroleum are highly oriented, causing often disruption and instability in both their supply and price. Although the current reserves seem immense, growing consumption will create a challenge for the world to replace fossil fuels with a new type of fuel. Under cold conditions; biodiesel performance is slightly lower than petroleum diesel performance, for biodiesel, the cloud and pour points are larger than for diesel oil. There are mainly classified into three generations:

1.1 First generation

Using traditional technologies, sugar, starch, vegetable oil or animal fats are used to produce biofuels. These are typically derived from grains high in sugar or starch fermented with bioethanol; or from seeds pressed with vegetable oil from biodiesel [15]. Biodiesel, bio-alcoholics, biogas and solid biofuels, vegetable oils, are all bio-fuels of the first generation.

1.2 Second generation

Biofuels are derived from non-food crops such as celluloid biofuels and waste biomass. Work continues on bio-fuels of the second generation, including bio-hydrogen, bio-methanol, bio-hydrogen diesel and mixed alcohol [7].

1.3 Third generation

Bio fuels are derived from the extraction of algae oil-also known as high yield "oilgae." Its production can be low cost.

Rising oil prices demand a report on biodiesel production. In India's industrial economy diesel fuels have an important role to play. The supply of fossil fuels is decreasing rapidly due to rising energy demand in the developing world, and the growing use of fossil fuels. Some alternative diesel fuel options usually include various vegetable oils, biogas producer gas and methanol, and ethanol.One of the Diesel engine's capable alternative fuels is from this sesame oil [30]. Sesame oil is a non-conventional, non-toxic, cheap and low-grade vegetable oil that can be produced from agricultural and plant resources locally. Their consumption is not associated with detrimental effects on the environment, since they produce less toxic emissions and greenhouse gases. The edible vegetable oil extracted from sesame seeds is sesame oil. In addition to being used as cooking oil, it is often used in many cuisines as a flavour enhancer. Sesame seed contains 50-60 percent high-quality oil rich in polyunsaturated fatty acids shown in Table1: (PUFA) and natural antioxidants, sesame and sesamoline. These bioactive components strengthen the stability and consistency of sesame oil along with many health benefits . The sesame seeds also contain two special substances, sesame and sesamoline, in addition to these basic nutrients. Both of these substances belong to a group of unique beneficial fibers called lingams, which have a diminishing effect of cholesterol in humans, preventing high blood pressure which increases vitamin E supplies in animals.Mass production of sesame oil is limited worldwide due to inefficient manual harvesting process needed to extract the oil.

IOP Conf. Series: Materials Science and Engineering **954** (2020) 012032 doi:10.1088/1757-899X/954/1/012032

Sesame oil is naturally light yellow in colour and is used in the manufacture of medicines as a solvent, and also in the preparation of soaps, paints, lubricants.

Table 1 Fatty acid percentage in sesame oil

Name of Fatty Acid	Formula	% by Weight
Linoleic acid	C18H32O2	41
Oleic acid	C18H34O2	39
Palmitic acid	C16H32O2	8
Stearic acid	C18H36O2	5
Others	-	7

I.Ors, S.Sarkoc et al. [1] Observed effects on performance , combustion and emission characteristics of the Direct Injection Compression Ignition engine fuelled with TiO_2 nano-particles applied to diesel ,biodiesel and n-butanol blends. Diesel, biodiesel (B100), B20, B20 + TiO_2 were used as test fuels. The addition of n-butanol to the fuel introduces density, kinematic viscosity and cold flow properties that have been greatly affected, though adding TiO_2 has little effect on these properties. Homa Hosseinzadeh-Bandbafha et al. [2] Effects of water-emulsified oil, biodiesel mixtures have been observed on the performance and emission parameters of the aqueous carbon nano-particle diesel engine as a novel nano-additive. Various sizes (38, 75 and 150μm) of nano-additive uel mixtures were combusted under four separate engine loads varying from 25 to 100 percent of the full load conditions in a diesel engine at a set engine speed of 1000 rpm. The most desirable performance characteristics were observed for the emulsified fuel mixture containing 38 μM carbon nano particles, which improved the brake power and thermal brake efficiency by 1.07 kW and 11.58 percent respectively, at full load operation, the brake specific fuel consumption decreases by approximately 107.3 g / Kwh. K.Kannan and Marappan [3] Efficiency and emission characteristics of the diesel engine checked with THEVETIA PERUVIANA Diesel and Diethyl Ether mixes 5 percent, 10 percent, 15 percent and 20 percent. It has been found from the comprehensive analysis that 20 percent DEE blend will result in a 5 percent rise in full-load biodiesel blend, a 14.63 percent reduction in harm smoke quality. It was concluded that better results would benefit from a 20 percent blend. Subrahmanyam. J. P et al. [4] Investigated the combustion and emission characteristics of the diesel engine using karanja oil methyl ester (KOME) blend in the 5 percent, 10 percent, 15 percent and 20 percent ratio with Di ethyl ester. Brake thermal efficiency with 15 percent KOME blend improved by 5.5 percent. Smoke opacity decreased to a minimum maximum load level with a 20 percent KOME- DEE blend. The optimum blend based on emission and efficiency characteristics was found to be the 15 per cent KOME DEE from the various KOME-DEE blend studies.V. Rambabu [5] Researched DI Diesel Engine 's efficiency, emission, and combustion characteristics along with Methanol Carburization with Linseed methyl ester. The findings and study of the $1/32^{th}$ throttle valve opening (3.73% of Methanol at maximum load) along with Bio-diesel activity provide improved thermal efficiency (23.546%) and low pollutants such as CO and HC. K. PrasadaRao [6] The output features of the diesel direct injection engine were investigated along with Methanol Carburization using Mahua methyl ester. From this study efficiency and emission analysis, HC, CO pollutants can be found to be low in scarifying small quantities of Brake Thermal Efficacy. In other words, the process of biodiesel MME along with improved braking power at the 1/16th throttle valve release. Mithun Das et al. [7] The refined castor oil for the production of biodiesel was used with special methanol and sodium hydroxide parameters. For analysis, biodiesel blends such as B5, B10 and B15 were considered, and research was conducted on the porous sphere. Diesel displayed a high rate of evaporation and the findings showed the high chemical reactivity of biodiesel mixtures. They concluded that the limited use of biodiesel in the blending of diesel biodiesel may help minimize reliance on fossil fuels. Nitin Shrivastava et al.[8] A 4-cylinder, 4-stroke, direct injection diesel engine with 50 and 150 gm / L alumina nano particles has been tested for Jatropha Oil Methyl Ester (JOME), resulted in the introduction in JOME of 50 and 150 mg / L nanoparticles, respectively, reducing BSFC by 9.1%, 12.2% (BTE). In JOME fuel, the same amount of nano particle addition was observed to increase by 2.4 percent and 6.2 percent. Reduces by 5.7 percent and 13.7 percent respectively the average HC emissions. In JOME fuel, The addition of 50 and 150 mg / l nanoparticles was demonstrated an average decrease of 8.4 percent and 16.7 percent in CO. JOME registered a reduction of 1.8 percent in NO_x and 4.7 percent in NO_x with the same amount of additional nanoparticles. Hossein Soukht Saraee et al. [9] For oil extracted from Pistacia Khinjuk using the SOx hlet extractor, a maximum yield of 35to40 percent was demonstrated. For DI Diesel engine research and testing, fuel samples such as B5, B10, B20, B40 and B100 have been considered. Biodiesel samples revealed a greater fuel consumption than petrol. Biodiesel blends were discovered with lower CO and HC smoke concentrations.

2. Fuel preparation

Biodiesel is composed of monoester alkyl fatty acids. It is extracted through the transesterification process from the tri-glycerides. Transesterification is the term used to describe organic reactions as the main class in which, by the exchange of alkyl groups, an ester is converted into another ester and is often referred to as alcoholysis. Transesterification is a reaction to equilibrium, and by combining the reactants, the process occurs. However, the presence of a catalyst significantly speeds up the transition in equilibrium. The fatty acid radicals of the triglyceride molecule break free from glycerin bonds, and the fatty acid radicals build new ester ties with the alcohol molecules, resulting in glycerin and fatty acid-free esters. Straight-chain radical fatty acids are described in the details below.
Transesterification reactions between triglyceride and alcohol:

CH_2COOR_1			CH_3COOR_1		CH_2 –OH
|		*Catalyst*			|
$CHCOOR_2$	+ 3 CH_3OH		CH_3COOR_2	+	CH –OH
|		*($NaOCH_3$)*			|
CH_2COOR_3			CH_3COOR_3		CH_2 –OH
Triglyceride (Vegetable oil)	Alcohol		Mixture of fatty esters		Glycerin

Preparing the biodiesel from sesame oil is performed in four steps.

2.1 Acidic Treatment

The process of transesterification of sesame oil was used directly to generate biodiesel from sesame oil because of the low acid value. During this process, with the help of a magnetic stirrer, the reactor was filled with 1 litre of refined sesame oil and heated to 60 ° C for 10 minutes.. Then, 250 ml of methanol was added to the sesame oil and added 2 -3 ml of H_2 SO_4 methanol- H_2 SO_4 after sometime. The preheated oil was then added to mixture. The reaction then kept for 1.5 h at a steady stirring rate of 1500 rpm. If the reaction stops, after one hour of watching the solution, pulp will calm down

IOP Conf. Series: Materials Science and Engineering **954** (2020) 012032 doi:10.1088/1757-899X/954/1/012032

Figure 1 one liter of raw Sesame oil in heater and settling

2.2 Basic Treatment

After extracting the pulp from the solution, it brings the solution into the conical flask and stirs with heat and adds 250 ml of methanol to the acidic solution and then adds 7 grams of NaOH pallets to the SOLUTION at 45 ° C after reaching a certain temperature. After the reaction was completed, almost all reaction products have been moved overnight to the separation funnel. To allow the separation of biodiesel from glycerine.

Figure 2 Base catalyzed stage

2.3 Water washing

Air agitation is used by the biodiesel final washing technique, the pH value of the oil no longer needs to be checked, as it is presumed that the biodiesel that comes out is essentially neutral. A third of the water is applied to the oil by volume and the bubble is cleaned to remove the bubbles. The mixture settles for one hour in a decanter, and then drains the water later.

Figure 3 Water washing

2.4 De-hydration

After water washing, the method of dehydration is carried out to remove the traces of water left in the residual oil. The oil is arranged in a beaker and heated to a heater that includes a magnetic stirrer for continuous stirring in order to eliminate traces of water present in the oil. As the water evaporates at 100 °C, the oil is heated to more than 100 ° C. After transesterification process the final Palm methyl ester (biodiesel) was obtained.

Figure 4 Oil is being heated and pure Sesame bio-diesel

Using the ultrasonic Hielscher ultrasonic UP400S, 400 watts at a frequency of 34 kHz for half an hour, the titanium oxide nanoparticles were dispersed into biodiesel mixtures. Weighted the nanoparticles with a mass fraction of 25 mg and 50 mg / l. Nano-particles of titanium oxide occur in the form of hexagonal white crystals, or even in small particles. Due to its clarity and very high refractive index the primary application of Tio_2 is as a white powder pigment. It is used in products such as paints, coatings, cosmetics and pharmaceuticals etc.

Table 2 The various physical and thermal properties of Tio_2

S.NO	Property	Value
1	Density	$4.23gm/cm^3$
2	Molar mass	79.93 gm/mol
3	Melting point	1843^0c
4	Boiling point	2972^0c

Figure 5 Hielscher ultrasonicator

The sample is then tested for flash & fire point using the Cleveland apparatus and the calorific value by the bomb calorimeter.

Table 3 Blended Diesel and Biodiesel properties

S. No	Property	Units	Blended sesame	Biodiesel
1	Density	kg/m^3	830	870
2	Specific Gravity	kg/m^3	0.830	0.824
3	Kinematic Viscosity	mm^2/sec	4.4	5.9
4	Calorific Value	kJ/kg	45500	46509.42
5	Flash point	^0C	75	184.6
6	Fire point	^0C	80	189.6

The average calorific value for blended sesame bio-diesel is 11019.36 Kcal / kg.
Thus, because 1 Kcal / kg = 4.18 KJ / kg, the calorific value = 46509.42 KJ / kg.

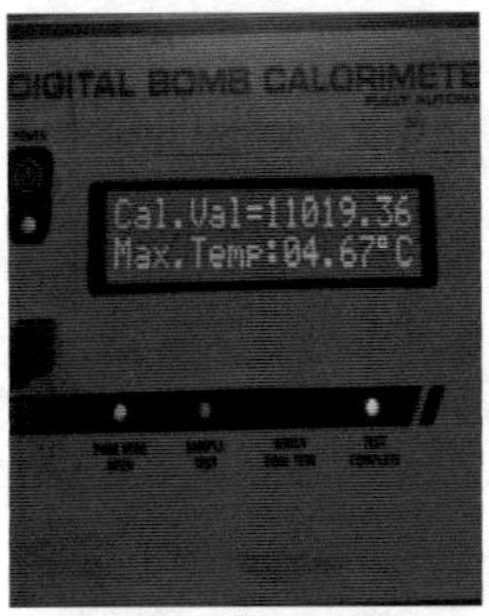

Figure 6 Calorific value reading

3. Experimental setup and procedure

Experiments have been performed on a Direct Injection Water cooled single-cylinder diesel engine with 4-stroke, the Table 4 indicates specifications of which are given. Figure 7 displays the schematics for the experimental set-up. To control the engine, a steady speed of 1500 rpm was used. Tests were carried on diesel engine by TiO_2 nano particulate substitute, Sesame oil. With load in 25 percent steps, from free load to full load. The motor was coupled to the electrical dynamometer to provide the brake load. For the diesel fuel and sesame oil, two different fuel tanks were used. For the volumetric fuel flow rate measurement, a 50 cm3 burette and a stop watch were used. Using the AVL-444 5 gas analyzer, emissions such as CO, HC and NOx were analyzed and smoke was measured by the Bosch smoke pump and the smoke meter is attached to the engine's exhaust manifold.

Figure 7 Experimental Test rig

Table 4 Specifications of test engine

Engine	AV-I, Kirloskar
Power (Kw)	4.4
Bore (mm)	87.50
Stroke (mm)	110
Compression ratio	17.5:01
Speed (Rpm)	1500
Injection pressure (Bar)	200
Injection timing	23^0 BDC

4. Results and discussions

4.1 Performance of the Engine

The intent of this study is to investigate the effects on engine performance parameters of brake torque, brake power (BP) and brake specific fuel consumption (BSFC), for which the metal based additives and oxygenated fuel mixtures are used.

The tests were performed at different engine speed levels' using the fuel blends of D100, B100, B20, B20 + TiO_2. Both test results for obtained and fuel properties and as regards confirmations, they were compared with one another.

4.2 Brake power and Engine torque

Figures.8 & 9 display the performance at different engine speeds of both the torque of the brake engine and the power of the test fuel blends. Max values for torque of 34.73 N-m, 28.89 N-m, 32.49 N-m, 35.59 N-m, for D100, B100, B20, B20 + TiO_2, fuel blends respectively, were registered at 1400 rpm. The average brake torque values as compared with diesel for Blend100 and B20 decreased by approximately 21%, 6.61% respectively, while for B20 + TiO_2 they increased by approximately 2.89%.

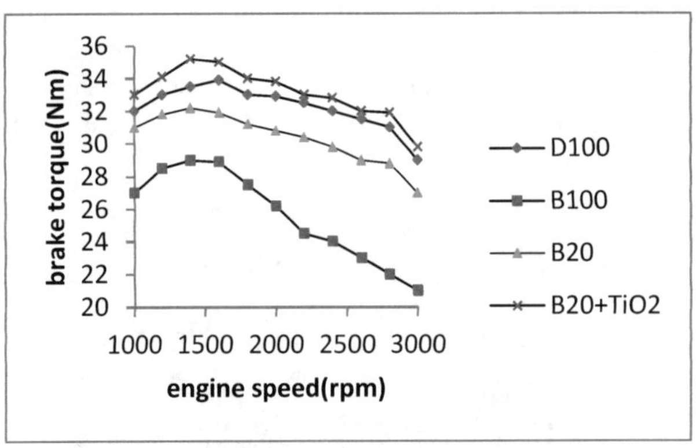

Figure 8 At different engine accelerations and full load conditions, brake torque values

For fuel mixtures D100, B100, B20, B20 + TiO_2, the maximum brake power values are 8.86 Kw, 8.16 Kw, 6.47 Kw and 5.7 Kw, respectively, recorded at 2800 rpm. Brake torque and engine speed are directly correlated with engine brake power, since improvements in brake torque have a direct effect on engine brake power at the same engine rpm. It's evident that the addition of TiO_2 to fuel mixes significantly increases the torque and power values of the engine brake. Nano-particles are also a high proportion of surface-volume and have greater oxidation of the fuel mixtures.

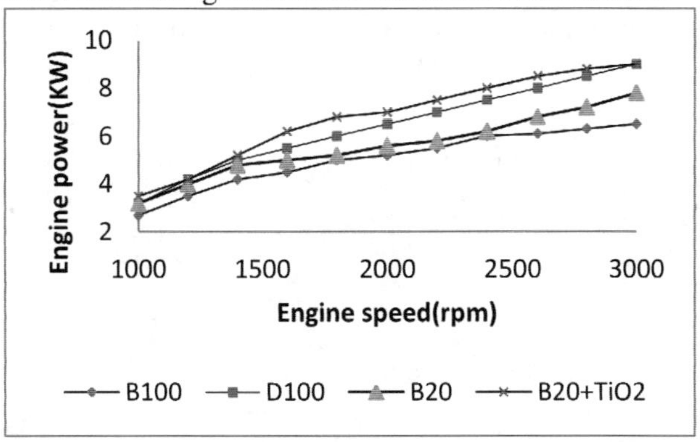

Figure 9 Brake power values and maximum load conditions at various engine speeds

4.3 Brake Specific Fuel Consumption (BSFC)

Figure 10 indicates the difference in Brake Specific Fuel Consumption (BSFC) at different engine speeds. The lowest BSFC values for D100, B20, B100, B20 + TiO_2 test fuels were 263.69 gr / Kwh, 271.99 gr / Kwh, 321.08 gr / Kwh and 250.18 gr / Kwh at 1800 rpm. Compared to euro diesel, B20 test fuels were 21.09 percent, 4.41 percent and 14.64 percent respectively, BSFC average increase for B100.

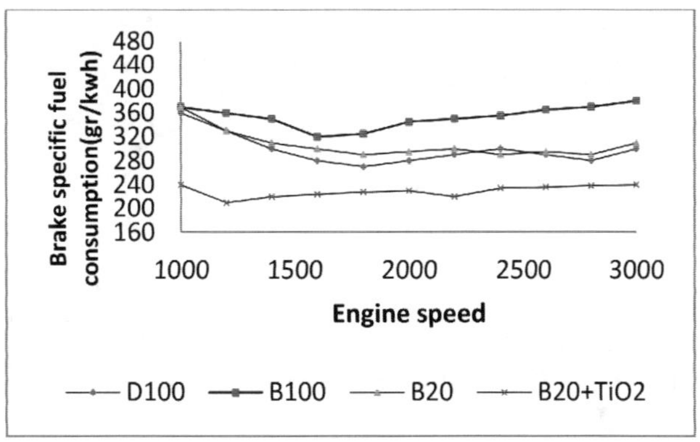

Figure 10 Brake-specific fuel consumption values at various engine speeds and with maximum load conditions

By comparisonB20 + TiO_2 BSFC decreased by 24.52 percent and 17.78 percent when adding TiO_2. It is clear that the BSFC was greatly impacted by the inclusion of biodiesel and TiO_2 in the fuel blends. The TiO_2 additive was reduced to BSFC. This is due to the high surface area of TiO_2, which, as a result of improved reaction surface area, leads to improved combustion efficiency.

4.4 Exhaust emissions

4.4.1 CO emissions

CO emissions Figure 11. The CO emissions outcomes are shown for both fully loaded test fuels and different engine speeds. Maximum values were registered for all fuel blends at 1000 rpm. Compared to diesel, B100, B20, B20 + TiO_2 average CO emissions decreased by 44.83%, 16.51%, 25.56%, 27.44% and 38.09% respectively. B100 fuel obtains the lowest CO emission values. It is also noted

IOP Conf. Series: Materials Science and Engineering **954** (2020) 012032 doi:10.1088/1757-899X/954/1/012032

that the additive fuel blend B20 + TiO_2 reduced the value of CO emissions by approximately 10.86 percent compared to B20. This can be caused by the fact that the reaction surface area is increased by nanoparticles (TiO_2) and hence the combustion reaction is enhanced, leading to a decrease in the formation of CO.

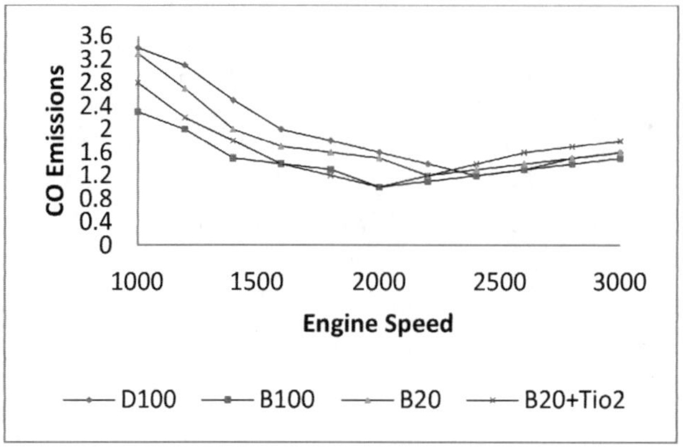

Figure 11 Values of CO emissions at different engine speeds

4.4.2 CO_2 emissions

Max CO_2 emission values for all test oils, as shown in Figure.12, have been achieved at 2800 rpm and full load engine condition. Increased average CO_2 emission levels were 13.82 percent, 3.88 percent, 9.94 percent compared to Euro diesel for B100, B20, B20 + TiO_2 test fuels.

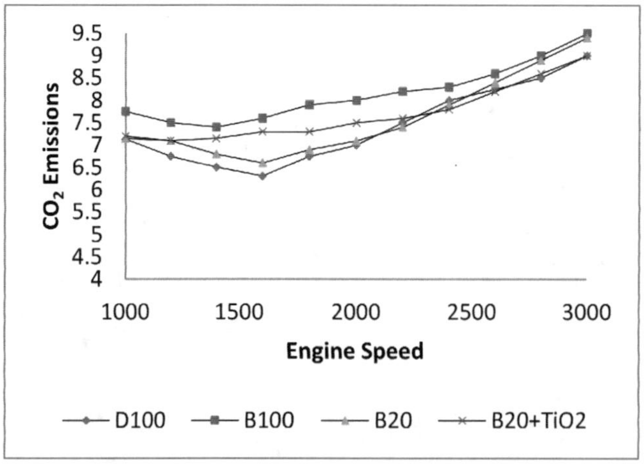

Figure 12 CO_2 emission rates at varying engine speeds and under optimal conditions for charging

The CO_2 emission values have raisen by around 5.88 percent and 5.79 percent, respectively, compared to B20 and B20+TiO_2. Adding TiO_2 to fuel blend increases average values for CO_2 emissions. This can be explained by the fact that CO_2 formation is rising as CO emissions fall, nano particles improve the combustion mechanism in such a way that the CO and CO_2 formation equilibrium occurs.

4.4.3 HC emissions

Figure.13 shows HC emissions for all test fuel at full load engine condition and varying engine speeds. Maximum emission values for HC were reported at 1000 rpm. While minimum values for all fuel mixtures were obtained at 3000 rpm, except for Euro diesel. Average reduction in HC emissions for B100, B20, B20 + TiO_2 test fuels was approximately 45.14 percent, 29.58 percent, 34.12 percent.

IOP Conf. Series: Materials Science and Engineering **954** (2020) 012032 doi:10.1088/1757-899X/954/1/012032

HC emission values decreased by 6.33 percent and 22.38 percent relative to the fuel blend without additives (B20) and TiO_2 additives (B20+TiO_2).

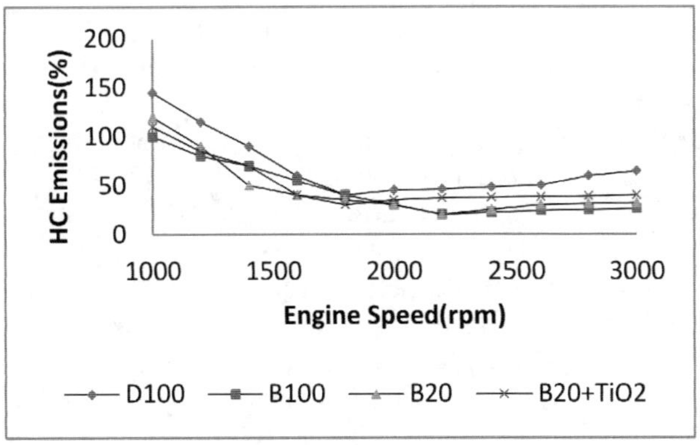

Figure 13 HC emission values and maximum load conditions at different engine speeds

4.4.4 NO_X Emissions

For all test fuels at full load and distinct engine speeds, the NOx emission results are shown in Figure 14. The maximum NOx emission values were recorded at 2800 rpm for all of the test fuels. Compared to euro diesel, the average rise in NOx values for test fuel B100, B20, B20 + TiO_2 was 18.90 percent, 4.73 percent and 11.99 percent respectively.

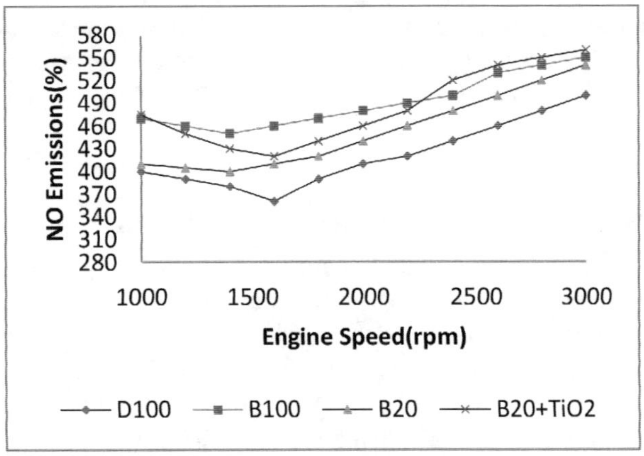

Figure 14: Emission values of NOx at different engine speeds and ideal charging conditions

By the addition of TiO_2 (B20 + TiO_2 and B20) causes NOx emission values to increase by approximately 6.98 percent and 3.97 percent relative to B20 and certain fuel mixes. For B100, the highest NOx emission value was recorded, while for B20+TiO_2 fuel blends, the lowest value was recorded. However, the combustion is improved by the TiO_2 additive.

6. Conclusions and discussions

In this research, biodiesel derived from sesame oil was mixed in different ratios with Euro diesel and nanoparticles of TiO_2. Thermo physical and chemical properties, such as density, fire and flash points and kinematic viscosity have been calculated and the calorific value has been calculated for different blends. The engine performance, combustion analysis and exhaust emission parameters were investigated. The following findings can be given below on the basis of the results of this report:

- Fossil-based production of biodiesel will greatly reduce the cost of manufacturing biodiesel. The properties of biodiesel follow specifications of both EN 14,214 and ASTM 6751. Thus, in diesel engines, biodiesel cooking waste may be used.
- Maximum brake torque and power values were reported at 1400 rpm and 2800 rpm respectively for all tested fuels. Compared to the B20 blend, the addition of TiO_2 to the B20 increased the engine brake torque and power by about 10.22%. Additionally, the B20 + TiO_2 brake engine's torque and power is about 9.76 percent higher than the B20 blend. However, brake-specific fuel consumption decreased by 27.73 percent. It can be concluded that the engine output was positively affected by the TiO_2 's contribution to the fuel mixture.
- The large nano-particulate surface area increases the reaction surface and the rate of heat transfer. The use of TiO_2 nano-particles in fuel mixtures has improved combustion during the combustion process due to a catalytic effect. As a result, the gross cylinder pressure and heat release rate for the TiO_2 additive are improved compared to Euro diesel.
- Although CO emissions, HC emissions and smoke opacity of sesame oil biodiesel have decreased, compared to Euro diesel emissions, NOx and CO_2 emissions have increased.
- Usage of TiO_2 nano-particles decreased emissions of CO, HC and smoke opacity. In comparison to fuel blends without TiO_2 additives, CO_2 and NOx emissions are improved in the combustion process.
- It can be concluded that biodiesel derived from sesame oil and nano-particles derived from TiO_2 can be regarded as promising alternative fuels for diesel engines, in view of all these findings. The torque and performance of the brake engine are not only significantly improved by TiO_2 nano-particles, but also reduces the basic fuel consumption of the fuel mixtures. It also improves the combustion method and the emissions of CO , HC and smoke opacity that have been positively affected. Further studies of various engine loads in multi-cylinder CIDI diesel engines are therefore suggested with a view to making a stronger conclusion on the effect of the TiO_2 additive. Additionally, it is possible to investigate the effects of other nano-particles on engine efficiency, combustion, and emission properties.

References

[1] I.Örsa, S. Sarıkoçb, A.E. Atabanic, S. Ünalanc, S.O. Akansuc, The effects on performance, combustion and emission characteristics of DICI engine fuelled with TiO2 nanoparticles addition in diesel/biodiesel/nbutanol blends, *Fuel* ***234*** *(2018) 177–188.*

[2] Homa Hossein zadeh-Bandbafhaa, Esmail Khalifeb, Meisam Tabatabaeic, Mortaza Aghbashloa, Majid Khanalia, Pouya Mohammadic,Taha Roodbar Shojaeia, Salman Soltanianc, Effects of aqueous carbon nano particles as a novel nano additive in water emulsified diesel/biodiesel blends on performance and emissions parameters of a diesel engine, *Energy Conversion and Management* ***196*** *(2019) 1153–1166.*

[3] T.K.Kannan and R.Marappan, Study of Performance and Emissions Characteristics of a Diesel Engine Using Thevetia Peruviana Biodiesel with Di-ethyl Ether Blends *European Journal, of Scientific Research Volume number* ***43*** *September2010.*

[4] J.P.Subrahmanyam and M.K.G.BabuSimultaneous Reduction of Smoke and NO_X Emissions of diesel and biodiesel fueled engines by using Diethyl Ether as Supplementary Oxygenated Fuel, presented at Indian Institute of Technology, Delhi.

[5] Dr.Rambabu. V, Experimental Investigation on the Performance,emission and combustion characteristics of DI Disel Engine with Linseed methyl Ester along with Methanal Carburization, *M.Tech. Thesis, Thermal Engg,NIT Warangal, 2011.*

[6] PrasadaRao, K., Experimental Investigation on the Performance of DIDisel Engine with Mahua methyl Ester along with Methanal Carburization, *M.E. Thesis, industrial engineering ,AU Visakhapatnam, 2011.*

[7] Mithun Das, Mouktik Sarkar, Amitava Datta, Apurba Kumar Santra, An experimental study on the combustion, performance and emission characteristics of a diesel engine fuelled with diesel-castor oil biodiesel blends, *Renewable Energy* ***119*** *(2018) 174e184,2018.*

[8] Nitin Shrivastava, Devanshu Shrivastava & Vipin Shrivastava, Experimental investigation of performance and emission characteristics of diesel engine using Jatropha biodiesel with alumina nanoparticles, *ISSN: 1543-5075 (Print)* ***1543-5083*** *(Online), 2018, 2018.*

[9] Hossein Soukht Saraee, Samad Jafarmadar, Mehran Sayadi, Aso Parikhani, Javad Kheyrollahi, Navid Pourvosoughi, Green fuel production from Pistacia Khinjuk and its engine test analysis as a promising alternative, *10.1016/j.jclepro.2017.04.053, JCLP* ***9407****, 2017*

[10] Venkata Ramesh and Mallikarjun, Biodiesel Production from palm oil by Transesterification Method, *I. Mech. Engg Sc.,****895****(8),0975-833X, 2012.*

[11] Pugazhvadivu. M and Sankaranarayanan G, Experimental studies on a diesel engine using mahua oil as fuel, *I. Mech. Engg Sc.,****787****(7),0974-6846, 2010.*

[12] G. Balaji & M. Cheralathan, Experimental investigation of varying the fuel injection pressure in a direct injection diesel engine fuelled with methyl ester of neem oil, *ISSN: 0143-0750 (Print)* ***2162-8246****, 2015.*

[13] Agnese Magno, Ezio Mancaruso, Bianca Maria Vaglieco a, Effects of a biodiesel blend on energy distribution and exhaust emissions of a small CI engine, *Energy Conversion and Management* ***96*** *(2015) 72–80, 2105.*

[14] Federico Millo, Biplab Kumar Debnath, TheodorosVlachos , Claudio Ciaravino, Lucio Postrioti, Giacomo Buitoni ,Effects of different biofuels blends on performance and emissions of an automotive diesel engine, *JFUE* ***9401****, 2015.*

[15] N. Panneerselvam, A. Murugesan, C. Vijayakumar & D. Subramaniam, Performance, emissions and combustion characteristics of CI engine fuel with watermelon (Citrullus vulgaris) methyl esters, *ISSN: 0143-0750 (Print)* ***2162-8246****, 2015.*

[16] T. Shaafi, R. Velraj, Influence of alumina nanoparticles, ethanol and isopropanol blend as additive with dieselesoybean biodiesel blend fuel: Combustion,engine performance and emissions, *Renewable Energy* ***80*** *(2015) 655e663, 2015.*

[17] C. Syed Aalam, C.G. Saravanan, M. Kannan, Experimental investigations on a CRDI system assisted diesel engine fuelled with aluminium oxide nanoparticles blended biodiesel, *doi.org/10.1016/j.aej.2015.04.009,****1110-0168****, 2015.*

[18] Özer Can, Erkan Öztürk , Hamit Solmaz , Fatih Aksoy , Can Çinar, H. Serdar Yücesu, Combined effects of soybean biodiesel fuel addition and EGR application on the combustion and exhaust emissions in a diesel engine, *Applied Thermal Engineering* ***95*** *(2016) 115–124, 2016.*

[19] M. Annamalai , B. Dhinesh , K. Nanthagopal , P. SivaramaKrishnan , J. Isaac Joshua, Ramesh Lalvani , M. Parthasarathy , K. Annamalai , An assessment on performance, combustion and emission behavior of a diesel engine powered by ceria nanoparticle blended emulsified biofuel, *Energy Conversion and Management* ***123*** *(2016) 372–380, 2016.*

[20] K. Khiari, S. Awad, K. Loubar, L. Tarabet, R. Mahmoud, M. Tazerout, Experimental investigation of pistacia lentiscus biodiesel as a fuel for direct injection diesel engine, *Energy Conversion and Management* ***108*** *(2016) 392–399, 2016.*

[21] A Parlak, H Karabas, I Ozsert, V Ayhan & I Cesur, Effects of tobacco seed oil methyl ester on performance and exhaust emissions of diesel engine, *ISSN: 1743-9671 (Print)* ***1746-0220*** *(Online), 2016.*

[22] Harish Venu, Venkataramanan Madhava, Effect of Al_2O_3 nanoparticles in biodiesel-diesel-ethanol blends at various injection strategies: Performance, combustion and emission characteristics, *Fuel* ***186*** *(2016) 176–189, 2016.*

[23] R.L. Krupakaran, T. Hariprasasd, A. Gopalakrishna & Pejjai Babu, The performance and exhaust emissions investigation of a diesel engine using γ-Al_2O_3 nanoparticle additives to biodiesel, *ISSN: 1758-3004 (Print)* ***1758-3012*** *(Online), 2016.*

[24] S. Karthikeyan, A. Elango & A. Prathima, The effect of cerium oxide additive on the performance and emission characteristics of a CI engine operated with rice bran biodiesel and its blends, *ISSN: 1543-5075 (Print)* ***1543-5083*** *(Online), 2016.*

[25] Seyyed Hassan Hosseini, Ahmad Taghizadeh-Alisaraei, Barat Ghobadian, Ahmad Abbaszadeh-Mayvan, Performance and emission characteristics of a CI engine fuelled with carbon nanotubes and diesel-biodiesel blends, *Renewable Energy* ***111*** *(2017) 201e213, 2017.*

[26] K. Nanthagopal, B. Ashok, A. Tamilarasu, Ajith Johny, Aravind Mohan, Influence on the effect of zinc oxide and titanium dioxide nanoparticles as an additive with Calophyllum inophyllum methyl ester in a CI engine, *Energy Conversion and Management* ***146*** *(2017) 8–19, 2017.*

[27] B. Ashok, K. Nanthagopal, R. Thundil Karuppa Raj, J. Pradeep Bhasker, D. Sakthi Vignesh, Influence of injection timing and exhaust gas recirculation of a Calophyllum inophyllum methyl ester fuelled CI engine, *Fuel Processing Technology* ***167*** *(2017) 18–30, 2017.*

[28] Akash Deep, Sarbjot Singh Sandhu, Subhash Chander, Experimental investigations on the influence of fuel injection timing and pressure on single cylinder C.I. engine fueled with 20% blend of castor biodiesel in diesel, *Fuel* ***210*** *(2017) 15–22, 2107.*

[29] Senthil Ramalingam, Silambarasan Rajendran, Pranesh Ganesan, Assessment of engine operating parameters on working characteristics of a diesel engine fuelled with 20% proportion of biodiesel diesel blend, *10.1016/j.energy.2017.09.134, EGY_EGY-D-**17-00564**, 2017.*

[30] J. Godwin John, V. Hariram, S. Sooryanarayanan, G. SivaSubramaniama, J. Reinhard Bonnke Osborn, A. Cyril Christo, S. Seralathan, Impact of varying the compression ratio on the combustion phenomenon of the diesel engine when fuelled with sesame biodiesel, materials today, j.matpr.2020.06.133, **2214-7853**, 2020 .

IOP Conf. Series: Materials Science and Engineering **954** (2020) 012033 doi:10.1088/1757-899X/954/1/012033

Optimization of Al-SiC-MW CNT metal composite by W-EDM using Taguchi method

Dr. C. Parswajinan*[1], Dr. B. Vijaya Ramnath[2], Aiswarya V[3], Aswini Subhatra M[5], S. Rajesh[5], J. Sriram[6]

*[1]Asst. Prof., Department of Mechanical Engineering, Sri Sai Ram Engineering College, Chennai – 600 044.
[2]Asst. Prof., Department of Mechanical Engineering, Sri Sai Ram Engineering College, Chennai – 600 044.
[3,4,5,6] Department of Mechanical Engineering, Sri Sai Ram Engineering College, Chennai – 600 044.

*parswajinan@gmail.com

Abstract. EDM process is primarily used in machining of Al6061 metal matrix composite reinforced with silicon carbide (SiC) and carbon nano-tube (CNT). This process can be used to measure the Metal Removal Rate (MRR), Tool Wear Rate (TWR) and Surface Roughness (SR) of the work piece material. Using Taguchi's parameter design, significant machining parameters affecting the performance of the process such as discharge current, pulse ON time and duty cycle are the input parameters. The performance measure for each combination of the parameters has been taken. In this work, the MRR, TWR and SR for each combination have been estimated. The results obtained are tabulated. The study demonstrates that the EDM process parameters can be adjusted to do better MRR, TWR and SR by optimizing the parameters using grey relational analysis (GRA).

1. Introduction

Nowadays, Electrical Discharge Machining (EDM) has been tremendously improving to meet the requirements in various manufacturing fields. It is a thermo electrical process in which the material is eroded from the workpiece by electric sparks between the workpiece and the electrode separated by a thin film of dielectric fluid, which is continuously fed to the machining zone to flush away the eroded particles. Several theories are proposed to explain the principle of EDM. The thermo-mechanical theory suggests that material removal in EDM operation is attributed to melting of material caused by flame jets that are formed as a result of various electrical effects of discharge. Thermo-electric theory suggests that the material removal in EDM operations take place as a result of generation of extremely high temperature due to the high intensity of discharge current. In EDM, it is mandatory to select machining parameters for achieving optimal machining performance. Usually, they are determined based on experience or handbook values. However, this does not ensure that this will result in optimal machining parameters due to the environment of machining or other parameters. To solve this issue, Taguchi method is used to determine the optimal machining parameters in the electrical discharge machining process. The Taguchi method optimizes the performance characteristics through the setting of process parameters and reducing the sensitivity of the system performance to sources of variation. As a result, the Taguchi method has become a powerful tool in the design of experiment methods. However, most published Taguchi applications to date have been concerned with the optimization of a single performance characteristic.

Jeykrishnan et al. [1 and 2] have done experiments on machining EN24 tool steel and D2 die steel by varying the parameters such as applied current, pulse on time and pulse off time in optimizing the MRR and TWR by employing Taguchi techniques and have concluded that the applied current plays a significant role in obtaining the best output characteristics. Vijaya Ramnath et al. [3] have done trials on drilling the fibre hybrid composites and have optimized the parameters using grey relational analysis (GRA) in obtaining the best output parameters. It is observed that composites reinforced with CNT showed improved mechanical properties [4] like strength and can be used as a better reinforcement for iron [5]. Vijaya Ramnath et al. inferred that increase in aluminium content increases the tensile strength [6]. Addition of CNT increases the compressive strength [7] but not hardness [8]. Mechanical properties are improved in reinforced iron compared to unreinforced iron [9]. Suresh rajan et al. [10] have employed EDM for machining inconel 825 alloys and have optimized the process parameters by using Taguchi techniques in-order to obtain the better surface roughness and have concluded that the pulse on time plays a significant role in getting the best output response. Akilesh et al. [11 and 12] have done electro-chemical machining on SKD-12 tool steel and D3 die steels and have optimized the process parameters using Taguchi techniques and have concluded that MRR and SR can be increased by varying the applied current, voltage and electrolyte concentration.

2. EDM PROCESS

This is the process used to remove metal from workpiece due to spark erosion. The input parameters such as pulse-on time, polarity, discharge current, dielectric fluid etc., which influences mostly the performance of machining process. Hence better machining performance in the EDM process can be achieved by proper selection of the machining parameters.

2.1 Selection of machining parameters

In this process, EDM machine (KT-200) was used for experimental purpose. In this method, the electrode is used as pure copper electrode with cylindrical cross section of diameter 14 mm is used for metal removal from workpiece. Here kerosene is used as dielectric medium. In this experiments, additive (aluminium powder or aluminium oxide powder) to the dielectric fluid (kerosene) is incorporated to mobilize the machining characteristics. The following machining parameters are taken into consideration such as pulse-on Current (A), time (T), and Duty Factor (DF).

2.2 Machining performance evaluation

Machining performance evaluation parameters can be calculated by using the following expression. EWR is defined as the ratio of Electrode Wear Weight (EWW) to Workpiece Removal Weight (WRW). It is usually expressed in percentage,

i.e. TWR (g/min) = (EWW / WRW) * 100 (1)

MRR is the ratio of WRW to a period of machining time in minutes (T),

i.e. MRR (g/min) = WRW / T (2)

By observing from the readings, the machining time for each workpiece can be optimized for 5 minutes. It is observed that, lower the TWR, better the machining performance. Higher the MRR, better the machining performance. Hence, EWR is lower the-better performance characteristic and MRR is higher-the-better performance characteristic.

2.3 Grey Relational Analysis (GRA)

The steps followed in Grey relational analysis were:

i. The input characteristics have been normalized for all the experimental trials;
ii. The Grey relational coefficient (GRC) have been generated;
iii. Upon taking the average of GRC, Grey relational grade (GRG) have been found out;
iv. To determine the significance of each parameter, Analysis of variance (ANOVA) has been incorporated along with GRG.
v. Optimum process parameters have been selected.

3. RESULTS AND DISCUSSION

The experimental layout for L16 orthogonal array (OA) has been given in Table- 3. 1 a and b.

Table - 3. 1. a. Experimental layout for L_{16} OA

MRR	S/N Ratio	TWR	S/N Ratio	SR
0.226	-12.9178	0.008	41.9382	3.91
0.244	-12.2522	0.009	40.91515	3.94
0.302	-10.4	0.002	53.9794	4.02
0.192	-14.334	0.002	53.9794	4.06
0.2	-13.9794	0.003	50.45757	4.15
0.226	-12.9178	0.004	47.9588	4.16
0.245	-12.2167	0.042	27.53501	4.25
0.215	-13.3512	0.051	25.8486	4.06
0.226	-12.9178	0.052	25.67993	4.08
0.236	-12.5418	0.052	25.67993	3.95
0.249	-12.076	0.054	25.35212	3.84
0.236	-12.5418	0.055	25.19275	3.74
0.246	-12.1813	0.056	25.03624	3.64
0.251	-12.0065	0.058	24.73144	3.85
0.259	-11.734	0.059	24.58296	3.95

Table - 3. 1. b. Experimental layout for L_{16} OA

S/N Ratio		GRC			GRG			Dev Seq			GRC		GRG	Rank
-11.7092	0.00	0.79	0.66	0.00	0.79	0.66	1.00	0.21	0.34	0.33	0.70	0.59	0.52	9
-11.8435	0.58	0.89	0.56	0.58	0.89	0.56	0.42	0.11	0.44	0.54	0.83	0.53	0.62	3
-11.9099	0.68	0.88	0.51	0.68	0.88	0.51	0.32	0.12	0.49	0.61	0.80	0.50	0.62	4
-12.0845	1.00	1.00	0.38	1.00	1.00	0.38	0.00	0.00	0.62	1.00	1.00	0.45	0.81	1
-12.1705	0.39	1.00	0.31	0.39	1.00	0.31	0.61	0.00	0.69	0.45	1.00	0.42	0.61	5
-12.361	0.43	0.98	0.16	0.43	0.98	0.16	0.57	0.02	0.84	0.47	0.97	0.37	0.60	6
-12.3819	0.58	0.96	0.15	0.58	0.96	0.15	0.42	0.04	0.85	0.54	0.93	0.37	0.60	7
-12.5678	0.68	0.30	0.00	0.68	0.30	0.00	0.32	0.70	1.00	0.61	0.42	0.33	0.48	13
-12.1705	0.52	0.14	0.31	0.52	0.14	0.31	0.48	0.86	0.69	0.51	0.37	0.42	0.43	15
-12.2132	0.58	0.12	0.28	0.58	0.12	0.28	0.42	0.88	0.72	0.54	0.36	0.41	0.47	14
-11.9319	0.63	0.12	0.49	0.63	0.12	0.49	0.37	0.88	0.51	0.58	0.36	0.50	0.51	11
-11.6866	0.71	0.09	0.67	0.71	0.09	0.67	0.29	0.91	0.33	0.63	0.35	0.60	0.58	8
-11.4574	0.63	0.07	0.84	0.63	0.07	0.84	0.37	0.93	0.16	0.58	0.35	0.75	0.64	2
-11.222	0.69	0.05	1.00	0.69	0.05	1.00	0.31	0.95	0.00	0.62	0.35	1.00	0.52	10
-11.7092	0.72	0.02	0.66	0.72	0.02	0.66	0.28	0.98	0.34	0.64	0.34	0.59	0.49	12
-11.9319	0.76	0.00	0.49	0.76	0.00	0.49	0.24	1.00	0.51	0.68	0.33	0.50	0.34	16

3.1 Effects of process parameters on MRR

The MRR increases, when the parameters such as Ton, current and the duty factor increases. This is mainly due to the fact that the more pulse on time and current results in increased machining exposure time, thereby more material gets removed. The Di-electric Current plays a vital role in MRR followed by duty factor and pulse on time. The response table and the ANOVA table for MRR have been given in Table 3.2 and 3.3. The S/N ratio graph for MRR has been given in Fig. 3.1.

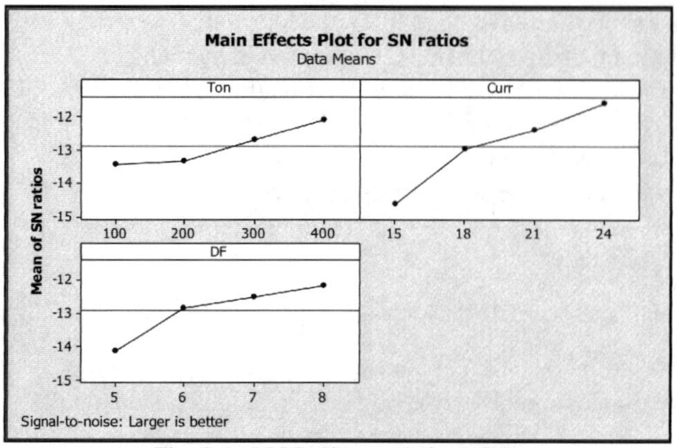

Fig. 3.1. S/N Ratio graph for MRR

Table- 3.2. Response table for MRR

Level	Ton	Curr	DF
1	-11.89	-11.88	-11.98
2	-12.37	-11.91	-11.85
3	-12.00	-11.98	-11.97
4	-11.58	-12.07	-12.03
Delta	0.79	0.19	0.18
Rank	1	2	3

Table- 3.3. ANOVA Table for MRR

Ton	Curr	DF	MRR	SNRA1
100	15	5	0.122	-18.2728
100	18	6	0.226	-12.9178
100	21	7	0.244	-12.2522
100	24	8	0.302	-10.3999
200	15	6	0.192	-14.334
200	18	5	0.2	-13.9794
200	21	8	0.226	-12.9178
200	24	7	0.245	-12.2167
300	15	7	0.215	-13.3512
300	18	8	0.226	-12.9178
300	21	5	0.236	-12.5418
300	24	6	0.249	-12.076
400	15	8	0.236	-12.5418
400	18	7	0.246	-12.1813
400	21	6	0.251	-12.0065
400	24	5	0.259	-11.734

3.2 Effects of process parameters on TWR

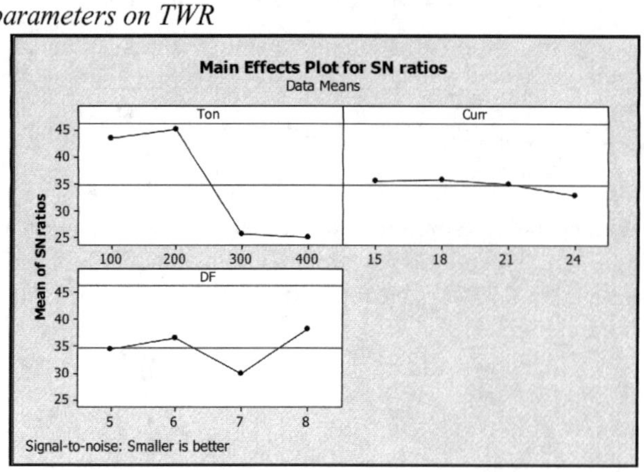

Fig. 2. S/N Ratio graph for TWR

The TWR decreases when the Ton, current and SR increases, as this happens due to the fact that the drop-in pressure makes the molecules to move towards the positively charged material, instead of cathode.

Pulse on Time plays an important role in determining the TWR followed by duty factor and current. The ANOVA table for TWR and Response Table for SR has been given in Table 3.4. The S/N ratio graph for TWR is given in Fig. 2.

Table -3.4. ANOVA for TWR and Response table for SR

T_{on}	Curr	DOF	TWR	S/N Ratio	SR	S/N Ratio
100	15	5	0.014	37.0774	3.85	-11.7092
100	18	6	0.008	41.9382	3.91	-11.8435
100	21	7	0.009	40.9151	3.94	-11.9099
100	24	8	0.002	53.9794	4.02	-12.0845
200	15	6	0.002	53.9794	4.06	-12.1705
200	18	5	0.003	50.4576	4.15	-12.361
200	21	8	0.004	47.9588	4.16	-12.3819
200	24	7	0.042	27.535	4.25	-12.5678
300	15	7	0.051	25.8486	4.06	-12.1705
300	18	8	0.052	25.6799	4.08	-12.2132
300	21	5	0.052	25.6799	3.95	-11.9319
300	24	6	0.054	25.3521	3.84	-11.6866
400	15	8	0.055	25.1927	3.74	-11.4574
400	18	7	0.056	25.0362	3.64	-11.222
400	21	6	0.058	24.7314	3.85	-11.7092
400	24	5	0.059	24.583	3.95	-11.9319

3.3 Effects of process parameters on SR

The surface roughness (SR) decreases, when the process parameters such as Ton, current and SR increases. This is due to the fact that the electrons emerge in the gap between the tool and the workpiece, thereby making the pressure to drop; as a result, the energy gathered between the tool and the workpiece increases. This helps in getting better SR.

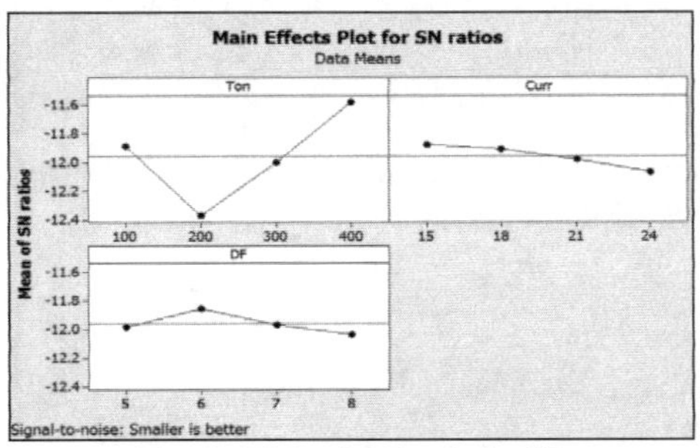

Fig. 3. S/N ratio Graph for SR

The pulse on time plays an important role in obtaining better SR, followed by the current and the DF. The response table and the ANOVA table for SR have been given in Table 3.5. The S/N ratio graph for SR has been given in Fig. 3.

Table- 3.5 ANOVA table for SR

MRR	TWR	SR		GRG			Dev Seq			GRC		GRĜ	Rank
0.122	0.014	3.85	0.00	0.79	0.66	1.00	0.21	0.34	0.33	0.70	0.59	0.52	9
0.226	0.008	3.91	0.58	0.89	0.56	0.42	0.11	0.44	0.54	0.83	0.53	0.62	3
0.244	0.009	3.94	0.68	0.88	0.51	0.32	0.12	0.49	0.61	0.80	0.50	0.62	4
0.302	0.002	4.02	1.00	1.00	0.38	0.00	0.00	0.62	1.00	1.00	0.45	0.81	1
0.192	0.002	4.06	0.39	1.00	0.31	0.61	0.00	0.69	0.45	1.00	0.42	0.61	5
0.2	0.003	4.15	0.43	0.98	0.16	0.57	0.02	0.84	0.47	0.97	0.37	0.60	6
0.226	0.004	4.16	0.58	0.96	0.15	0.42	0.04	0.85	0.54	0.93	0.37	0.60	7
0.245	0.042	4.25	0.68	0.30	0.00	0.32	0.70	1.00	0.61	0.42	0.33	0.48	13
0.215	0.051	4.06	0.52	0.14	0.31	0.48	0.86	0.69	0.51	0.37	0.42	0.43	15
0.226	0.052	4.08	0.58	0.12	0.28	0.42	0.88	0.72	0.54	0.36	0.41	0.47	14
0.236	0.052	3.95	0.63	0.12	0.49	0.37	0.88	0.51	0.58	0.36	0.50	0.51	11
0.249	0.054	3.84	0.71	0.09	0.67	0.29	0.91	0.33	0.63	0.35	0.60	0.58	8
0.236	0.055	3.74	0.63	0.07	0.84	0.37	0.93	0.16	0.58	0.35	0.75	0.64	2
0.246	0.056	3.64	0.69	0.05	1.00	0.31	0.95	0.00	0.62	0.35	1.00	0.52	10
0.251	0.058	3.85	0.72	0.02	0.66	0.28	0.98	0.34	0.64	0.34	0.59	0.49	12
0.259	0.059	3.95	0.76	0.00	0.49	0.24	1.00	0.51	0.68	0.33	0.50	0.34	16

4. CONCLUSION

From various observations from the experiments, the following points can be concluded. The performance characteristics of MRR, TWR and SR is significantly improved through this research work. This research has been studied and optimized the parameters for enhancing the performance machining characteristics of EDM process. From this research work the optimized results are found that MRR, TWR and SR were 0.302 g/mm^3, 0.002 g/mm^3 and 4.02 μm, respectively.

REFERENCE

1. J. Jeykrishnan, B. Vijaya Ramnath, S. Akilesh, R. P. Pradeep Kumar, "Optimization of process parameters on EN24 tool steel using Taguchi technique in electro-discharge machining", *Materials science and Engineering*, 2016, Vol. 149, pp. 01022.

2. J. Jeykrishnan, B. Vijaya Ramnath, A. Jude Felix, C. Rupan Pernesh, S. Kalaiyarasan, "Parameter optimization of electro-discharge machining (EDM) in AISI D2 die steel using Taguchi technique", Indian journal of science and technology, Vol. 9, Issue: 43, 2016.

3. B. Vijaya Ramnath, S. Sharavanan, J. Jeykrishnan, "Optimization of process parameters in drilling of fibre hybrid composite using Taguchi and grey relational analysis", *Materials science and Engineering,* 2017, Vol. 183, pp. 012003.

4. B. Vijaya Ramnath, C. Parswajinan, C. Elanchezhian, S. V. Pragadeesh, P. R. Ramkishore, V. Sabarish, "A Review on CNT Reinforced Aluminium and Magnesium Matrix Composites", *Applied Mechanics and Materials*, 2014, Vol. 591, pp. 120-123.

5. C. Parswajinan, B. Vijaya Ramnath, M. Vetrivel, P. Ramanarayanan, S. Bharath, T. Ajay, R. Raghav Chander, "A Review on Composite Materials with Ferrous, CNT and Powder Metallurgy", *Applied Mechanics and Materials*, 2015, Vol. 813-814, pp. 9-13.

6. B. Vijaya Ramnath, C. Elanchezhian, M. Jaivignesh, S. Rajesh, C. Parswajinan, A. Siddique Ahmed Ghias, "Evaluation of mechanical properties of aluminium alloy–alumina–boron carbide metal matrix composites", *Materials and Design*, 2014, Vol. 58, pp. 332–338.

7. C. Parswajinan, B. Vijaya Ramnath, M.Vetrivel, C. Elanchezhian, K. Loganathan, R. Sarvesh, C. Rohit Prasanna, R. N. Karthik Babu , "Experimental Investigation of Mechanical and Chemical Properties of Aluminium reinforced with MWCNT", *Applied Mechanics and Materials*, 2015, Vol. 766-767, pp 287-292.

8. B. Vijaya Ramnath, C. Parswajinan, C. Elanchezhian, S. V. Pragadeesh, C. Kavin, P. R. Ramkishore, V. Sabarish, "Experimental Investigation on Compression and Chemical Properties of Aluminium Nano Composite", *Applied Mechanics and Materials*, 2014, Vol. 680, pp. 7-10.

9. C. Parswajinan, B. Vijaya Ramnath, C. Elanchezhian, S. V. Pragadeesh, P. R. Ramkishore, V. Sabarish , "Investigation on Mechanical Properties of Nano Ferrous Composite", *Procedia Engineering,* 2014, Vol. 97, PP. 513 – 521.

10. J. Jeykrishnan, B. Vijaya Ramnath, G. Sureshrajan, M. Siva Bharath, X. HervinSavariraj, S. Akilesh, "Effects of die-sinking electro-discharge machining on surface roughness in inconel 825 alloy", Indian journal of science and technology, Vol. 9, Issue: 41, 2016.

11. J. Jeykrishnan, B. Vijaya Ramnath, C. Elanchezhian, S. Akilesh, "Parametric analysis on Electro-chemical machining of SKD-12 tool steel", Materials today: Proceedings, 2017, Vol. 4(2(A)), pp. 3760-3766.

12. J. Jeykrishnan, B. Vijaya Ramnath, C. Elanchezhian, S. Akilesh, "Optimization of process parameters in electro-chemical machining (ECM) of D3 die steels using Taguchi technique", Materials today: Proceedings, 2017, Vol. 4(8), pp. 7844-7891.

IOP Conf. Series: Materials Science and Engineering **954** (2020) 012034 doi:10.1088/1757-899X/954/1/012034

Multi response Optimization of Welding Parameters in Dissimilar Plasma Arc Welded Joints Using RSM Based GRA Coupled with PCA

K Palani[1] and S Arunprasad[2]

[1] Department of Mechanical Engineering, Sri Chandrasekharendra Saraswathi Viswa Mahavidyalaya (Deemed to be University), Kanchipuram, 631561, Tamilnadu, India

[2] Department of Production Engineering, Sri Sairam Engineering College, Chennai, 600 044, Tamil Nadu, India

E-mail: palani.k@kanchiuniv.ac.in

Abstract. The quality of the joints is essential to achieve customer satisfaction and earning profit in manufacturing industries. This work investigates the multi-response optimization of the plasma arc welded (PAW) process parameters of the dissimilar Ti-3Al-2.5V titanium alloy and AA8011-H24 aluminium alloy joints using RSM based GRA coupled with PCA. The principal component analysis is incorporated into the response surface methodology based grey relational analysis to achieve the optimum responses viz., tensile strength and tensile elongation with the welding parameters, viz., welding current, torch angle, gas flow rate and welding speed of the PAW joints. The response surface and contour plots are used to study the effect of parameters of the joints. The welding speed is the most significant parameter on the composite welding quality index (CWQI) followed by welding current, gas flow rate and torch angle to decide the quality of the joints at the optimal levels.

Keywords: *Titanium alloy; Aluminium alloy; Grey relational analysis; Principal component analysis; Composite welding quality index*

1. Introduction

In aerospace, aviation and transportation industries, the joining of dissimilar alloys, viz., titanium alloys and aluminium alloys are focused due to its unique properties, viz., improved strength in Ti alloys and also, better corrosion resistance and high strength to weight ratio in aluminium alloys [1, 2]. It is the difficult task in joining of high and low melting materials due to their various thermal properties viz., coefficient of heat transfer coefficient and melting temperature in fabrication industries [3]. The formation of the intermetallic compounds (IMCs) are occurred in the weld interface especially in the heat-affected zone (HAZ) of the dissimilar Al and Ti alloys by the variation in melting and solidification temperatures in the welding process and these brittle IMCs deteriorates the strength properties of the dissimilar joints [4, 5]. The dissimilar alloys such as Ti alloys and Al alloys are conventionally fusion welded using the tungsten inert gas welding [6], spot welding [7], and laser welding [8] in the aviation and automobile industries. The solid-state joining processes such as friction welding [9], diffusion bonding [10], friction stir welding processes [11-14] and friction stir processing [15, 16] are used in joining of dissimilar alloys.

The formation of the weld failures, viz., porosity, improper fusion, hot cracking, and solidification shrinkage are reduced in the fabricated dissimilar welded joints using the high energy fusion welding techniques such as electron beam welding and plasma arc welding processes [17]. Lakshminarayanan et al. [18] examined that the higher tensile strength and the toughness of the FSW joints whereas 24% and 39% higher tensile strength and the toughness of the GTAW process were examined due to the appearance of fine and equiaxed grains in the weld nugget zone of thee interstitial free steels.

Mvola et al. [19] investigated that the dissimilar magnesium and aluminium alloys were successfully welded with cold metal transfer and the appearance of fractures were occurred in the magnesium alloy fusion zone (FZ) due to the formation of Al-Cu IMCs in the welded joints. Esmaily et al. [20] successfully butt welded joints on Ti-6Al-4V alloy using friction stir welding and tungsten inert gas welding processes and reported that the broader fusion zone and heat-affected zone were observed in gas tungsten arc welding compared to friction stir welding process at 650 rpm rotational speed. The multi-response optimization methods are applied to identify the effective responses of the PAW process in the dissimilar alloy materials.

Rajakumar et al. [21] examined that the optimum shear strength of 87 MPa, the microhardness of 163 HV and interface layer thickness of 7 μm for the optimum process settings of bonding temperature of 510 °C, bonding pressure of 17 MPa and holding time of 37 min using the developed mathematical models of diffusion bonded dissimilar Cp-Ti with AA7075 aluminium alloys. The hybrid optimization technique viz., response surface methodology and grey relational analysis coupled with principal component analysis were applied on the Electro discharge machining process on dissimilar alloy materials by Sahu et al. [22]. Palani et al. [23] applied the hybrid fuzzy-based response surface optimization of welding parameters on Vickers microhardness and impact strength of FSWed AA8011-H24 aluminium alloy joints. Also, they suggested that the reduction of the multiple responses into a single effective response with significant contribution of the machining process were effectively used in determining with the grey relational grade.

The various hybrid optimization techniques were applied to find the very effective parameters which were helpful to decide the quality of the welded joints. Siddiquee et al. [24] applied the hybrid grey relational analysis with principal component analysis on the centreless cylindrical grinding process. They reported that the reflection of the relative importance of the responses was found by applying grey analysis whilst the weightage of the responses is improved the quality of the grinding process by using PCA. The optimization of multiple responses of submerged arc welding parameters, viz., bead width, bead penetration and bead reinforcement successfully made using response surface methodology based fuzzy logic with trapezoidal and triangular membership functions (MF) by Sharma et al. [25]. Also, they reported that the optimized results using a trapezoidal membership function showed the better-optimized results than the triangular membership functions. In this investigation, the work has been carried out on the multi-response optimization of welding parameters in dissimilar plasma arc welded joints using RSM based GRA coupled with PCA.

2. Experimental work

In this investigation, the Ti3Al2.5V alloys and AA8011 aluminium alloys [26] (Procured from Sri Kataria Steels in Mumbai) were used with the dimensions of 150 mm x 150 mm x 6 mm for butt joint configuration using the plasma arc welding process (Model PWH/-2A with the thermal arc 2A plasma welding torch) and are presented in Figure 1. The specifications of the PAW machine setup are listed in Table 1. This setup carries the PAW machine control module, the direct current power source, gas supply (Plasma gas cylinder, Shielding gas cylinders with 80% Argon gas and 20% Hydrogen gas and, Coolant supply. The shielding gas environment is provided for the fabrication of the aluminium alloys and titanium alloy joints due to the oxidation behaviour of the alloy materials.

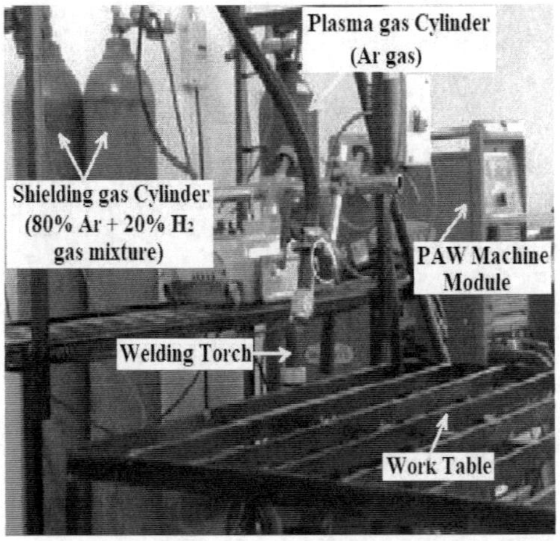

Figure 1. Schematic arrangement of PAW Machine setup

Table 1. Specifications of the PAW Machine (Model: PWH/-2A)

Sl. No.	Specifications	Details
1	Operation type	Automatic
2	Plasma gas	Argon gas
3	Shielding gas	80% Ar and 20% H_2
4	Nozzle to Plate distance	1 mm
5	Plasma gas flow rate	15 – 20 Lpm
6	Torch Angle	85° – 95°
7	Welding Speed	25 – 50 mm/min
8	Welding Current	90 – 120 Amp
9	Polarity	DCEN
10	Filler rod	ER4043 standard
11	Electrode	3% Thoriated 'W' electrode
12	Electrode Diameter	1.2 mm

The chemical composition of the AA8011 aluminium alloy and Ti3Al2.5V alloy are presented in Table 2. The plates for investigation initially washed with an ultrasonic bath sonicator and the butt surfaces were cleaned with acetone.

Table 2. Chemical composition of the AA8011-H24 Aluminium alloy and Ti3Al2.5V alloy

Chemical composition (Wt.%) of AA8011-H24 Aluminium alloy				
Fe – 0.74	Mn – 0.10	Si – 0.56	Cu – 0.12	Mg – 0.10
Zn – 0.08	Cr – 0.12	Ti – 0.05	Remaining Al	
Chemical composition (Wt.%) of Ti3Al2.5V Ti alloy				
Fe – 0.33	Si – 0.15	Al – 2.9	V – 2.5	O – 0.15
C – 0.06	H – 0.02	N – 0.02	Remaining Ti	

3. Central Composite Design

In this present investigation, the effects of the welding parameters on the mechanical properties of the dissimilar AA8011-H24 Al alloy and Ti3Al2.5V Ti alloy joints were discussed using the four factors, five levels of the central composite rotatable design with the thirty-one different welding experiments for the PAW process. The design of experiments was conducted using the Design Expert® 11.0 software to examine the optimal responses on the superior responses of PAW process.

3.1 Identifying the Important FSW and PAW Process Parameters

The fishbone diagram is applied to find the various parameters, viz., work and tool material properties, dimensions of the workpiece and tools, filler materials and machining parameters involved in dissimilar PAW process to decide the good quality of the welded joints. In this work, the independent welding process parameters, welding current (Iw), torch angle (θ), plasma gas flow rate (Fr) and welding speed (S) are selected based on the previous literature related to the fusion welding process [27], machine capability and preliminary trial experiments which affects the tensile properties of the fabricated joints.

3.2 Finding the working limits of parameters

The feasible working limits of the input parameters are found by conducting a large number of trail experiments for PAW welding process using the dissimilar AA8011-H24 Al alloy and Ti3Al2.5V Ti alloys with 6 mm thick plates. The trial experiments are conducted by keeping one of the parameters as variable and other parameters were kept constant. The range of each welding parameter is selected based on by inspecting the presence of defects, viz., hot cracking, solidification shrinkage, tunnel defect, wormhole, piping defect, and kissing bond through the macrostructure across the weld direction. The upper and lower limits of the four-factor, five levels of the design matrix are coded as +2 and -2 respectively and the intermediate coded values are determined by the equation (1),

$$X_i = \frac{2\,[2X - (X_{\max} + X_{\min})]}{(X_{\max} - X_{\min})} \tag{1}$$

Where,

Xi is the required coded value of a variable *X; X* is any value of the variable lies between X_{min} to X_{max}; X_{min} is the lower limit of the variable, and X_{max} is the upper limit of the variable.

The selected welding parameters with their limits, units and notations of the PAW process are presented in Table 3.

Table 3 PAW process parameters and their levels

Process parameters	Unit	Notation	Levels				
			-2	-1	0	+1	+2
Welding current	Amps	I_w	94	98	102	106	110
Torch angle	deg	θ	86	88	90	92	94
Gas flow rate	Lpm	F_r	16	17	18	19	20
Welding speed	mm/min	S	25	30	35	40	45

3.3 Development of Design Matrix

The various combinations of each process parameters were formed which carries the different experimental trails in the rows and the levels of each welding parameters in the columns of a design matrix [28]. In this work, the four factors, five levels of CCD matrix were used to conduct the 31 sets of coded conditions for the experiments. The non-linear quadratic models were developed from the design matrix which has 31 experimental conditions due to the welding behaviour on the alloy materials on the tensile properties of PAW butt joints using the response surface methodology.

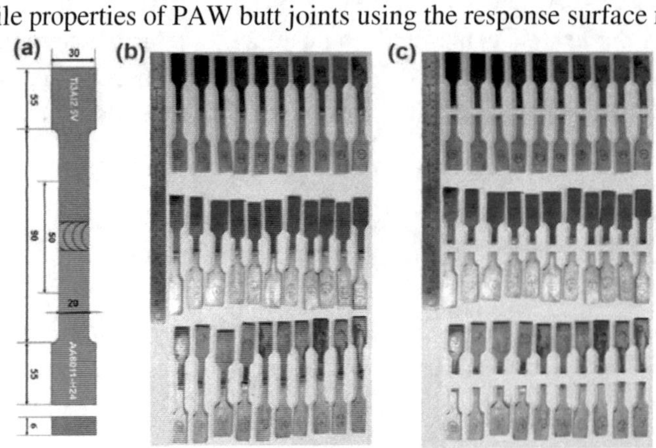

Figure 2 (a) Standard Tensile Specimen (b) Fabricated Tensile Specimens and (c) Fractured Tensile Specimens

Table 4. Design Matrix for PAW Process (Coded Values)

Std	Welding Current (Amp)	Torch Angle (Deg)	Gas Flow rate (Lpm)	Welding Speed (mm/min)	Tensile Strength (Mpa)	Tensile Elongation (%)
1	-1	-1	-1	-1	132.4	3.21
2	+1	-1	-1	-1	143.1	3.48
3	-1	+1	-1	-1	130.1	4.10
4	+1	+1	-1	-1	138.9	4.02
5	-1	-1	+1	-1	135.0	3.56
6	+1	-1	+1	-1	136.6	3.66
7	-1	+1	+1	-1	130.2	4.12
8	+1	+1	+1	-1	128.4	3.88
9	-1	-1	-1	+1	133.8	3.35
10	+1	-1	-1	+1	141.2	3.84
11	-1	+1	-1	+1	138.0	3.92
12	+1	+1	-1	+1	142.3	4.08
13	-1	-1	+1	+1	139.6	3.55
14	+1	-1	+1	+1	137.2	3.87
15	-1	+1	+1	+1	141.2	3.75
16	+1	+1	+1	+1	136.2	3.78
17	-2	0	0	0	134.0	3.44
18	+2	0	0	0	138.8	3.72
19	0	-2	0	0	140.0	3.16
20	0	+2	0	0	136.3	3.91
21	0	0	-2	0	135.4	3.95
22	0	0	+2	0	132.1	4.00
23	0	0	0	-2	136.4	3.95
24	0	0	0	+2	144.2	3.96
25	0	0	0	0	145.4	4.39
26	0	0	0	0	146.4	4.42
27	0	0	0	0	144.4	4.38
28	0	0	0	0	145.7	4.42
29	0	0	0	0	143.3	4.41
30	0	0	0	0	145.1	4.43
31	0	0	0	0	143.8	4.42

The tensile specimens were sliced perpendicular to the weld direction using a power hacksaw and machined as per ASTM-E08 standards and were tested in the Omega Inspection and Analytical Laboratory, Chennai [29]. The fabricated plasma arc welded tensile specimens and fractured specimens are presented in Figure 2. The recorded experimental tensile test responses, namely Tensile Strength and Tensile Elongation for PAW processes are presented in Table 4 in the design matrix using response surface methodology.

4. Result and Discussions

This work was carried the optimisation of multi responses of the PAW process variables, namely welding current (Iw) in Amp, torch angle (θ) in deg, gas flow rate (Fr) in LPM and welding speed (S) in mm/min. The range of the above independent variables should be considered as $102 \leq Iw \leq 110$; $86 \leq \theta \leq 94$; $16 \leq Fr \leq 20$; and $25 \leq S \leq 45$. Furthermore, the grey relational analysis was applied to convert the multi-response optimisation problem into the single response optimisation problem using the calculated grey relational grade [30].

4.1 Normalised sequence of PAW responses

The normalisation of the measured responses of the PAW process is the primary step for grey relational analysis based on the Larger the better concept using Eqn. (2) and this step prevents the large variation in the measured responses, viz., tensile strength and tensile elongation. Usually, the normalised responses are considered in the range 0 to 1 [31]. Table 5 shows the normalised values of the responses of the dissimilar plasma arc welded Ti-3Al-2.5V Ti alloy and AA8011-H24 aluminium alloy joints.

For higher-the-better

$$x_i(k)^* = \frac{x_i(k) - x_i(k)^-}{\left[x_i(k)^+ - x_i(k)^-\right]} \qquad For\ \left(i = 1,\ 2\ldots\ m\ and\ k = 1,\ 2\ldots\ n\right) \qquad (2)$$

Where xi (k)* is the normalised response, x0 (k) is the desired responses, xi (k)+ is the maximum of xi (k), and xi (k)- is the minimum of xi (k).

Table 5 Normalised responses, Grey relational coefficient and Grey relational grade of the Responses

Std	Welding Current (Amp)	Torch Angle (deg)	Gas Flow rate (LPM)	Welding Speed (mm/min)	Normalised PAW Responses		Grey Relational Coefficient		Grey Relational Grade
					Tensile Strength (MPa)	Tensile Elongation (%)	Tensile Strength (MPa)	Tensile Elongation (%)	
1	98	88	17	30	0.2222	0.0394	0.3668	0.3423	0.3668
2	106	88	17	30	0.8167	0.2520	0.5662	0.4006	0.5662
3	98	92	17	30	0.0944	0.7402	0.5069	0.6580	0.5069
4	106	92	17	30	0.5833	0.6772	0.5766	0.6077	0.5766
5	98	88	19	30	0.3667	0.3150	0.4316	0.4219	0.4316
6	106	88	19	30	0.4556	0.3937	0.4654	0.4520	0.4654
7	98	92	19	30	0.1000	0.7559	0.5146	0.6720	0.5146
8	106	92	19	30	0.0000	0.5669	0.4346	0.5359	0.4346
9	98	88	17	40	0.3000	0.1496	0.3935	0.3703	0.3935
10	106	88	17	40	0.7111	0.5354	0.5761	0.5184	0.5761
11	98	92	17	40	0.5333	0.5984	0.5359	0.5546	0.5359
12	106	92	17	40	0.7722	0.7244	0.6659	0.6447	0.6659
13	98	88	19	40	0.6222	0.3071	0.4944	0.4191	0.4944
14	106	88	19	40	0.4889	0.5591	0.5130	0.5314	0.5130
15	98	92	19	40	0.7111	0.4646	0.5584	0.4829	0.5584
16	106	92	19	40	0.4333	0.4882	0.4815	0.4942	0.4815
17	94	90	18	35	0.3111	0.2205	0.4057	0.3908	0.4057
18	110	90	18	35	0.5778	0.4409	0.5072	0.4721	0.5072
19	102	86	18	35	0.6444	0.0000	0.4589	0.3333	0.4589
20	102	94	18	35	0.4389	0.5906	0.5105	0.5498	0.5105
21	102	90	16	35	0.3889	0.6220	0.5098	0.5695	0.5098
22	102	90	20	35	0.2056	0.6614	0.4913	0.5962	0.4913
23	102	90	18	25	0.4444	0.6220	0.5216	0.5695	0.5216
24	102	90	18	45	0.8778	0.6299	0.6892	0.5747	0.6892
25	102	90	18	35	0.9444	0.9685	0.9204	0.9407	0.9204
26	**102**	**90**	**18**	**35**	**1.0000**	**0.9921**	**0.9923**	**0.9845**	**0.9923**
27	102	90	18	35	0.8889	0.9606	0.8726	0.9270	0.8726
28	102	90	18	35	0.9611	0.9921	0.9562	0.9845	0.9562
29	102	90	18	35	0.8278	0.9843	0.8567	0.9695	0.8567
30	102	90	18	35	0.9278	1.0000	0.9369	1.0000	0.9369
31	102	90	18	35	0.8556	0.9921	0.8802	0.9845	0.8802

4.2 Calculation of the grey relational coefficient and grey relational grade

The calculation of grey relational coefficient (GRC) is to express the relationship between the predicted and the actual experimental responses of the experiments in Table 5. The grey relational coefficient ζi (k) can be calculated based on Eqn. (3),

$$\zeta_i(k) = \frac{\Delta_{\min} + \psi\Delta_{\max}}{\Delta_{oi}(k) + \psi\Delta_{\max}} \quad (3)$$

In this, the distinguishing coefficient, (ζ) of 0.5 is considered for the determination of the grey relational coefficient [32].

4.3 Determination of the eigenvalues, eigenvectors and principal component scores

The eigenvalues and eigenvectors are determined from the correlation coefficient and the Variance-Covariance matrix array is represented in Eqn. (4) as follows:

$$(N_{kl} - \lambda_k I_m)V_{ik} = 0 \quad (4)$$

Where N is the correlation coefficient matrix form of N_{kl} and λ_k is the kth eigenvalue and $\sum_{k=1}^{n} \lambda_k = n$, k = 1, 2… n and $V_{ik} = [a_{k1} a_{k2} \ldots\ldots a_{kn}]^T$ is the eigenvector corresponding to the eigenvalue λ_k.

The principal component score, ψj corresponding to each experiment is formulated based on Eqn. (5) as follows:

$$\Psi_n = \sum_{j=1}^{n} X_{ij} \cdot V_{jk} \quad (5)$$

Where $\psi_1, \psi_2, \psi_3, \ldots$ represents the first principal component, second principal component, third principal component and so on. X_{ij} is the corresponding normalised measured response matrix, and V_{jk} is the eigenvector corresponding to the eigenvalues λ_k.

The principal component analysis is incorporated into the response surface methodology based grey relational analysis for optimising the welding process parameters, namely tensile strength, yield strength, tensile elongation and joint efficiency to achieve the quality of the fabricated joints effectively and the weights of each response are determined. The correlation coefficient array was applied to determine the eigenvalues and eigenvectors and these are used to find the principal components of the responses of the PAW welded joints is represented in the Eqn. (4).

The Principal components and Eigenvalues of the responses for dissimilar plasma arc welded AA8011-H24 aluminium alloys and Ti-3Al-2.5V titanium alloy joints are given in Table 6. The weights of the responses are found by squaring of the eigenvalues of the principal components assigned for the experiments. The cumulative contribution of each response of the friction stir welded joints are calculated based on the weights determined for the responses using principal component analysis is listed in Table 7.

IOP Conf. Series: Materials Science and Engineering **954** (2020) 012034 doi:10.1088/1757-899X/954/1/012034

Table 6. Principal component and Eigen Values of the responses

Items	Principal Components	
	PC(1)	PC(2)
Eigenvalues	1.6944	0.3056
Proportion	0.8472	0.1528
Accumulated Proportion	0.8472	1.0000
Responses	**Eigen Vectors**	
Tensile Strength, (MPa)	0.707	0.707
Tensile Elongation,(%)	-0.707	-0.707

Table 7. Contribution of the responses

PAW Responses	Contribution	Cumulative Contribution (%)
Tensile Strength, (MPa)	0.7323	73.23
Tensile Elongation, (%)	0.2677	100

The quality of the joints is decided with above 85% of the contribution of the parameters on the responses. This work has the combined first two principal components, PC (1) and PC (2) has 100% of the variation of the responses [33]. The composite welding quality index (CWQI) is determined from the first two principal component scores which are used to optimise the responses of the PAW joints. This CWQI is calculated by the sum of the products of the first two principal component scores multiplied by their corresponding eigenvalues is represented in Eqn. (6).

$$CWQI = \sum_{k=1}^{2}(\lambda_k * \mathrm{PCS(k)}) \quad (6)$$

The higher CWQI value of 4.7613 was examined on the 26th experimental runs which provide the optimum responses of the PAW process. This optimum CWQI value shows the higher tensile properties, viz., tensile strength and tensile elongation of the dissimilar plasma arc welded AA8011-H24 aluminium alloy and Ti-3Al-2.5V titanium alloy joints.

The mathematical model is developed for the composite welding quality index from the coefficients of the PAW responses based on the second-order polynomial model using the Design Expert 8.0® software at 95% of confidence intervals [34, 35]. The backward elimination procedure is applied to eliminate the insignificant parameters for the CWQI values to adjust the fitted quadratic model. The final quadratic model of composite welding quality index in terms of coded and actual PAW parameters are presented in Eqn. (7) as follows:

$$\begin{aligned} CWQI \text{ (Coded)} = {} & +21.77+(1.81*Iw)-(0.31*\theta)-(0.85*Fr)+(2.15*Sw) \\ & -(1.01*Iw*\theta)-(2.42*Iw*Fr)-(0.58*Iw*S) \\ & -(1.17*\theta*Fr)+(1.26*\theta*S)+(0.69*Fr*S) \\ & -(3.30*Iw^2)-(2.89*\theta^2)-(3.46*Fr^2)-(1.97*S^2) \end{aligned} \quad (7)$$

Table 8. Four principal component scores and the composite welding quality index

Exp No	Principal Components Scores PCS(1)	PCS(2)	(λ1*PCS(1))	(λ_2*PCS(2))	CWQI	Rank
1	1.0718	0.0020	0.78488	0.00054	0.78541	30
2	4.3206	0.0037	3.16398	0.00099	3.16497	10
3	1.4082	0.0046	1.03122	0.00123	1.03246	28
4	3.6735	0.0063	2.69010	0.00169	2.69179	13
5	2.1903	0.0029	1.60396	0.00078	1.60473	24
6	2.8500	0.0015	2.08706	0.00040	2.08746	20
7	1.3618	0.0063	0.99725	0.00169	0.99893	29
8	0.6841	0.0040	0.50097	0.00107	0.50204	31
9	1.5184	0.0041	1.11192	0.00110	1.11302	27
10	4.2239	0.004	3.09316	0.00107	3.09423	11
11	3.3621	0.0052	2.46207	0.00139	2.46346	17
12	4.7775	0.0051	3.49856	0.00137	3.49993	9
13	3.4661	0.0031	2.53823	0.00083	2.53905	14
14	3.3870	-0.0001	2.48030	-0.00003	2.48027	16
15	3.8546	0.0083	2.82272	0.00222	2.82495	12
16	2.8122	0.0029	2.05937	0.00078	2.06015	21
17	1.5652	0.0062	1.14620	0.00166	1.14786	26
18	3.4053	0.0040	2.49370	0.00107	2.49477	15
19	3.1164	0.0024	2.28214	0.00064	2.28278	18
20	2.7660	0.0069	2.02554	0.00185	2.02739	23
21	2.8060	0.0029	2.05483	0.00078	2.05561	22
22	1.8611	0.0043	1.36288	0.00115	1.36403	25
23	2.8587	0.0068	2.09343	0.00182	2.09525	19
24	5.1131	0.0059	3.74432	0.00158	3.74590	8
25	6.1687	0.0030	4.51734	0.00080	4.51814	3
26	**6.5009**	**0.0027**	**4.76061**	**0.00072**	**4.76133**	**1**
27	6.0520	0.0000	4.43188	0.00000	4.43188	5
28	6.2459	0.0038	4.57387	0.00102	4.57489	2
29	5.8095	-0.0005	4.25430	-0.00013	4.25416	7
30	6.0926	0.0037	4.46161	0.00099	4.46260	4
31	5.9272	0.0001	4.34049	0.00003	4.34052	6

The analysis of variance test is conducted for composite welding quality index which is applied to predict the model at the 95% confidence level [36]. From Table 9, the composite welding quality index has the R^2 value of 0.9948 and the model for CWQI shows a reasonable agreement with 99.48% of the variability. The Adjacent R^2 value of 0.9902 is confirmed the model which is highly significant. The Predicted R^2 value of 0.9848 is in reasonable agreement with the corresponding Adjacent R^2 value. The lack of fit F-value of 94.41% is examined for CWQI which measures the failure of a model not considered in the regression. The most significant parameters are identified from the quadratic model for composite welding quality index [37]. The main effects of the process parameters of the PAW process on CWQI are examined in Table 10 and Figure 3 respectively.

Table 9. Analysis of Variance on Composite Welding Quality Index

Source	Sum of Squares	df	Mean Square	F Value	p-value Prob > F	Remarks
Model	1153.81	14	82.42	217.47	< 0.0001	Significant
I_w-Welding Current	78.91	1	78.91	208.22	< 0.0001	Significant
θ-Torch Angle	2.30	1	2.30	6.08	0.0254	Significant
F_r – Gas Flow rate	17.33	1	17.33	45.73	< 0.0001	Significant
S -Welding Speed	111.37	1	111.37	293.87	< 0.0001	Significant
$I_w * \theta$	16.42	1	16.42	43.32	< 0.0001	Significant
$I_w * F_r$	93.73	1	93.73	247.32	< 0.0001	Significant
$I_w * S$	5.31	1	5.31	14.02	0.0018	Significant
$\theta * F_r$	21.81	1	21.81	57.56	< 0.0001	Significant
$\theta * S$	25.41	1	25.41	67.06	< 0.0001	Significant
$F_r * S$	7.63	1	7.63	20.12	0.0004	Significant
I_w ^2	311.32	1	311.32	821.49	< 0.0001	Significant
θ ^2	238.23	1	238.23	628.63	< 0.0001	Significant
F_r ^2	342.01	1	342.01	902.49	< 0.0001	Significant
S ^2	110.66	1	110.66	292.00	< 0.0001	Significant
Residual	6.06	16	0.38			
Lack of Fit	2.13	10	0.21	0.32	0.9441	Not Significant
Pure Error	3.94	6	0.66			
Cor Total	1159.88	30				

Table 10. Main Effects of PAW Process Parameters on CWQI

Process Parameters	Level 1	Level 2	Level 3	Level 4	Level 5	(Max-min)	Rank
Welding Current	22.5904	16.1059	41.7938	20.9034	30.1589	25.6879	2
Torch Angle	30.2374	17.7918	41.2318	19.2175	29.8175	23.44	4
Gas Flow rate	30.9467	17.3524	41.2762	19.6569	28.5319	23.9238	3
Welding Speed	33.1426	12.2888	40.0508	24.7205	42.2656	29.9768	1

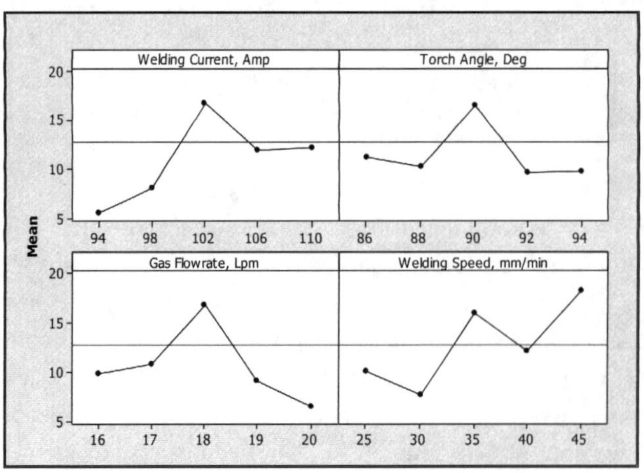

Figure 3. Main Effects Plot for CWQI

4.4 Confirmation experiments for dissimilar PAW joints

The optimum welding parameters are found through the determination of the composite welding quality index using the hybrid approach of RSM based GRA coupled PCA analysis. Besides, the five different confirmation tests are conducted at the optimal process settings, such as welding current of 102 Amp, torch angle of 90°, gas flow rate of 18 mm and welding speed of 35 mm/min on the responses of the dissimilar PAW joints. The average values of the tensile properties are listed in Table 11. It is used to determine the maximum value of the CWQI which relates the better quality performance of the fabricated joints. Finally, it is concluded that the hybrid optimisation enhances the tensile properties of the dissimilar plasma arc welded AA8011-H24 aluminium alloys and Ti-3Al-2.5V titanium alloy joints compared to the optimisation of the parameters using the RSM approach.

Table 11. Confirmation Experiment for Dissimilar PAW Joints

Welding Current (Amp)	Torch Angle (Deg)	Gas Flow rate (LPM)	Welding Speed (mm/min)	Tensile Strength (MPa)	Tensile Elongation (%)
102	90	18	35	144.87	4.41

5. Conclusions

- The hybrid approach of RSM based GRA coupled PCA analysis was applied to optimise the tensile properties, viz., tensile strength and tensile elongation of the dissimilar plasma arc welded AA8011-H24 aluminium alloys and Ti-3Al-2.5V titanium alloy joints.
- The developed mathematical models are analysed to correlate the most significant welding process parameters, viz., 102 of Amp welding current, 90° of torch angle, 18 LPM of gas flow rate and 35 mm/min of welding speed on the responses of the dissimilar plasma arc welded joints and these optimum parameters are helpful to optimise the responses in deciding the quality of the welded joints.
- The analysis of variance was applied on the main effects of PAW process parameters on the responses, it was showed that welding speed is the dominant factor to decide the good quality of the welded joints followed welding current, gas flow rate and torch angle based on the CWQI.
- The confirmation tests showed that the experimental results were much closer to the predicted optimum results with minimum variations on the measured responses of the dissimilar PAW joints.

6. References

[1] Yakabu DY, Macedo BD, Resende HB and Batalha M H F 2018 *Int J Adv Manuf Technol* **95** 1339–1355
[2] Aonuma M and Nakata K 2011 *Mater. Trans* **52** (5) 948-52
[3] Malikov AG, Orishich AM, Vitoshkin IE 2020 *J Appl Mech Tech Phy* ***61*** 307–317
[4] Palani K, Elanchezhian C, Ramnath BV, Bhaskar GB Jagadeesh JS and Kumar GM 2017 *Int. J. Res. App. Sci. Eng. Tech* **5** 437–442
[5] Rostami H, Nourouzi S and Jamshidi Aval H 2018 *J Mech Sci Technol* ***32*** 3371–3377

[6] Ramkumar T, Selvakumar M, Narayanasamy P, Ayisha Begam A, Mathavan P and Arun Raj A 2018 *J Manuf Proc* ***30*** 290-298
[7] Mousavi Anijdan S H, Sabzi M, Ghobeiti-Hasab M and Roshan-Ghiyas A 2018 *Mater. Sci. and Eng: A* ***726*** 120-125
[8] Prabakaran MP and Kannan GR 2019 *Opt Las Technol* ***112*** 314-322
[9] Paventhan R, Lakshminarayanan PR and Balasubramanian V 2012 *Int J Iron & Steel Res* ***19*** (1) 66-71
[10] Fernandus MJ, Senthilkumar T, Balasubramanian V and Rajakumar S 2012 *Mater and Des* ***33*** 31–41
[11] Wei Y, Aiping W, Guisheng Z and Jialie R 2008 *Mater. Let* **62** 2836-39
[12] Palani K, Elanchezhian C, Vijaya Ramnath B, Bhaskar GB and Naveen E 2018 *Mater Today: Proc* ***5*** 24515–24524
[13] Ilangovan M, Boopathy SR and Balasubramanian V 2015 *Def. Tech* **11** (2) 174–184
[14] Vijayan S, Raju R and Rao SRK *Mater. Manuf. Proc* **25** 1206-12
[15] Palani K, Elanchezhian C, Sai Prakash K H V, Deepak Kumar, Dayanand, Keshav Kumar and Sreekanth K 2018 *IOP Conf. Series: Mater Sci Engg* **390** 012072
[16] Palani K, Elanchezhian C, Avinash K, Karthik C, Chaitanya K, Yugandar K and Karthik S 2018 *IOP Conf. Series: Mater Sci Engg* ***390*** 012108
[17] Vaidya WV, Horstmann M, Ventzke V, Petrovski B, Koc M, Kocik R and Tempus G 2010 *J. Mater. Sci* **45** 6242-54
[18] Lakshminarayanan AK and Balasubramanian V 2011 *J. Mater. Eng. Perf* **20** 82-89
[19] Mvola B, Kah P and Martikainen J 2014 *Int J Mech Mater Eng* ***9*** 21
[20] Esmaily M, Mortazavic SN, Todehfalah P and Rashidi M 2013 *Mater & Des* ***47*** 143-150
[21] Rajakumar S and Balasubramanian V 2015 *Int. J. Adv. Manuf. Tech* **86** (1-4) 1095-1112
[22] Sahu PK, Kumari K and Pal S 2016 *Adv. Manuf 4* 237–247
[23] Palani K, Elanchezhian C, Vijaya Ramnath B and Ramadoss R 2020 *Mater Today: Proc* ***23*** (2) 573-582
[24] Siddiquee AN, Khan ZA and Mallick Z 2010 *Int J Adv Manuf Technol* ***46*** 983–992
[25] Sharma A, Chaudhary AK, Arora N and Mishra BK 2009 *Int J Adv Manuf Technol* ***45*** 1096–1103
[26] Palani K and Elanchezhian C 2015 *Appl. Mech. Mater* **813** 451-55
[27] Şefika Kasman 2013 Int. J. Adv. Manuf. Tech **68** 795–804
[28] Datta S, Bandyopadhyay A and Pal PK 2008 *Int. J. Adv. Manuf. Technol.* **39** (11-12) 1136-43
[29] Palanikumar K, Karunamoorthy L and Karthikeyan R 2006 *Mat. Manuf. Proc* **21** (8) 846-52
[30] Rajyalakshmi G and Venkata Ramaiah G 2013 *Int. J. Adv. Manuf. Technol.* **69** (5-8) 1249-62
[31] Palani K and Elanchezhian C 2015 *Appl. Mech. Mater* **766** 921-27
[32] Kimura M, Nakamura S, Kusaka M, Seo K and Fuji A 2005 *Sci. Tech. Weld. Join* **10** (6) 666-73
[34] Sundaram NS and Murugan N 2010 *Mater. Des* ***31*** (9), 4184-93
[35] Palani K and Elanchezhian C 2015 *Appl. Mech. Mater* **813** 446-450
[36] Ranganathan S and Senthilvelan T 2011 *Int. J. Adv. Manuf. Technol.* (2011) **56** 455-62
[37] Gagliardi F, Ciancio C and Ambrogio G 2017 *Int. J. Adv. Manuf. Technol.* **94** (1-4) 719-28

IOP Conf. Series: Materials Science and Engineering **954** (2020) 012035 doi:10.1088/1757-899X/954/1/012035

Study of the G-ratio of aluminium silicon carbide nano particles reinforced metal matrix composites

A.Nandakumar[1] T. Rajmohan[2] R.Sundararajan[3] K.Venkatesh[4] N.Muni Deepak[5] K.Venkateshwaran[6]

Department of Mechanical Engineering, Sri Chandrasekharendra Saraswathi Viswa Maha Vidyalaya University, Enathur, Kanchipuram – 631561, India

E-mail: [1]nandakumar.a@kanchiuniv.ac.in and [2]rajmohanscsvmv@yahoo.com

Abstract: Grinding plays an essential role in obtaining high dimensional accuracy and surface finish. Three types of work pieces were used with different ratios of SiC (1%,2%,3%). When the composition of SiC increases, the G-ratio decreases. This type of material is highly used in aerospace applications due to its low density and good wear and corrosion resistance. It is genuinely believed that Al-SiC has tremendous potential for application in the aerospace industry. A critical role is played by grinding in achieving good surface finish and improved dimensional accuracy. Therefore, a strategy to improve the grinding rate and surface finish is important for economic production. In this research, the various grinding ratio was found out and results show that G-ratio is high when coolant is used and G-ratio is low when coolant is not used.

Key words: Grinding; Nano TiO_2; Nano SiC particles; G Ratio

1. Introduction

Composites are materials in which two phases are combined, usually with strong interfaces between them. They usually consist of a continuous phase called the matrix and discontinuous phase in the form of fibers, whiskers or particles called the reinforcement. Considerable interest in composites has been generated in the past because many of their properties can be described by a combination of the individual properties of the constituent phases and the volume fraction in the mixture. Composite materials are gaining wide spread acceptance due to their characteristics of behavior with their high strength to weight ratio [1]. Metal Matrix Nano Composites (MMNCs) are being used increasingly in both the aeronautical and automotive industries [2]. The successful and cost-effective machining of MMCs is essential in order to get defect free components having a required dimension and improved surface finish. The grinding operation might participate a significant role for finishing and durable machining, where a single grinding process symbolizes a cost-effective choice, as it removes require for prior conventional machining process [3]. Numerous ceramic reinforcements are recognized for MMCs, but Silicon carbide (SiC) has expanded interest due to its high rigidity, high specific strength, solidity, wear resistance and dimensional strength [4].

The machining of SiC reinforced composites presents significant challenges to industries due to the high abrasiveness of the reinforcing particles and the anisotropic & non homogenous structure. Precipitate collapse of tool is accounted as the main difficulty in machining of MMC [5]. It is significant to extend quantitative analytical models for sustainable manufacturing and product design in order to appreciate the vital role of the components of sustainability in manufacturing of products [6]. Increasing quantities of metal matrix composites (MMCs) are being used to replace conventional materials in many applications, especially in the automobile and recreational industries, owing to increasing performance requirements [7]. The most popular types of MMCs are aluminium alloys reinforced with ceramic particles. These low-cost composites provide higher strength, stiffness and fatigue resistance [8], with a minimal increase in density over the base alloy. The superior mechanical properties achieved by the reinforcements in MMCs, on the other hand, significantly influence their "machinability" [9]. This term which describes the operational characteristics of a cutting tool in machining a material is rather ill-defined and is normally judged in terms of tool life, surface finish and power required for material removal [10].

Aluminium-based metal matrix composites (MMCs) reinforced with ceramic particles are advanced materials known for their good damping properties, high specific strength, and high wear resistance [11]. Methods to produce these composites and studies on their mechanical properties have gained popularity. MMCs are increasingly used in astronautic, automobile, and military industries [12]. In addition, the sporting goods industry has also been in the forefront of MMCs development capitalizing on the materials' high specific properties [13]. Reports on machining of Aluminium-based MMCs reinforced with ceramic particles [14] are still scarce. Despite many advantages, full implementation of MMCs is cost-prohibitive. This is partially due to the poor machinability of the materials. Although near-net-shape MMC components can be produced, final finishing is still required for obtaining designed final dimensions and required surface finish [15].

Significant cost and fabrication problems, including machining, must be overcome for the successful application of these composites. Surface finish and surface integrity are important for surface sensitive parts subjected to fatigue. Subsurface damage due to the machining of MMCs results from conventional and unconventional processes, therefore, finishing processes such as grinding are used to improve the surface integrity of machined MMCs [16]. The objective of this paper is to study the G ratio of aluminium silicon carbide nano particles reinforced metal matrix composites. Grinding plays an essential role in obtaining high dimensional accuracy and surface finish. Three types of workpieces were used with different ratios of SiC (1%, 2%,3%). This type of material is highly used in aerospace applications due to its low density and good wear and corrosion resistance.

2. Materials and Methods

Aluminium billet obtained from M/s Micro Fine chemicals, India. and SiC Nanoparticles procured from M/S US Research Nanomaterials Inc, USA are employed as base material for synthesizing the Nanocomposites. Aluminium reinforced with SiC Nanoparticles are fabricated by Vacuum based solidification processing [17]. SAE20W40, Cashew nut-based vegetable oil and Nano TiO_2 were procured from M/s Ganapathy traders, India. The average diameter and length of the Nano tubes vary from 10-20 nm and 10-30 μm respectively and Nano TiO_2 were mixed with Cashew nut-based vegetable oil using ultrasonic processor.

IOP Conf. Series: Materials Science and Engineering **954** (2020) 012035 doi:10.1088/1757-899X/954/1/012035

The preparation of the Nano fluids includes dispersion of 10 gm of Nano TiO_2 in 500 ml of vegetable oil-based fluid. The processed samples were kept ideal for 24 hours earlier to the property measurements. The Scanning Electron Microscopy (SEM) (Quanta FEG 200, Japan) analysis was hired for defining the size and morphology of the nanoparticles. It is clear from the Figure 1 that morphology of nanoparticles is nearly spherical in shape and it is also confirmed that the sizes of the dispersed nanoparticles are within the specified range. The properties of NanoTiO_2 mixed fluids were tested and the results were presented and compared with the SAE20W40 and Cashew nutshell Oil and are presented in the Table 1.

Table 1. Rheological properties of the Lubricants

Properties	SAE20W40	Vegetable Oil (Cashew nut shell Oil)	Vegetable Oil (Cashew nut shell Oil + TiO2)
Flash point (°C) [ASTM D92]	200	214.27	190.2
Thermal Conductivity Watt/mK	0.152	0.161	0.169
Viscosity @100°C (cSt) [ASTM D445]	15.2	15.48	16
Viscosity index [ASTM D2270]	120	126	158

Figure 1. SEM image of Nano TiO_2

From the rheological properties it is found that a considerable improvement in the properties was observed in TiO_2 filled cashew nut-based vegetable oil due to the enhanced heat cropping ability of Nano fluid [18]. The gravity die casting technique was used to prepare the Nano SiC reinforced Al matrix composites and microstructure of prepared samples were presented in Figure 2.

IOP Conf. Series: Materials Science and Engineering **954** (2020) 012035 doi:10.1088/1757-899X/954/1/012035

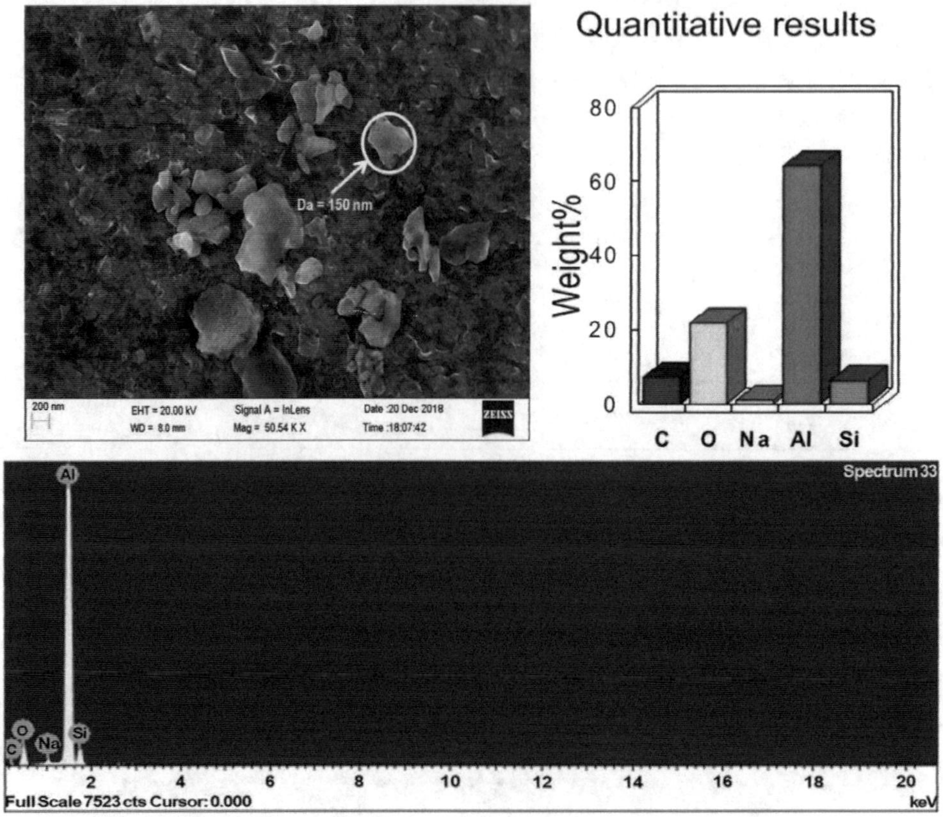

Figure 2. Microstructure and EDAX of fabricated specimen

3. Experimental Work

The cylindrical grinding machine is a type of grinding machine used to shape the outside of an object. The cylindrical grinder can work on a variety of shapes; however, the object must have a central axis of rotation. This includes but is not limited to such shapes as a cylinder, an ellipse, a cam, or a crankshaft. Roughness is an important parameter when trying to find out whether a surface is suitable for a certain purpose. Rough surfaces often wear out more quickly than smoother surfaces. A roughness tester is used to quickly and accurately determine the surface texture or surface roughness of a material.

4. Results and Discussions

The Grinding experiments are carried out on a horizontal spindle cylindrical grinding machine (Type G13P, HMT make). In present study the cylindrical specimens of MMNC were ground with the selected aluminium oxide grinding wheel (AA60K5V8). The G Ratio of the experimentation is provided in the following Tables 2 and 3. The G ratio used by the operation is determines, volume of metal removed divided by volume of wear of tool.

IOP Conf. Series: Materials Science and Engineering **954** (2020) 012035 doi:10.1088/1757-899X/954/1/012035

Table 2. Grinding ratio (G-Ratio) without coolant

% OF NANO SiC	1%		2%		3%	
	Before	**After**	**Before**	**After**	**Before**	**After**
WORKPIECE DIAMETER	16.12	15.92	16.18	16.03	17.06	16.96
WHEEL DIAMETER	189.18	189.14	188.18	188.14	188.10	188.06
LENGTH OF WORKPIECE	122.18		136.18		126.18	
WHEEL THICKNESS	21.04		21.04		21.04	
G-RATIO	**2.45**		**2.07**		**1.35**	

Table 3. Grinding ratio (G-Ratio) with coolant

% OF NANO SiC	1%		2%		3%	
	Before	**After**	**Before**	**After**	**Before**	**After**
WORKPIECE DIAMETER	15.92	15.72	15.98	15.88	16.86	16.76
WHEEL DIAMETER	188.08	188.06	188.04	188.02	188.02	188.00
LENGTH OF WORKPIECE	122.18		136.18		126.18	
WHEEL THICKNESS	21.04		21.04		21.04	
G-RATIO	**4.88**		**2.74**		**2.68**	

4.1 Analysis of Grinding Ratio

The bar chart describes the variation of G-Ratio with and without coolant is shown in Figure 3. Y-axis represents the grinding ratio and X-axis represents the percentage of Nano SiC in Al. The G ratio gradually decreases when there is an increase in the percentage of Nano SiC. There is a sharp drop in G-Ratio between 1% and 2% and a slight drop between 2% and 3% when no coolant is used. The G Ratio is less all conditions of MMNC for providing coolant in grinding operations.

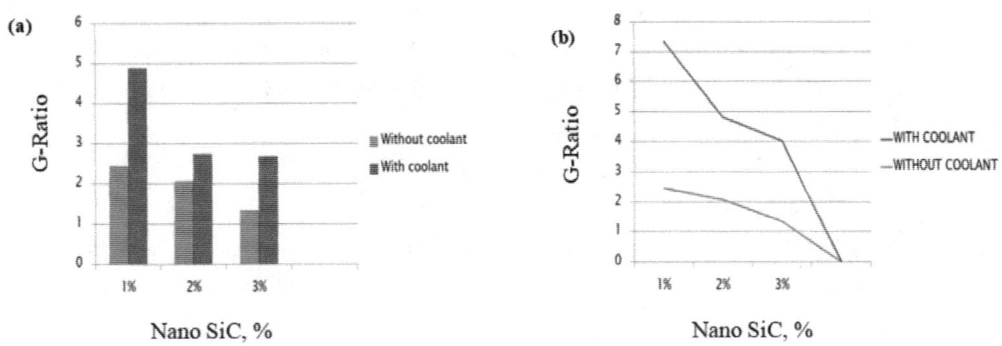

Figure 3. Grinding specimen: (a) Bar chart and (b) Line diagram

5. Conclusions

Based on the results of the present experimental investigation, the following conclusions for MQL grinding of MMNC can be drawn.

- The large apertures are a feature of good cutting capability and show that Nano fluids are reserved on the grinding wheel surface. The fissure networks on the grinding wheel are like micro-reservoirs that can accumulate Nano fluids for the time being. This property guarantees that micro-reservoirs can carry TiO_2 Nano particles into the grinding zone.
- Nano fluids on abrasive grain surfaces can form a layer of TiO_2 Nano fluid lubricating film through physical and chemical reactions; this layer can inhibit the generation of wear flats and cause them to fall off of the abrasive grains effectively.
- The TiO_2 nano particles with a nano size and spherical appearance serve as good "bearings"in the grinding Zone.
- TiO_2 nano particles prevent direct contact in the friction pair, change the sliding friction between the abrasive grains and the workpiece into rolling friction, and reduce the wear flats of grains by narrowing the contact area between them and the workpiece.
- The G-Ratio gradually decreases when there is an increase in the percentage of Nano SiC. There is a sharp drop in G-Ratio between 1% and 2% and a slight drop between 2% and 3% when no coolant is used.
- The G-Ratio is less for all conditions of MMNC for providing coolant in grinding operations.

6. References

[1] Chennakesare Reddy A 2010 *Ind J Sci and Tech* ISSN: 0974- 6846
[2] Vijayabhaskar S and Rajmohan T 2019 *Silicon **11***, 1701–1716
[3] Ponappa K, Aravindan S and Rao PV 2011 *Instit Mech Engs. B – j. Eng. **226*** 1675–1683
[4] Rajmohan T, Palanikumar K and Prakash S 2016 *Else. Compo. Part B - Eng. **50*** 297-308
[5] Rajmohan T and Palanikumar K 2013 *Int J Adv Manuf Tech **64*** 1249–1261
[6] Kaynak Y, Lu T and Jawahir IS 2014 *Mach. Sci. Tech. **18*** (2), 149–198
[7] Hung NP, Zhong ZW and Zhong CH 1997 *Proceed Compos Engg*. 459–460
[8] Hung NP, Zhong ZW, Zhong CH 1997 *J Mater Manuf Proc **12*** (6) 1075–1091
[9] Rao DN and Srikant RR 2006 *J Engg Manuf. P. I. Mech. Eng. B – J. Eng **220*** 1803–1806
[10] Grzesik W 2008 *Else Sci USA,* 226–245
[11] Wen D and Ding Y 2004 *Int J Heat Mass Trans. **48*** 5181–5188
[12] Nandakumar A, Rajmohan T and Vijayabhaskar S 2019 *Silicon. **11*** 2987–2999
[13] Jaswinder Singh 2017 *Int J Appl Engg Res **12*** 2963-2968
[15] Majumdar S, Kumar S, Roy D, Chakraborty S and Das S 2018 *Mater. Manuf. Proc. **33*** 1459-1465
[16] Nandakumar A and Rajmohan T 2018 IOP Conf Series: Mater Sci and Engg. ***390*** (1) 012 -033
[17] Rajmohan T 2019 *Partic Sci Tech **37*** 21-30
[18] Su Y 2016 *Int J Adv Manuf Tech* **83** 2083–2089

IOP Conf. Series: Materials Science and Engineering **954** (2020) 012036 doi:10.1088/1757-899X/954/1/012036

Application of Evolutionary Algorithms to Predict Mechanical Properties on Natural Fibers – A Review

Prasanth kumar. R[1], Rajmohan.T[2], Vijayan. D[3]

[1] Postgraduate Student. Department of Mechanical Engineering, Sri Chandrasekharendra Saraswathi Viswa Mahavidyalaya University.
[2] Associate Professor. Department of Mechanical Engineering, Sri Chandrasekharendra Saraswathi Viswa Mahavidyalaya University.
[3] Assistant Professor. Department of Mechanical Engineering, Sri Chandrasekharendra Saraswathi Viswa Mahavidyalaya University.

✉ [1]prasanthrentachinthala@gmail.com

Abstract. Natural fibers are an abundant source in nature, having an inherent property in which include biodegradable, decomposable and cost efficient. Moreover, they exhibit extensive mechanical properties such as tensile modulus, flexural modulus than synthetic fibers when reinforcing polymer resins. However, adding inappropriate levels of quantity of resins misleads on obtaining desired mechanical properties. Several approaches have been utilized in the past decades to get desired mechanical properties on the natural fibers, in which applying evolutionary algorithms are the most recent and familiar in the composite research community. Therefore, in this study an application of evolutionary algorithms on natural fibers are presented. This objective of the review is to provide detailed information of the natural fibers along with its recent algorithmic approaches. Moreover, the effect of various treatments and its changes in mechanical, thermal and electrical properties of matrix.

Keywords Natural fibers. Evolutionary Algorithms. Mechanical Properties.

1. Introduction

In recent years, natural fibers have gaining much attention by the composite research community due to their low density, bio degradability, sustainability and good thermal, acoustic properties. Specifically, high performance applications such as automobile, aircraft, and marine applications are being replaced from synthetic fiber to natural fibers by combining appropriate levels of quantities of polyester and polymer resins. Usually natural fibers are classified into different types as represented in Figure. Impressive investigations have been carried out using many natural fibers such as kenaf, hemp, flax, sisal, coir and jute using various polymer matrices. In general, the development of polymer matrices was composed based on the earlier studies conducted based on the composite or by trial and error methods. Despite the advantages of trial and error method, there are more setbacks such as more computation time and requires numerous databases to understand fiber/matrix ratio for fabricating the composites. Alternatively, by applying soft

IOP Conf. Series: Materials Science and Engineering **954** (2020) 012036 doi:10.1088/1757-899X/954/1/012036

computing techniques the complexity in finding these values can be minimized. Numerous soft computing techniques are being utilized in recent days for optimizing the natural fiber composites. They are Genetic Algorithm (GA), Simulated Annealing (SA) algorithm, Particle Swarm Optimization (PSO), Teaching Learning Based Optimization (TLBO) algorithm, etc. therefore, taking these applications of soft computing techniques into consideration an in-depth review has been presented on the application of multi objective evolutionary approach on natural fibers.

2. Fiber Flowchart

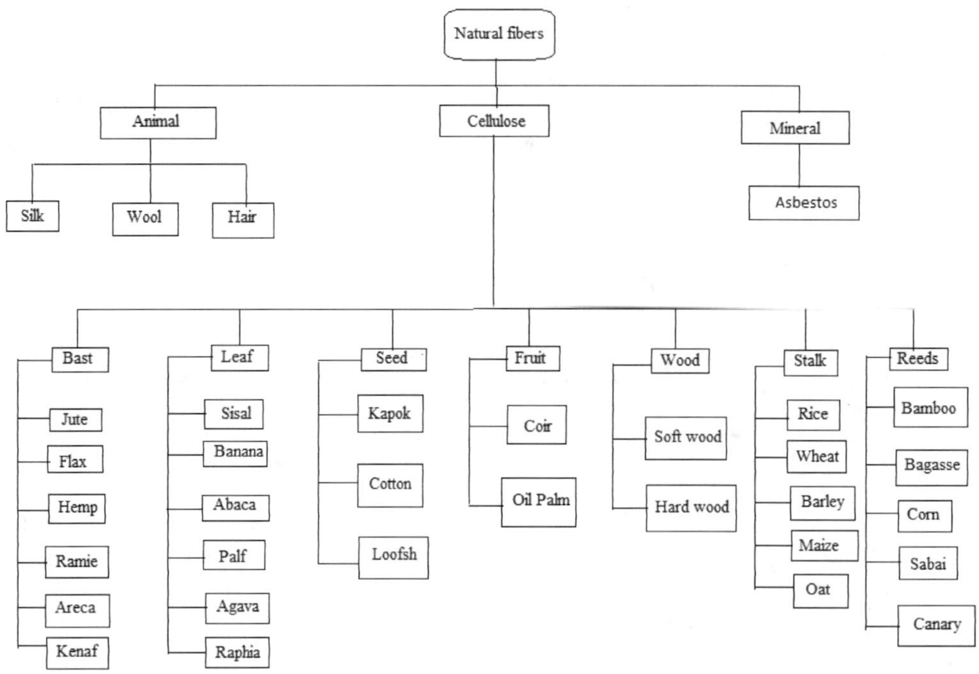

3. Evolutionary Algorithms

3.1 Genetic Algorithm

S Jayabal et al. [1] investigated studies on optimization of drilling characteristics in fiber reinforced composites was carried out using Nelder-Mead and Genetic algorithms. They selected the composites with natural fiber known as coir for reinforcing material. They determined optimum values of machining parameters and process parameter through experiments planned, conducted and analyzed using the Box-Behnken design, Nelder-Mead and Genetic algorithm methods. A optimized mathematical model of drilling machining parameters such as thrust force, torque and tool wear were developed and they improved tool life by selecting optimal cutting parameters such as drill bit diameter, speed and feed rate so that they reduced tool wear.

J Lilly Mercy et al. [2] conducted experimental studies on genetic optimization of machining parameters affecting thrust force during drilling on composite plates. In this work pineapple unidirectional fiber mats were used as reinforcement material and epoxy resin as matrix material to fabricate composite plates.

Drilling experiments were conducted under different conditions of machining by Taguchi L_{27} orthogonal array and established mathematical relationship between input parameters and output response by using statistical modelling. They achieved drilling operation on pineapple reinforced composite plates by using HSS drill bits with varied cutting parameters and optimization of thrust force done by using genetic algorithm so that they achieved increase in feed rate with lower thrust force and speed.
Velumani S et al. [3] investigated studies on mathematical modeling and optimization of mechanical properties polymer-based matrix composite. They used short coir fibers as reinforcement and vinyl ester polymer as matrix material. The second order mathematical models were developed by using statistical packages like regression modelling were developed using experimental database and optimization was performed by using genetic algorithm. They noticed the results such that genetic algorithm optimization indicated the combination of fiber length and fiber content influenced improvement in values of mechanical properties such as tensile, flexural and impact strengths.

3.2 Simulated Annealing Algorithm
Kailainathan S et al. [4] conducted studies on hybrid particulate fiber polymer composites, they used industrial bio-waste teak wood powder and natural sisal fiber as reinforcements for polymer matrix. Hybrid composite fabrication was done by using compression molding technique, process parameters were developed by using mathematical models and further these models were optimized by simulated annealing algorithm to get better mechanical properties. They noticed improved mechanical properties such as flexural and tensile behavior over the specified range of conditions.
Bharathiraja G et al. [5] investigated studies on optimization of mechanical behaviors of bio particulates filled coir-polyester composites. In this work they used red mud and termite mound soil as bio particulates, green husk coir fibers as reinforcement material and bio particulates are mixed in proportions with polyester resin system so that resin system consist of unsaturated orthophthalic polyester resin. They conducted Analysis Of Variance for understanding mechanical behaviors caused by the effects of fiber length and particulate content on coir polyester composites, over the specified range of conditions to predict mechanical behavior they developed nonlinear regression models and fabrication parameters, optimization of mechanical behavior was analyzed by using simulated annealing algorithm. They obtained improved optimal values of mechanical properties such as tensile, flexural and impact strengths was determined using simulated annealing algorithm by optimization nonlinear regression models and through ANOVA analysis they concluded by using bio particulates mechanical performance greatly improved in coir-polyester composites.

3.3 Teaching Learning Based Optimization Algorithm
Sivakumar M et al. [6] conducted studies on studies on shear strength of CNT/ coir fiber/ fly ash-reinforced epoxy polymer composites. Through design of experiment approaches they used technique called response surface methodology and prepared experimental specimens by varying contents of reinforcements with epoxy, for understanding reinforcement percentage on the shear modulus of composite they carried out study known as ANOVA and two models were predicted shear modulus by using regression models. Out of two models one was based on ANN and other through multiple linear regression model and optimization of reinforcement parameters were done by using TLBO algorithm. They implemented TLBO algorithm for both models to get optimum parameters with increased shear modulus and by using ANN model they obtained the set of proportionate results of reinforcement contents along with increased shear modulus.
Mangat [7] conducted experimental investigations on bio degradable structure for bio medical application based on natural fiber embedded additive manufacturing, their intention is to explore the behavior of mechanical and bacterial characteristic of chemically treated waste natural fiber inserted with three dimensional structures produced with fused filament deposition for biomedical application. They used raw

animal fiber and chemically treated for the usage in biomedical structures, silk and sheep wool fibers used as reinforcements and polylactic acid used as matrix material. The laminated composites are prepared by using a low-cost desktop time additive manufacturing method. Through design of experiments technique they obtained results of output characteristics such as dimensional accuracy, hardness, three-point bending strength and bacterial test which was influenced by input parameters obtained through ANOVA and signal/noise tests, optimization of input parameters for the design of embedded structures for scuffled based biomedical applications done using different optimization techniques like Taguchi L9 orthogonal array method, single parametric optimization, TLBO and multi parametric optimization.

3.4 Crow Search Algorithm

Sultana N et al. [8] investigated studies on modeling approach of surface response coupled with crow search algorithm for optimizing the properties of jute fiber reinforced concrete, natural fiber jute is reinforced with concrete in-order to get greater strengths in JFRCC. A mathematical models was developed using response surface methodology with box-behnken design, response surface methodology is integrated with crow search algorithm in-order to find global optimal solution. For making standard dimensions of cylinders and cubes depends on three independent factors such as water-cement ratio, length and volume of jute fiber. For finding second order polynomial models for both compressive and tensile strengths of JFRCC they used RSM-based on BBD method initially and later RSM integrated with crow search algorithm to find global optimal solution and results stated the maximum compressive and tensile strengths were found along with the optimal set of independent factors.

3.5 Particle Swarm Optimization Algorithm

M Megahed et al. [9] conducted studies on optimization of hybrid natural laminated composite beams for a minimum weight and cost design, the fibers are carbon/flax used as reinforcement with epoxy resin as matrix material to get hybrid natural laminated composite beams and compared with hybrid carbon/glass/epoxy, neat glass/epoxy and neat flax/epoxy laminated beams. The main objective of study is to lower the weight and cost of the laminated composite beams with a specified lower bound constraint of natural frequency. The modeling and analysis of the problem were performed based on the Euler-Bernoulli beam theory, the beam is composed of many layers and the design variables such as fiber material, fiber volume fractions, thicknesses and fiber orientations. Layer thickness and fiber orientations are the manufacture limitations known as discrete variables, so that PSO algorithm is used to deal with constrained problem with discrete variables and optimal stacking sequence design is investigated. The results stated that optimal hybrid composite that satisfy the lower frequency limits such as high and low stiffness skin core layers and PSO determines the optimum layer thickness and fiber volume fractions. They noticed with appropriate stiffness to weight ratio out of all hybrid laminated beams, carbon/flax/epoxy is cheaper and lighter.

3.6 Response Surface Methodology

Boumaaza et al. [10] conducted studies on natural fibers reinforced plasters and the effects on mechanical properties when natural fiber is treated with alkaline solution and optimization technique carried out using RSM. They used natural fibers such as sisal, flax and jute fibers and treated with different concentrations of NaOH alkaline solution, with varied fiber lengths reinforced with plaster mortals. They determined experimental parameters using ANOVA test, for optimizing output responses they used RSM and desirability function DF. They stated that the reinforcement of plasters with alkali treated fibers possess greater flexural strength, displacement and flexural modulus, fibers like flax lowered crack propagation and bridged the voids so that behavior of plaster was change from brittle to non-linear in nature. They stated the RSM is a good modelling tool that identifies insignificant and interaction factors with reduced complexity and given effective predictions for given problem and they noticed fiber lengths plays a pre dominant role when compared with alkali dosage variable over the response of process.

3.7 Artificial Neural Network

Khazi et al. [11] conducted predictive studies on ANN models for varying filler content for cotton fiber/PVC composites based on experimental load. Here they use cotton fiber as filler content to improve the mechanical properties of the PVC composite material. A series of experiments carried out in laboratory by varying filler content in the composites depending experimental data, ANN models are developed, trained and tested on tensor flow using python library keras to implement back-propagation method. They predicted load displacement curves by using different methods such as grid search hyper parameter turning method and k-fold cross validation methods, ANN models are also used to determine load displacement curves. Six ANN models was developed using tensor flow with backend keras library they observed developed models predicted the load-displacement curves for varying content of fiber in composites.
Okoye et al. [12] investigated studies on multi objective optimization of newbouldia leaves fiber and recycled high density polyethylene composites for the application of fireboard. Optimization of fire board properties was carried out by using MOR, MOE, IB and WA fireboard in response to the variations in process parameters and experimental design was developed using CCD which was used to develop RSM model for variation in physical and mechanical properties of fireboard. ANN model is also used to predict properties of fireboard moreover it was used as fitness function for multi objective optimization. They stated results such that ANN and RSM models produced significant properties of fireboard, the optimized Pareto front of the non-dominated sorting genetic algorithm over desirability function as a multi-optimization tool for the design and practical applications of the composite.

4. Conclusion

Evolutionary algorithms plays a vital role in solving natural fibers because it's combinatorial problem which has many solutions of similar performance. Evolutionary algorithms are predominant in successfully solving discrete variables and can get optimal solutions. Design engineers will have a choices of set of solutions so that they can select optimal solution. By mastering the specialized evolutionary algorithms it can reduce the optimization cost when compared with other evolutionary techniques. From the review, it can be understood that, the evolutionary algorithms can be used in solving complex multi-objective problems with in less duration and cost effective.

5. References

[1] S Jayabal, U Natarajan 2010 Optimization of thrust force, torque, and tool wear in drilling of coir fiber-reinforced composites using Nelder–Mead and genetic algorithm methods, DOI 10.1007/s00170-010-2605-7.
[2] J Lilly Mercy, P Sivashankari, M Sangeetha, K R Kavitha, S Prakash 2020 Genetic Optimization of Machining Parameters Affecting ThrustForce during Drilling of Pineapple Fiber Composite Plates – an Experimental Approach, https://doi.org/10.1080/15440478.2020.1788484
[3] S Velumani, P Navaneetha Krishnan, S Jayabal 2011 Mathematical Modeling and Optimization of Mechanical Properties of Short Coir Fiber-Reinforced Vinyl Ester Composite Using Genetic Algorithm Method, DOI: 10.1080/15376494.2012.699599.
[4] Kailainathan S, R Muralikannan, K Nijandhan, Srisaran Venkatachalam 2019 High-strength hybrid particulate-fibre polymer composites: The role of process temperature on the mechanical strength, https://doi.org/10.1088/2053-1591/ab54a0

[5] G Bharathiraja, S Jayabal, R Prithivirajan, S Sathiyamurthy 2014 Optimization of mechanical behaviors of bio particulates filled coir-polyester composites using simulated annealing.
[6] Sivakumar Mahalingam et al. 2019 Studies on shear strength of CNT/coir fibre/fly ash-reinforced epoxy polymer composites, https://doi.org/10.1680/jemmr.19.00098.
[7] Amarveer Singh Mangat 2018 Experimental investigations on natural fiber embedded additive manufacturing-based biodegradable structures for biomedical applications https://doi.org/10.1108/RPJ-08-2017-0162
[8] N Sultana et al. 2020 An experimental investigation and modeling approach of response surface methodology coupled with crow search algorithm for optimizing the properties of jute fiber reinforced concrete, Construction and Building Materials 243 (2020) 118216
[9] M Megaheda et al. 2020 Optimization of hybrid natural laminated composite beams for a minimum weight and cost design, Composite Structures 239 (2020) 111984
[10] Messaouda Boumaaza, Ahmed Belaadi, Mostefa Bourchak 2020 The Effect of Alkaline Treatment on Mechanical Performance of Natural Fibers-reinforced Plaster: Optimization Using RSM, https://doi.org/10.1080/15440478.2020.1724236
[11] Monzure Khoda Kazi et al. 2020 Predictive ANN Models for Varying Filler Content for Cotton Fiber/PVC Composites based on Experimental Load Displacement Curves, https://doi.org/10.1016/j.compstruct.2020.112885
[12] Chidozie Chukwuemeka Nwobi-Okoye 2019 RSM and ANN Modeling for Production of Newbouldia Laevies Fibre and Recycled High Density Polyethylene Composite: Multi Objective Optimization Using Genetic Algorithm, DOI 10.1007/s12221 020 9597-1.

IOP Conf. Series: Materials Science and Engineering **954** (2020) 012037 doi:10.1088/1757-899X/954/1/012037

Review on Thermal analysis of Polymer Matrix Composites

G.Venkatakoteswararao[1,*], R.Vinayagamoorthy[2]

[1, 2] Department of Mechanical Engineering, Sri Chandrasekharendra Saraswathi Viswa Mahavidyalaya (Deemed to be University), Kanchipuram, 631561, Tamilnadu, India

Corresponding author: [*, 1]venkatakoteswararao02@gmail.com, [2]vin802002@gmail.com

Abstract. Natural fibers are widely used for reinforcement in composite materials and well tried to be effectively exchange fibre strengthened chemical compound composites to some extent in applications like domestic, automotive and lower finish region elements. A few explores have been investigated in last decades to research the conduct of natural fiber composites. The natural fibre strengthened composites are atmosphere friendly, have high strength to weight quantitative relation like artificial optical fiber strengthened composites. Deferential scanning calorimetry (DSC) and Thermogravimetric analysis (TGA) were used to evaluate the thermal properties. TGA is used to measure temperature change and mass loss of samples and DSC enables the measurement of glass transition temperature. This article presents an extensive study on the thermal properties of polymeric composites; the examination on polymeric composites is consistently in the phase of improving the properties.

Keywords: Natural Fibers, Thermal analysis, DSC, TGA

1. Introduction

Polymers and polymer composites are used presently in a wide variety of industrial applications ranging from construction materials to electronics, and in our everyday lives from chopsticks to trash bins due to their versatility, light weight, low cost and excellent chemical stability. In some fields, low thermal conductivity of polymers limits their applications. [1-4]. For instance, Due to the limited heat spreading power and low thermal conductivity, polymers can be one of the major technical barriers for flexible electronics based on polymer. The composite properties improved the lignocellulose fibres and they obviously contribute to the individual properties of their components but also to the interface of the fibre matrix. Mechanically these composites fail mainly due to fibre pullout, fibre debonding and fibre breakage. In reality, the matrix's function in the fibre-reinforced composites used to move the load to the rigid fibres through Interface Shear stresses. That process calls for a Composite cannot be optimised and leaves it vulnerable to impacts on the environment that can degrade it, thereby reducing its lifespan. Insufficient adhesion between hydrophobic polymers and hydrophilic fibres contributes to weak natural mechanical properties in composites lined with silicone fibre. Recently, several studies have reported lignocellulose fibre chemical modifications has reasonable attachment of polymer matrix to fibres. Bad interface adhesion means full interface functionality. [5-13]. Several researchers studied the thermal effects and hybrid composite materials reinforced with various vegetable fibres (i.e., jute, sisal, ramie, coconut, sugar palm, etc.) or natural fibres mixed with glass fibres. FRP is also reinforced with various types of fibres to boost basic mechanical and thermal properties by means of different processing methods. Therefore, it is important to research any hybrid composite's thermal conductivity, specific heat, transition step, thermal stability and thermal degradation. The DSC is also one of the analyses required to assess the materials chemical properties. Glass transition reveals somehow the transition from glass to elastic Polymer-state [14-24].

2. Studies on Thermal Analysis

Literature studies of various natural fibers thermal properties are studied the main thermal analysis are TGA and DSC .TGA is to measure the temp changes, mas loss of sample is measured and any material is heated at particular temp of the material is starting to decompose to particular mass. It is mainly focused on the thermal stability of material. DSC enables the measurements of the transition such as glass transition melting and crystallization. The thermal properties of various natural fibers values are listed in table1.

2.1Thermal degradation of Lignocellulose fiber

In DSC and TGA analysis, lignocellulose fibres exhibit related patterns. Illustrates the three major stages of natural fibre thermal decomposition in which Main decomposition that occurs between 200-400 ° C. The thermal analysis relies on the respective atmosphere. Since Instance, the DSC test for cellulose under inert atmosphere changes the endothermic peak to oxidative (air or oxygen) exothermic peak.

2.2 Thermal degradation of Jute fiber

The jute composites at about 1000C indicating the elimination of moisture from the fibre. All composites showed the precise thermal decomposition inside the temperature range of 210^0 C to280^0 C. The degradation temperature of the untreated and dealt with Jute/Epoxy composites takes location [2] at 210 to 265^0 C respectively. However, inside the case of watermelon stuffed composite, the decomposition takes region at 275^0C. A small lift within the degradation temperature can be visible from the above composite suggests a huge discount in weight reduction. In the direction of higher temperatures, the weight loss is extensively reduced for treated composite while compared to untreated one. The better degradation temperature of the treated composite may be attributed to the presence of filler content, which led to multiplied thermal balance [20].

2.3 Thermal degradation of maize fiber

The maize fibers are presume that it is essential need to get great grip between fiber and network, to get a decent composite material strands should change from hydrophilic to hydrophobic characters. The maize fiber and polyester tar covered maize fiber gives a valuable data on warm corruption estimations of composites. Properties of normal filaments rely upon developing conditions and handling conditions. This variety makes more troublesome to break down the impacts of the strands and their interfaces on the warm properties of the composite material, these challenges call for improvement of new techniques [7].

2.4 Thermal degradation of Borassus fiber

In this study, examine the variation in particular heat, thermal conductivity of composite with honour to temperature, fiber and fly ash. The fly ash varies from 10 to 30% and the temperature range between 30 to 120°c. The composites with and without fly ash were keen with Borassus Flabellifer strengthened Composites in polyester matrix. Borassus Flabellifer Reinforce Composites acts as a thermal insulating component and it doesn't require any corrosion so, the longer styles is finished and besides prefabricate to totally shapes, set up time is very less, for this reason they're friendly reinforcing materials for the advance of load bearing.

IOP Conf. Series: Materials Science and Engineering **954** (2020) 012037 doi:10.1088/1757-899X/954/1/012037

2.5 Thermal degradation of Palm fiber

Thermal conductivity and thermal diffusivity of date-palm tree fiber reinforcement gypsum were estimated. The goal is to make use of the composite supplies with manufacture of thermal insulation for buildings. Thermo physical property measurements of composites were bought utilizing the method of a periodic temperature slope. The implications have shown that the incorporation of date palm tree fibers induces a lower in each thermal conductivity and thermal diffusivity. This demonstrates that the thermal conductivity of the insulating plate realized in our laboratory to be to be more important in palm fibers more essential than date palm tree fibers [13].

2.6 Thermal degradation of kenaf and pineapple fiber

The finding of this investigation demonstrated that NaOH treated fiber explicitly the kenaf and mengkuang delivered higher greatest pinnacle temperature than the UT strands. Notwithstanding, for glass progress temperature, the treated PALF, kenaf and mengkuang fiber, the result demonstrated lower an incentive when contrasted with the untreated fiber. Further examination needs to investigate on physical properties testing, for example, the water assimilation, softening point and thickness. The DSC examination results that can just give glass progress temperature. The examination of warm properties for PALF, kenaf and mengkuang strands could bring an improvement of utilizing characteristic fiber composites for the utilization of creating the games gear [12].

2.7 Thermal degradation of hair fiber

The effects of the loading of fibres and the alteration of the surface on the HF-/HDPE-reinforced composites have thermal properties been investigated. The thermal characteristics (TGA, DTG and DSC) is enhanced by the application of the surface modification of AT fibres 0.25-N. Via the implementation of more therapies (0.5-N AT) on HF, the fibre properties were reduced because of The Fiber Surface Fracture. Compared to untreated and other levels of AT-reinforced composites, the degradation of the 0.25-N AT 15 percent HF-/HDPEreinforced composite showed optimal performance. It has been observed that the exothermic peaks have moved to a higher temperature of the 0.25-N treated fibre compared with the untreated and another stage of AT fibre in DTA analysis [18].

2.8Thermal degradation of sisal fiber

DSC studies have shown that the mode of sisal degradation Fiber is a feature of the environment involved. In inert ambience, The DSC curves showed distinct peaks for each portion of the Sisal fibres, and two exothermic peaks in the air are watched. Cellulose degradation in inert environment whereas in the oxidative atmosphere, endothermic processes occur, It happens by exothermic processes. The withdrawal of lignin Holocellulose and cellulose thermally degraded from fibre occurring in both inert and oxidant at lower temperatures, Atmospheres, since the thermal usability of these constituents is simpler. The TGA studies have corroborated DSC's results the degradation actions of sisal fibre and its degradation help to summarise for constituents the finding has been that sisal thermal degradation fibre has better results compare to other fibers [21].

Table 1 Thermal properties of natural fiber

Fiber	Peak Temperature of Derivative Weight Change (°C)	Residue (wt. %)
Jute	335-350	14-19
Kenaf	380-430	10-13
Pine apple	220-280	26-62
Sisal	300-375	20-45
Palm	220-270	18-55
Borassus	350-400	22-65
Lignocellulose	200-400	18-40

3. Conclusions

- The work made on a review of thermal properties of natural fibers such as jute, kenaf, maize, palm, pineapple, and Lignocellulose and Borassus fibers.
- Apparently, the thermal activity of natural fibres brings a Correlation with their chemical constituents, such as cellulose, lignin, and hemicellulose.
- Weight loss initially in natural Fibers are correlated with temperatures between 50 and 100 ° C Hydroxyl groups evaporate from the fibre surface.
- Thermogravimetric isothermal and non-isothermal Study shows how pyrolysis acts in both the process.
- In advanced polymer composites, enhancing thermal stability in natural fiber-reinforced polymer composites is of great concern about materials.
- There are several works mentioned above in Different conditions that could be helpful for a particular reason in choosing products.
- However, the composites of natural fiber-reinforced polymers are being studied to improve their thermal resistance properties.

4. References

[1] Vinayagamoorthy R 2020 *Polym. Polym. Compos. doi:* 10.1177/0967391120943461.
[2] Vinayagamoorthy R 2020 *Glass. Fiber. Reinf. Polym. Compos. doi: 10.1515/9783110610147-001.*
[3] Vinayagamoorthy R 2020 *Polym. Polym. Compos. doi: 10.1177/0967391119883163.*
[4] Vinayagamoorthy R, Vikas Konda, Parikshit Tonge, Tumuluri Nikhil Koteshwar, Premkumar M 2019 *Mater. Today: Proc.* **16** 567.
[5] COCK F, Cuadri A and Partal P 2013 *Polymer Testing doi.org/10.1016/j. 2013.03.015.*
[6] Vinayagamoorthy R, Tumuluri Nikhil Koteshwar, Vajhala Venu Madhav, Twarakavi Karthik Sai, Vikas Konda 2019 *Mater. Today: Proc.* 16 277.
[7] Vinayagamoorthy R 2019 *J. Nat. Fibers. doi: 10.1080/15440478.2019.1598916.*
[8] Vinayagamoorthy R, Venkatakoteswararao G 2020 *J. Thermoplast. Compos. Mater.* 33 1429.

[9] Vinayagamoorthy R 2020 *J. Thermoplast. Compos. Mater*. 33 828.

[10] Aji IS, Zainudin ES, Khalina A, Sapuan SM, Khairul MD 2012 *Journal of Thermal Analysis and Calorimetry volume 109, 893–900900.*

[11] Vinayagamoorthy R, Ankur Sharma, Vignesh Iyer, Navneeth G 2019 *Adv. Manuf. Processes., Lect. Notes. Mech. Eng. 521.*

[12] Rajmohan T, Vinayagamoorthy R, Mohan K 2019 *J. Thermoplast. Compos. Mater. 32 1282.*

[13] Vinayagamoorthy R, Manoj I. V, Narendra Kumar G, Sai Chand I, Sai Charan Kumar G. V, Suneel Kumar K 2018 *IOP Conf. Series: Mater. Sci. Eng. 390 012029.*

[14] Vinayagamoorthy R 2019 *J. Nat. Fibers.* 16 163.

[15] Poddar P, Islam MS, Sultana S, Nur HP and Chowdhury AMS 2017. *J Material Sci Eng 2016, doi; 10.4172/2169-0022.1000270.*

[16] F.A. Fauzi, Z. Ghazalli, J.P. Siregar, C. Tezara 2016 *MATEC Web of Conferences 78:01042 doi: 10.1051/matecconf/20167801042.*

[17] Shraavana Bavan D and Mohan Kumar G C 2013 *Vol 3 World Congress on Engineering.*

[18] Prashant Srivastava, Shishir Sinh 2017 *Science and Engineering of Composite Materials doi; 10.1515/secm-2017-0035.*

[19] A. Oushabi 2017 *South African journal of chemical engg 23 doi; 10.1016/j.sajce.2017.04.005.*

[20] Madhusudhan T, SenthiL Kumar M. 2016. *IOSR Journal of Pharmacy vol 6 pp; 23-27.*

[21] Adriana R. Martina, Maria Alice Martinsb, Odilon R.R.F. da Silva c, Luiz H.C. Mattoso 2010 *Thermochimica Acta 506(s 1–2):14–19 doi: 10.1016/j.tca.2010.04.008.*

[22] Vinayagamoorthy R, Sivanarasimha S, Vijay Padmanabhan, Vedula Ganesh, Karthikeyan S 2015 *Int. J. Appl. Eng. Res* 10 663.

[23] Vinayagamoorthy R 2018 *J. Reinf. Plast. Compos.* 37 49.

[24] Vinayagamoorthy R 2017 *J. Reinf. Plast. Compos.* 36 1577.

IOP Conf. Series: Materials Science and Engineering **954** (2020) 012038 doi:10.1088/1757-899X/954/1/012038

Analysis of Power Consumption in the drilling of Nano SiC reinforced Aluminium matrix composites

V. V. Kalyan Chakravarthy[1] T. Rajmohan[2]

Department of Mechanical Engineering
Sri Chandrasekharendra Saraswathi Viswa Maha Vidyalaya University,
Enathur, Kanchipuram – 631561, India

kalyanchakravarthy@kanchiuniv.ac.in, rajmohanscsvmv@yahoo.com

Abstract. Nanoceramic particles reinforced Metal Matrix Composites (MMNC) have gained immense attention from both the scientific and industrial community. The employment of these advanced class materials in the industries has given many advantages. These Nanocomposites are very difficult to machine, and the number of holes is to be drilled on their surface for many applications as drilling is considered as one of the significant applications in the industry. The researchers and industrial experts are striving very hard to attain green manufacturing techniques in drilling, which creates less impact on the environment and consume lower power. The main aim of current research is to analyse the power consumption in the drilling of Nano silicon carbide (SiC) reinforced Aluminium (Al) matrix composite under various machining conditions, cutting speeds, feed rates. For this purpose, a series of experiments have been planned using Taguchi's L18 orthogonal array. The drilling tests were performed using a vertical machining centre (VMC) which is controlled by computer numeric control (CNC) employing carbide drills of 10mm dia with cutting point angles of 90°, 118° and 135°. The machining operation is performed under Dry cutting and Cryogenic cutting condition. On successful completion of the experiments, the surface morphology is studied using Field Emission Scanning Electron Microscopy (FESEM).

Key Words: Nano SiC particles; Drilling; Cryogenic; LN_2; Taguchi L18 array; FESEM.

1. Introduction

The Metal Matrix Nano Composites (MMNCs) are remarkably focused over the recent years mostly due to their wider functional, structural characteristics and their applications [1]. These advanced class of materials own superior mechanical strength, and best distortion resistance, the product manufacturing with these materials remain a significant challenge for the industrial sector [2]. Henceforth, various researchers had been conducted on the viability study of green manufacturing of MMNCs, to reduce the undesirable influence of machining on the environment, and also to reduce the power consumption in the machining of these advanced class of materials [3]. Many researchers have studied the fabrication of Nano SiC composites using stir casting process, and these composites are fabricated using Al as matrix materials and Nano SiC as the reinforcing material. The ceramic Nano SiC helped in improving the hardness, yield strength and ultimate tensile strength of the composites

IOP Conf. Series: Materials Science and Engineering **954** (2020) 012038 doi:10.1088/1757-899X/954/1/012038

[4-6]. Suresh et al. had investigated the influence of nanoparticles in the Aluminium matrix composites fabricated using stir casting furnace. The authors had concluded that the inclusion of nanoparticles had produced products superior to the base material in terms of hardness and tensile strength. [7] Thakre et al. has carried research in the burr height produced in the drilling of Al- SiC composites. The researchers have employed RSM based BBD to analyse the burr pattern [8]. Rajmohan et al. had researched Aluminium (Al) based composites and had employed Taguchi design to analyse the drilling characteristics of the Al matrix composites [9]. Suresh et al. had studied the turning performance of Al-based hybrid matrix composites. The experiments are planned using Taguchi l_{27} orthogonal array, and the optimisation of the experimental results are carried out using Grey Fuzzy algorithm. [10]. Sahu et al. had investigated the machining performance of Al-TiC matrix composites. The authors had planned the experiments using the Taguchi l_{16} orthogonal array and had optimised the machining performance using the multiobjective optimisation. [11].

Wang et al. [12] have proposed a model for energy consumption in the drilling process, authors have proposed the model based on equation consisting idle power which is calculated based on the spindle speed, the cutting power, calculated from the cutting force. Franco et al. [13] had focused their research in the analysis of energy consumption in the micro drilling using EDM and ultrashort pulse laser machining. Rajemi et al. [14] developed a methodology of optimisation by analysing energy use in turning operations for a machined product. Pervaiz et al. [15] had studied the surface roughness and energy consumption in performing drilling on Al 6061. Authors had found the power consumption using power logger and had analysed the energy consumption in the drilling.

An extensive and comprehensive review of recent articles exposes that not much effort has been shown towards investigating the power consumption in the drilling of Nano SiC-Al composites. This investigation presents the experimental results and its power consumption with Dry machining and Cryogenic drilling. The drilling tests are conducted on VMC as per Taguchi L_{18} orthogonal array. Scanning electron microscope (SEM) was utilised for inspecting the subsurface morphology of the machined specimen.

2. Materials and Methods

In the current investigation, the materials utilised are shown in table 1.

Table 1 materials used

Sl.no	Material	Size	Supplier	Purity
1	Aluminium	Billet	M/s Micro Fine chemicals, India	99.99%
2	Nano SiC	50-80 nm	M/S US Research Nanomaterials Inc.	99%
3	Cryogenic lubricant (LN_2)		M/s Cryo air products Chennai, India	99%

The MMNC of Nano SiC reinforced Al matrix composites are fabricated using ultrasonic-assisted stir casting furnace [5]. The specimen is fabricated in the form of rectangular blocks of 150x100x100 mm. Nano SiC reinforced Al matrix composites are examined with metallographically to confirm the presence of nanoparticles using FESEM and is presented in figure 1.

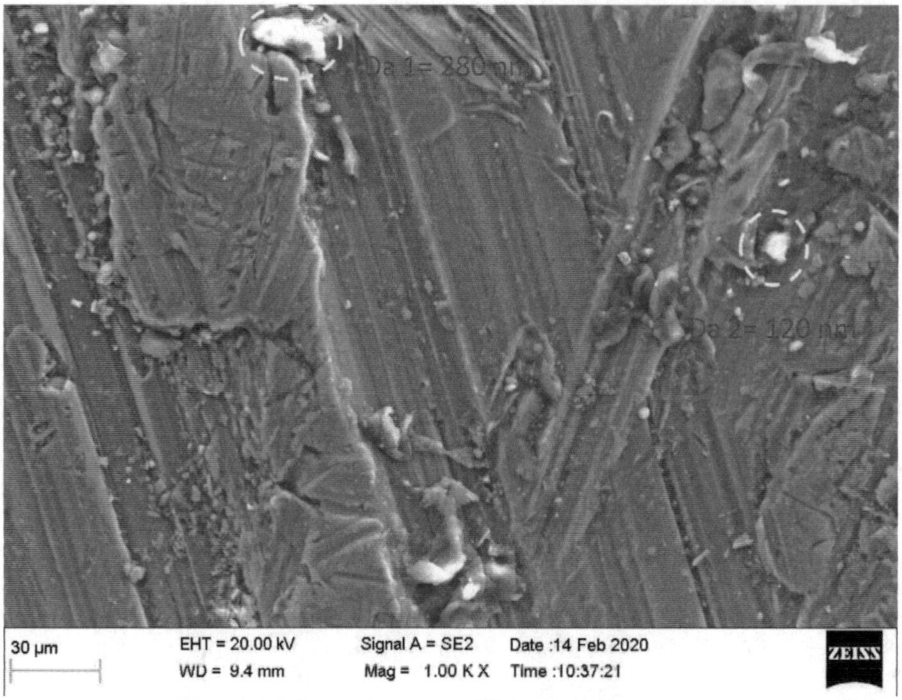

Figure 1 Microstructure of Fabricated specimen

The metallogrphical examination provides the confirmation of the inclusion of nano SiC particles in the fabricated specimen.

2.1 Experimental details

The experiments are conducted using a CNC assisted Vertical Machining Centre (VMC) of 'ARIX Machines Co. Ltd' Dynamometer of 'Kistler' make is attached to the VMC to record the variations in Torque produced while machining. Carbide drills of 10 mm diameter with different point angles (90°, 118°, 135°) are employed for the drilling operation.

For the current research, the experiments are planned using Taguchi L_{18} orthogonal array and the parameters opted are Spindle Speed, Feed rate, Cooling condition, Wt% of Nano SiC. The parameters opted for the machining are shown in Table 2. The outputs considered here is Torque

Table 2 Drilling Parameters and their levels

Sl.no	Parameters	Notation	Units	Levels		
				1	2	3
1	Speed of Spindle	N	rpm	500	1000	1500
2	Feed rate	*f*	mm/min	50	100	150
3	Point angle	*Ø*	Degree	90	118	135
4	Wt% of Sic	*w*	%	1	2	3
5	Type of Cooling			Dry		Cryogenic

Machining is carried out in varying lubricative conditions which are Dry Machining (DM), and Cryogenic Machining (CM). In DM the heat generated is allowed to reduce under the atmospheric conditions, and in CM a cryogenic system has been employed to supply the LN_2 on the machining zone with a delivery pressure of 3 bar [16]. The experimental layout is provided in figure 2.

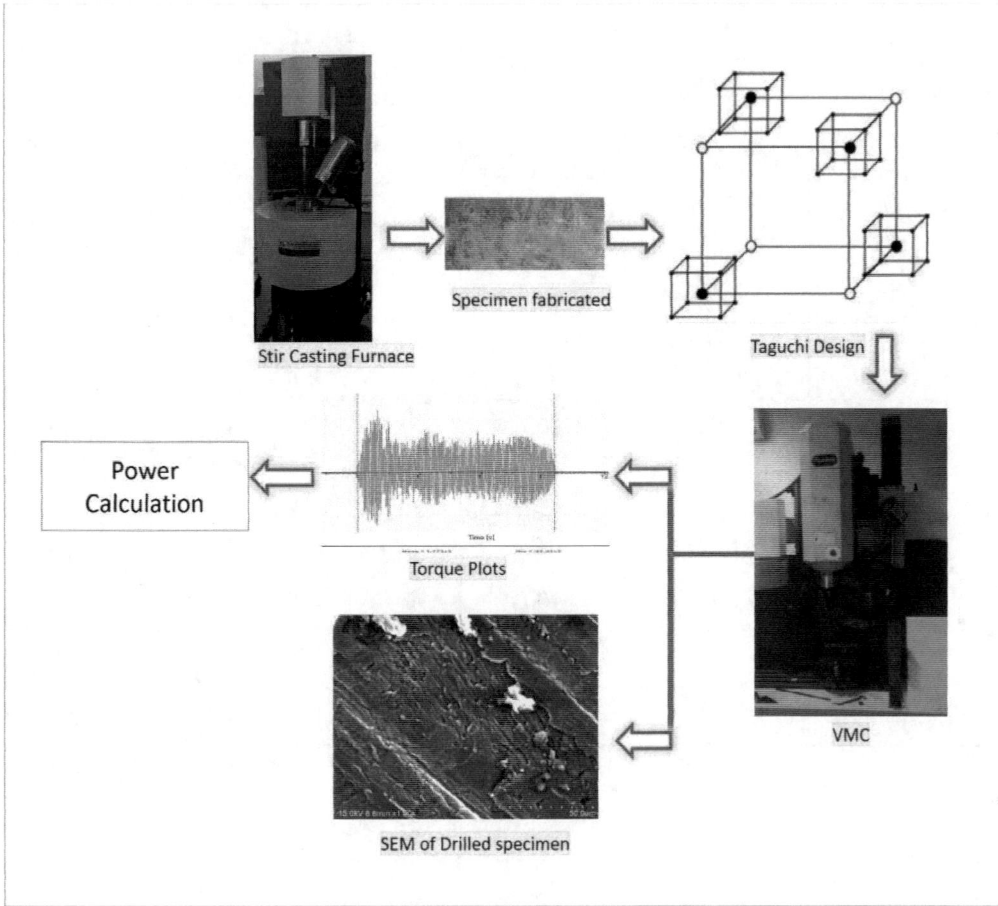

Figure 2 Experimental Layout

The experimental design and the results of the experimentation are provided in table 3. The power has been calculated form the Torque generated in the experimental procedure. The equation 1 is used for the calculation of power consumption and is provided

$$power = \frac{2\pi NT}{60} \text{-------------------}(1)$$

Equation 1 for calculating power
Where,
N= Spindle speed (rpm)
T= Torque (N-m)

IOP Conf. Series: Materials Science and Engineering **954** (2020) 012038 doi:10.1088/1757-899X/954/1/012038

Table 3 Experimental design

Sl. No	Cooling Condition	Spindle speed (rpm)	Feed Rate (mm/min)	Wt% of SiC	Point angle	Torque (N-m)	Power (w)
1	Dry	500	50	1	90	1.773	92.79
2	Dry	500	100	2	118	1.192	62.38
3	Dry	500	150	3	135	1.707	89.33
4	Dry	1000	50	1	90	1.046	109.48
5	Dry	1000	100	2	118	1.453	152.08
6	Dry	1000	150	3	135	1.987	207.97
7	Dry	1500	50	2	118	1.446	227.02
8	Dry	1500	100	3	135	1.508	236.76
9	Dry	1500	150	1	90	1.796	281.97
10	Cryogenic	500	50	3	135	1.584	82.90
11	Cryogenic	500	100	1	90	1.829	95.72
12	Cryogenic	500	150	2	118	1.309	68.50
13	Cryogenic	1000	50	2	118	1.364	142.77
14	Cryogenic	1000	100	3	135	1.905	199.39
15	Cryogenic	1000	150	1	90	1.938	202.84
16	Cryogenic	1500	50	3	135	1.18	185.26
17	Cryogenic	1500	100	1	90	1.785	280.25
18	Cryogenic	1500	150	2	118	1.905	299.09

3. Results and Discussion

3.1 Effect of Spindle speed on power consumption

The variations recorded are analysed and provided in figure 3. The results illustrate that the power consumption tends to increase from 81.33w to 248.7w as the spindle speed increases from 500rpm to 1500 rpm in the case of DM and the in the contrary the power consumption is found to increment from 82.4w to 254.9w as the spindle speed increases from 500rpm to 1500 rpm in CM. This raise in the power consumption is majorly due to the strain hardening of the material in the case CM.

IOP Conf. Series: Materials Science and Engineering **954** (2020) 012038 doi:10.1088/1757-899X/954/1/012038

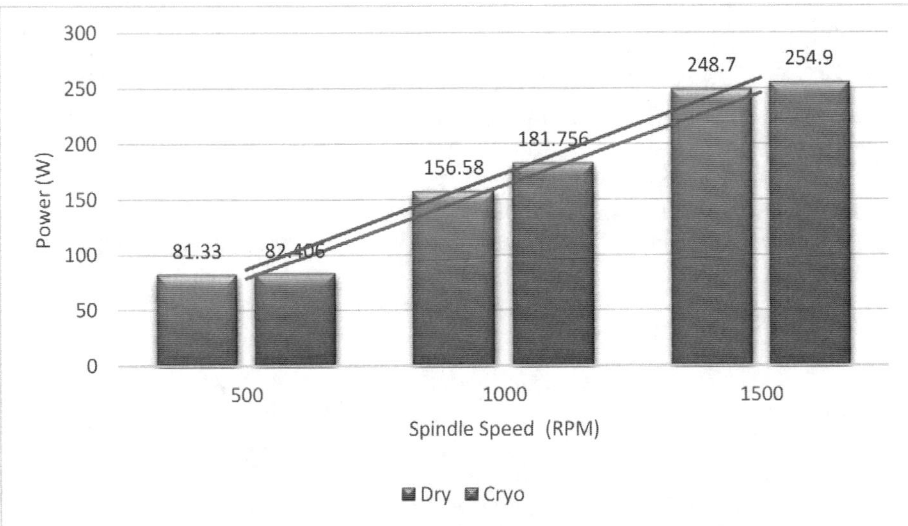

Figure 3 Effect of Spindle speed on Power consumption

3.2 Effect of Feed rate on power consumption

The variations recorded are provided in figure 4. The results indicate that the power consumption tends to increase from 143.1w to 193.09w as the spindle speed increases from 50 mm/min to 1500 mm/min in the case of DM and the, on the contrary, the power consumption is found to increment from 136.9 w to 190.1w as the spindle speed increases from 500rpm to 1500 rpm in CM. From the observation, it is found that as the feed rate increases the power consumption also increases, this is majorly due to the development of thrust forces at higher feeds which results in higher power consumption.

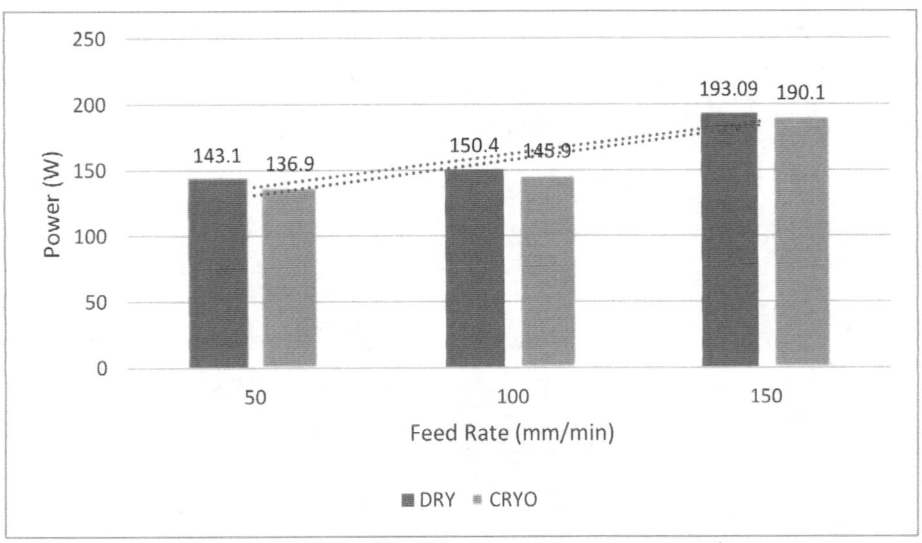

Figure 4 Effect of feed rate on power consumption

3.3 Effect of wt% of Nano SiC on power consumption

The variations in power consumption are recorded, and analysed is provided in figure 5. The results illustrate that the power consumption tends to follow a typical structure which shows that the power consumption had reduced from 161.14 w to 147.16 w when the wt % of Nano SiC increased from 1% to 2% in DM and the power consumption had increased from 147.16 w to 177.97 w when the wt % of Nano SiC increased from 2% to 3% in DM. The CM has followed a decrement of power consumption in the whole cycle as the wt % of Nano SiC increased from 1% to 3% the power consumption found to be reduced from 192.29 w to 155.85 w

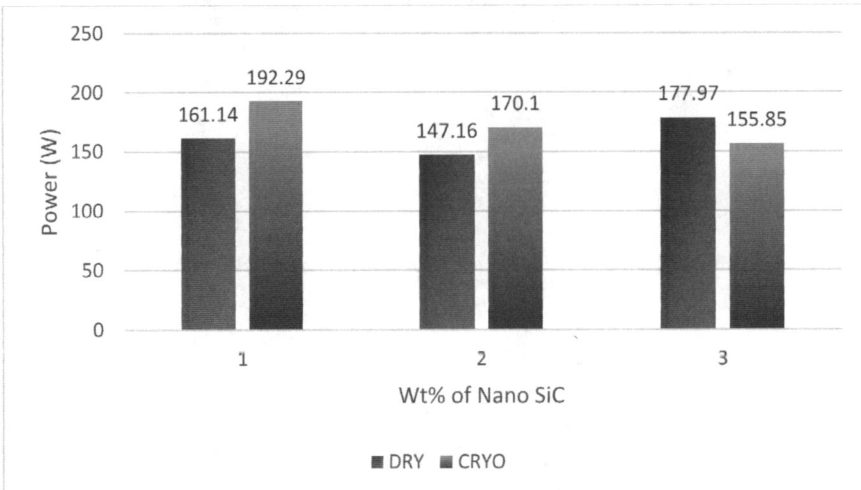

Figure 5 Effect of Nano SiC wt % on Power consumption

3.4 Effect of cooling condition on power consumption

The variations in the experimental procedure are recorded and are provided in figure 6. The results illustrate that the power consumption is found to be higher in CM when compared with DM. The results indicate that the power consumption of DM is 162.9 w and the power consumption in the CM is 172 w. The power consumption in the CM is found to be high due to the employment of Cryogenic LN_2 as the material gets hard when cooled rapidly.

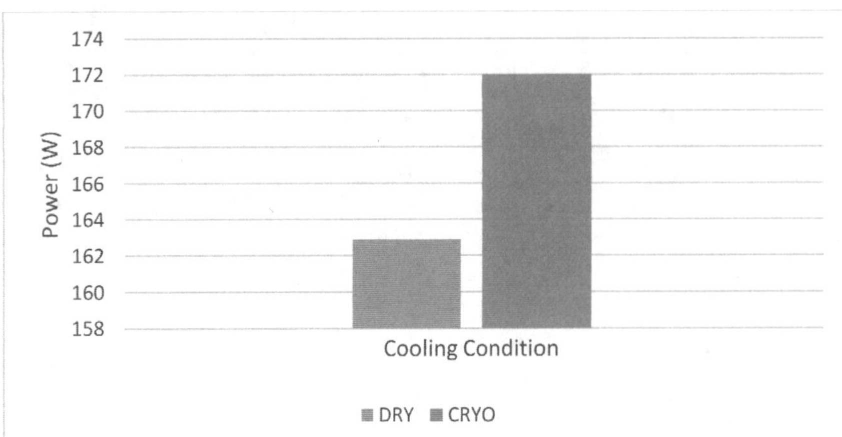

Figure 6 Effect of Cooling Condition on Power consumption

3.5 Analysis of Machined Surface

The machined Nano SiC reinforced Al matrix composites is examined for the microstructural changes employing FESEM. The micrographic studies indicate that Nano SiC is found to be evenly distributed in the material. The FESEM micrographs are presented in the figures 7-12. These micrographs of the Nano SiC reinforced Al matrix composites indicates the presence of voids, continuous and discontinues tracks; adhered debris, craters in the machined zone. When the drilling process starts, the drill bit tends to crush the workpiece, causing surface tearing. These tracks were formed because of the strain hardening of the specimen. It can be concluded that the utilisation of cryogenic LN2 in the experimentation generates better surface finish and also helps in increasing the tool life [16].

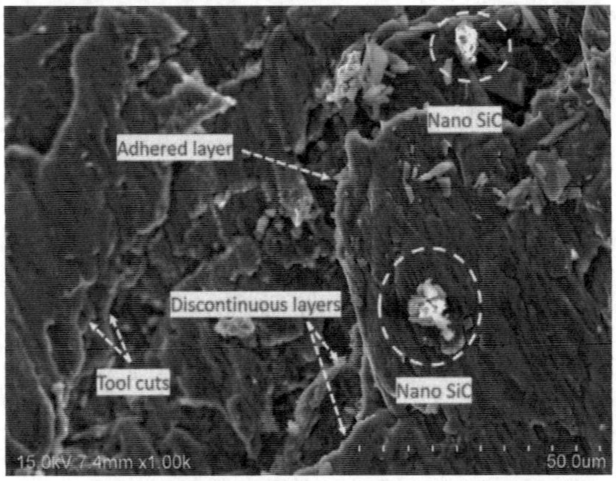

7(a)

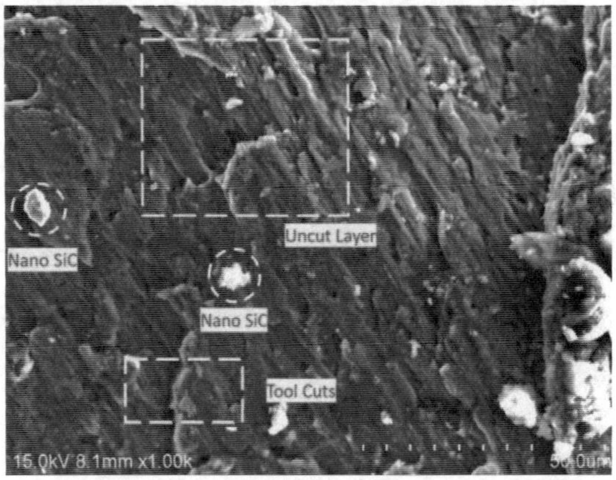

7(b)

Figure 7(a) Dry drilling at Spindle speed = 1500, Feed rate = 150mm/min, Point angle = 90, Wt= 1%
7(b) Cryogenic drilling at Spindle speed = 1500, Feed rate = 150mm/min, Point angle = 90, Wt= 1%

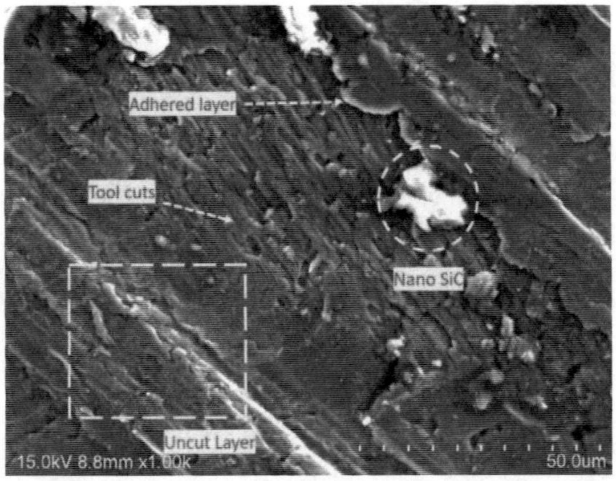

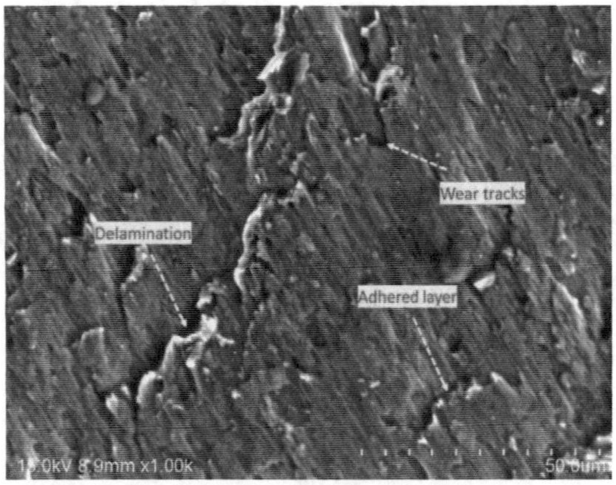

8(a) 8(b)

Figure 8(a) Dry drilling at Speed of Spindle = 500, Feed rate = 100mm/min, Point angle = 118, Wt= 2%
8(b) Cryogenic drilling at Spindle speed = 500, Feed rate = 100mm/min, Point angle = 118, Wt= 2%

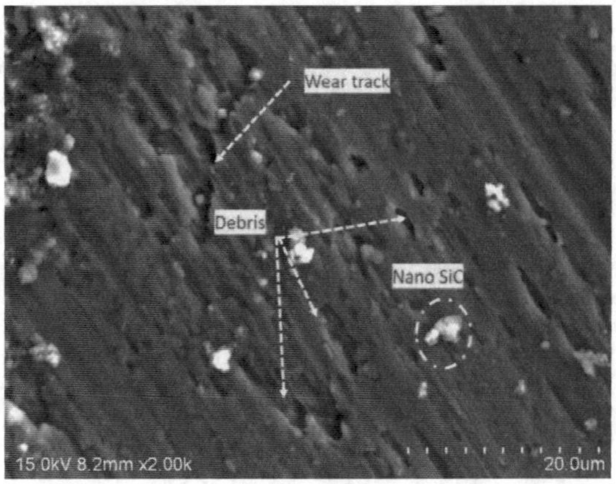

9(a)

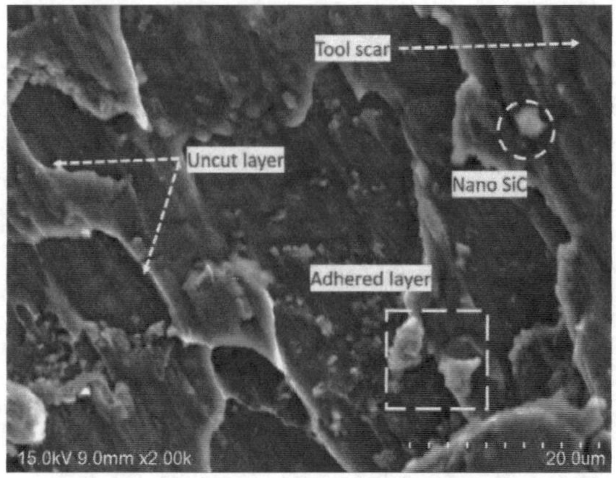

9(b)

Figure 9(a) Dry drilling at Speed of Spindle = 1500, Feed rate = 100mm/min, Point angle = 135, Wt= 3%
9(b) Cryogenic drilling at Spindle speed = 1500, Feed rate = 100mm/min, Point angle = 135, Wt= 3%

4. Conclusions

- In this paper, an analysis for power consumption in the drilling of Nano SiC reinforced Al matrix composites had been carried basing upon the spindle speed utilised in the experimental procedure and the Torque generated while machining.
- The power consumed for the machining process was found using an empirical calculation.
- Higher feeds tend to consume higher power.
- Employment of cryogenic LN_2 consumes higher power in drilling, which is mainly due to the hardening of materials at low temperatures.
- The application of advanced cryogenic LN_2 helped in providing a better surface finish.
- It is also observed that the tool life has also increased by employing cryogenic LN_2.
- FESEM metallographic studies indicate the distribution of Nano SiC particles evenly in the composite
- The metallographic studies indicate that the formation of cracks on the surface of composite, which are majorly due to the strain hardening of the material.

References

[1] Rajmohan, T., Palanikumar, K., Arumugam, S. (2014) Synthesis and characterisation of sintered hybrid aluminium matrix composites reinforced with nano copper oxide particles and micro silicon carbide particles. COMPOS PART B-ENG. https://doi.org/10.1016/j.compositesb.2013.10.060
[2] Vijayabhaskar, S., Rajmohan, T. (2018) Experimental Investigation and optimisation of Machining parameters in WEDM of Nano-SiC Particles Reinforced Magnesium Matrix Composites SILICON. https://doi.org/10.1007/s12633-017-9676-0

[3] Nandakumar, A., Rajmohan, T. (2018) Grinding of MMC using MQL based vegetable oil-Review IOP Conference Series: Materials Science and Engineering 390 (1), 012033

[4] Mazahery, A., & Shabani, M. O. (2012). Characterisation of cast A356 alloy reinforced with nano SiC composites. Transactions of Nonferrous Metals Society of China, 22(2), 275–280. doi:10.1016/s1003-6326(11)61171-0

[5] Nandakumar, A., Rajmohan, T. & Vijayabhaskar, S. (2019) Experimental Evaluation of the Lubrication Performance in MQL Grinding of Nano SiC Reinforced Al Matrix Composites Silicon. https://doi.org/10.1007/s12633-019-0088

[6] Ashok Kumar, R., & Krishnakumar, T . (2018) Casting and characterisation of Al6063/SiC nano composites produced using stir casting method. Materials Today: Proceedings, 5(11), 23853–23862. doi:10.1016/j.matpr.2018.10.177

[7] Suresh, S., Gowd, G. H., & Deva Kumar, M. L. S. (2019). Mechanical Properties of AA 7075/Al2O3/SiC Nano-metal Matrix Composites by Stir-Casting Method. Journal of The Institution of Engineers (India): Series D. doi:10.1007/s40033-019-00178-1

[8] Thakre, A. A., & Soni, S. (2016) Modeling of burr size in drilling of aluminum silicon carbide composites using response surface methodology. Engineering Science and Technology, an International Journal, 19(3), 1199–1205. doi:10.1016/j.jestch.2016.02.007

[9] Rajmohan, T., Palanikumar, K. and Kathirvel, M. (2012) Optimisation of machining parameters in drilling hybrid aluminium metal matrix composites. Transactions of Nonferrous Metals Society of China, 22(6), pp.1286-1297.

[10] Suresh, P., Marimuthu, K., Ranganathan, S. and Rajmohan, T. (2014) Optimisation of machining parameters in turning of Al-SiC-Gr hybrid metal matrix composites using grey-fuzzy algorithm. Transactions of Nonferrous Metals Society of China, 24(9), pp.2805-2814.

[11] Sahu, M.K., Valarmathi, A., Baskaran, S., Anandakrishnan, V. and Pandey, R.K., (2014) Multiobjective optimisation of upsetting parameters of Al–TiC metal matrix composites: a grey Taguchi approach. Proceedings of the Institution of Mechanical Engineers, Part B: Journal of Engineering Manufacture, 228(11), pp.1501-1507.

[12] Wang, Q., Zhang, D., Chen, B., Zhang, Y. and Wu, B., (2019) Energy Consumption Model for Drilling Processes Based on Cutting force. Applied Sciences, 9(22), p.4801.

[13] Franco, A., Rashed, C.A.A. and Romoli, L., 2016. Analysis of energy consumption in micro-drilling processes. Journal of Cleaner Production, 137, pp.1260-1269.

[14] Rajemi, M.F., Mativenga, P.T. and Aramcharoen, A., (2010) Sustainable machining: selection of optimum turning conditions based on minimum energy considerations. Journal of Cleaner Production, 18(10-11), pp.1059-1065.

[15] Pervaiz, S. and Deiab, I., (2015) Surface roughness and energy consumption analysis of conventional and peck drilling approaches. Proceedings of the Institution of Mechanical Engineers, Part B: Journal of Engineering Manufacture, 229(12), pp.2180-2195.

[16] Kalyan Chakaravarthy, V.V., Rajmohan, T., Vijayan, D., Palanikumar, K. and Latha, B., (2020) Sustainable drilling performance optimisation for Nano SiC reinforced Al matrix composites. Materials and Manufacturing Processes, pp.1-9.

IOP Conf. Series: Materials Science and Engineering **954** (2020) 012039 doi:10.1088/1757-899X/954/1/012039

Effect of Nano Particles on Tribological Behavior of Reciprocating Air Compressor Oil Using Fourball Tribometer: An Experimental Investigation

P. Chenga Reddy[1]*and S. Arumugam[2]

[1, 2] Department of Mechanical Engineering, Sri Chandrasekharendra Saraswathi Viswa Mahavidyalaya, Enathur, Kanchipuram - 631561, Tamil Nadu, India.

Chengareddy.p@kanchiuniv.ac.in

Abstract. This paper presents an experimental investigation of tribological properties of pentaerythritol ester (PE) as biolubricant with the addition of nanoparticles i.e. copper oxide (CuO). The nano particle was added with pentaerythritol ester by wt.% as 0.1, 0.3 and 0.5%. The experimental work was conducted on a fourball tribometer under various conditions such as applied load of 40 kg, speed of 1200 rpm, experimental duration of 1 hr and a temperature of 75°C under the standard of ASTM 4172. The 0.3% PE based nanoparticle lubricant shows the low coefficient of friction (COF), frictional torque (FT) and wear scar diameter (WSD) as compared to other nano based lubricants and with base oil. The tested worn surface was examined with SEM analysis and results are reported that the 0.3% nanoparticle lubricant shows the smoother and low surface roughness.

1 Introduction

In current scenario, the crude (mineral) oil plays a significant role in automobile and other applications. At the same time there is a huge demand in fuel and lubrication area. But in other side, these mineral oils create many problems such as emission of pollutants, toxic gases, degradability problems which damages environment, increasing of crude oil cost due to the depletion of its reserves. With this connection, world needs an environmentally friendly lubricants to avoid such type drawbacks with the crude oil. One such alternative lubricant is vegetable based biolubricant which fulfill the requirement of mineral oil [1]. But the usage of straight vegetable oil having some problems like poor thermo-oxidative stability and cold flow properties [2]. This can be rectified by modifying the vegetable oil via chemically to improve the cold flow properties and thermo-oxidative stability by conventional methods [3]. These conventional methods are taking more reaction time leads more power consumption and produces low yields which increases the cost of final product [4]. So, the researchers are attempted a few experimental works on the production of biolubricant with high yield percentage with high conversion rate at low reaction time using ultrasonic assisted transesterification process.

Many researchers are produced high yield biodiesel from various vegetable oils (namely castrol oil, coconut oil, rapeseed oil, soybean oil, etc.) using ultrasound assisted transesterification process [5-9]. But very rare attempts were completed to produce biolubricants by using ultrasonic irradiation process. Arumugam et al [10] produced pentaerythritol ester as biolubricant using ultrasonic assisted transesterification process under nitrogen atmosphere, results are revealed that the ultrasonic irradiation process produces the high yield percentage of pentaerythritol ester (PE) with low reaction time and PE shows low coefficient friction and low wear depth. Another work carried out by optimizing the ultrasonicator process parameters like ultrasonic amplitude, ultrasonication time, catalyst concentration and ultrasonic pulse for high yields. Reported that the yield about 81.4% of pentaerythritol ester was obtained at optimal conditions i.e. 15 s ultrasonic pulse, 60% of ultrasonic amplitude, 1.5 wt% of catalyst concentration, and 100 °C reaction temperature [11]. However, the addition of nanoparticles are shows significant role on tribological properties in terms of reduction in friction and wear rate. Baskar et al [12] was investigated the tribological characteristics of chemically modified rapeseed oil by addition of with and without different nanoparticles like CuO, WS_2 and TiO_2. From the experimental results the addition of CuO nanoparticle lubricant shows improved tribological properties in terms of wear and friction reduction, also it shows smoother wear scar images than that of other nanoparticles. Gulzar et al [13-14] was used nano based chemical modified palm oil as lubricant to investigate the anti wear properties. They reported that the nano lubricants are shown better anti wear properties than that of without nano addition lubricants. Rajaganapathy et al [15] suggested that the addition of 0.5% CuO nanoparticle with vegetable-based lubricant shows improved viscosity, thermal conductivity and reduction in friction and wear than that of other lubricants. Another work was carried out by Kashyap and Harsha [16] on oxide nanoparticle with the addition of chemically modified rapeseed oil and the nano-based biolubricants are shown better wear reduction properties than that of raw rapeseed oil and mineral based oils. Tribological tests were conducted using four-ball tribometer and pin-on-disc tribometer under nano-based biolubricants and it shows remarkable results in terms of reduction in friction and wear [17-18].

From the literature studies, ultrasonic irradiation method is plays most significant role on production of biolubricants with high yields at low reaction time. Also, the CuO nanoparticles having good antiwear properties in terms of reduction in wear and friction. This present study is to investigate the synthesis of pentaerythritol ester as bio-based compressor oil using ultrasonic irradiation process and also tribological studies were to be carried out under nano-based biolubricants using four-ball tribometer.

2. Experimental

2.1 Ultrasonic Irradiation Transesterification Process

A two-stage transesterification process was used to synthesis the pentaerythritol ester from rapeseed oil using ultrasonicator. An ultrasonic sonicator of Oscar make, Model PR-1000, 750W Ultrasonic Power, 1 min of pulse duration, adjustable frequency of upto 25 kHz was used for this transesterification process. In first stage: the products of raw rapeseed oil (1500ml), catalyst of sodium hydroxide (13.38g), alcohol as methanol (267.6ml) was taken in three necked flask and kept in ultrasonicator for the duration of 20 min reaction time at 55°C temperature (not to be exceed 60°C). The excess alcohol was removed from the obtained product by warm water wash. Afterwards the methyl ester was separated from glycerol by using separating funnel. Then the obtained methyl ester was heated up at a temperature of 55°C to evaporate the presence of excess water. Second stage: the obtained quantity of final product of methyl ester of 960m1, p-TSA of 1.5 g as catalyst, 224.5 g of pentaerythritol and solvent of xylene as 20 ml was taken in a three necked flask and placed in a ultrasonicator chamber followed by transesterification process. The whole reaction was operated under nitrogen atmosphere with 1h reaction time, 20 kHz of frequency and at a temperature of 100°C. Later, the reactants were heated up at a temperature of 70°C to evaporate the presence of rest of the water in the reactants and warm water is used to take off the excess alcohol by wash from the products [10]. The obtained final product is pentaerythritol ester (PE).

2.2 Preparation of Nano Lubricants

The nanoparticles of CuO was added with formulated pentaerythritol ester on wt.% (0.1%, 0.3% and 0.5%) basis using ultrasonic bath as shown in Fig 1. The anti-wear properties of the material were improved by the addition of nano additives with the pentaerythritol ester. For this purpose, an ultrasonic sonicator (Make: LMUC-2A, Model: LABMAN, Ultrasonic frequency: 40±3 kHz, Ultrasonic wattage: 50W) was used to ensure homogenous mixing and well scattering of nano particles into the formulated lubricant i.e. pentaerythritol ester [19]. The CuO nanoparticles was supplied by Nano labs, India. 99% of mass purity, the average diameter of the nanoparticles is about 30-50 nm. The weight of the nanoparticles is measured by precision digital electronic weighing balance machine.

Fig. 1. Ultrasonic Bath

The measured nanoparticles are mixed with pentaerythritol ester on wt.% basis i.e. 0.1%, 0.3%, and 0.5% (prepared samples are shown in Fig. 2.) and placed the mixture on a magnetic stirrer for 30 min. Afterwards, the mixture was set down in ultrasonic sonicator with a duration of 1 h. The prepared sample properties are shown in Table 1. PE with 0.3wt.% of CuO nanoparticle was shown improved physico-chemical properties in terms of viscosity, viscosity index, flash point, pour point and total acid number than that of other lubricants.

Fig. 2. Nano-based compressor oil samples

Table 1. Properties of compressor oil samples

Sl.No.	Properties	Standards	SAE 30	PE	PE+0.1wt.% of CuO	PE+0.3wt.% of CuO	PE+0.5wt.% of CuO
1	Kinematic viscosity at 100° C, cSt	ASTM D445	9.16	9.02	10.6	10.8	10.2
2	Viscosity index	ASTM D2270	95	124	118	126	112
3	Flash point, ° C	ASTM D92	210	208	252	254	242
4	Pour point, ° C	ASTM D97	< -10	< -10	-11	-11	-11
5	Total acid number (TAN)	ASTM D664-04	0.99	1.87	2.54	2.42	2.74

2.3 Tribological tests

The four ball tribometer (Make: M/S Ducom Instruments Pvt. Ltd., Bangalore; Model: TR-30L-IAS) was used to test the lubricity performance of the prepared lubricant samples i.e. SAE30, PE, PE with 0.1wt.% of CuO nanoparticle, PE with 0.3wt.% of CuO nanoparticle and PE with 0.5wt.% of CuO nanoparticle. For this test a 12.7mm diameter of ball is made with AISI E52100 chrome alloy steel and its hardness has 65 HRC. For each test, 10 ml of prepared compressor oil sample was filled in a ball pot for the evaluation of lubricity properties. The tribological properties of the lubricants are tested as per ASTM D4172. The testing conditions are as follows: sample oil temperature is 75 °C, load as 40 kg, 1200 rpm of ball rotational speed with 1 h operational time for each test setup. The wear scar spots of the tested balls are analyzed using optical microscope was shown in Fig. 3. Frictional torque was measured with a load cell and co-efficient of friction (μ) was determined using given Eq. (1) [20].

$$\mu = \frac{T\sqrt{6}}{3wr} \quad (1)$$

Where T= frictional torque in Nm, w=applied load in N and r=distance from the center of the contact point on the lower balls to the axis of rotation in m.

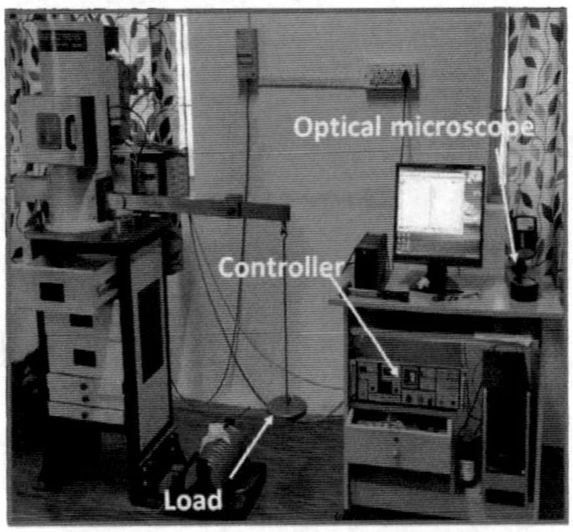

Fig. 3. Four-ball tribometer

3 Results and Discussions

3.1 Frictional Torque Analysis

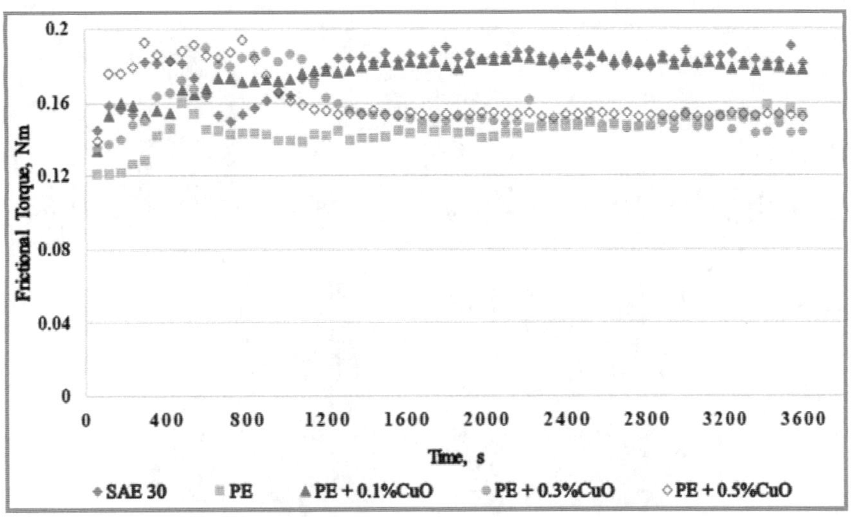

Fig. 4. Frictional torque

Fig. 4 demonstrates the frictional torque for various compressor biobased nano lubricants. The frictional torque value of all the nano-based lubricants are closer to each other but PE and PE based nano lubricants are shows low mean frictional torque than that of petroleum-based lubricant i.e. SAE 30. Among all the lubricants PE with 0.3 wt.% of CuO nano lubricant shows minimal frictional torque than that of other biolubricant samples and with base oil, this is due to the presence of nano particles with the pentaerythritol ester. The turbo layer forms between the moving ball and stationary ball which minimize the friction, this is only due to the presence of ester groups and dissolved nano particles in the pentaerythritol ester. However, PE with 0.1 wt.% CuO nano lubricant and SAE30 mineral lubricant shows the high frictional torque [20 & 21].

3.2 Analysis of co-efficient of friction

Fig. 5 illustrates the co-efficient of friction for various lubricants. The least coefficient of friction (COF = 0.0761) was observed with the lubricant of PE with the addition of 0.3 wt.% of CuO nano particle as compared with the other lubricants. The highest COF was noted as 0.1009 and 0.09934 for SAE30 and PE with 01 wt.% CuO nano lubricant this is due to the oil layer lost between the contact area of the tested ball. Pentaerythritol ester along with nano particles has a good enough to maintain a stable film formation between the contact area [10].

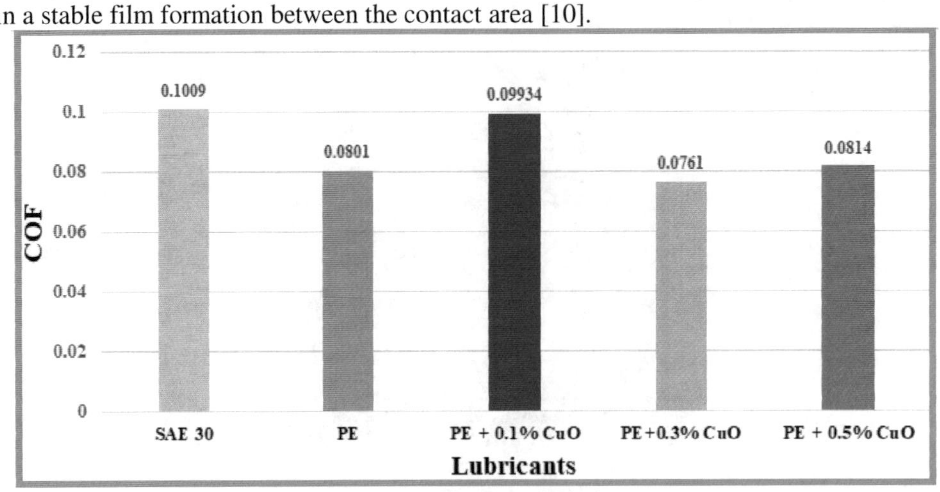

Fig. 5. Co-efficient of friction

3.3 Wear analysis

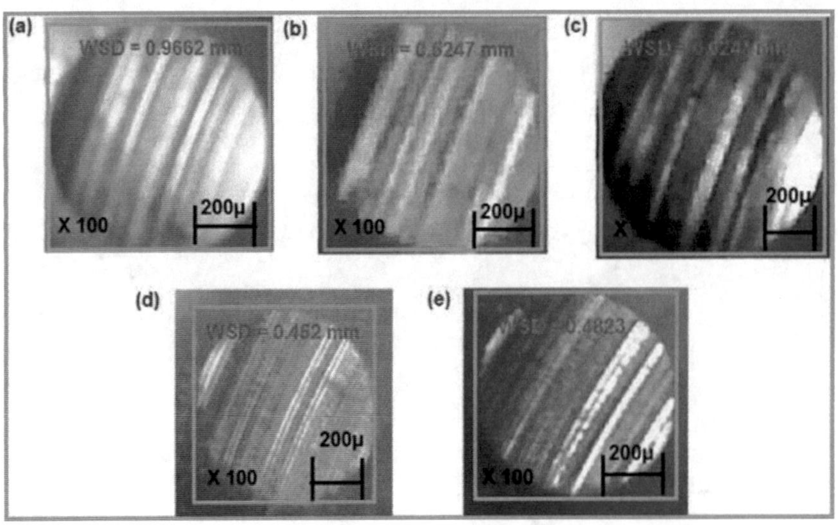

Fig. 6 Wear scar images of various lubricants (a) SAE30, (b) PE, (c) PE with 0.1 wt.% (d) PE with 0.3 wt.% and PE with 0.5 wt.% CuO nanoparticles

Fig. 6 and 7 shows the wear scar images and its morphology. From Fig. 6, it is noted that the wear scar diameters (WSD) are decreased drastically between 0.9662 to 0.452 with the addition of nano CuO with pentaerythritol ester. Among all the lubricants 0.3 wt.% nano CuO lubricant shows the minimal WSD is about 0.452 mm than that of other lubricants i.e. SAE30, PE, PE with 0.1 wt.% and PE with 0.3 wt.%. This is strongly believed that the oil film formed between the moving ball and stationary ball to protect from the friction and wear. This oil film was maintained till the end of experiment is ony due to its high viscosity of the PE with 0.3 wt.% CuO nano lubricant. Fig. 7 illustrates the morphology of the tested wear samples under various compressor oil samples i.e. SAE30 and PE with nano lubricants. PE with 0.3 wt.% nano lubricant shows the smoother surface as compared to the other lubricants i.e. base oil, pure pentaerythritol ester and other nano based lubricants. The other tested samples are showed as rough surfaces, pits, grooves, scuffing and deep pits [19].

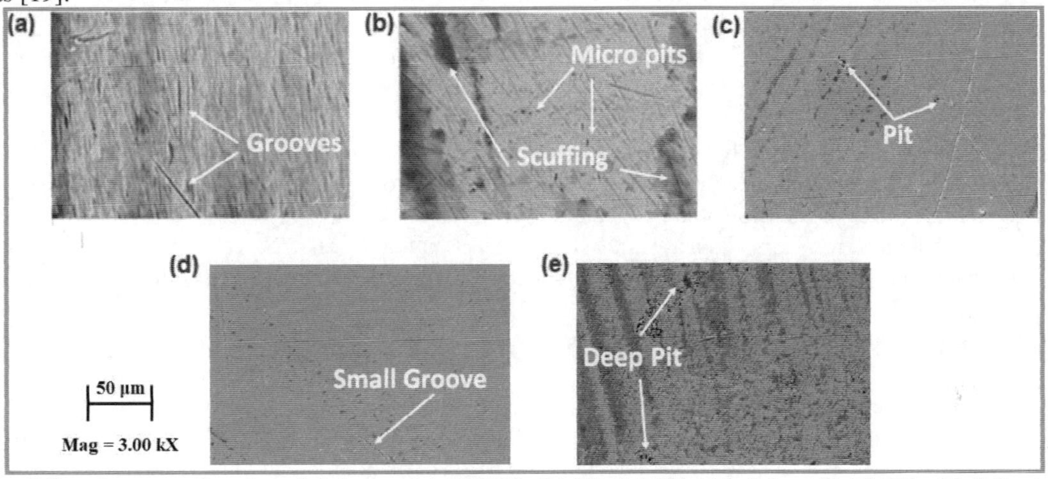

Fig. 7 Sem images of various lubricants (a) SAE30, (b) PE, (c) PE with 0.1 wt.% (d) PE with 0.3 wt.% and PE with 0.5 wt.% CuO nanoparticles

4 Conclusions

In this study, the tribological characteristics were carried out using four ball tribometer under various PE based nano lubricants. The following conclusions were made from the obtained results:

- The ultrasonic irradiation process is most effective and efficient method to produce high yield percentage of pentaerythritol ester than that of conventional method i.e. mechanical stirring.
- The improved physico-chemical properties (i.e. viscosity, viscosity index, flash point, pour point and total acid number) of the prepared nano-based lubricants are meets the ASTM standards.
- PE with 0.3 wt.% CuO nano lubricant shows the low frictional torque and least co-efficient of friction than other nano based lubricants and with base oil.
- Low wear scar diameter was observed with the PE with 0.3 wt.% CuO nano lubricant.
- SEM images of the tested specimens confirmed that the PE with 0.3 wt.% CuO nano lubricant specimen shows the flatter surface.

From the tribological study, it can be suggested that the CuO nano lubricant have great anti friction stability than that of other lubricants.

References

1. Mobarak, H. M., Niza Mohamad, E., Masjuki, H. H., Kalam, M. A., Al Mahmud, K. A. H., Habibullah, M., & Ashraful, A. M. (2014). The prospects of biolubricants as alternatives in automotive applications. Renewable and Sustainable Energy Reviews, 33, 34–43.
2. Reeves, C. J., Menezes, P. L., Jen, T.-C., & Lovell, M. R. (2015). The influence of fatty acids on tribological and thermal properties of natural oils as sustainable biolubricants. Tribology International, 90, 123–134.
3. McNutt, J., & He, Q. (Sophia). (2016). Development of biolubricants from vegetable oils via chemical modification. Journal of Industrial and Engineering Chemistry, 36, 1–12.
4. Heikal, E. K., Elmelawy, M. S., Khalil, S. A., & Elbasuny, N. M. (2017). Manufacturing of environment friendly biolubricants from vegetable oils. Egyptian Journal of Petroleum, 26(1), 53–59.
5. Mahamuni, N. N., & Adewuyi, Y. G. (2009). Optimization of the Synthesis of Biodiesel via Ultrasound-Enhanced Base-Catalyzed Transesterification of Soybean Oil Using a Multifrequency Ultrasonic Reactor. Energy & Fuels, 23(5), 2757–2766.
6. Encinar, J., Pardal, A., Sánchez, N., & Nogales, S. (2018). Biodiesel by Transesterification of Rapeseed Oil Using Ultrasound: A Kinetic Study of Base-Catalysed Reactions. Energies, 11(9)
7. Thanh, L. T., Okitsu, K., Sadanaga, Y., Takenaka, N., Maeda, Y., & Bandow, H. (2010). Ultrasound-assisted production of biodiesel fuel from vegetable oils in a small-scale circulation process. Bioresource Technology, 101(2), 639–645.
8. Choedkiatsakul, I., Ngaosuwan, K., & Assabumrungrat, S. (2013). Application of heterogeneous catalysts for transesterification of refined palm oil in ultrasound-assisted reactor. Fuel Processing Technology, 111, 22–28.
9. Choudhury, H. A., Malani, R. S., & Moholkar, V. S. (2013). Acid catalyzed biodiesel synthesis from Jatropha oil: Mechanistic aspects of ultrasonic intensification. Chemical Engineering Journal, 231, 262–272.
10. Arumugam, S., Chengareddy, P., & Sriram, G. (2018). Synthesis, characterisation and tribological investigation of vegetable oil-based pentaerythryl ester as biodegradable compressor oil. Industrial Crops and Products, 123, 617–628.
11. Arumugam, S., Chengareddy, P., Tamilarasan, a., and Santhanam, V. (2019). RSM and Crow Search Algorithm-Based Optimization of Ultrasonicated Transesterification Process Parameters on Synthesis of Polyol Ester-Based Biolubricant. Arabian Journal for Science and Engineering. 44, 5535–5548.
12. Baskar, S., Sriram, G., & Arumugam, S. (2016). Experimental analysis on tribological behavior of nano based bio-lubricants using four ball tribometer. Tribology in Industry, 37, 449–454.
13. Gulzar, M., Masjuki, H., Varman, M., Kalam, M., Mufti, R. A., Zulkifli, N., … Zahid, R. (2015). Improving the AW/EP ability of chemically modified palm oil by adding CuO and MoS2 nanoparticles. Tribology International, 88, 271–279.
14. Gulzar, M., Masjuki, H. H., Kalam, M. A., Varman, M., Zulkifli, N. W. M., Mufti, R. A., & Zahid, R. (2016). Tribological performance of nanoparticles as lubricating oil additives. Journal of Nanoparticle Research, 18(8).
15. Rajaganapathy, C., Vasudevan, D., & Murugapoopathi, S. (2020). Tribological and rheological properties of palm and brassica oil with inclusion of CuO and TiO2 additives. Materials Today: Proceedings. doi:10.1016/j.matpr.2020.05.032
16. Kashyap, A., & Harsha, A. (2016). Tribological studies on chemically modified rapeseed oil with CuO and CeO2 nanoparticles. Proceedings of the Institution of Mechanical Engineers, Part J: Journal of Engineering Tribology, 230(12), 1562–1571.

17. Chaurasia, S. K., Sehgal, A. K., & Singh, N. K. (2020). Improved Lubrication Mechanism of Chemically Modified Mahua (Madhuca indica) Oil with Addition of Copper Oxide Nanoparticles. Journal of Bio- and Tribo-Corrosion, 6(3).
18. Singh, Y., Sharma, A., Singh, N. K., & Chen, W.-H. (2020). Development of bio-based lubricant from modified desert date oil (balanites aegyptiaca) with copper nanoparticles addition and their tribological analysis. Fuel, 259, 116259.
19. Arumugam, S., & Sriram, G. (2014). Synthesis and characterization of rapeseed oil bio-lubricant dispersed with nano copper oxide: Its effect on wear and frictional behavior of piston ring–cylinder liner combination. Proceedings of the Institution of Mechanical Engineers, Part J: Journal of Engineering Tribology, 228(11), 1308–1318.
20. Aziz, N. A. M., Yunus, R., Rashid, U., & Zulkifli, N. W. M. (2016). Temperature effect on tribological properties of polyol ester-based environmentally adapted lubricant. Tribology International, 93, 43–49.
21. Zulkifli, N. W. M., Azman, S. S. N., Kalam, M. A., Masjuki, H. H., Yunus, R., & Gulzar, M. (2016). Lubricity of bio-based lubricant derived from different chemically modified fatty acid methyl ester. Tribology International, 93, 555–562

IOP Conf. Series: Materials Science and Engineering **954** (2020) 012040 doi:10.1088/1757-899X/954/1/012040

Electro-chemical corrosion Behavior of Vegetable oil-Based Engine oil

R. Ellappan[1*], S. Arumugam[1], V. Santhanam[1]

[1] Department of Mechanical Engineering, Sri Chandrasekharendra Saraswathi Viswa Mahavidyalaya, Kanchipuram-631561. Tamilnadu, India.

Correponding E-mail: [1,*]ellsraaj@gmail.com

Abstract. Progressive development of bio-lubricants is the need of the day for replacing petroleum-based lubricants in terms of dependency and degradation. Though the developed lubricants meeting out the properties of petroleum-based lubricants, the hygroscopic nature of the lubricants derived from vegetable oils are suspected to deteriorate the metal surfaces by corrosion. The vegetable oil-based engine oil was formulated via chemical alteration method. The mobility of ions is the key factor for electro-chemical corrosion. In this work we executed the corrosion study of engine oils on metallic coupons through electro-chemistry technique. therefore, vegetable oil-based engine oil and synthetic engine oil were tested in conductivity test using Galvanic cell. In this Galvanic cell, Tetra butyl ammonium perchlorate is used as a catalyst to enhance the mobility of ions in engine oils (electrolyte). The test coupons of ferrous and non-ferrous metal are considered as electrodes. The Tafel plot and impedance curve showed that the corrosion rate for the given current potential is not in the measurable range. Therefore, the electro-chemical corrosion study is not a suitable method to evaluate the corrosion rate of metals on viscous fluids like vegetable oil-based engine oil.

1. Introduction

Lubricants are mainly focused in the tribological aspects of contact surfaces. Subsequently, the development of vegetable oil-based bio-lubricants is the best alternate for synthetic lubricants in the aspects of degradation and sustainability. Wagnar et al.[1] identified the alternative methods to improve the biodegradability of high oleic sunflower oil by chemical modification and also suggested that vegetable oil lubricants could be produced in higher volume through this route. Erhan et al. [2] reported that the total production and consumption rate of non-renewable petroleum-based lubricants must be replaced by vegetable oil-based lubricants due to their better rheological properties and bio-degradability. Arumugam et al. [3] reported the chemical modification of vegetable oil and found that the biodegradability of base stock doesn't affect even after the chemical modifications.

Bekal et al. [4] recorded the mixture of synthetic engine oil and vegetable oil as crankcase lubricant in diesel engine fuelled with bio-diesel. They concluded that the blended engine offers better performance as compared with synthetic engine oil. Savita kaul [5] conducted the static immersion test for different metallic coupons in biodiesel environment and showed that the biodiesel induced the traces of corrosion on the metallic surfaces. Chew et al. [6] investigated the comparative corrosion study of magnesium and aluminium metals in biodiesel environment and SEM, XRD, FTIR results showed that the considerable amount of corrosion is identified in magnesium. Singh et al. [7, 8] observed the formation of new phase in the biodiesel immersed surfaces and concluded that the presence of moisture in biodiesel promoting oxidation of metals. Sergio Luiz de Assis et al. [9] studied the electrochemical corrosion behaviour of titanium alloys in hank's solution, and the polarization curve as well as EIS (Electro chemical Impedance Spectroscopy) test results showed the formation of passive film on the surfaces to act as barriers against corrosion. Jun Qu et al. [10] suggested that the addition of ionic liquids in vegetable oils showed no indication of corrosion on cast iron surface at room temperature.

IOP Conf. Series: Materials Science and Engineering **954** (2020) 012040 doi:10.1088/1757-899X/954/1/012040

Bommersbach et al. [11] identified that the amine group inhibitor forms thick film at the interface of steel surface and surrounded fluid. This thick film prevents the oxidation of metallic ions. Wang et al. [12] analysed biodiesel corrosion behavior in sea water environment through electrochemical studies and found that the ion concentration in biodiesel exposed area is more than sea water exposed area and hence, the biodiesel protect the metal surface against corrosion. Tian et al. [13] conducted the electrochemical corrosion study of steel in oil-water emulsion and observed that the anodic dissolution of steel is inhibited by oil, therefore the lower solubility of ionic spices found in oil. Zhang et al. [14] concluded that the presence of oil in the electrolyte solution decreases the anodic dissolution of metal due to the oil phase formation. Bommau et al. [15] used the different concentrations of natural oil as corrosion inhibitor on tin plates, observed the better inhibitor efficiency at moderate temperatures. Wang et al. [16] investigated the corrosiveness and stability of oil-brine mixed solutions through electrochemical studies and found that corrosion could be retarded by the addition of oil in brine solution.

2. Materials and Methods

2.1 Test coupon selection and preparation

To study the electrochemical corrosion behaviour of lubricants, the material selection is the important parameter to correlate with the real time component interfacing surfaces which are exposed with lubricants as electrolyte in this electrochemical corrosion study. A cast-iron test coupons of size 10x10x1 mm are polished in 500-1200 grit emery papers and polished surfaces are cleaned with distilled water and chemical reagents are separately used for the further cleaning of silicon carbide abrasive deposits over ferrous and non-ferrous coupons and were prepared as per ASTM G1 standard as shown in Figure 1(a). Teflon is surrounded over the test coupon to so as to expose minimum area of contact in the electrolyte solution, to avoid the over loading in the corrosion cell as shown in Figure 1(b).

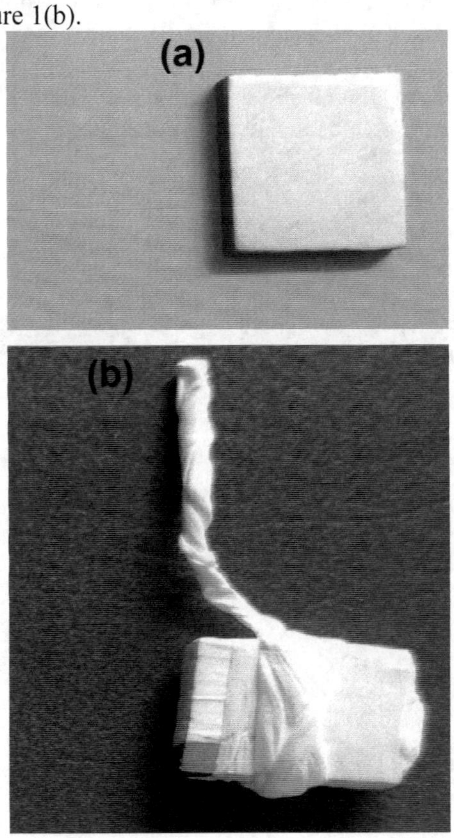

Figure 1. Metallic coupon in electrochemical corrosion study (a) before and (b) after

2.2 Conductivity analysis of lubricants

A 50 ml aliquot of the biolubricant was taken in a neutral glass beaker for the conductivity measurement. The lubricant sample was subjected to ultrasonication for ten minutes to ensure homogeneity and conductivity was measured with a high sensitivity conductivity meter with cell constant of $1 cm^{-1}$(Make:EQUIP-TRONICS; Model:EQ-667; India) and its corresponding setup is shown in Figure 2. The specific conductance of biolubricant was compared with a standard synthetic lubricant i.e., SAE20W40. For this, also the same test was repeated for synthetic lubricant.

Figure 2 Conductivity test set up

2.3 Experimental set up

The electro-chemical corrosion studies were done in a three-electrode corrosion cell, with different components of IC engine material as working electrodes, Saturated calomel electrode (SCE) as reference electrode and a platinum foil as counter electrode. Tetra butyl ammonium perchlorate, a catalyst was added to test lubricant to improve the ion mobilities. In this study, bio-lubricant and synthetic lubricants are used as electrolyte. The potentiodymanic polarization curves were recorded between +0.1 V to -0.1 V windows. Earlier to this, vegetable oil (Rapeseed oil) based biolubricant was synthesised through epoxidation, hydroxylation and esterification reaction by adopting the detailed procedure of Arumugam et al. [3]. The Figures 3(a) and 3(b) show the three-electrode galvanic cell and electro-chemical corrosion test setup.

Figure 3. (a) Three-electrode galvanic cell and (b) Electro-chemical corrosion test setup

3. Results and Discussion

3.1 Conductivity of lubricants

Table 1 shows the specific conductance of synthetic and biolubricant. The conductivity study shows that the biolubricant is having higher specific conductance than synthetic lubricant-SAE20W40. The specific conductance of lubricants is played a major role to accelerate corrosion in the electrochemical corrosion.

Table 1. Conductivity of lubricants

Parameter	Synthetic Lubricant	Biolubricant
Specific conductance (mS cm^{-1})	1.204	1.232

3.2 Electro-chemical corrosion analysis

A cast iron in synthetic lubricant and as well as biolubricant samples was studied via the polarization curves as depicted in Figure 4. As seen from Figure 4, there was a slight improvement in the anodic current density, and a synchronised negative shift of the corrosion potential of the cast iron material against biolubricant was detected. The ester molecules correspond to vegetable oil-based lubricant led to the establishment of a defensive corrosion scale, which resulted in diminished cast iron dissolution. The noticeable anodic dissolution increases and a negative shift in corrosion potential recommended a higher activity of the cast iron material under biolubricant atmosphere as compared with SAE20W40. The slight shift in the anodic current density of the cast iron sample, signifying a lesser formation of iron oxides over the electrode surface.

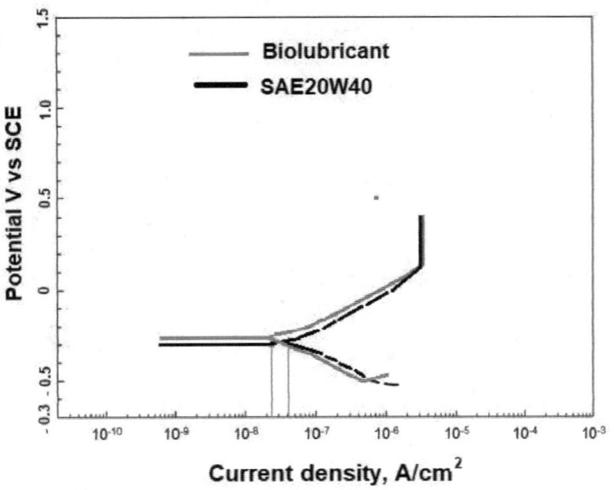

Figure 4. Potentio dynamic polarization curve

4. Conclusions

The corrosion behaviour of various engine components in presence of newly synthesized bio-lubricant was investigated electrochemically.

- The higher specific conductance value for biolubricant confirmed the presence of more ionic species when compared to synthetic counter-part.
- The potentio dynamic polarization studies were done to investigate the nature of corrosion.
- The results of these studies revealed that the corrosion of cast iron material vegetable oil-based biolubricant behaves almost an identical behavior with synthetic lubricant in terms of electro-chemical corrosion.
- This study further opens up an idea to execute tribo-corrosion analysis of vegetable oil-based lubricant in addition to electro-chemical corrosion analysis.

References

[1] Wagner, H., Luther, R. and Mang, T., 2001. Lubricant base fluids based on renewable raw materials: their catalytic manufacture and modification. Applied Catalysis A: General, 221(1-2), pp.429-442.

[2] Erhan, S.Z., Sharma, B.K., Liu, Z. and Adhvaryu, A., 2008. Lubricant base stock potential of chemically modified vegetable oils. Journal of Agricultural and Food Chemistry, 56(19), pp.8919-8925.

[3] Arumugam, S., Ellappan, R., Sangavi, S., Sriram, G. and Ramakrishna, P., 2018. Feasibility Analysis of Biodegradable Automotive Lubricant: An Evaluation of Material–Lubricant Compatibility in a Corrosion Perspective. Arabian Journal for Science and Engineering, 43(3), pp.1345-1368.

[4] Bekal, S. and Bhat, N.R., 2012. Bio-lubricant as an alternative to mineral oil for a CI engine—an experimental investigation with pongamia oil as a lubricant. Energy Sources, Part A: Recovery, Utilization, and Environmental Effects, 34(11), pp.1016-1026.

[5] Kaul, S., Saxena, R.C., Kumar, A., Negi, M.S., Bhatnagar, A.K., Goyal, H.B. and Gupta, A.K., 2007. Corrosion behavior of biodiesel from seed oils of Indian origin on diesel engine parts. Fuel processing technology, 88(3), pp.303-307.

[6] Chew, K.V., Haseeb, A.S.M.A., Masjuki, H.H., Fazal, M.A. and Gupta, M., 2013. Corrosion of magnesium and aluminum in palm biodiesel: A comparative evaluation. Energy, 57, pp.478-483.

[7] Singh, B., Korstad, J. and Sharma, Y.C., 2012. A critical review on corrosion of compression ignition (CI) engine parts by biodiesel and biodiesel blends and its inhibition. Renewable and Sustainable Energy Reviews, 16(5), pp.3401-3408.

[8] ASTM G:1.Standard Practice for Preparing, Cleaning, and Evaluation Corrosion Test Specimens

[9] ASTM G5 – 13ˊ2 Standard Reference Test Method for Making Potentiodynamic Anodic Polarization Measurements

[10] de Assis, S.L., Wolynec, S. and Costa, I., 2006. Corrosion characterization of titanium alloys by electrochemical techniques. Electrochimica Acta, 51(8-9), pp.1815-1819.

[11] Qu, J., Bansal, D.G., Yu, B., Howe, J.Y., Luo, H., Dai, S., Li, H., Blau, P.J., Bunting, B.G., Mordukhovich, G. and Smolenski, D.J., 2012. Antiwear performance and mechanism of an oil-miscible ionic liquid as a lubricant additive. ACS applied materials & interfaces, 4(2), pp.997-1002.

[12] Bommersbach, P., Alemany-Dumont, C., Millet, J.P. and Normand, B., 2005. Formation and behaviour study of an environment-friendly corrosion inhibitor by electrochemical methods. Electrochimica Acta, 51(6), pp.1076-1084

[13] Wang, W., Jenkins, P.E. and Ren, Z., 2012. Electrochemical corrosion of carbon steel exposed to biodiesel/simulated seawater mixture. Corrosion Science, 57, pp.215-219..

[14] Tian, B.R. and Cheng, Y.F., 2008. Electrochemical corrosion behavior of X-65 steel in the simulated oil sand slurry. I: Effects of hydrodynamic condition. Corrosion Science, 50(3), pp.773-779.

[15] Zhang, G.A. and Cheng, Y.F., 2009. Electrochemical corrosion of X65 pipe steel in oil/water emulsion. Corrosion Science, 51(4), pp.901-907.

[16] Bammou, L., Chebli, B., Salghi, R., Bazzi, L., Hammouti, B., Mihit, M. and Idrissi, H., 2010. Thermodynamic properties of Thymus satureioides essential oils as corrosion inhibitor of tinplate in 0.5 M HCl: chemical characterization and electrochemical study. Green Chemistry Letters and Reviews, 3(3), pp.173-178.

[17] Wang, Z.L., Zhang, J., Wang, Z.M., Zhou, L., Han, X., Li, Q.F. and Gao, S., 2014. Emulsification reducing the corrosion risk of mild steel in oil–brine mixtures. Corrosion science, 86, pp.310-317.

Hybrid WCMFO algorithm for the optimization of AWJ process parameters

A Tamilarasan[1*], T Rajmohan[1], KG Ashwinkumar[1], B Dinesh[1], M Praveenkumar[1], R.Dinesh Reddy[1], K V V Surya Kiran[1],Elangumaran R[1]and S Krishnamoorthi[1]

[1]Department of Mechanical Engineering, Sri Chandrasekharendra Saraswathi Viswa Mahavidyalaya, Enathur, Kanchipuram – 631561, Tamilnadu, India.

[1*]Corresponding Author: tamilrj2010@gmail.com

Abstract. Selecting optimal and proper cutting parameters for achieving desired results in the AWJ process is a demanding task. Therefore,the main contribution of this work is to apply a new reliable hybrid WCMFO algorithm for process optimization in the AWJ process. Based on the box-behnken design technique, 17 experimental runs were conducted and quadratic model for surface roughness was developed to fit with experimental data. Then, the WCMFO algorithm is implemented in the consideration of surface roughness as a fitness function. The key advantage of this algorithm is that it does not accumulate to any local optima, and the existence of a hybrid algorithm allows it to store the best solutions available so far. The predicted optimal settings were verified through confirmatory experiments, and the results validated.

1. Introduction

Composite materials are the most advanced engineering materials[1]. A composite is a heterogeneous material produced by the synthetic assembly of two or more components in order to obtain unique characteristics and properties[1]. Concern over preserving natural sources and recycling has led to renewed interest in biomaterials focused on sustainable raw materials. As a result , new forms of plant fibre composites have been produced in recent years. Natural-fibre reinforced composites offer a good mechanical performance and eco-friendliness. The use of composites based on natural fibres is growing rapidly[2]. Natural fibres, such as sisal, flax , banana, hemp, oil palm and jute, have range of techno-economic and ecological advantages over synthetic fibres (i.e. fibres of glass). The combination of fascinating mechanical and physical properties and their environmentally friendly character has given rise to interest in a number of industrial sectors, in particular the automotive industry[2]. However, there are some drawbacks in natural fibres composites that include inferior mechanical properties and high moisture absorption behaviour compared to their traditional synthetic counterparts. Mechanical properties also depend on processing stage, especially concerning the degree of fibre adhesion matrix, physical integrity and overall structural consistency[2].

Machining composite material causes significant problems due to the anisotropic and non-homogeneous nature of the material[2]. The process of machining composite material is distinct and complicated from that of metals. In traditional machining methods, such as turning, milling and drilling, the poor surface finish of the composite was poor and the removal of fibre, subsurface damage, delamination, and removal of bulk matrix are the major failures typically observed[2]. Technology advancement creates the use of non-conventional energy sources instead of the conventional process. Abrasive water-jet machining is such an non-traditional cutting process that uses a water-abrasive mixture as a high-speed water jet to extract material from the workpiece surface[3]. The high-speed water jet propels the abrasive mixing chamber to pass through a nozzle which enhances the jet's speed and leads it in a narrow zone to reach the work surface. The abrasive jet strikes at very high speed on the workpiece and extracts the substance through the erosive action of abrasives[3]. The brittle fracture is caused by the hammering action of the abrasives on the workpiece, and the residual wear fragments are separated from the machining region by the water jet[4]. One of the key benefits of the AWJM process over the AJM process is that no dust is created since a combination of water and abrasives is used to clean the material. The performance of an AWJM process depends on its various machining parameters, like stand-off distance, the diameter of the abrasive water-jet nozzle, abrasive concentration, nozzle pressure, abrasive grain size and feed rate of the nozzle, etc[4]. Kerf geometry is an important trait for abrasive water jet cutting. Kerf geometry is characterised by surface topography such as roughness and roughness along with kerf width and taper angle. It has wide access and, as the jet cuts through the material, its width decreases and the kerf is formed. Variation by millimetre of penetration as much as half of the kerf width can be described as kerf taper. Kerf tapers are formed as the jet loses its strength, piercing from the top to the bottom. In the present work, the AWJ cutting is proposed to assess the required input process parameters conditions for the improvement on the reduction of the surface roughness in cutting of natural fibre(pineapple) based composites. By conducting cutting experiments on the basis of the BBD design matrix. Further , a hybrid WCMFO algorithm is implemented to obtain the best AWJ optimal process parameters.

2. Experimental work

Pineapple Leaf Fiber (PALF) used as reinforcement fibre in most of matrices has shown its significant role in being inexpensive, having superior properties compared to other natural fibres, as well as encouraging agricultural economy[2]. PALF is described as multi-cellular and lignocelluloses materials, which is extracted from the leaves of the Ananas cosomus (belong to Bromeliaceae family) via retting process[2-3].The high impact polystyrene mats were cut into rectangular pieces as per the dimension of mould cavity. Pineapple based fibre-reinforced polyester composite was fabricated, using the compression moulding method. In the compression moulding method starts by placing the mixture containing matrix and reinforcing materials on the bottom half of the preheated cavity mould. This method also allows the processing of short fibre laminates to be carried out. The mould is then closed and the top half is lower and the pressure is increased to the preset level. With increasing pressure, the materials begin to melt and flow into the cavity. Then the mould is cooled down and the product is removed from the mould. The schematic view of the AWJ cutting zone as shown in the Fig. 1. In this w mainly on three controllable parameters such as feed rate(4-10 m/min), stand-off distance (1-3mm)and water jet pressure (150-300Mpa) and response variable (surface roughness), each factor has three levels designed by box-behnken design methodology and shown in Table 3. Surface quality is the intrinsic observation of every machined surface. In the same way, surface roughness is produced in the water jet machine due to the cutting mechanism involved. Mitutoyo make, surface roughness tester SJ 401 of measure collection/motion:80micron/0.001micron was used to calculate surface roughness values . The roughness significance was considered in three places at unlike locations chosen randomly and the standard value was directly measured for investigation.

Figure 1. Cutting Zone of AWJ process

Table 1. Experimental plan and collected responses

Exp.No	Process parameter			Response
	Feed rate	Stand-off Distance	Water jet pressure	Surface Roughness(SR)
	m/min	mm	Mpa	Micron
1	4	1	225	1.354
2	10	1	225	1.787
3	4	3	225	1.553
4	10	3	225	2.111
5	4	2	150	1.165
6	10	2	150	1.596
7	4	2	300	1.493
8	10	2	300	2.043
9	7	1	150	1.236
10	7	3	150	1.521
11	7	1	300	1.627
12	7	3	300	1.904
13	7	2	225	1.638
14	7	2	225	1.654
15	7	2	225	1.697
16	7	2	225	1.637
17	7	2	225	1.682

3. Regression model representation

The DOE results were statistically analysed using the variance analysis (ANOVA) to assess the influence of the three variables tested for the response. Polynomial equations have been generated to create a relationship between significant factors and surface roughness. The proposed model in uncoded unit values was realised by the effect and interaction of each variable as shown in the following equation.

SR(Micron)=-0.35011+0.031583×Feed Rate+0.062583×Stand of distance+0.010156×Water jet pressure+0.010417×Feed Ratex Stand off Distance+1.32222E-004×Feed Rate×Water jet pressure-1.88840E-005×Water jet pressure^2 (R^2=0.9939) (1)

Predicted response values have been determined by the equations described above. The positive sign of the coefficients in the regression equation indicated a synergistic effect, while the negative sign indicated an antagonistic effect on the response.It has been observed from Eq. (1) that the all the process parameter variables are positive effects on the Ra. Further, the significance and adequacy of the developed model have been determined using ANOVA with 95 % of CI.The value of the correlation coefficient (R^2) was 0.9538, which shows that 95.38% of total variations can be clarified by the model. In order to estimate the adequacy of the regression model, the diagnostic plots are given in Figures 2 and 3. The normal percentage probability plot is presented in Figure 6(b). The data points show that neither there is any apparent problem with normality nor the response transformation is required. The plot of the residuals versus the fits was used to examine the reliability of the model, as illustrated in Figure 6(c).As a result, no series of the increasing or decreasing points, of the patterns such as the increasing residuals with increasing fits and of a predominance of the positive or negative residuals should be found which is fortunately met in the work.

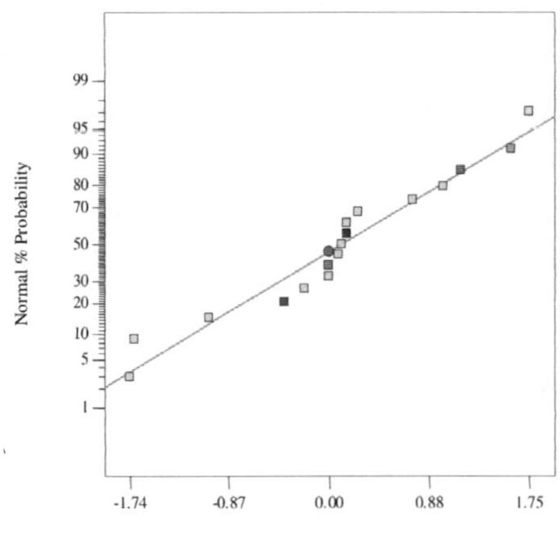

Figure 2. Normal % probability plot of the residuals

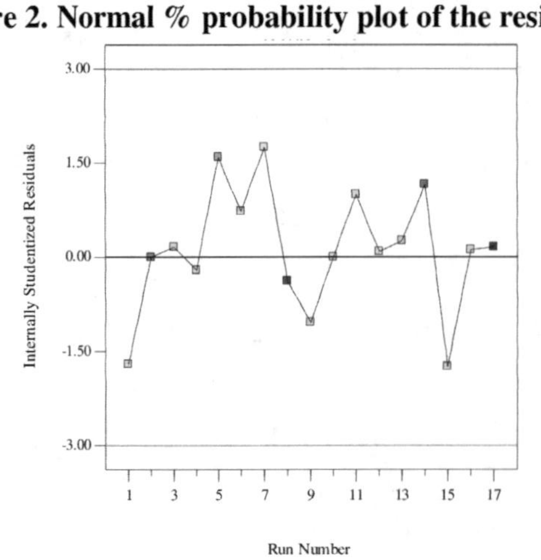

Figure 3. Plot of residuals versus run number

4. Hybrid WCMFO Algorithm and implementation

Hybrid WCMFO algorithm, which imitates the behavior of the nature flow of water cycle and moth navigation in the nature, has been introduced by the Khalilpourazari and Khalilpourazary in the year 2018[5]. Therefore, it constitutes a hybrid between the two techniques, comprising the water cycle and moth-flame optimization algorithms[6]. In order to avoid the trapping of the local Optima and the premature convergence of the population, the proposed algorithm incorporates the advantages of the water cycle algorithm(WCA) and the moth flame algorithm (MFA)[5]. The WCMFO that integrates the Levy operator with the MFO spiraling movement into the water cycle process, respectively, to increase the capacity of the WC for exploration and exploitation capability[6]. Ideally, WCA is able to explore the problem space with great capacity. The WCA maps streams and rivers towards the sea and allows search agents to change their role with regards to the best solution. But, the WCA tends to suffer from the lack of an efficient operator capable of performing exploitation. Instead, MFO makes very good use of its spiral movement capacity, but cannot effectively explore the solution area[6]. As every moth updates its location to its respective flame. The MFO therefore does not share information about the best solution obtained up to this point. This inspires the implementation of an efficient WCA and MFO hybridisation that can benefit from the advantaged effects of both algorithms[5]. The WCA is regarded as the fundamental algorithm in the proposed WCMFO algorithm[5-7]. In the WCA, the first step in changing the location of the streams and rivers is by using the spiral motions of the moths. The simple WCA update process only takes into account the space from the stream to the river when its location is changed. In short, in the space between the stream and the respective river is the next position of the stream[7]. Alternatively, the MFO algorithm updating procedure enables moths to change their location around their flames. The spiral system movement enables streams and rivers to improve their position, significantly improves the hybrid WCMFO's exploitability. The second change in the WCA is to improve the process of rainfall. In the metaheuristic algorithms, randomization plays an important role. There are two mechanisms are known in order to improve randomization in the WCMFO algorithm. As in fundamental WCA, the first is the raining cycle. During the cycle, WCMFO creates new solutions if the distance between a river or a stream and the sea is less than d_{max}. The second is for a random walk (i.e.Levy Flight) to allow stream to flow alone in the solution area[5] .Take into account the WCA iteration, if the rivers are updated and no better solution is found, then rivers and sea position should not be changed until next iteration. In hybridized algorithm, streams should change their location using the following equation in order to increase the randomness of the algorithm[5]

$$X_{i+1} = X_i + Levy(\dim) \otimes X_i \tag{2}$$

If x_{i+1} is the next position on the stream, x_i is the current position on the stream and dim is a decision variable. Then, the levy flight is determined with the following formula.

$$Levy(x) = \frac{0.01 \times \sigma \times r_1}{|r_2|^{\frac{1}{\beta}}} \tag{3}$$

where r_1 and r_2 between zero and one, and are generated at random. The parameter σ is computed as follows by the below equation

$$\sigma = \left(\frac{\Gamma(1+\beta) \times \sin\left(\frac{\pi\beta}{2}\right)}{\Gamma\left(\frac{1+\beta}{2}\right) \times \beta \times 2^{\left(\frac{\beta-1}{2}\right)}} \right)^{\frac{1}{\beta}} \tag{4}$$

IOP Conf. Series: Materials Science and Engineering **954** (2020) 012041 doi:10.1088/1757-899X/954/1/012041

The WCMFO's pseudo - code is presented as follows[5]:

set the parameters of WCMFO such as N_{pop}, N_{sr}, a, and maximum number of iterations
***for** i=1:N_{pop}*
Create a random stream
Calculate the objective function value of the stream
end for
sort the streams from best to worst based on their objective function value
Sea← the first stream
Rivers←n_{sr}-1
Stream←n_{pop}-n_{sr}
Determine the intensity of flow for rivers and sea
i=0;
***While** i < maximum number of iterations*
i=i+1;
***for** streams*
Update the position of stream using spiral movement
Stream_objective=objective function value of the new stream
***if** stream_objective < river_objective*
River_position= the new stream
***if** stream_objective < sea_objective*
Sea_position= the new stream
end if
end if
***if** river_objective < sea_objective*
Sea_position =River_position
end if
end for
***for** rivers*
Update the position of rivers using spiral movement
river_objective =objective function value of the new river
***if** river_objective < sea_objective*
Sea_position =River_position
end if
end for
***for** streams*
Update the position of the streams using Levy flight
end for
***for** Rivers and streams*
d=calculate the distance between each river or stream and the sea
***if** $d < d_{max}$*
raining process (for both rivers and streams)
end if
end for
Linearly decrease the parameter max *d*
Linearly decrease the parameter a
end while

The empirical equation (Eqn. 1) for surface roughness in the AWJ process formed using the RSM method was used as the objective or fitness function for optimizing using a hybrid WCMFO algorithm. A Matlab code has been compiled for executing the WCMFO. The corresponding code of WCMFO is subsequently run on MATLAB R2013a, 5 GB RAM and 2.30 GHz processor operating platform. The main goal of the process optimization on AWJ is to decide on the process parameters, namely the feed rate, stand-off distance and water jet pressure, in order to optimize the objective function which is surface roughness(SR). The problem was formulated as an unconstrained optimization problem where the objective was to minimize the surface roughness.

i.e. Minimize surface roughness, with the limits,

$$4 \text{ m/min} \leq \text{feed rate} \leq 10 \text{ m/min}$$
$$1\text{mm} \leq \text{stand-off distance} \leq 3\text{mm}$$
$$150 \text{ Mpa} \leq \text{water jet pressure} \leq 300\text{Mpa}$$

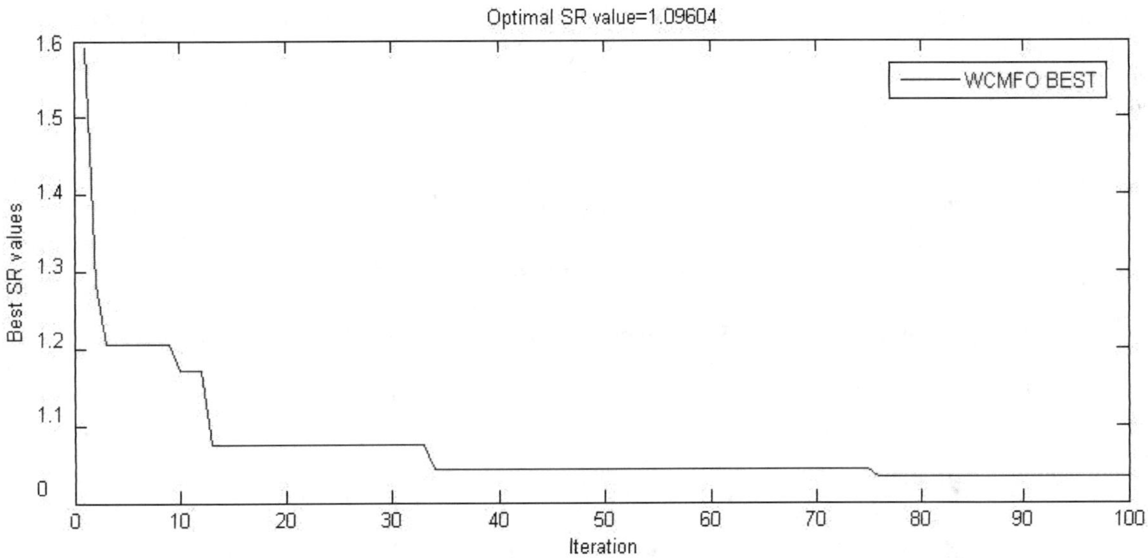

Figure 4. Performance of the WCMFO algorithm across 100 generations

The optimization process configuration is shown in Figure 4 illustrates the resulting values of the optimal SR values for the 100 iterations. There is a significant decrease in the best SR values until the global best solution was found in 77th iteration. The optimal values of the AWJ optimal conditions along with the respective surface roughness(SR) values are listed in Table 2.

Table 2. Comparison between the initial and optimal conditions achieved by CS algorithm

Model Summary		Initial condition	Optimal condition	% of improvement
Process parameters	Feed rate(m/min)	4	5.245	
	Stand-off Distance (mm)	1	1.454	
	Water jet pressure(Mpa)	225	151.547	
SR value		1.354	1.09604	19.05

Table 3. Results of confirmation experiment

Feed rate (m/min)	Stand-off Distance (mm)	Water jet pressure (Mpa)	Predicted	Confirmation Experiment	Relative Error(%)
5.245	1.454	151.547	1.09604	1.098	0.1788

As seen from the Table 2, the percentage of improvement of surface roughness value to corresponding optimal condition is achieved by 19.05% . Further, the confirmation experiments were carried out to verify the corresponding surface roughness predicted by WCMFO algorithm. Table.5 shows that the percentage of error between the predicted and experimented values. From this analysis, it is observed that the calculated error(i.e. 0.1788) is very small and it proves the predictability and effectiveness of the proposed WCMFO algorithm.

5. Conclusion

The current work proposes an effective process parameter optimization approach that integrates RSM and a hybrid WCMFO algorithm on engineering optimization concepts. The proposed approach can effectively assist engineers in determining the optimal process parameter settings for AWJ under consideration of surface roughness. The following conclusions can be drawn.

1. Effective application of the Box – Behnken approach used for the experimentation.
2. The chosen response was modelled using the response surface methodology in terms of the AWJ process parameters.
3. A novel hybrid WCMFO algorithm is implemented to achieve the best optimal surface roughness value and the algorithm output showed that for achieving minimal surface roughness 1.09604 Microns.
4. The results of confirmation experiment performed and agree well with the predicted optimum values by WCMFO with minimum relative errors.
5. The results obtained would be useful and serve as a technical database for AWJ cutting industries for achieving better quality .

References

[1] Rajmohan T, and Palanikumar K 2013 Application of the central composite design in optimization of machining parameters in drilling hybrid metal matrix composites *Measurement* 46(4)1470-1481.
[2] Vinayagamoorthy R and Rajmohan T 2018 Machining and its challenges on bio-fibre reinforced plastics: A critical review *Journal of Reinforced Plastics and Composites* 37(16) 1037-1050.
[3] Vigneshwaran S, Uthayakumar M and V Arumugaprabu 2017 Abrasive water jet machining of fiber reinforced composite materials *Journal of Reinforced Plastics and Composites* 37(4) 1-8.

[4] Tamilarasan A, Arumugam S, Rajamani D, Changareddy P, Balasubramanian E and Pranay P Multi-response Optimization of AWJ Process Parameters in Cut Quality Characteristics of Hastelloy C-276 Lecture Notes in Mechanical Engineering7 381-388.

[5] Khalilpourazari S, Khalilpourazary S 2019 An efficient hybrid algorithm based on Water Cycle and Moth-Flame Optimization algorithms for solving numerical and constrained engineering optimization problems. Soft Computing 23 1699-1722.

[6] Renugambal A and Selva Bhuvaneswari K 2020 Image Segmentation of Brain MR Images Using Otsu's Based Hybrid WCMFO Algorithm *Computers, Materials & Continua* 64(2) 681-700.

[7] Navard MHGN and Lesani H 2019 Optimal Allocation of Electric Parking and Distributed Generation in distribution system Based on Hybrid Water Cycle-Moth Flame Optimizer Algorithm *Signal processing and Renewable Energy* 3(1) 67-84.

IOP Conf. Series: Materials Science and Engineering **954** (2020) 012042 doi:10.1088/1757-899X/954/1/012042

Effect of engine oil dilution by waste ayurvedic oil biodiesel on tribological behavior of liner-ring tribo pair material

R. Balakumar [a,*], **G. Sriram** [b], **S. Arumugam** [c]
[a, b, c] Department of Mechanical Engineering,
Sri Chandrasekharendra Saraswathi Viswa Mahavidyalaya,
Kanchipuram-631561. Tamilnadu, India
Corresponding author : [a,] *balavhd@gmail.com, [b]drg.sriram@gmail.com,
[c]aru_amace@yahoo.co.in

Abstract. The fastest exhaustion of petroleum-based fossil fuels and its ever-ending environmental issues which scares the researchers to search for the alternate source for commercial diesel fuel. In this way, the biodiesel as an alternate fuel plays an important role in a diesel engine. There are many vegetables were used as a biodiesel feedstock, the utilization of waste feedstock as a biodiesel resource will be the prime importance in the reduction of waste management followed by its dumping processes in the earth surface. The present experimental investigation is used to examine the utilization of waste ayurvedic oil (WAO) biodiesel blends diluted in the synthetic oil and its tribological study using pin on disc tribometer under standard test conditions. The ultrasound irradiation assisted transesterification process method is employed in the conversion of waste ayurvedic oil in to WAO methyl ester (WAOME). Then the WAOME was added with diesel fuel to form the biodiesel blends and named as B10, B15 and B20. In addition to this, the blends were contaminated with synthetic lubrication oil (SAE20W40) to study the dilution effects of the biodiesel. The real-time cylinder liner-ring tribopair materials were used for the investigation. The experimental results reveal that, the tribological behavior i.e., lower co-efficient of friction and wear was observed with WAO20 biodiesel blend than other biodiesel blends and diesel fuel.

Keywords: waste ayurvedic oil biodiesel; ultrasonication; fuel from waste; pin on disc.

1. Introduction

At present, the cost of biodiesel is high as compared to diesel fuel as biodiesel is produced from refined vegetable oils. The best method to cut down the cost of biodiesel is the use waste feedstock such as waste cooking oil, waste fish oil, waste transformer oil, waste plastic oil, waste mango seed oil and wastes from animal fat. Many researchers reported the use of biodiesel as a diesel engine fuel from such waste feed stocks [1-10]. Oil feed stock is the prime cost involved in the synthesis of biodiesel that accounts for more than 60% of the overall biodiesel processing cost. If any waste source is used as biodiesel feedstock, the economics of biodiesel could be greatly promoted. Additionally, the use of biodiesel from waste feedstocks also cut down the cost of waste treatment which is a prime importance of waste management system.

Knothe & Razon [11] found that, the biodiesels were derived from many sources which comprise of fatty acid methyl ester or fatty acid ethyl ester. The biodiesels have high flash point, biodegradability, low sulfur content, good burning efficiency, minimized emission characteristics and it has 10 wt % oxygen content. This concluded that, the biodiesel was partial substitute for fossil fuels. Pelitli et al. [12] investigated the waste oils from the various vehicle engine crankcases and waste gearboxes removed while the vehicles maintenance and analyze the oil managements. This study leads the fitness for recycling the oil, amount of quantity can be recovered or disposal of the oils used. The investigations based on the various analysis of cadmium, lead, total halogens, flash points, chlorine, arsenic, chromium etc. From the various metals analyzed from the various engine crankcase oils the lead plays the higher quantity and arsenic, cadmium and chromium play lesser than lead. The more amount of removed metals was noticed with the vehicles around 15 to 35 years old. From the investigation findings the engine crankcase oils and gearbox oils can be used for recycling [13].

Out of various feedstocks used by many of the researchers, none of the investigators walked through in the studies of utilizing the Waste Ayurvedic Oil. The various authors investigated the various biodiesel production methods on different food stock and found that influencing parameter as time and temperature are 120min and 50°-60°C. As well as the molar ratio between the alcohol and oil is 6-12:1 with the suitable catalyst as 1% by weight [14,15]. In the transesterification and pyrolysis-based biodiesel process, more amount of power is needed and larger amount of reaction time is wasted. To overcome the difficulties the ultrasonic based transesterification process method is used for the conversion of biodiesel.

In IC engines, the various engine components produce friction between the metal-metal contact, which in-turn reduces the engine life and hence lubricity behavior of fuel is one of the most predominant factors in extending engine reliability [16]. Masarof et al. [17] investigated the friction and wear characteristics of palm and calophyllum inophyllum biodiesel blends in addition to performance and emission behavior. They confirmed that 20% of palm oil biodiesel blend displayed the favorable engine performance, lower emissions together with better lubricating performance. Shanta et al. [18] studied the tribological effects of various biodiesels from peanut oil, soybean oil, canola oil and chicken fat in a pin on disk tribometry test setup. The test was conducted on steel disk with stainless steel ball material. The investigation result analysis clears that, the frictional force and specific wear rate were minimum with the mixture of animal fat with lubricating oil as a feed stock than the other mixtures of lubricating oil with biodiesel blends. The main influencing parameter of low distillation temperature of biodiesel improves the lubricating oil dilution. Arumugam and Sriram [19] analyzed the engine cylinder liner and piston ring tribo pair material in a pin on disc tribometer. The various lubricating oils were rapeseed oil-based lubricant, rapeseed oil biodiesel based lubricating oil, commercially available lubricating oil and diesel-based lubricant. From the various studies like coefficient of friction, frictional force and wear depth it is concluded that, the rapeseed oil-based biodiesel contaminated with the commercially available synthetic lubricant shown a better performance and better impact in the life of the engine than all other lubricants used.

From the above literatures, it is proven that rare attempts were made to efficiently exploit waste ayurvedic oil as diesel engine fuel, in spite of it being as waste and sparingly feasible resource. Many literatures reported the usefulness of waste cooking oil, waste fish oil, waste transformer oil, waste plastic oil and wastes from animal fat as biodiesel. Apart from the various waste feed stocks to be used as alternative engine fuels which are listed above, one such important waste feedstock for the synthesis of biodiesel is waste ayurvedic oil.

2. Materials and Methods

2.1 Synthesis of waste ayurvedic oil biodiesel using ultrasonic irradiation-based transesterification method

In this study, biodiesel is formulated from WAO via ultrasonic irradiation assisted transesterification process [20, 21]. For this, WAO was collected from in and around ayurvedic hospitals located in southern region of Tamilnadu, India and it was filtered using filtering cloth in order to remove sediments and other foreign bodies present in it. The other chemical substances such as methanol, sodium hydroxide pellets, hydrochloric acid and anhydrous sodium sulphate was purchased from M/S Maruthi traders, Chennai.

The ultrasonic irradiation-based transesterification process setup was used to convert the WAO in to WAO methyl ester. The 1000ml triple neck glass flask was used for this process among three neck the centre neck is used for ultrasonic probe and remaining two is used for thermocouple and addition of rection mixtures. 400g filtered WAO, 80g methanol and 4g of NaOH pellets were added in the flask. The ultrasonic irradiation was applied for this process. After the process, the mixture was left to 12 h to settle down then, the top layer was separated using separating funnel and allowed to bubble wash to attain purified WAO methyl ester (WAOME). At final, the WAOME was blended with diesel fuel in the volume basis to prepare the blends such as,

WAO10: 10% of WAOME+90% of diesel.
WAO15: 15% of WAOME+85% of diesel.
WAO20: 20% of WAOME+80% of diesel.

To prepare the homogeneous mixing the blends were mixed thoroughly using mechanical stirrer at 60°C with 500rpm for 15min. Table 1 shows the properties of WAO biodiesel blends and diesel fuel. Finally, the WAO biodiesel blends and the diesel fuel were contaminated by 10% with synthetic lubrication oil (SAE 20W40) and the mixture was mixed thoroughly with mechanical stirrer in the glass flask.

Table 1 Properties of WAO biodiesel blends and diesel fuel

Properties	WAO10	WAO15	WAO20	Diesel
Density, kg/m^3	837.7	842.1	854.1	830.2
Kinematic Viscosity @40°C, cSt	2.69	3.02	3.41	2.71
Flash point, °C	61	65	71	56
Fire point, °C	56	60.5	64	48.5

2.2 Tribological test configuration

To study the tribological characteristics of WAO biodiesel such as coefficient of friction and wear, Wear and friction monitor (Make: Ducom, India: Model: TR20LE-400PHM) was used in accordance with ASTM G99. The tribometer was connected to a personal computer with data accusation system. The pins used for the tribotest were made out of engine's piston ring material i.e., AISI-304 and the disc were made out of engine's cylinder liner material AISI-316 steel with a hardness of 60 HRC. The specifications of the pin and disc are as follows:

Length of the pin : 30mm
Diameter of the pin : 10mm
Thickness of the disc : 8mm
Diameter of the disc : 165mm

The details of the test matrix are as follows:

Load : 80N
Temperature : ambient
Disc speed : 300rpm
Sliding velocity : 0.7835m/s
Duration : 1 hour
Sliding distance : 2850m
Fuels used : WAO10, WAO15, WAO20 and diesel.

In general, the lubricity behavior of any fuel is evaluated using fuel dilution method [22-24] for which commercial engine oil i.e. SAE20W40 was diluted with 10% of WAO biodiesel such as WAO10, WAO15, WAO20 and also with commercial diesel and accordingly four different lubricant samples were prepared. The wear and coefficient of friction (COF) was measured using linear variable differential transformer and electronic sensors respectively.

3. Results and discussion

When biodiesel is used as fuel for diesel engine, it may lead to some degree of fuel dilution problem with engine oil [19]. The measurement of tribological behavior of WAO biodiesel is much more important in order to study the fuel lubricity and anti-wear properties. The variation of co-efficient of friction (COF) and wear between pin-disc tribo pair under the influence of various ayurvedic oil biodiesel fuels and diesel fuel are shown in Fig. 1. (a) & (b).

From Fig. 1(a) it is depicted that all the three blends of WAO biodiesel showed the reduced COF as compared to diesel fuel. Under similar operating conditions, WAO20 had the lowest COF followed by WAO15 and WAO10. Out of the four-test fuel, diesel fuel is showed highest COF. The lowest COF is mainly attributed by the presence of polar groups and long-chain fatty acids in WAO biodiesel. Further, the existence of oxygen moieties and double bonds present in WAO20 biodiesel is responsible for the overall improvement of lubricity behavior of WAO biodiesel. The Fig. 1(b) shows the wear of pin material under the influence of various WAO biodiesel and as well as diesel. The wear of pin was increased with increase in experimental time. The variation in pin wear was only considered in this tribological study due to the hardness of the pin material was much lower than disc material.

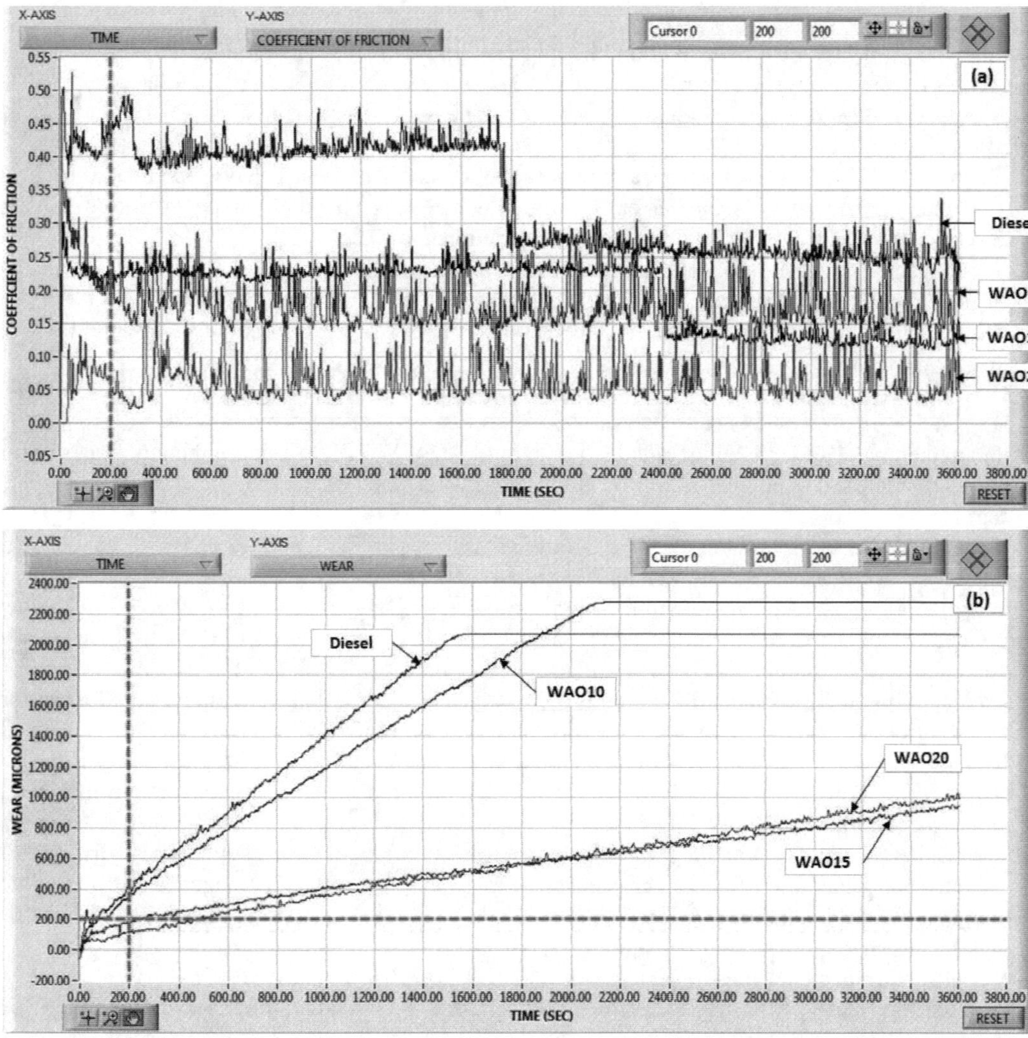

Fig. 1. Tribological behavior as a function of time (a) co-efficient of friction (b) wear

The wear of pin was increased with increase in experimental time. The variation in pin wear was only considered in this tribological study due to the hardness of the pin material was much lower than disc material. It is depicted that the wear was dramatically reduced for WAO20 it was nearly about 56% lower than diesel fuel. Similarly, the reduction in wear was about 9% and 58% respectively for WAO15 and WAO10 as compared to diesel fuel. This wear reduction trend confirms the anti-wear behavior of WAO20 biodiesel. The adsorption of polar groups that is present in the WAO20 biodiesel on the pin-disc tribo pair surfaces during the rubbing, providing them good anti-wear properties. Typically, the interactions between ZDDP (Zinc dialkyl dithiophosphate) additive molecules present in the SAE20W40 and WAO20 biodiesel which are responsible for the formation of wear protective layer are much stronger than that of WAO10 and WAO15. The similar studies by Maleque et al. [25], reported the better anti-wear behavior of palm oil-based biodiesel is due to the formation of hydrocarbon protective layer by the presence of long-chain fatty acid in biodiesel.

4. Conclusions

In the current tribological investigation, the lubricity behavior of WAO biodiesel was instrumented using pin on disc tribometer. The outcome of the experimental study is summarized as follows:

- It is achievable to originate the biodiesel from one such waste source is waste ayurvedic oil (WAO).
- The use of WAO biodiesel improved the anti-friction and anti-wear behavior compared to diesel fuel and in particular WAO20 was outperformed which suggested that higher fraction of WAO in biodiesel may play a vital role on reducing the friction and wear.
- As compared to diesel, the wear was reduced significantly of about 56% for WAO20 due to the adsorption of polar groups present in WAO20 biodiesel.
- Under similar tribological conditions, WAO20 had the lowest COF of about 0.04 while the COF for the diesel is 0.13.
- The reduced COF with WAO based biodiesel might be due to the existence of oxygen moieties and double bonds which is not a part of diesel fuel.
- From the above investigation results it is observed that, WAO20 performs better wear scuffing protection to the engine liner-ring tribo pair material and best alternate for diesel fuel.

References

[1] D. Capuano, M. Costa, S. Di Fraia, N. Massarotti, L. Vanoli, (2017) Direct use of Waste vegetable oil in internal combustion engines, *Renew. Sustain. Energy Rev.* 69 759-770.

[2] Thi Tuong Vi Tran, Sunanta Kaiprommarat, Suwadee Kongparakul, Prasert Reubroycharoen, Guoqing Guan, Manh Huan Nguyen, Chanatip Samart, (2016) Green biodiesel production from waste cooking oil using an environmentally benign acid catalyst, *Waste Manag.* 52 367-374.

[3] Hao Yang, Xinghu Li, Mingfei Mu, Guiyue Kou & Lijun Li, (2017) Performance and emissions analysis of a diesel engine directly fueled with waste cooking oil biodiesel, *Int. j. of Ambient Energy*, 38 428-434.

[4] R. Yahyaee, B. Ghobadian, G. Najafi, (2013) Waste fish oil biodiesel as a source of renewable fuel in Iran, *Renew. Sustain. Energy Rev.* 17 312–319.

[5] Balakumar, R.; Sriram, G.; Arumugam, S. (2019) Feasibility Study of Biodiesel Synthesis from Waste Ayurvedic Oil: Its Evaluation of Engine Performance, Emission and Combustion Characteristics. *Arabian Journal for Science and Engineering* 44:8067–8079

[6] Ayatallah Gharehghani, Mostafa Mirsalim, Reza Hosseini, (2017) Effects of waste fish oil biodiesel on diesel engine combustion characteristics and emission, *Renew. Energy* 101 930-936.

[7] S. Prasanna Raj Yadav, C.G. Saravanan, R. Vallinayagam, S. Vedharaj, William L. Roberts, (2015) Fuel and engine characterization study of catalytically cracked waste transformer oil, *Energy Convers. Manag.* 96 490–498.

[8] Md Nurun Nabi, Md Shamim Akhter, Md Atiqur Rahman, (2013) Waste transformer oil as an alternative fuel for diesel engine, *Procedia Eng.* 56 401 – 406.

[9] M. Mani, G. Nagarajan, S. Sampath, (2010) An experimental investigation on a DI diesel engine using waste plastic oil with exhaust gas recirculation, *Fuel* 89 1826–1832.
[10] Ivana B. Bankovic-Ilic, Ivan J. Stojkovic, Olivera S. Stamenkovic, Vlada B. Veljkovic, Yung-Tse Hung, (2014) Waste animal fats as feedstocks for biodiesel production, *Renew. Sustain. Energy Rev.* 32 238–254.
[11]. Knothe, G & Razon, LF (2017), 'Biodiesel fuels', *Progress in Energy and Combustion Science*, vol. 58, pp. 36-59.
[12] V. Pelitli, O. Dogan, H. J. Koroglu, (2017) Waste oil management: Analyses of waste oils from vehicle crankcases and gearboxes, *Global J. Environ. Sci. Manage.*, 3(1): 11-20.
[13] Abd Allah S. A. El-Gharabawy, (2017) Cost Analysis for Biodiesel Production from Waste Cooking Oil Plant in Egypt, *International Journal of Smart Grid, ijSmartGrid A.A.S.A. El- Gharabawy*, Vol.1, No.1,.
[14] Puneet Verma , M.P.Sharma , (2016) Review of process parameters for biodiesel production from different feed stocks", *Renewable and Sustainable Energy Reviews*. 62 1063–1071.
[15] Umer Rashid and Farooq Anwar, (2008) Production of Biodiesel through Base-Catalyzed Transesterification of Safflower oil using an optimized protocol, *Energy & Fuels*, 22 1306–1312
[16] Fazal M, Haseeb A, Masjuki H, (2011) Biodiesel feasibility study: an evaluation of material compatibility; performance; emission and engine durability, *Renew. Sustain. Energy Rev.* 15 1314–1324.
[17] M.H. Mosarof, M.A. Kalam, H.H. Masjuki, Abdullah Alabdulkarem, A.M. Ashraful, A. Arslan, H.K. Rashedul, I.M. Monirul, (2016) Optimization of performance, emission, friction and wear characteristics of palm and Calophyllum inophyllum biodiesel blends, *Energy Convers. Manag.* 118 119–134.
[18] Shanta, S. M., G.J.Molina and V. Soloiu (2011). Tribological effects of mineral-oil lubricant contamination with biofuels: a pin-on-disk tribometry and wear study. *Advances in Tribology*, Vol. 2011, Article ID 820795, 7 pages.
[19] Arumugam, S., and G. Sriram (2012). Effect of bio-lubricant and biodiesel contaminated lubricant on tribological behavior of cylinder liner–piston ring combination. *Tribology Transactions*, Vol. 55, No. 4, pp. 438-445.
[20] Leonardo S.G.Teixeira, Julio C.R.Assis, Daniel R.Mendonca, Iran T.V.Santos, , Luiz A.M.Pontes, Josanaide S.R.Teixeira, (2009) Comparison between conventional and ultrasonic preparation of beef tallow biodiesel. *Fuel Process. Technol.* 90 1164–1166.
[21] Huong LTT, Tan PM, Hoa TTV, (2011) Biodiesel production from fat of tra catfish via heterogeneous basic-catalyzed transesterification using ultrasonic mixing. *E-J Surf. Sci. Nanotechnol.* 9 477–481.
[22] Balakumar, R.; Sriram, G.; Arumugam, S. (2018) Effect of lubricant contaminated with waste ayurvedic oil biodiesel on tribological behavior of cylinder liner-piston ring tribo pair material. *Mater. Today Proc.* 5, 13220–13226
[23] Balakumar, R.; Sriram, G.; Arumugam, S. (2018) Assessment on tribological characteristics of waste ayurvedic oil biodiesel blends using high-frequency reciprocating rig tribometer. *IOP Conf. Ser. Mater. Sci. Eng.* 390, 012047
[24] Balakumar, R.; Sriram, G.; Arumugam, S.; Shree Navin, A. (2018) Analysis on tribological characteristics of waste ayurvedic oil biodiesel blends using four ball tribometer. *IOP Conf. Ser. Mater. Sci. Eng.* 390, 012080
[25] Maleque, M. A., Masjuki, H. H., and Ishak, M, (1998) Bio-Fuel Contaminated Lubricant and Hardening Effects on the Friction and Wear of AISI 1045 Steel, *Tribol. Trans.* 41(1) 155– 159.

IOP Conf. Series: Materials Science and Engineering **954** (2020) 012043 doi:10.1088/1757-899X/954/1/012043

Heavy metals Assessment of Temple Ponds in Kanchipuram, Tamil Nadu, India

P Meenakshi[1], G Sriram[2]

[1]Assistant Professor, Department of Civil & Structural Engineering, Sri Chandrasekharendra Saraswathi Viswa Mahavidyalaya, Enathur, Kanchipuram - 631 561, Tamilnadu, India

[2]Professor, Department of Mechanical Engineering, Sri Chandrasekharendra Saraswathi Viswa Mahavidyalaya, Enathur, Kanchipuram - 631 561, Tamilnadu, India

Email: [1]meenathavamani82@gmail.com an [2]drg.sriram@gmail.com

Abstract. Heavy metals are a major environmental pollutant, they are slow, non-degradable material it affects the aquatic ecosystem. Heavy metal pollutant molecules arrive the pond water through anthropogenic activities, domestic waste, hospital wastes and temple wastes. The investigation was designed to study the seasonal spreading of specific heavy metals Iron (Fe), Copper (Cu), Lead (Pb), Arsenic (As) and Zinc (Zn)) in pond water. Seventeen pond water samples were collected during pre- monsoon (March 2017) and post monsoon (January 2018) period. The concentration of Iron in the pre monsoon greater than class A limit. The concentration of other prescribed heavy metal was less than central pollution control board standards in both the season. Hence pond water having a negligible amount of heavy metal pollution. This environmental observing and valuation research useful for managing and planning for the protection of pond.

Keywords: *Heavy metals; Temple ponds; pre monsoon.*

1. Introduction

The world is in the greatest problem due to lack of quality and amount of water. Water pollution is one of the universal problems, touching both the industrialized and the emerging nations [1]. The water pollution problems are different in many aspects. Toxic metals, acids, sediment, animal and human wastes, and synthetic organic compounds deposited in the waterway that causes the water pollution of developing nations [2]. Pollution changes the water chemistry which affects the water quality and upset the ecological balance in the aquatic ecosystem and disturbs the productivity. Most pollutants that move in the aquatic environment are chemical in nature and range from completely toxic substance such as agricultural pesticides, cyanide and salt of various heavy metals to nutrient such as phosphate, nitrate, fertilizer and organic matter of domestic and industrial origin [3]. The major role for water pollution is the presence of heavy metal. There are two main causes of heavy metals into the environment that are natural process and anthropogenic. Recent days urban and industrial actions are growing heavy metals into the aquatic ecosystem, when they exceed standard concentration, they have toxic effects on living organisms also they decrease survival, growth and species diversity[4,5].In certain condition heavy metal is high concentration they may have a negative impression on the receiving environment. Ponds are very significant water bodies in the country [6]. The various resources of the pollutions of the pond are washing the clothes, animals, dumping of the waste by unauthorized small scale units functioning in the surrounding of the pond [7,8]. These resources, create a passage for mixing heavy metal into the ponds [9]. The increased pollution level decreases the quality and quantity of water. Kanchipuram is holiest city having more temples and festivals. Many festivals conducted in ponds thepam, thirthavari, nadavavi etc. Ponds also act as a reservoir for storing rainwater, it will increase the water table level for surrounding area. At present it is the main target to measure the water pollution level and its impression on living organisms in the pond water. The present research work is the assessment of heavy metal components during pre-monsoon and post-monsoon period (March 2017 and January 2018) in temple ponds in Kanchipuram.

2. Study Area

This study is focused in the temple ponds present in the Kanchipuram. It is one of the eminent cities of temples in South India. Kanchipuram is a town in the northern portion of Tamil Nadu state. It is a traditional and visitor spot, with the population more than 170,000 people. The geographic location lies between 12°46'30'' - 12°52'00'' North Latitude and 79°39'00'' - 79°46'20'' East longitude. The sampling points in Kanchipuram cover residential, industrial, spiritual, low and high population areas. The study area shown in Figure 1.

Figure 1. Study Area

3. Sampling of Temple pond water

The pond water samples were collected at seventeen temples in Kanchipuram at pre monsoon (PRM), post monsoon(POM). The level of water in pond varies from low, medium and high. The samples were collected in well cleaned polythene bottles of one litter capacity. The pond water samples marked as P1, P2 up to P17. Then the sample was immediately taken to a laboratory for testing. The parameters pH, Total Dissolved Solids, Electrical conductivity, heavy metals Iron, Arsenic, Lead, Copper, Zinc were tested in the laboratory. The standard (CPCB & BIS) permissible limit for each measured parameter shown in Table 1.

Table 1. Standard permissible limits for each measured parameters

Sl.No	Water Quality Parameters	Characteristics of water				
		Drinking after disinfection (A)	Bathing (B)	Drinking with treatment infection (C)	Fish Culture (D)	Irrigation (E)
1	pH	6.5 – 8.5	6.5 – 8.5	6.5 – 8.5	6.5 – 8.5	6.0 – 8.5
2	Electrical Conductivity	-	-	-	1000	2250
3	Total dissolved solids	500	-	1500	-	2100
4	Iron	0.3	-	50	-	-
5	Arsenic	0.05	0.2	0.2	-	-
6	Lead	0.1	-	0.1	-	-
7	Copper	1.5	-	1.5	-	-
8	Zinc	15	-	15	-	-

4. Result and Discussion

4.1 Water Quality Parameters

The values of water quality parameters measured in selected sampling points are presented in Table 2.

Table 2. Water Quality Parameters values in sampling points

Sl. No.	Name of the Temple	Sampling Point	pH		TDS		EC	
			PRM	POM	PRM	POM	PRM	POM
1	Lakshmi Narayana Temple	P1	7.44	7.03	230	159	354	244
2	Kamakshi Amman Temple	P2	8.42	7.32	382	744	588	1144
3	Kailasanathar Temple, Sevilimedu	P3	7.47	9.31	161	164	248	252
4	Kasi Viswanatha Temple	P4	7.42	8.26	618	436	950	670
5	Astabhujakoram Temple	P5	7.40	7.91	573	148	882	228
6	Punniyakoteeswarar Temple	P6	7.47	7.68	305	183	469	282
7	Kosalakottam	P7	7.59	8.04	306	744	470	1144
8	Kachapeshwarar Temple	P8	7.25	6.86	159	114	244	135
9	Varadharaja Perumal Temple, Front side	P9	7.85	7.58	165	172	254	265
10	Varadharaja Perumal Temple, Back side	P10	7.48	7.62	424	740	653	1139
11	Sonnavannam Saitha	P11	7.94	7.80	932	1146	1434	1763
12	Dharmalingeswarar Temple	P12	9.07	8.86	1326	1160	2040	1785
13	Ekambareswarar Temple	P13	7.45	7.52	132	117	203	180
14	Santhaleeswarar Temple	P14	7.76	7.46	676	1076	1040	1656
15	Vilakoli Perumal Temple	P15	7.85	7.45	835	750	1284	1154
16	Kailasanathar Temple	P16	7.96	7.66	574	547	883	842
17	Thamarai Kulam	P17	7.4	7.42	425	1204	654	1853

The pH value varies between 7.4 to 9.07 in PRM and 6.86 to 9.31 in POM. pH values of sixteen temple ponds are within the permissible limits, P12 in PRM, P3 in POM having greater pH value and it shows it is in alkaline nature. It is due to melted chemical composites and biological process in the solution [10]. The seasonal variation of pH values in all sampling points shown in Figure 2. TDS values vary between 159 mg/l to 1326 mg/l in PRM and 114 mg/l to 1204 mg/l in POM, both seasons all measured values were within the standard limit [11]. The greater value of Total dissolved solids reduces the purity of water, affects the photosynthesis process. The seasonal variation of TDS values in all sampling points shown in Figure 3. Electrical conductivity values varies between 203 mg/l to 2040 mg/l in PRM and 135 mg/l to 1853 mg/l in POM. EC depends on the presence of solid compounds, it may not cause a major effect on pond water [12]. The seasonal variation of EC values in all sampling points shown in Figure 4.

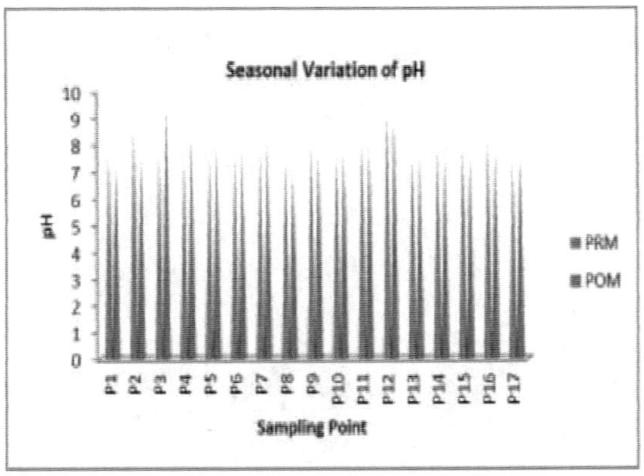

Figure 2. Seasonal Variation pH in Sampling Points

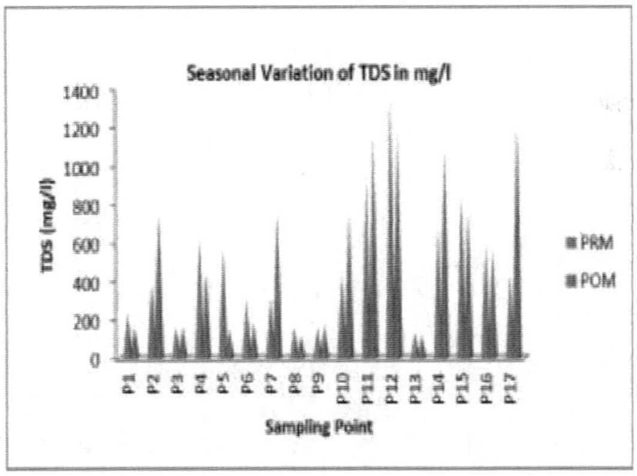

Figure 3. Seasonal Variation TDS in Sampling Points

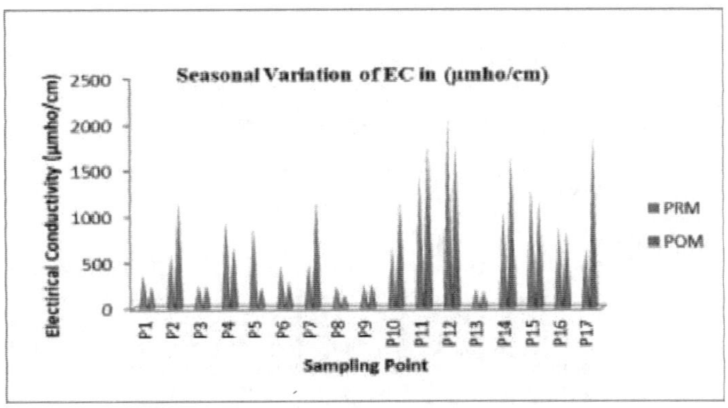

Figure 4. Seasonal Variation EC in Sampling Points

4.2 Heavy metal distribution

Five heavy metals Fe, As, Pb, Cu, Zn was assessed for the concentrations in seventeen selected stations. The observed results were presented in Table 3.

IOP Conf. Series: Materials Science and Engineering **954** (2020) 012043 doi:10.1088/1757-899X/954/1/012043

Table 3. Heavy metal values in Sampling Points

Sl. No.	Sampling Point	Fe (mg/l)		As (mg/l)		Pb (mg/l)		Cu (mg/l)		Zn (mg/l)	
		PRM	POM	PRM	POM	PRM	POM	PRM	POM	PRM	POM
1	P1	0.54	0.07	< 0.1	< 0.1	< 0.1	< 0.1	< 0.1	< 0.1	< 0.1	< 0.1
2	P2	0.67	0.09	< 0.1	< 0.1	< 0.1	< 0.1	< 0.1	< 0.1	< 0.1	< 0.1
3	P3	0.72	0.04	< 0.1	< 0.1	< 0.1	< 0.1	< 0.1	< 0.1	< 0.1	< 0.1
4	P4	0.82	0.03	< 0.1	< 0.1	< 0.1	< 0.1	< 0.1	< 0.1	< 0.1	< 0.1
5	P5	0.68	0.08	< 0.1	< 0.1	< 0.1	< 0.1	< 0.1	< 0.1	< 0.1	< 0.1
6	P6	0.53	0.03	< 0.1	< 0.1	< 0.1	< 0.1	< 0.1	< 0.1	< 0.1	< 0.1
7	P7	0.46	0.08	< 0.1	< 0.1	< 0.1	< 0.1	< 0.1	< 0.1	< 0.1	< 0.1
8	P8	0.37	0.02	< 0.1	< 0.1	< 0.1	< 0.1	< 0.1	< 0.1	< 0.1	< 0.1
9	P9	0.42	0.07	< 0.1	< 0.1	< 0.1	< 0.1	< 0.1	< 0.1	< 0.1	< 0.1
10	P10	0.54	0.07	< 0.1	< 0.1	< 0.1	< 0.1	< 0.1	< 0.1	< 0.1	< 0.1
11	P11	0.74	0.03	< 0.1	< 0.1	< 0.1	< 0.1	< 0.1	< 0.1	< 0.1	< 0.1
12	P12	0.82	0.02	< 0.1	< 0.1	< 0.1	< 0.1	< 0.1	< 0.1	< 0.1	< 0.1
13	P13	0.65	0.03	< 0.1	< 0.1	< 0.1	< 0.1	< 0.1	< 0.1	< 0.1	< 0.1
14	P14	0.65	0.05	< 0.1	< 0.1	< 0.1	< 0.1	< 0.1	< 0.1	< 0.1	< 0.1
15	P15	0.58	0.06	< 0.1	< 0.1	< 0.1	< 0.1	< 0.1	< 0.1	< 0.1	< 0.1
16	P16	0.46	0.04	< 0.1	< 0.1	< 0.1	< 0.1	< 0.1	< 0.1	< 0.1	< 0.1
17	P17	0.67	0.07	< 0.1	< 0.1	< 0.1	< 0.1	< 0.1	< 0.1	< 0.1	< 0.1

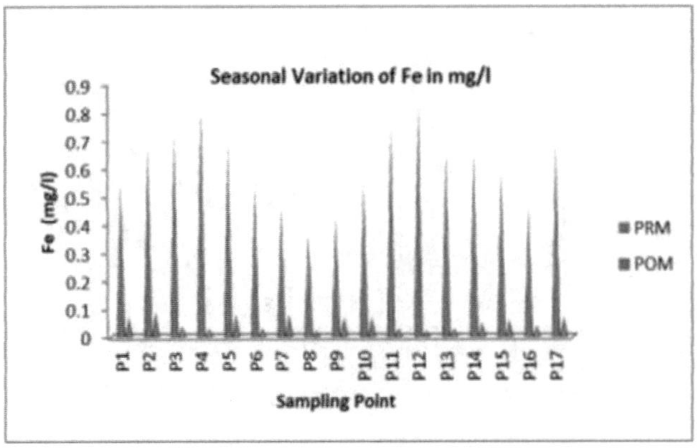

Figure 5. Seasonal Variation EC in Sampling Points

IOP Conf. Series: Materials Science and Engineering **954** (2020) 012043 doi:10.1088/1757-899X/954/1/012043

4.2.1 Iron, (Fe)

The heavy metal iron is a nutrient for human consumption. Maximum metabolic processes are going with the presence of iron [13]. It is a necessary component of hemoglobin and myoglobin, important for doing the movement of peroxides, cytochromes and other hemoprotein and flavoprotein enzymes [14,15]. In the sampling points, the variation of iron shown in Figure 5. The value of iron in PRM period ranges from 0.37 mg/l to 0.82 mg/l and POM period ranges from 0.02 mg/l to 0.09 mg/l. The present data show fluctuations of values in PRM and POM period. In PRM period considering Class A the values greater than the standard limit, According to WHO (1984),the safe condition for iron value 0.3 mg/l in drinking water and it varied between 1.94-5.76 mg/l in pond water. Considering other classes the values are below the standard limit. In POM all the sampling points having less value than the prescribed standard limit.

4.2.2 Arsenic, (As)

The heavy metal Arsenic is an essential compound for animal species; it plays a major role in protein synthesis. It is a dietary mineral for humans, it can take from food the limited concentration is 15-25μg. Arsenic mixing in the environment through mining, burning of fossil fuels, mixing in ground water through the dissolving of weathered rock, soil. Humans may absorb arsenic from water, food, air and skin contact with soil, water. Plants absorb arsenic easily it creates the presents of arsenic in food. The concentration of inorganic arsenic present in surface water is high, it can consume for plants in water, plant eating organisms, fish. Consumption of excessive inorganic arsenic cause health effects, like less production of white, red blood cells, stomach irritation, lung irritation, skin changes, development of cancer cells and damage DNA. In the study area, Arsenic concentration for PRM and POM period was less than the prescribed standard limit.

4.2.3 Lead, (Pb)

Lead is a non- necessary component. Deposition of leaded gasoline and industrial sources contaminate soil. Lead is brought from certain jobs welding, cutting, painting, demolition of buildings, industrial mineral activities mining, and extraction [16]. It found in some foods, traditional medicines. In drinking water lead mix from lead pipes, fittings. The contaminated drinking water is linked to hip fracture, damage to kidneys and it creates hypertension for human beings. The natural organic modules, like garlic oil and vitamin E, act as a depleting agent to lead in the human body .The acceptable limit for lead is 0.01 mg/l (BIS 2012). In the study area, lead concentration for PRM and POM period was less than the prescribed standard limit.

4.2.4 Copper, (Cu)

The heavy metal copper is an important portion of various enzymes, it is necessary for the synthesis of hemoglobin. Copper in the water due to heavy mounting vehicle traffic. In the stagnant water, the copper ions interact with organic species from industrial wastes having the potential appearance, ability to precipitate a complex system, deposit on the river basin and percolate to the water table. The maximum concentration of copper in the water is poisonous to the human health and causes hypertensions and produces pathological variations in brain tissues [17].In the study area, copper concentration for PRM and POM period was less than the prescribed standard limit.

4.2.5 Zinc (Zn)

Zinc is an essential nutrient, it shows a significant role in the physiological and metabolic progression of many organisms. Melting of Zinc ores is the leading source of contamination of Zinc [18]. Other sources are municipal waste, automobiles and farming use of pesticides and fungicides having the zinc sulfate are the supplementary sources of ecological pollution due to Zinc. The highest concentrations of zinc toxic to health. Humans need a required level of Zn, less concentration causes infections, anemia and birth defects in pregnant women. In the study area, Zinc concentration for PRM and POM period was less than theprescribed standard limit.

5. Conclusion

The concentration of heavy metals and water quality parameters from seventeen sampling points was determined. The results showed that there was a less concentration of Arsenic, Lead, Copper, and Zinc in pre monsoon and post monsoon period. The concentration of Iron in the pre monsoon greater than standard limit comparatively to class A and below the limit for the remaining classes, in the postmonsoon below the standard limit for all classes due to dilution of pond water by rain. Analyzing all the parameters the pond water was not directly used for drinking, it can useful for bathing, irrigation, fish culture etc.

6. References

[1] Mercy M and Dhanalakshmi B 2017 *Int Res J Pharm* ***8*** 41-45
[2] Banerjee P and Prasad B 2018 *J of App Chem* ***11*** 01- 05
[3] Sridhar S G D, Sakthivel A M, Sangunathan U, Balasubramanian M, Jenefer S, Rafik M M and Kanagaraj G 2017 *Appl Water Sci* ***7*** 4651–4662
[4] Silambarasan K, Senthilkumaar P and Velmurugan K 2012 *Eur J of Exper Bio* ***2*** 2192-2198
[5] Ramachandran A, Krishnamurthy R R, Jayaprakash M and Balasubramanian M 2018 *J of App Geo and Geophy* ***6*** 29-35
[6] Maurya A, Negi T and Negi R K 2018 *Asia j Anim Sci* ***12*** 16-22
[7] Giri S and Singh A K 2014 *Water Qua. Expo Health* ***5*** 173–182
[8] Bhatnagar A and Devi P 2013 *Int J of Env Sci* ***3*** 1980-2009
[9] Tapashi Gupta and Mrinal Paul 2013 Curr. World Env. **8** 127-131
[10] Bhuyan MDS, Bakar MA, Rashed-Un-Nabi M D, Senapathi V, Chung SY and Islam MDS 2019 *Appl Water Sci* ***9*** 1-13
[11] Goher ME, Hassan AM, Ibrahim B, Abdel-Moniem, Fahmy AH and El-sayed SM 2014 *Egyp J of Aquatic Res* ***40*** 225–233
[12] Sisira S. Withanachchi, Ghambashidze G, Kunchulia I, Urushadze T and Ploeger A2018 *Int J Envir Res Pub Health* ***15*** 1-37
[13] Flowra FA, Ghosh JK, Sayed Jewel MDA, Tumpa AS and Hussain MA 2012 *J Life Earth Sci* ***7*** 115-117
[14] Adeyemi MM and Ugah IA 2017 *J of Env Sc. Toxi and Food Tech* ***11*** 39-43
[15] Armid A, Shinjo R, Ruslan R and Fahmiati 2017 *IOP Conf. Ser Mater Sci and Engg* ***172*** 012002
[16] Nguyen NTT and Volkova IV 2018 *IOP Conf Ser. Mater Sci and Engg* ***451*** *012203*
[17] Goher ME, El-Rouby WMA, El-Dek SI, El-Sayed SM and Noaemy SG 2018 *IOP Conf Series Mater Sci and Engg* ***464*** 012003
[18] Jiao Z, Li H, Song M and Wang L 2018 IOP *Conf Ser Mater Sci and Engg* **394** 052055

IOP Conf. Series: Materials Science and Engineering **954** (2020) 012044 doi:10.1088/1757-899X/954/1/012044

Eco Friendly Machining Processes for Sustainability - Review

T. Rajmohan[1*] V.V. Kalyan Chakravarthy[2] A. Nandakumar[3] S.D. Satish Kumar[4]

Department of Mechanical Engineering
Sri Chandrasekharendra Saraswathi Viswa Maha Vidyalaya University,
Enathur, Kanchipuram – 631561, India

[1]rajmohanscsvmv@yahoo.com [2]kalyanchakravarthy@kanchiuniv.ac.in [3]nandakumar.a@kanchiuniv.ac.in
[4]sathishsd.2@gmail.com

Abstract: Green manufacturing is a style for industrialized that minimizes dissipate and contamination. The pollution prevention resolves in manufacturing industries to expand and execute various environmentally-friendly strategies. The main purpose of green machining is to hold up future generations by attaining sustainability. Cutting fluids are discoloured with metal particles and foulness products which reduce the efficiency of cutting fluids. To minimize the unpleasant environmental effects coupled with the use of cutting fluids, the harmful components from their formulations have to be eliminated or reduced to the acceptable level. In addition, mineral based cutting fluids are going to be replaced with vegetable based cutting fluids since they are environmentally friendly. The present review primarily determined on examining different aspects of machining process from an environmental sensitivity. This paper presents the environmental green machining such as dry machining, machining with minimum quantity lubricant, Cryogenic machining, hot machining, vegetable oil-based cutting fluids and nano filled cutting fluids. The Performance of Eco-friendly machining are discussed in terms of cutting forces, tool flank wear, cutting temperatures and surface finish are also discussed.

Key words: Eco friendly machining; Cutting fluid; MQL; Cryogenics; Vegetable oil; Nano filled fluid

1. Introduction

In 1868 researchers have used water as cutting fluid in machining and observed that speeds have increased by 30–40 %. The quality, productivity and cost of the product mainly depend on performance of machining operations. In machining the various properties of parameters affecting the cutting performances are cutting fluids work piece, cutting tool, machining conditions and machine tool [1]. The appropriate function of cutting fluid must not only develop the performance of machining, but also complete the needs such as non-toxic, non-harmful to health for operators, not a fire hazard, not smoke and easy disposal [2]. Veera Bhadra Rao et al [3] have concluded that flood cooling uses heavy coolant which increases cost of coolants, but has lots of health issues due to physical contact to skin and mist generated has serious effects such as respiratory disorders and other diseases. Cutting fluids have grown to be added difficult in requisites of both workers health and environmental pollution, with severe concerns ensuing from their use, behaviour and high disposal costs. Therefore, researchers have focused to minimize the use of cutting fluids and cutting energy due to important environmental links. [4]. Benedicto et al [5] have made a review on the analysis of the

IOP Conf. Series: Materials Science and Engineering **954** (2020) 012044 doi:10.1088/1757-899X/954/1/012044

applicability of cutting fluids in machining operations. They have identified main alternatives in machining like minimum quantity lubrication technique thereby meeting the industrial needs in terms of economy, environmental and technical effects.

The cutting fluids are classified into cutting oils, soluble oils, synthetic fluids and semi-synthetic fluids. Mineral based cutting fluids are even though reasonably priced the application is restricted due to the limited resource, poor biodegradability and the accessibility of mineral oil is highly dependent on political considerations. Generally, the cutting fluids are diluted with water, such that about 95 % of the cutting fluid, by volume, is water [6]. The biodegradable cutting fluids such as soybean oil, canola oil, sunflower oil, coconut oil, sesame oil, castor oil etc. as environmentally friendly lubricants as an option to mineral based cutting fluids. Modern advancement of high oleic vegetable oils as lubricants is stimulated by growing environmental and economic concerns to protect endangered species to the next generation. Vegetable oils perform better than the other oils and the reasons are illustrated in Figure 1. [7].

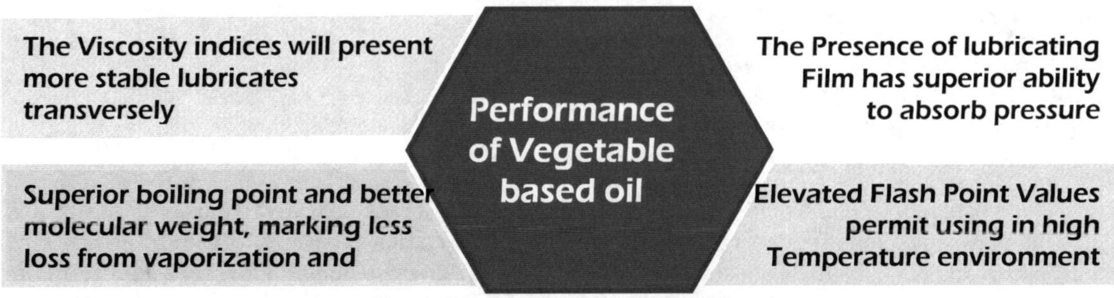

Figure 1 Properties of vegetable oil

Researchers have proved that vegetable-based oil and nano filled oil shows superior lubricating property than conventional oil [8&9]. Cryogenic machining is a modern method of cooling the chip-tool interface, to decrease the tool wear, to modify the features of material, consequently which develop the machining performance and product excellence. This method uses non-combustible and non-corrosive gas liquefied by cooling to −196 °C as coolant [10]. The sustainable products, processes, and systems are the functional elements of sustainable manufacturing. The reduced material utilization, minimizing energy consumption and waste-free manufacturing processes are the major challenges to the industry to manufacture sustainable products [11]. The conceptual layout of sustainable machining is provided in figure 2.

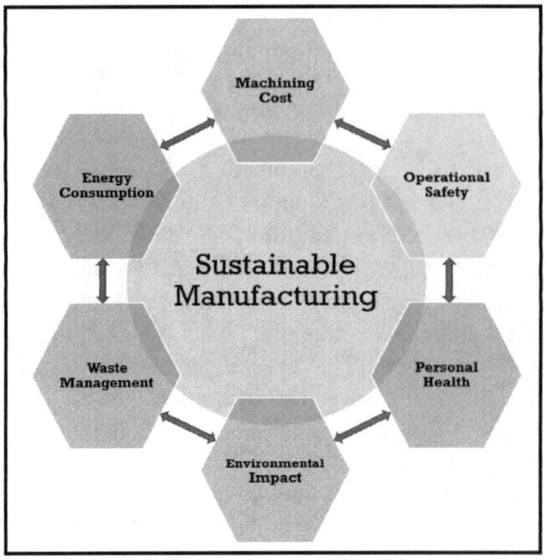

Figure 2. Schematic diagram of Sustainable manufacturing

IOP Conf. Series: Materials Science and Engineering **954** (2020) 012044 doi:10.1088/1757-899X/954/1/012044

Careful examination of literature review reveals that very limited review articles are available in the area of sustainability of the machining process i.e. energy-efficient manufacturing Processes. Therefore, the paper presents the various environmental cognisant machining techniques and discusses the performance of eco-friendly machining by considering cutting forces, tool wear, cutting temperatures and surface roughness.

2. Sustainable Machining Techniques

2.1 DRY Machining (DM)

The machining process conceded in the nonexistence of cutting fluid is termed as dry machining (DM). High performance cutting tools enables dry machining for sustainable expertise [12]. Wu et al (2018) have proved that the feasibility of dry machining of the high-throughput drilling of compacted graphite iron for 3150 holes [13]. Thomas et al. have stated that dry machining cause high temperature in the machining zone which results in tool life reduction and inaccurate dimensions of the work piece. [14] Goindi and Sarkar have performed a comprehensive review of dry machining based on sustainability aspects meeting major research objectives. They have concluded that dry machining is environmentally friendly which eliminates hazardous cutting fluids during machining operations resulting in sustainable machining [15]. The extremely difficult to machine metals such as nickel alloys, titanium alloys, Co-Cr alloys have machined by cemented carbide (WC/Co) tools filled with solid lubricant. The performance of dry machining is enhanced by self-lubricating tool with the employ of conservative tools [16].

2.2 Minimum Quantity Lubrication (MQL)

The machining process sprints in the presence of an extremely small amount of cutting fluid, in the assortment of 10–150 ml/h, distributed in the cutting zone in a vapour technique is called MQL [17]. Inconel was machined under MQL and found that machinability is improved for MQL as compared to dry or wet machining [18]. Liu et al. attempted to prove that dry machining and MQL could be the most appropriate processes for machining titanium. Tool coating was one key factor to realize high speed machining and green manufacturing [19]. Khan et al showed that an improved surface roughness of concerning 31.6% was achieved for MQL over wet cutting and about 18% was obtained for dry cutting [20]. Boubekri and Shaikh has concluded that MQL serves as an alternate to flood cooling but without significant health concerns. Dry and MQL gave comparable results for turning, milling, drilling and grinding for all materials. [21]. Nandakumar et al. has employed MQL machining condition and has developed superior surface finish at lower wear of tool in grinding Al-SiC composites [22]. Most of the authors have demonstrated perceptible enhancement in performance with MQL machining compared to dry machining, more significant results could have been obtained. Table 1 explored the difficulties in machining of materials under dry conditions [23-33].

Table 1. Machining of materials under dry conditions [Addona et al., 2017; Thakur et al., 2014; Ramanujam et al., 2014; Thakur et al., 2014; Singarvel et al., 2014; Rajaguru and Arunachalam, 2017; Fernandez-Abia et al., 2013; Thakur et al., 2014; Nexhat Qehaja et al., 2015; Torres et al., 2015]

SI.NO.	MATERIAL USED	CUTTING TOOL USED	REMARKS
1	Inconel 718	Carbide CVD coated insert	During the investigation no major signs of tool damages are observed while machining Inconel 718 at low speeds i.e. at 60 and 90m/min. Comparatively at high cutting speeds tool gets worn out at faster rate with major tool failure patterns like heavy notching.

2	Inconel 825	Uncoated cemented carbide insert and CVD multilayer coated tool	CVD multilayer coating has synergistic influence on the improvement of the machined surface integrity-based super alloy.
3	Inconel 718	PVD coated tungsten carbide cutting insert	As cutting speed increased from 30m/min to 70m/min, the predominant tool wear mechanisms are abrasion, oxidation and edge chipping due to the friction and thermal softening.
4	Inconel 825	Uncoated and PVD multilayer coated cemented carbide inserts	The average flank wear increased with both cutting speed as well as with machining duration. However use of PVD multilayer coated improved the resistance to flank wear.
5	EN25 Steel	CVD and PVD coated carbide tools	The results of ANOM shows that a combination of machining parameters for this investigation are cutting speed of 244m/min, feed rate of 0.10mm/rev and depth of cut of 1.0mm with CVD coated tool. The result of ANOVA shows that the coated tool is the most significant parameter followed by cutting speed.
6	Super Duplex Stainless Steel	CVD and PVD coated tungsten carbide cutting inserts	MT-TiCN-Al_2O_3 coating provides relatively better performance in-terms of tool wear, cutting force, cutting temperature and surface integrity
7	AISI 304L austenitic stainless steel	PVD advanced tools	The best coatings for turning of difficult to machine materials as austenitic stainless steels are nACo and AlTiN coatings
8	Inconel 825	Uncoated and CVD coated cemented carbide tool	Micro hardness increased with increase in cutting speed. Coated tool resulted in reduction in micro hardness in the surface and sub-surface region.
9	Cold rolled steel C62D	SNMM coated tungsten carbide inserts	The results revealed that feed rate seems to influence surface roughness more significantly than nose radius and cutting time.
10	2030-T4 aluminium alloy	Inserts of hard metal with a coating of titanium nitride	The most significant factor was the interaction effect between the depth of cut and the feed rate.

2.3 Cryogenic Machining

High abrasive materials, super alloys and novel machine tools have utilised cryogenic machining for safe and environmental-friendly method [34]. In machining of AISI 304 stainless steel by oblique cryogenic cooling, tool life was improved more than four times [35]. Joshi et al have made an experimental study to assess the effects of dry, MQL and cryo-MQL on grinding force, temperature, surface roughness and chip morphology on Inconel 751. The experimental results indicate that liquid nitrogen and MQL has improved the machining characteristics as compared with MQL and dry grinding [36]. Pereira et al [2015] have presented a new cooling system which consists of CO_2 cryogenic and MQL technique. The new system provides an environmental and economic efficient alternative. The results indicate that CO_2 + MQL system is more effective in improving the machining characteristics of Inconel 718 during milling operation when compared to conventional machining techniques [37]. Gholap and Mohod had stated that cryogenic fluids such as liquid nitrogen are not reusable, sometimes may not be feasible [38]. Kaynak and Gharibi have made an experimental study on the progression of tool wear in the turning operation of AISI 4140 steel under CO_2 assisted MQL, CO_2 assisted and LN_2 assisted cryogenic machining conditions. The experimental results indicate that CO_2 assisted machining is not performed better in turning of AISI 4140 steel and is not recommended [39]. Tapoglou et al [2017] have investigated the performance of CO_2 cryogenic coolant in milling of titanium alloy Ti-6Al-4V to analyse tool life under different coolant delivery options. They have performed machining under cryogenic CO_2, flood cooling, dry machining and cryogenic CO_2 combined with MQL environments. The results indicate that cryogenic CO_2 plus MQL has achieved a tool life of 18.5minutes better than other machining conditions [40].

Jerold and Pradeep Kumar have investigated to study the influence of cryogenic coolants in machining of titanium alloy Ti-6Al-4V and its effects on cutting temperature, cutting forces, surface roughness, chip morphology and tool wear. They have compared the results of cryogenic machining with dry and wet machining. The results indicate that the use of CO_2 coolant has improved the machinability characteristics of titanium alloy when compared to other cryogenic coolants [41]. Fengbiao Wang et al [2017] have investigated the performance of Nickel-based super alloy on tool wear during milling operation under cryogenic machining conditions. They have found that the application of liquid nitrogen coolant has reduced the tool wear [42]. Chetan et al have done an experimental study to analyse the performance of liquid nitrogen in cryogenic machining of Nimonic 90 alloy by using uncoated tungsten carbide inserts. The experimental results indicate that the liquid nitrogen has suppressed the notch wear and edge fracture for the uncoated tools during the machining [43]. Shokrani et al have investigated the role of cryogenic cooling in enhancing the machinability characteristics of Ti-6Al-4V titanium alloy during CNC end-milling operation. The experimental results indicated that the cryogenic cooling improved the machinability by reducing surface roughness by 40% when compared to conventional machining conditions [44]. The table 2 shows the machining of materials under cryogenic cooling and it is determined that substantial improvement in tool life and surface finish-dimensional accuracy [45-49].

Table 2. Machining of Hard materials under Cryogenic cooling [Sunil Magadum et al., 2014; Alborz Shokrani, 2016; Iturbe et al., 2016; Ibrahim Sadik et al., 2016; Aramcharoen and Shaw, 2014]

SI.NO.	MATERIALS USED	CUTTING TOOL	CUTTING FLUID USED	CRYOGENIC PARAMETERS	REMARKS
1	EN-24 hardened steel	TiAlNi coated carbide tool inserts	Liquid nitrogen	Liquid nitrogen at -185°C was used as a cryogenic coolant.	There is an improvement in tool life by 22.46%, 35.74% and 38.60% in cryogenic machining over flood coolant machining at cutting speed 125m/min, 160m/min and 200m/min respectively. The influence of cryogenic cooling reduces the cutting force by 15.42%, 8.22% and 7.58% at 125, 160 and 200m/min respectively.
2	Cobalt chromium (CoCr) alloy	Solid carbide cutting tool	Neat oil is used for MQL system. Liquid nitrogen is used for cryogenic cooling system.	Liquid nitrogen at 1.5bar pressure and 20kg/h flow rate was sprayed along the cutting tool into the cutting zone.	Results indicated that 71% reduction in surface roughness and 96% reduction in flank wear can be achieved using cryogenic cooling when compared to conventional machining practice.

3	Ti-6Al-4V Titanium alloy	Coated carbide inserts	Oil based coolant is used for flood machining. Liquid nitrogen(LN_2) is used for cryogenic system	The flow rate for cryogenic media was controlled and regulated at 0.35 l/min while pressure was set at 4bars.	The results showed that tool wear resistance increases when using the developed cryogenic modular compared to oil-based coolant.
4	Inconel 718	CVD TiCN-Al_2O_3 –TiN coated carbide inserts.	The MQL oil used was the KLUBERTCUT CO 6-150 oil. Liquid nitrogen (LN_2) is used for cryogenic cooling system.	Liquid nitrogen was delivered to the rake face of the cutting tool, with a jet cooling system in order to reduce the temperature on the tool-chip interface and facilitate chip evacuation.	Reported that the combination between cryogenic and MQL machining improves the machining performance for short machining times in comparison with dry and MQL machining.
5	Ti-6Al-4V Titanium alloy	A cutter equipped with two PVD coated inserts were used.	Blasocut BC 25 flood emulsion is used for wet milling. Liquid carbon dioxide (CO_2) is used for cryogenic cooling system.	The supply pressure was constant at 50bar. The different flow rates of CO_2 for cryogenic cooling are 0.65, 0.19 and 0.15kg/min respectively.	The flow rate of coolant has limited influence on the rate of flank wear development. Cryogenic cooling with CO_2 reduces the tendency of thermal cracks to propagate laterally, delaying the chipping of the cutting edge.
6	Inconel 718	Two-flute end mills of TiN coated tungsten carbide were used.	Liquid nitrogen (LN_2) is used as the cryogenic media.	The liquid nitrogen was supplied through nozzle in order to deliver the cooling to the cutting zone.	The cryogenic cooling significantly reduces the temperature in the cutting zone and improves the cooling accessibility through the tool-chip interface compared to conventional cooling method.

Kalyan Chakaravarthy et al. has performed drilling operation on Al-SiC composite under cryogenic LN_2 and MQL condition. The experimental results show that the thrust force, surface roughness, burr height and temperature generated are reduced to a great extent [50].

2.4 Hot Machining

Hot Machining is novel technique in which the temperature of the work piece is elevated to several hundred Celsius over ambient, which sources reduction in the shear strength of the material. Tigham first innovated the process of hot machining in 1889, since then it has fashioned much interest among various investigators. Researchers observed that at an optimum temperature the tool life increases to a highest value and after that it reduces in hot machining of stainless steel, S-816alloy, X-alloy and Inconel-X [51-52]. Many researchers have proved that hot machining is a very useful method for machining of hard-to-cut materials and to reduce tool wear; power consumed and increase surface finish. [53-54] The effect of temperature of work-piece is found to be the most significant on tool life [55]. Authors have simulated the hot machining of Inconel 718 using finite element analysis. They concluded that use of heat in secure immediacy to the surface of the work piece can considerably reduce the force. The increase of process zone temperature reduces heat generation which reduce flow stress and cutting force in hot machining. The previous studies confirm that the performance of hot machining is improved by extended tool life. Improved surface finish and higher metal removal rate [56].

2.5 Machining using nano filled cutting fluid

In recent times scientists engaged nano particles in conventional lubricants since its extraordinary development in thermo-physical, and heat transfer capabilities, decrease the coefficient of friction and wear effect to develop the efficiency and consistency of machine tools [57]. Kalyan Chakaravarthy et al. have employed Nano CuO filled rapeseed oil in the drilling of Al- SiC composites. The experimental results have shown that the surface finish of the machined specimen is found to be superior which is majorly due to the nano particles in the vegetable oil [50]. Sahu et al have evaluated the machining performance of MWCNT filled Nano fluid in turning of Ti-6Al-4V alloys and found that nano fluids are effective in overcoming the problems faced by the conventional machining conditions [58]. Venkatesan et al have made an experimental study to improve the machinability characteristics of Nimonic 90A alloy during machining operation by using Nano fluid MQL technique. They have prepared Nano fluids by dispersing copper oxide Nano particle with groundnut oil through ultra-sonication process. They have obtained optimum response values at 40m/min cutting speed, 0.14rev/min and concentration 0.25 by Taguchi methodology [59]. Ahsan Ul Haq et al have done an investigation to analyse the effects of feed, depth of cut and flow rate on the temperature in face milling of D2 tool steel. They have done machining by MQL and Nano fluid MQL conditions. RSM based central composite design was used to design the experiments. The results indicate that 25% temperature reduction is observed during NFMQL technique as compared to MQL [60].

Conclusions

Sustainable manufacturing turns out to be a global concept, substance important elements on all the fields as well as machining processes. The technique to sustainable manufacturing through environmentally-friendly machining begins from the steps that must be taken to implement ecological machining methods in order to make these technologies consistent, environmentally friendly and cost efficient. The review focuses on the performance and environmental blow of environmentally-friendly machining for various materials and machining conditions. The salient features are as follows:

- High performance cutting tools enables dry machining for sustainable expertise. The dry machining lessen costs associated with cutting fluid which is approximately 7–17% of total manufacturing costs, dry machining technique will unavoidably be considered
- MQL gave comparable results for turning, milling, drilling and grinding for all materials.
- MQL based nano fluid lubrication has clearly demonstrated significant decrease in consumption of cutting fluid as well as the cutting energy or cutting power while improving surface quality, this technique has been established as a environmentally friendly green manufacturing
- The cryogenic cooling application in machining operations could be firm as enabling considerable progress in tool life and surface finish-dimensional accuracy through reduction in tool wear through control of machining temperature desirably at the cutting zone.
- Cryogenic LN_2 coolant reduced the cutting temperature to an extent about 9–34% and 3–17% when compared to wet and CO_2 machining conditions respectively. Cryogenic CO_2 cooling reduced the cutting forces efficiently than wet and cryogenic LN2 machining to a range of about 17–38% and 2–12% respectively (Dilip Jerold and Pradeep Kumar 2012)
- The economical machining can be carried out using low costly cutting tools by implementing hot machining technique to machine difficult to cut materials.

- The introduction of vegetable oil-based metalworking fluids in machining applications has made it possible to achieve better performance as reported by all researchers. Researchers proved that vegetable oil assisted machining made it credible to increase assembly volume by 10% and also reducing tool cost by 40 to 50%
- Nano fluids assisted machining exhibits the energy-efficient heat transfer due to superior thermal conductivity and heat transfer coefficient

References

[1] Srikant, R.R.; Ramana, V.S.N.V. (2015) Performance evaluation of vegetable emulsifier based green cutting fluid in turning of American Iron and Steel Institute (AISI) 1040 steel – an initiative towards sustainable manufacturing. *Journal of Cleaner Production,* Volume 108, Part A, Pages 104-109.

[2] Rao, D.N.; Srikant, R.R. (2006) Influence of emulsifier content on cutting fluid properties. *Proc Inst Mech Eng, Part B: Eng Manuf,* 220: 1803–1806

[3] Miriyala Veera Bhadra Rao; Vasim A Shaikh; Bhushan T Patil. (2018) Recent trends in the effective utilization of minimum quantity lubrication (MQL) in turning low carbon steels. *Industrial Engineering Journal,* 11:2, 29-33.

[4] Rajmohan, T.; Sathishkumar, S.D.; Palanikumar, K. (2017) Effect of a nanoparticle-filled lubricant in turning of AISI 316L stainless steel (SS). *Particulate Science and Technology,* 35 (2): 201-208.

[5] Benedicto, E.; Carou, D.; Rubio, E.M. (2017) Technical, Economic and Environmental Review of the lubrication/cooling systems used in machining processes. *Procedia Engineering,* 184:99-116.

[6] Padmini Rapeti; Vamsi Krishna Pasam; Krishna Mohana Rao Gurram; Rukmini Srikant Revuru. (2018) Performance evaluation of vegetable oil based nano cutting fluids in machining using grey relational analysis-A step towards sustainable manufacturing. *Journal of Cleaner Production,* 172: 2862-2875.

[7] Khan, M.M.A.; Dhar, N.R. (2006) Performance evaluation of minimum quantity lubrication by vegetable oil in terms of cutting force, cutting zone temperature, tool wear, job dimension and surface finish in turning AISI-1060 steel. *J Zhejiang UnivSci A,* 7(11): 1790–1799

[8] Baskar, S.; Sriram, G.; Arumugam, S. (2016) The Use of D-optimal Design for Modeling and Analyzing the Tribological Characteristics of Journal Bearing Materials Lubricated by Nano-Based Biolubricants. *Tribology Transactions,* 59(1): 44-54, DOI: 10.1080/10402004.2015.1063179.

[9] Arumugam, S.; Sriram, G.; Ellappan, R. (2014) Bio-lubricant-biodiesel combination of rapeseed oil: an experimental investigation on engine oil tribology, performance, and emissions of variable compression engine. *Energy,* 72: 618-627.

[10] JianchaoZhang.,ChangheLi;YanbinZhang;MinYang;DongzhouJia;GuotaoLiu;YaliHou;RunzeLi;NaiqingZhang;QidongWu;HuajunCao. (2018) Experimental assessment of an environmentally friendly grinding process using nanofluid minimum quantity lubrication with cryogenic air. *Journal of Cleaner Production,* 193: 236-248.

[11] Jawahir, I.S.; Wanigarathne, P.C.; Wang, X. (2006) Product design and manufacturing processes for sustainability, mechanical engineers. In: Kutz M (ed) Handbook: manufacturing and management, third edn, vol 3, John Wiley & Sons, Inc.,Delmar, 414–439.

[12] Davim, J.P.; Sreejith, P.S.; Silva, J. (2007) Turning of brasses using minimum quantity of lubricant (MQL) and flooded lubricant conditions. *Mater Manuf Processes,* 22(1):45–50.

[13] Wenwu Wu; Ali Kuzu; David Stephenson; Jun Hong; Mustafa Bakkal; Albert Shih. (2018) Dry and minimum quantity lubrication high-throughput drilling of compacted graphite iron. *Machining science and Technology,* 22(4): 652-670

[14] Jacob Thomas; Keyur Kunte; Vaibhav Arote. (2016) Review on machining techniques: Dry machining and cryogenic machining. *International Journal of Advance Research in science and Engineering,* 5:2, 188-194

[15] Gyanendra Singh Goindi; Prabir Sarkar. (2017) Dry machining: A step towards sustainable machining – challenges and future directions. *Journal of cleaner production,* 165:1557-1571.

[16] Ahmad-Yazid, A.; Taha, Z.; Almanar, I.P. (2010) A review of cryogenic cooling in high speed machining (HSM) of mold and die steels. *Sci Res Essays,* 5(5):412–427.

[17] Grzesik, W. (2008) Dry and semi-dry machining, advanced machining processes of metallic materials. 226–245.

[18] Vasu, V.; Pradeep Kumar Reddy, G. (2011) Effect of minimum quantity lubrication with Al2O3 nanoparticles on surface roughness, tool wear and temperature dissipation in machining Inconel 600 alloy. *Proceedings of the Institution of Mechanical Engineers, Part N: Journal of Nanoengineering and Nanosystems,* 225: 3–16.

[19] Liu, Z.; An, Q.; Xu, J.; Chen, M.; Han, S. (2013) Wear performance of (nc-AlTiN)/(a-Si3N4) coating and (nc-AlCrN)/(a-Si3N4) coating in high-speed machining of titanium alloys under dry and minimum quantity lubrication (MQL) conditions. *Wear,* 1–11.

[20] Khan, A.A.; Ahmed, M.I. (2008) Improving tool life using cryogenic cooling. *Journal of Materials Processing Technology,* 196(1–3): 149–154.

[21] Nourredine Boubekri; Vasim Shaikh. (2015) Minimum Quantity Lubrication (MQL) in machining: Benefits and Drawbacks. *Journal of Industrial and Intelligent Information,* 3(3): 205-209.

[22] Nandakumar, A., Rajmohan, T. and Vijayabhaskar, S., 2019. Experimental Evaluation of the Lubrication Performance in MQL Grinding of Nano SiC Reinforced Al Matrix Composites. *Silicon, 11*(6), pp.2987-2999.

[23] Addona, D.M.D.; Sunil J Raykar; Narke, M.M. (2017) High speed machining of Inconel 718: tool wear and surface roughness analysis. *Procedia CIRP,* 62: 269-274.

[24] Thakur, A.; Dewangan, S.; Patnaik, Y.; Gangopadhyay, S. (2014) Prediction of Work Hardening during Machining Inconel 825 using Fuzzy Logic Method. *Procedia Materials Science,* 5: 2046-2053.

[25] Thakur, A.; Gangopadhyay, S.; Maity, K.P. (2014) Effect of cutting speed and tool coating on machined surface integrity of Ni-based super alloy. *Procedia CIRP,* 14: 541-545.

[26] Thakur, A.; Mohanty, A.; Gangopadhyay, S.; Maity, K.P. (2014) Tool Wear and Chip Characteristics during Dry Turning of Inconel 825. *Procedia Materials Science,* 5: 2169-2177.

[27] Thakur, D.G.; Ramamoorthy, B.; Vijayaraghavan, L. (2009) Optimization of Minimum Quantity Lubrication Parameters in High Speed Turning of Superalloy Inconel 718 for Sustainable Development. *World Academy of Science, Engineering and Technology,* 54: 224–226.

[28] Ramanujam, R.; Venkatesan, K.; Vimal Saxena; Philip Joseph. (2014) Modeling and Optimization of Cutting Parameters in Dry Turning of Inconel 718 using Coated Carbide Inserts. *Procedia Materials Science,* 5: 2550-2559.

[29] Singarvel, B.; Selvaraj, T.; Jeyapaul, R. (2014) Multi Objective Optimization in Turning of EN25 Steel Using Taguchi Based Utility Concept Coupled With Principal Component Analysis. *Procedia Engineering,* 97: 158-165

[30] Rajaguru, J.; Arunachalam, N. (2017) Coated tool Performance in Dry Turning of Super Duplex Stainless Steel. *Procedia Manufacturing,* 10: 601-611.

[31] Fernandez-Abia, A.I.; Barreiro, J.; Fernandez-Larrinoa, J.; Lopez de Lacalle, L.N.; Fernandez-Valdivielso, A.; Pereira, O.M. (2013) Behaviour of PVD coatings in the turning of austenitic stainless steels. *Procedia Engineering,* 63: 133-141.

[32] Nexhat Qehaja; Kaltrine Jakupi; Avdyl Bunjaku; Mirlind Bruci; Hysni Osmani. (2015) Effect of Machining Parameters and Machining Time on Surface Roughness in Dry Turning Process.

Procedia Engineering, 100: 135-140.
[33] Torres, A.; Puertas, I.; Luis, C.J. (2015) Surface roughness analysis on the dry turning of an Al-Cu alloy. *Procedia Engineering,* 132: 537-544.
[34] Pusavec, F.; Krajnik, P.; Kopac, J. (2010) Transition to sustainable production—part I: application on machining technologies. *J Clean Prod,* 18(2): 174–184.
[35] Khan, A.A.; Ahmed, M.I. (2008) Improving tool life using cryogenic cooling. *Journal of Materials Processing Technology,* 196(1–3): 149–154.
[36] Joshi, S., Rawat, K. and Balan, A.S.S., 2018. A novel approach to predict the delamination factor for dry and cryogenic drilling of CFRP. *Journal of Materials Processing Technology*, *262*, pp.521-531.
[37] Pereira, O.; Catala, P.; Rodriguez, A.; Ostra, T.; Vivancos, J.; Rivero, A.; Lopez-de-Lacalle, L.N. (2015) The use of hybrid CO2+MQL in machining operations. *Procedia Engineering,* 132:492-499.
[38] Tushar Anil Gholap; Suprabhat A Mohod. (2015) Review on experimental analysis of cryogenic cooling on various machining processes. *International Journal of Advance Research in science and Engineering,* 4(9): 140-149
[39] Yusuf Kaynak; Armin Gharibi. (2018) Progressive tool wear in cryogenic machining: The effect of liquid nitrogen and carbon dioxide. *Journal of Manufacturing and Materials Processing,* 2(2):1-12.
[40] Nikolaos Tapoglou; Maureen I Aceves Lopez; Ian Cook; Christopher M Taylor. (2017) Investigation of the influence of CO2 cryogenic coolant application on tool wear. *Procedia CIRP,* 63: 745-749.
[41] Dilip Jerold, B.; Pradeep Kumar, M. (2013) The influence of cryogenic coolants in machining of Ti-6Al-4V. *Journal of Manufacturing Science and Engineering,* 135:031005-1-8.
[42] Fengbiao Wang; Lili Li; Jingkai Liu; Qilin Shu. (2017) Research on tool wear of milling nickel-based super alloy in cryogenic. *Int J Adv Manuf Technol,* 91:9(12), 3877-3886.
[43] Chetan, Habtamu, A.; Behera, B.C.; Ghosh, S.; Rao, P.V. (2017) Machining of Nimonic 90 Alloy under dry and LN2 environment using AlTiN coated and uncoated tungsten carbide inserts. Proceedings of 10th International Conference on Precision, Meso, Micro and Nano Engineering (COPEN 10) December 07-09, 2017, Indian Institute of Technology Madras, Chennai – 600 036, India.
[44] Alborz Shokrani; Vimal Dhokia; Stephen T Newman. (2016) Comparative investigation on using cryogenic machining in CNC milling of Ti-6Al-4V titanium alloy. *Machining science and Technology,* 20(3): 475-494.
[45] Sunil Magadum; Arun Kumar, S.; Yoganath, V.G.; Srinivasa, C.K.; GuruMurthy, T. (2014) Evaluation of Tool life and Cutting forces in Cryogenic Machining of Hardened Steel. *Procedia Materials Science,* 5: 2542-2549.
[46] Alborz Shokrani.; Vimal Dhokia.; Stephen T Newman. (2016) Cryogenic high speed machining of cobalt chromium alloy. *Procedia CIRP,* 46: 404-407.
[47] Iturbe, A.; Hormaetxe, E.; Garay, A.; Arrazola, P.J. (2016) Surface integrity analysis when machining Inconel 718 with conventional and cryogenic cooling. *Procedia CIRP,* 45: 67-70.
[48] Ibrahim Sadik, M.; Simon Isakson; Amir Malakizadi; Lars Nyborg. (2016) Influence of coolant flow rate on tool life and wear development in cryogenic and wet milling of Ti-6Al-4V. *Procedia CIRP,* 46: 91-94
[49] Ampara Aramcharoen. (2016) Influence of cryogenic cooling on tool wear and chip formation in turning of titanium alloy. *Procedia CIRP,* 46: 83-86.
[50] Kalyan Chakaravarthy, V.V., Rajmohan, T., Vijayan, D., Palanikumar, K. and Latha, B., 2020. Sustainable drilling performance optimization for Nano SiC reinforced Al matrix composites. *Materials and Manufacturing Processes*, pp.1-9.
[51] Tour, S.; Fletcher, L.S. (1949) Hot spot machining. Iron Age 78, 164.
[52] Krabacher, E.J.; Merchant, M.E. (1951) Basic factors in hot machining of metals. *Trans. ASME,*

73: 761.

[53] Ranganathan, S.; Senthilvelan, T. (2011) Multi-response optimization of machining parameters in hot turning using grey analysis. *int j adv manuf technol,* 56:455– 462.

[54] Tosun, N.; Ozler, L. (2004) Optimization for hot turning operations with multiple Performance characteristics. *Int J Adv Manuf Technol,* 23: 777– 782.

[55] Asit Kumar Parida; Kalipada Maity. (2017) Effect of nose radius on forces, and process parameters in hot machining of Inconel 718 using finite element analysis. *Engineering Science and Technology, an International Journal,* 687-693.

[56] Ozler, L.; Inan, A.; Ozel, C. (2000) Theoretical and experimental determination of tool life in hot machining of austenitic manganese steel. *Int. J. Mach. Tool Manuf,* 41: 163–17.

[57] Nor Azwadi Che Sidik; Syahrullail Samion; Javad Ghaderian; Muhammad Noor Afiq Witri Muhammad Yazid. (2017) Recent progress on the application of nanofluids in minimum quantity lubrication machining: A review. *International Journal of Heat and Mass Transfer,* 108: 79–89.

[58] Neelesh Kumar Sahu; Atul B Andhare; Roja Abraham Raju. (2017) Evaluation of performance of Nano fluid using multiwalled carbon Nano tubes for machining of Ti-6AL-4V. *Machining science and Technology,* 22(3): 476-492.

[59] Venkatesan, K.; Devendiran, S.; Abhijeet Thakur; Ashish Chauhan, V.; Pavan Kalyan, P.V. (2018) Study on influence of machinability characteristics in machining of Nimonic 90A alloy using copper oxide Nano fluid in MQL mode. *International Journal of Mechanical Engineering and Technology,* 9(3):978-986

[60] Muhammad Ahsan Ul Haq; Aqib Mashood Khan; Le Gong; Tao Xu; Longhul Meng; Salman Hussain. (2018) A comparative study of face milling of D2 steel using AL2O3 based Nano fluid Minimum Quantity Lubrication and Minimum Quantity Lubrication. *Advances in Science and Technology Research Journal,* 12(1): 99-105.

IOP Conf. Series: Materials Science and Engineering **954** (2020) 012045 doi:10.1088/1757-899X/954/1/012045

Synthesis and Tribological characteristics of a nano CuO particle-filled water-based lubricants

T. Rajmohan[1] D. Hemanth kumar[2] R. Nivithetha[3] Elamburthi[4] V.V. Kalyan Chakravarthy[5]

Department of Mechanical Engineering
Sri Chandrasekharendra Saraswathi Viswa Maha Vidyalaya University,
Enathur, Kanchipuram – 631561, India

rajmohanscsvmv@yahoo.com[1] hemanthdeena@gmail.com[2]
kalyanchakravarthy@kanchiuniv.ac.in[5]

Abstract. The oil-based lubricating system tends to produce high cutting temperatures owning to the low cooling capability of oil lubricants. The addition of lubricating additives in nanoscale could increase the tribological and lubrication performance of the water-based lubricants. In this work, nano CuO particle-filled water-based lubricants as cutting fluids were developed using ultrasonication process. The Thermophysical Properties of synthesized nano filled water-based lubricant was tested and analyzed. The tribological characteristics were fairly investigated using four-ball and pin on disc apparatus testing units. The results indicated that nano CuO filled water-based lubricant shows enhanced tribological performance in terms of coefficient of friction (COF) and wear scar diameter (WSD) compared with base coolant. The superior performance of CuO filled water-based lubricant was credited to the detail that CuO nanoparticle additives arrived the rubbing interface and made a tribo-thin layer of CuO. Water-based lubricant showed the highest surface deformation compared to nano-based lubricants which indicate that the damage was due to adhesive wear.

Key Words: Nano CuO particles; Wear Scar; Tribology; water-based lubricants; ultrasonication.

1. Introduction

The entire purpose of international standard 14001 is to guard environment in balance with socio-economic requirements. The materials selection, energy utilize, solid waste, liquid residues, and gaseous filtrates are the environmental stressors in different stages of product life cycle. The work pieces, tools, metalworking fluids, cleaning fluids and coatings are eliminated as wastes to the environment in manufacturing process [1]. The sustainable products, processes, and systems are the functional elements of sustainable manufacturing. The reduced material utilization, minimizing energy consumption and waste-free manufacturing processes are the major challenges to the industry to manufacture sustainable products [2].The manufacturing process consists of several fundamentals such as Material, energy, data, machining, lubrication and chips [3]. The employ of cutting fluids in machining process is the most significant environmental exposure [4]. It has been approximated that the utilization of cutting fluid can account for more than 10-15 % of machining costs [5]

IOP Conf. Series: Materials Science and Engineering **954** (2020) 012045 doi:10.1088/1757-899X/954/1/012045

In 1868 researchers have used water as cutting fluid in machining and observed that speeds have increased by 30–40 %. The quality, productivity and cost of the product mainly depend on performance of machining operations. In machining the various properties of parameters affecting the cutting performances are cutting fluids work piece, cutting tool, machining conditions and machine tool [6]. The appropriate function of cutting fluid must not only develop the performance of machining, but also complete the needs such as non-toxic, non-harmful to health for operators, not a fire hazard, not smoke and easy disposal [7]. The cutting fluids are classified into cutting oils, soluble oils, synthetic fluids and semi-synthetic fluids. Mineral based cutting fluids are even though reasonably priced the application is restricted due to the limited resource, poor biodegradability and the accessibility of mineral oil is highly dependent on political considerations. Generally the cutting fluids are diluted with water, such that about 95 % of the cutting fluid, by volume, is water [8]. The remaining 5 % is a mixture of oil, emulsifiers, and additives [9]. The additives are utilized to minimize the corrosion, to limit the acidity, to manage the microbial growth, to get better the lubricity, and to avoid the foaming. Some of these additives are toxic and harmful to human health, wildlife and environment [10].

Tao Lv et al [11] found that the cutting fluids of high thermal conductivity were needed to attain a better machiningperformance.One of the most active methods to improve the coolingability of lubrication is to use water-based cutting fluids [12-13].But, water itself could not form real lubricating films in the impression area since of its low viscosity and lubricity[13], which thus directed to a weak load--carrying capacity and highfriction.Nevertheless, recent studies [14] have confirmed that the addition of lubricating additives in nanoscale, such asSiO_2, Al_2O_3, CuO, TiO_2, and MoS_2, could progress the tribologicalproperties of the base lubricants.Wang et al.[15] reported that the addition of SiO_2 nanoparticles into deionizedwater could improve its tribological performance andextreme pressure properties significantly. Setti et al. [16] showed that Al_2O_3 nanoparticle based water-lubricant led to a significant reductionin the tangential forces and grinding temperatures comparedwith the base lubricant.

2 Experimental details

2.1 Preparation of nano CuO filled water-based lubricants

Copper oxide (Cuo) nano particles of 40-80nm in diameter (purchased M/s US Nano Research.,USA) was used as additives. Lauryl sodium sulfate (SDS, purchased from Ganapathy traders, India) was engaged as a surface-active agent in the research of water-based lubricants. Polyethylene glycol 300 (PEG300) was firstly diffused into purified water at a mass ratio of 3:2 at ambient temperature to prepare base lubricant [11]. Cuo nanoparticles in powder was then directly added into this base lubricant (water PEG solution) at weight ratios of 0.5%and 1% and SDS was consecutively plunged at a concentration of 10 wt% of the CuO nano particles content. Consequently, the Cuo nano particles was ultrasonically dispersed in water bath for 1 h by means of an ultrasonic unit of 20KW. For comparison, water-based lubricants without CuO nano particles was separately formulated using the same process. The synthesized water-based lubricants are highly steady without precipitation and stratification after 40 days of aging, as shown Figure. 1.

Figure 1. Samples of Lubricant prepared

IOP Conf. Series: Materials Science and Engineering **954** (2020) 012045 doi:10.1088/1757-899X/954/1/012045

2.2 Rheological and physical properties of water based lubricants

The properties of nano CuO filled water lubricants were tested and the results were shown in Table 1. The viscosity of the base-lubricant and the nano-lubricants are measured using a Redwood Viscometer with ASTM D-445 standards.

Table 1. Properties of water based lubricants

Properties	0.5 % nano CuO+Water	1% wt nano CuO+Water	Water based lubricant without additive
Flash point (°C) [ASTM D92]	493	476	422
Thermal Conductivity Watt/mK	0.736	0.924	0.607
Viscosity @100°C (cSt) [ASTM D445]	1.649	1.649	1.03
Fire point	517	521	473

Three trials are conducted for each sample and average is taken for tabulation. Flash and fire points are measured by standard methods using a Clevel and open cup flash and fire point apparatus in accordance with ASTM D-92.Transient hot wire liquid thermal conductivity method was used for determination of thermal conductivity under ASTM D7896. Three trials are conducted for each test and average is taken for tabulation.

2.3 Friction and wear tests

a. Four-ball tester

The friction and wear performances of water-based lubricant using nano CuO as additives were performed using the four-ball testing on TR30L 1AS Ducom India, as shown in Figure. 2.

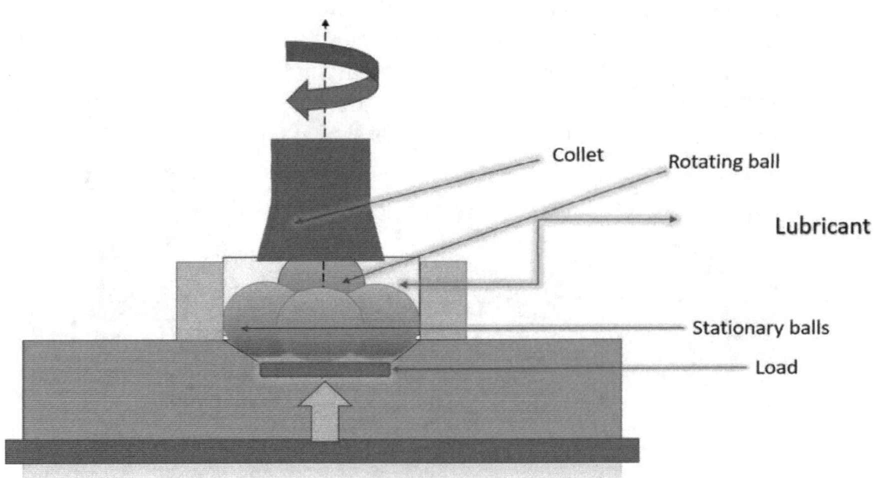

Figure 2. Schematic layout of Four ball Tribometer

The steel balls used in four-ball tests were 12.7 mm in diameter and made of AISI high chromium steel, which is having hardness of 65 HRC [17-18]. The tests were conducted at normal load of 400 N, rotational speed of 1200 rpm and room temperature of 25 °C for 30 minutes. Prior to testing, the balls were ultrasonically cleaned in acetone for 15 min. The wear scar diameters (WSDs) of the balls after testing were measured by using CCD camera 100X system. Each test was repeated three times and average values of coefficient of friction (COF) and wear scar diameter (WSD) were recorded.

b. Pin-On-Disc Tribometer

The tribological studies have been conducted using a pin-on-disc tribometer in accordance with ASTM G 99-05 standards. The pin material used is aluminium alloy (Al-86%, Si-12% and other elements-2%) and the disc is of steel (EN-32) with a hardness of 65 HRC. The layout of the Pin on disc tribometer is shown in figure 3.

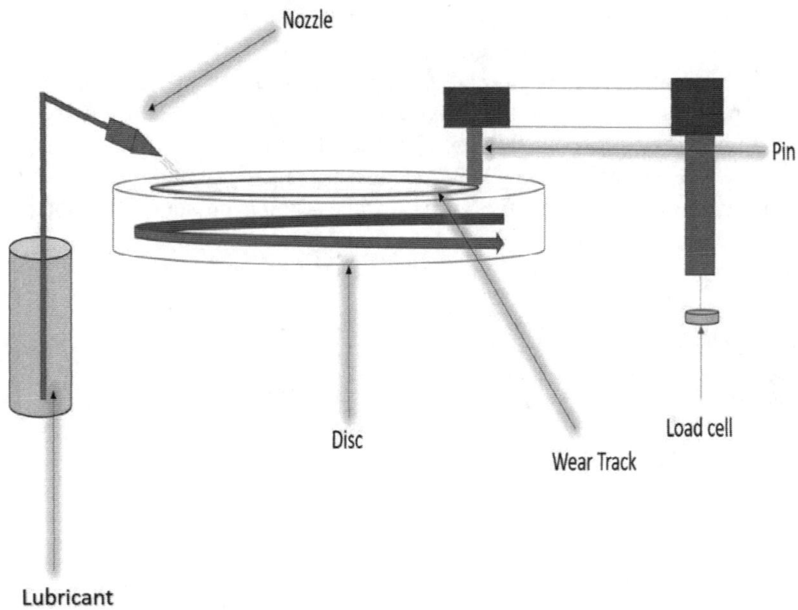

Figure 3. Schematic layout of Pin on Disc Tribometer

The diameter and length of the pin are 8 mm and 10 mm respectively. After the Machining operation of the aluminium pin, its sliding surface is polished using emery paper of 800 grit size; the same polishing method is also adopted for steel disc. The surfaces are then cleaned by acetone and heptanes respectively. The operating conditions for the test is shown in Table 2.

Table 2. Operating conditions for friction and wear test

Four ball wear test	Normal load-400N Rotational speed-1200rpm Temperature-25°C Time -30sec
Pin on disc test apparatus	load-5-50N Rotational speed-100-1500rpm Temperature-120°C Time -300sec

Water based lubricant is supplied at the sliding interface in small quantity between aluminium alloy pin and steel disc on the pin-on-disc tribometer in order to keep boundary/thin film lubrication conditions. Coefficient of friction is measured directly by an electronic display.

3. Results and discussions

3.1Properties of water based lubricants

Significant enhancement in the properties was observed in nano CuO filled water based lubricants due to the better heat yielding capability of nano fluid. Viscosity, flash and fire point of nano CuO filled water based lubricants were improved up to 40% compared to base lubricant. The thermal conductivity of nano CuO filled water lubricant is enhanced up to 50%.The increase in thermal conductivity of the nano fluid is due to dynamic interaction of the flow field and the nano CuO

3.2 Tribological properties of nano CuO filled water-based lubricant

Figure 4,5 shows the effect of nano CuO concentration on the tribological performance of water-based lubricants. It is noticed that the least COF and WSD were attained with 1wt% nano CuO particles filled lubricant, which confirmed a COF reduction of 15.8% and wear reduction of 10.4% in comparison with the base water lubricant. As seen in Figures 4 and 5, the COF curves of the nano CuO filled water lubricant are more constant in contrast with that of the base lubricant, and the worn surfaces of the balls shown in Fig. 4 are smoother as well. These may be credited to the fact that CuO nanoparticles acted as ball bearings during the sliding [19], which could diffuse the stress between the friction pairs, and so resulted in lesser COF and WSD. Meanwhile, the CuO nanoparticles might serve as the spacers to fill the furrows and pits on the worn surfaces which prevented the interface from direct contact, and hence led to lower COF and WSD too [20].

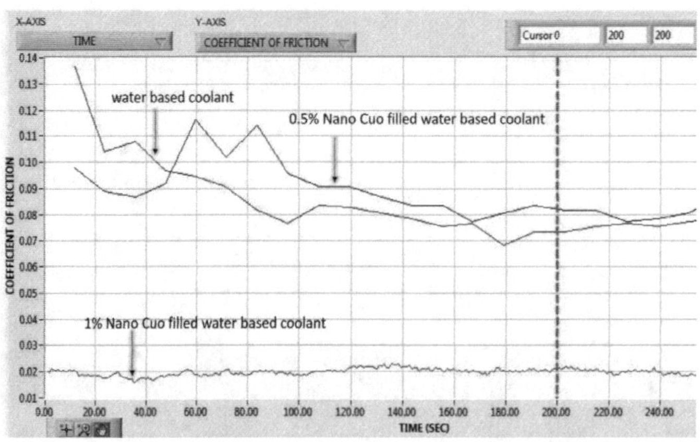

Figure 4. Effect of Lubricants on COF at 10N load

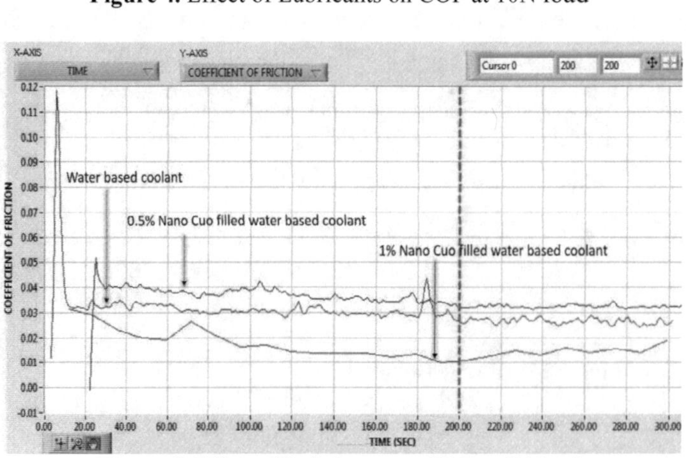

Figure 5. Effect of Lubricants on COF at 5N load

3.3 Analysis of Worn Surfaces by CCD and SEM

The worn scar surface investigation is significant for ruling the sharp failures that occurred due to the effect of friction and wear between the contact surfaces. The SEM analysis of the stationary steel balls and Pins of the Pin of Disc tribometer are used in this study in higher load conditions is illustrated in Figures 6 and 7 respectively.

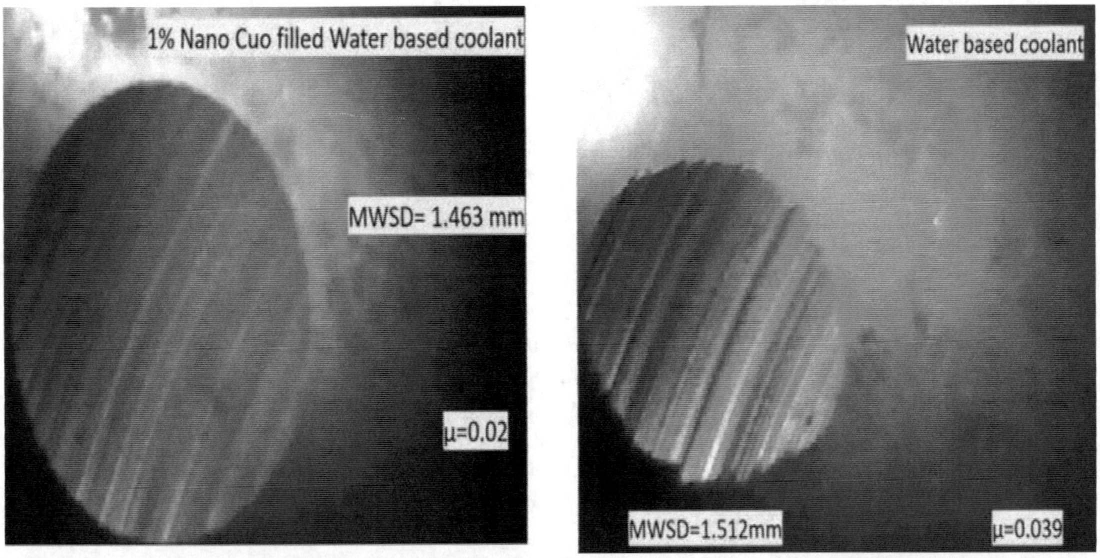

Figure 6. Surface Morphology of Ball employed in Four Ball Tribometer using CCD

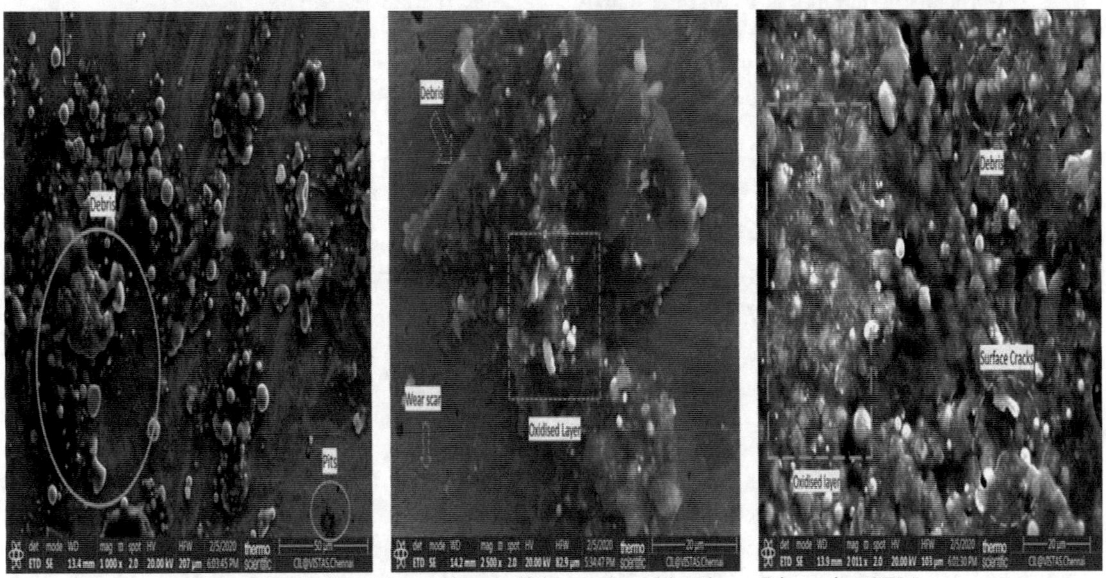

Figure 7. Surface Morphology of Pin employed in Pin on Disc using SEM

The worn scar surface areas of the stationary ball for base water lubricant was higher than nano CuO filled water based lubricants. Water based lubricant showed the highest surface deformation compared to nano based lubricants which indicates that the damage was due to adhesive wear. Apart from this, the circular, tiny and steady scuffs on the ball characterize abrasive wear[21] It is seen that CuO nano particles were easy to formulate up the nanosized furrows and pits, and provided as spacers to effectively prevent the contact of the friction peak. Temporarily, the CuO nanoparticles might act as rolling elements between the moving parts, which helped to reduce the friction at the contacting surface[20]. Furthermore, because of the high contact pressure and rotating speed, a smooth protecting film which consists CuO would form on the rubbing interface[22]. This film could not only bear the load but also separate the rubbing surfaces, which thus resulted in lower friction with the reduced adhesion of work piece materials[11]. Therefore, the tribological characteristics of nano filled water based lubricant were much better than base lubricant.

4. Conclusions

In this project, a comparative performance analysis of nano Cuo filled water-based lubricant under different weight ratios with base lubricant was systematically examined through triboligical experiments. The major conclusions drawn from the present investigation as follows

- Nano CuO filled water based lubricant was successfully synthesized through ultrosonication process.
- Rhological and physical properties of water based lubricants was tested**.** The significant enhancement in the properties was observed in nano CuO filled water based lubricants up to 50% due to the better heat yielding capability of nano fluid.
- The friction and wear performances of water-based lubricant using nano CuO as additives were performed using the four-ball testing and pin on disc apparatus.
- The promising results were obtained from the experiments that the 1 wt% CuO filled water lubricant showed up a COF reduction of 18.8 and wear reduction of 17.6% in comparison with the base water lubricant
- Water based lubricant showed the highest surface deformation compared to nano based lubricants which indicates that the damage was due to adhesive wear.
- The CuO nanoparticles might act as rolling elements between the moving parts, which helped to reduce the friction at the contacting surface.Furthermore, because of the high contact pressure and rotating speed, a smooth protecting film which consists CuO would form on the rubbing interface

In conclusion, CuO nanoparticles could effectively progress the tribological properties performances of the water-base lubricant. It is understood that this novel water-based lubricant suggests an efficient approach and capable industrial application for metal cutting.

References

[1] Goindi, G.S. and Sarkar, P., (2017). Dry machining: a step towards sustainable machining–challenges and future directions. *Journal of cleaner production*, *165*, pp.1557-1571.
[2] Pusavec, F., Deshpande, A., Yang, S., M'Saoubi, R., Kopac, J., Dillon Jr, O.W. and Jawahir, I.S., (2015). Sustainable machining of high temperature Nickel alloy–Inconel 718: part 2–chip breakability and optimization. *Journal of Cleaner Production*, *87*, pp.941-952.
[3] Rajemi, M.F., Mativenga, P.T. and Aramcharoen, A., (2010). Sustainable machining: selection of optimum turning conditions based on minimum energy considerations. *Journal of Cleaner Production*, *18*(10-11), pp.1059-1065.
[4] Pervaiz, S., Kannan, S. and Kishawy, H.A., (2018). An extensive review of the water consumption and cutting fluid based sustainability concerns in the metal cutting sector. *Journal of Cleaner Production*, *197*, pp.134-153.
[5] Debnath, S., Reddy, M.M. and Yi, Q.S., (2014). Environmental friendly cutting fluids and cooling techniques in machining: a review. *Journal of cleaner production*, *83*, pp.33-47.
[6] Srikant, R.R. and Ramana, V.S.N.V., (2015). Performance evaluation of vegetable emulsifier based green cutting fluid in turning of American Iron and Steel Institute (AISI) 1040 steel–an initiative towards sustainable manufacturing. *Journal of Cleaner Production*, *108*, pp.104-109.

[7] Amiril, S.A.S., Rahim, E.A. and Syahrullail, S., (2017). A review on ionic liquids as sustainable lubricants in manufacturing and engineering: Recent research, performance, and applications. *Journal of Cleaner Production*, *168*, pp.1571-1589.
[8] Padmini, R., Krishna, P.V. and Mohana Rao, G.K., (2016). Experimental evaluation of nano-molybdenum disulphide and nano-boric acid suspensions in vegetable oils as prospective cutting fluids during turning of AISI 1040 steel. *Proceedings of the Institution of Mechanical Engineers, Part J: Journal of Engineering Tribology*, *230*(5), pp.493-505.
[9] Mia, M., Gupta, M.K., Singh, G., Królczyk, G. and Pimenov, D.Y., (2018). An approach to cleaner production for machining hardened steel using different cooling-lubrication conditions. *Journal of Cleaner Production*, *187*, pp.1069-1081.
[10] Ghosh, S. and Rao, P.V., (2015). Application of sustainable techniques in metal cutting for enhanced machinability: a review. *Journal of Cleaner Production*, *100*, pp.17-34.
[11] Tao, Lv., Shuiquan Huang., Xiaodong Hu., Yaliang Ma., Xuefeng Xu. (2018) Tribological and machining characteristics of a minimum quantity lubrication (MQL) technology using GO/SiO_2 hybrid nanoparticle water-based lubricants as cutting fluids. *Int J Adv Manuf Technol* **96,** 2931–2942. https://doi.org/10.1007/s00170-018-1725-3
[12] Itoigawa, F., Childs, T.H.C., Nakamura, T. and Belluco, W., (2006). Effects and mechanisms in minimal quantity lubrication machining of an aluminum alloy. *Wear*, *260*(3), pp.339-344.
[13] Mao, C., Zhang, J., Huang, Y., Zou, H., Huang, X. and Zhou, Z., (2013). Investigation on the effect of nanofluid parameters on MQL grinding. *Materials and Manufacturing Processes*, *28*(4), pp.436-442.
[14] Sharma, A.K., Tiwari, A.K. and Dixit, A.R., (2015). Improved machining performance with nanoparticle enriched cutting fluids under minimum quantity lubrication (MQL) technique: a review. *Materials Today: Proceedings*, *2*(4-5), pp.3545-3551.
[15] Wang, J.H., Song, M., Li, J.L. and Wang, X.B., (2011). The preparation and tribological properties of water-soluble nano-silica particales. *Tribology*, *31*(2), pp.118-123.
[16] Setti, D., Sinha, M.K., Ghosh, S. and Rao, P.V., (2015). Performance evaluation of Ti–6Al–4V grinding using chip formation and coefficient of friction under the influence of nanofluids. *International Journal of Machine Tools and Manufacture*, *88*, pp.237-248.
[17] Peña-Parás, L., Taha-Tijerina, J., Garza, L., Maldonado-Cortés, D., Michalczewski, R. and Lapray, C., (2015). Effect of CuO and Al2O3 nanoparticle additives on the tribological behavior of fully formulated oils. *Wear*, *332*, pp.1256-1261.
[18] Cozza, R.C., (2014). Influence of the normal force, abrasive slurry concentration and abrasive wear modes on the coefficient of friction in ball-cratering wear tests. *Tribology International*, *70*, pp.52-62.
[19] Gupta, B., Kumar, N., Panda, K., Dash, S. and Tyagi, A.K., (2016). Energy efficient reduced graphene oxide additives: Mechanism of effective lubrication and antiwear properties. *Scientific reports*, *6*, p.18372.
[20] Sayuti, M., Sarhan, A.A. and Salem, F., (2014). Novel uses of SiO2 nano-lubrication system in hard turning process of hardened steel AISI4140 for less tool wear, surface roughness and oil consumption. *Journal of Cleaner Production*, *67*, pp.265-276.
[21] Song, H.J. and Li, N., (2011). Frictional behavior of oxide graphene nanosheets as water-base lubricant additive. *Applied Physics A*, *105*(4), pp.827-832.
[22] Peng, D.X., Kang, Y., Hwang, R.M., Shyr, S.S. and Chang, Y.P., (2009). Tribological properties of diamond and SiO2 nanoparticles added in paraffin. *Tribology international*, *42*(6), pp.911-917.

Friction Stir Welding and Friction Stir Processing of Dissimilar Alloys: A Review

K Palani[1] and C Elanchezhian[2]

[1] Department of Mechanical Engineering, Sri Chandrasekharendra Saraswathi Viswa Mahavidyalaya (Deemed to be University), Kanchipuram, 631561, Tamilnadu, India

[2] Department of Mechanical Engineering, Sri Sairam Engineering College, West Tambaram, Kanchipuram, 600044, Tamilnadu, India

E-mail: palani.k@kanchiuniv.ac.in

Abstract. The present work is focused on friction stir welding and friction stir processing of dissimilar alloys due to its emerging technique applied in the wide area of applications viz., aviation, ship building, defence industries etc. In recent years, the novel solid-state joining techniques viz., friction stir welding and friction stir processing are applied in joining of dissimilar alloys even though there is the difference in mechanical properties and chemical composition of the alloys. Due its eco-friendly, versatile and energy-efficient process, these are implemented to reduce the harmful effects and formation of defects in joining of dissimilar alloys compared to the fusion welding techniques. The specially designed non-consumable rotating tools are used in joining process by generating frictional heat between the tool surface and alloys to produce the sound welds. This work discusses the status, opportunities and challenges of above processes in joining of dissimilar alloys, viz., aluminium, steels, titanium etc.,

Keywords: *Friction stir welding; Friction stir processing; Mechanical properties; Optimization techniques; Material characterization.*

1. Introduction

The quality of the welded joints was studied in joining of the various similar and dissimilar alloy materials in the aerospace, fabrication and aviation industries. Initially, the fusion welding process was tried to fabricate the dissimilar material viz., steel and its alloys. Besides, it can be extended into the FSW process due to avoid the unnecessary defects produced by the fusion welding processes. The FSW process was applied to produce the similar and dissimilar joints with considerable amount of frictional heat between the tool circumferential surface and base materials. The various designed non-consumable tools were used to fabricate the dissimilar joint efficiently compared to the fusion welding processes. The researchers have focused on the material characterization, material flow, heat transfer, grain boundary formation, temperature distribution in the dissimilar joints. The various optimization techniques viz., response surface methodology, Taguchi method, grey relational analysis, fuzzy approach and artificial intelligence techniques were applied to study the significant parameters which decide the quality of the joints.

2. Mechanical and Metallurgical Characterization on dissimilar FSW Joints

The various welding process parameters, such as different combinations of materials, tool materials, geometry of the FSW tools, different grades of electrodes were carried out in the similar and dissimilar welded joints of friction stir welding process. These above independent factors are mainly applied based on the mechanical and metallurgical characterization of the joints.

IOP Conf. Series: Materials Science and Engineering **954** (2020) 012046 doi:10.1088/1757-899X/954/1/012046

2.1 Mechanical Properties

The defect-free joints were achieved by the suitable welding parameters in assisting to enhance the mechanical properties of the fabricated welded joints. The investigation was made on the dissimilar friction stir butt welding of pure titanium and AA5083-H321 aluminium alloy by Amini et al. [1] and they examined that there was no region for the thermo-mechanically affected zone (TMAZ) due to the higher melting temperature of the pure titanium compared to the aluminium alloy. Also, they reported that the highest hardness was predicted due to the dynamic recrystallization and material flow in the nugget zone (NZ) whereas the interface has the brittle intermetallic compounds (IMCs) for the welded joints. Aonuma et al. [2] found that the titanium with AA7075 aluminium alloy has the lower tensile strength relate to the titanium alloy with AA2024 aluminium alloy. The lower tensile strength of the welded joints was observed for pure titanium with aluminium alloy welded joints by increasing the travel speed whereas the titanium alloy and aluminium alloy combinations has the lower tensile strength of the welded joints.

The dissimilar butt joining between AA 2024-T3 and AA 7050-T7651 aluminium alloys of 2 mm thickness was successfully done by friction stir welding (FSW) and stationary shoulder friction stir welding (SSFSW) processes by Barbini et al. [3]. They found that the welded samples along the thickness, the frictional heat was not properly generated due to the non-uniform material flow and plastic deformation and presented in Figure 1. The half of the stir zone width was examined at the bottom of the weld relate to the crown of the weld due to the variation in microstructure with non-uniform heat distribution in FSW joints.

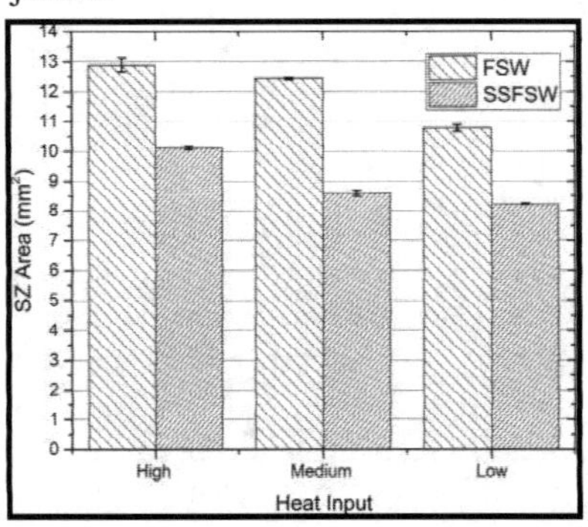

Figure 1. Comparison of the SZ area in SSFSW and conventional FSW for different Heat Input

Azizi et al. [4] used the square pin tool to investigate the microstructures and mechanical properties with 10 mm thick pure copper plates of four different welding speeds of 50, 100, 150, and 200 mm/min and constant rotational speed of 700 rpm of the butt joints. The defects with finer grains were examined due to the less heat input and high peak temperature at the higher welding speeds of the tool. The more ductile fracture surfaces were observed at the lower welding speeds, whereas the lower tensile elongation was found at the higher welding speeds of the tool. Charit et.al [5] improved the tensile properties of the as-cast alloy using the friction stir processing in addition with Zn, Mg and Sc particles. They reported that the different grain sizes 1.8 μm, 5.5 μm and 0.68 μm were found for the artificial aged, solution heat treated and no post heat treaded cast alloy by variations in plunge depth conditions.

Dehghani et al. [6] reported that the formation of fine equiaxed grains at the stir zone was found whereas the coarse grains with precipitates occurred at the heat affected zone of the friction stir welding of AA7075-O aluminium alloys. The maximum strength and hardness was found at the stir zone of the FSW butt joints which were more than the strength and hardness of the base materials and also it indicates the there is no formation of defects at the welded joints. Delijaicov S et al. [7] reported that the irregular mixing of materials was found at the low penetration depth due to the more tilt angle of the tool with its vertical position. The transverse speed and tool tilt angle are the most influencing parameters on the joint strength of the fabricated FSW joints. Zhang et al. [8] made an investigation on the microstructure and mechanical properties of the dissimilar ultrasonic spot welded AA2139 aluminium alloy to Ti-6Al-4V titanium alloy with 1 mm thick sheets. Due to the less fracture at the interface of the lap joints, the maximum shear strength was found owing to other dissimilar combinations of the Lap joints.

Cao et al. [9] investigated that the effects of the FSW process parameters, viz., rotational speed and the pin length on the microstructure and mechanical properties of the lap welded 2 mm thick AZ31B-H24 magnesium alloy sheets. The variation in probe length of the tool increased the tensile shear load, while these load initially increased by increasing the rotational speed and further increment of the tool rotational speed induced the decrement of the load for magnesium alloy joints and presented in Figure 2. Also, they have studied that the grain sizes, heat distribution rate were increased by increasing the tool rotational speed due to the development of the grain growth after recrystallization.

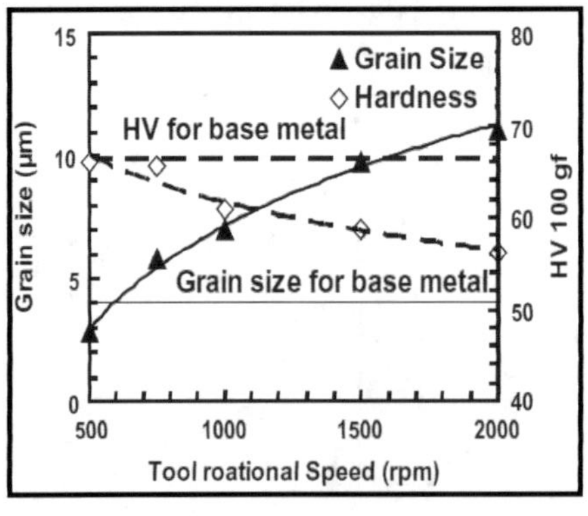

Figure 2 Effects of tool rotational speed on grain size and hardness in the stir zone of the top sheet at the mid-Thickness obtained at a welding speed of 20 mm/s and a pin length of 2.75 mm

Dorbane et al. [10] investigated effects of the welding parameters on the microstructural, mechanical, and fracture properties of dissimilar friction stir welded AZ31B magnesium alloy and AA6061-T6 aluminium alloy joints using the various combinations of the rotational speeds and welding speeds of the tool and found the superior properties at optimal FSW process conditions. They observed that the grain refinement of the both Mg-rich and Al-rich alloys were occurred at the stir zone (SZ) with the grain sizes of 2.43±0.87 μm and 0.94±0.24 μm respectively. From the Figure 3, they examined the formation of the thin layer of intermetallic compounds (IMC) at the interface near the SZ which deteriorates the weld properties. The discontinuity of microhardness was found at the weld interface due to the presence of IMCs in the joints.

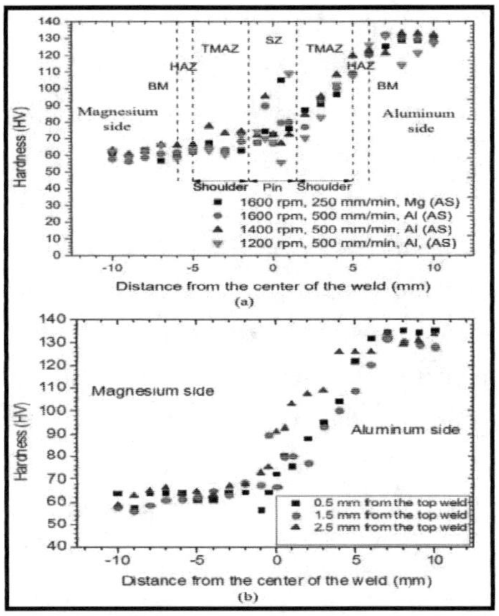

Figure 3 (a) Microhardness values obtained at 2 mm from top of weld for the various parameters, (1)1600 rpm and 250 mm/min, (Mg on AS) (2) 1600 rpm and 500 mm/min, (Al on AS) (3) 1400 rpm and 500 mm/min, (Al on AS) (4) 1200 rpm and 500 mm/min, (Al on AS) and (b) Microhardness obtained on welded joints using 1400 rpm, 500 mm/min parameters at three different distances from top of the weld.

3. Various Optimisation techniques used in FSW and FSP Processes

The various welding process parameters, such as different combinations of materials, tool materials, geometry of the FSW tools and the optimisation of the process parameters were important to achieve the objectives for the good quality of the welded joints. The different optimization techniques were applied on the similar and dissimilar aluminium alloy materials to achieve the quality of the joints [11 -13].

3.1 Similar and Dissimilar FSW process

Farzadi et al. [14] used the effective statistical technique, namely response surface methodology for the optimization of the parameters of the FSW of AA7075 aluminium alloy butt joints by proper selection of parameters, viz., rotational speed, transverse speed, shoulder diameter, and pin diameter of the tool on the tensile strength of the welded joints. They found the maximum tensile strength and joint efficiency with the optimum process parameters settings 510 rpm of rotation speed, 95 mm/min of welding speed, and 16 mm of shoulder diameter with 5 mm of pin diameter of the tool.

Sundaram et al. [15] successfully welded the hot worked AA2024 aluminium alloy with the cold worked AA5083 aluminium alloy with five different tool pin profiles using the FSW process. They developed the mathematical models which give the good correlation between the process parameters and the responses of the friction stir butt welded joints. They examined that the highest tensile elongation and tensile strength of the joints was found for tapered hexagon pin profile due to its higher pulsating effect and smooth material flow, whereas the straight cylinder pin profile gave the lower strength and elongation of the joints.

Palani et al. [16] successfully optimized the friction stir welding parameters of the dissimilar AA8011-H24 aluminium alloys and Ti3Al2.5V titanium alloys using the Taguchi based grey relational analysis with principal component analysis. They reported that the drastic reduction of the strength of the dissimilar welded joints with the optimum process settings of 1700 r/min tool rotational speed, 60mm/min welding speed, 6 kN axial load using the square pin profiled tool. Winiczenko et al. [17] optimized the process parameters on the tensile strength and microstructural properties of the dissimilar AISI 1020-ASTM A536 joints using the hybrid RSM based Genetic Algorithm. They developed the mathematical model was used to predict the tensile strength of the joints and also, they studied the effects of the welding parameters on the fabricated joints. Okuyucu et. al. [18] applied the artificial neural network algorithm on the friction stir welded aluminium plates and the better correlation results have been observed by comparing the measured and predicted results of tensile strength, hardness of weld metal and hardness of HAZ compared to elongation and yield strength of the FSW weld specimens.

Sahu et al. [19] made a feasibility study of the optimization of the FSW parameters of AM20 magnesium alloy using Taguchi grey relational analysis. They reported that the analysis by response surface methodology showed the better results compared to the taguchi method at the optimal process parameters of the welded joints. Palani et al. [20] applied hybrid fuzzy approach to optimize the process parameters, such as rotational speed of 2050 rpm, tilt angle of 90o and transverse speed of 40 mm/min with the responses, Vickers microhardness of 57.7 HV, and Impact strength of 12.4 J for the AA8011-H24 aluminium alloy FSW joints. They showed that the tool tilt angle is the most significant parameter for the Vickers microhardness followed by the welding speed, whereas welding speed is the most significant parameter for Impact strength of the joints followed by the tool tilt angle.

Shojaeefard et al. [21] made the friction stir welding process successfully on dissimilar AA 5083 aluminum alloys and brass plates and optimized the process parameters of the welded joints using Taguchi method. They reported that the sound welded joints were occurred based on the variation in rotational speed of the tool with the formation of fine and equiaxed grains at the stir zone, while the elongated grains were examined at the thermo mechanically affected zone. Also, they examined that the tensile shear force initially increases by increasing the rotational speed and further increment of rotational speed reduced the tensile shear force due to the variations of heat input between the tool and workpiece contact surfaces.

3.2 Similar and Dissimilar FSP process

Dadashpour et al. [22] investigated the friction stir processing of cast AZ91C magnesium alloy with and without SiO2 Nano particles and also studied the effect of SiO2 Nano particle addition and the fracture mechanism on the mechanical properties of the fabricated joints. Also, they reported that the enhanced mechanical properties are found in the fabricated joints with nanoparticles addition related to the base cast alloys and joints without nanoparticles due to the evenly dispersion of the alloying elements in the weldment. Due to the evenly dispersion of the alloys, it deteriorates the fracture behaviour of the fabricated joints.

IOP Conf. Series: Materials Science and Engineering **954** (2020) 012046 doi:10.1088/1757-899X/954/1/012046

Khodabakhshi et al. [23] studied the deformation behavior of a hybrid aluminium matrix nanocomposite reinforced with TiO2, MgO and Al3Ti nanoparticles using FSP with threaded tool pin. They reported that the enhanced tensile properties are observed at the atmospheric temperature for the reinforcement of Al3Ti and MgO nanoparticles due to the dynamic strain ageing with a negative strain rate compared to the TiO2 nano particles reinforcement in the fabricated joints. Cioffi F et al. [24] investigated that the lateral off-set was the dominant parameter for the fracture location and resistance of the dissimilar friction stir welded AA2024 alloy and 17%SiC/2124Al composites which improves the bonding quality of the welded joints. They reported that the boundary of the alloy-composite mostly affects the strength of the welded joints through the variation in hardness.

Bodaghi et al. [25] aimed to make the joints of AA5052-O aluminium sheets with the addition of SiC Nano powder and investigated the effect of SiC nanoparticles on the mechanical properties and microstructural behaviour of the welded joints. The grain size of the welded joints at stir zone was reduced due to addition of SiC Nano powder during FSW process whereas the tensile strength was decreased. The microhardness was considerably increased compared to the welded specimen without Nano powder and higher loss of weight and poor wear resistance was observed due to improper distribution of Nano powder in stir zone.

4. Effect of Process Parameters on the Mechanical properties of the joints

4.1 Tool Pin Profile

The effect of the microstructure and mechanical properties of the of the friction stir welded AA6061-T6 aluminium alloy joints using various tool tilt angles and two different profiled pin tools were discussed by Banik et al. [26]. They found that the tapered threaded pin tool produced the stacked onion ring pattern, whereas there was no onion ring formation in the welded region. Also, they observed that the more stable flow of material using the tapered threaded tool with higher tilt angle and it enhances the tensile properties of the welded joints. Ramachandran et al. [27] studied the effect of geometry of the tool pin profile on the dissimilar friction stir welded AA5052 aluminium alloy and HSLA Steel joints using the various profile angles of the tapered tools. The tapered tool produced the significant penetration in steel side by the formation of the intermetallic compound which enhances the strength properties.

Palani et al. [28] discussed that the variations of welding speed affected the strength of the joints due to the lack of diffusion of the parent materials. Also, they examined that the tensile strength was improved by keeping constant rotational speed of the tool (1250 rpm) and welding speed (50 mm/min) with variations of plunge depth range of 2.25 to 2.5 mm. Further, they suggested that the hexagonal tool pin produced the higher tensile strength and higher hardness of the joints compared to other profiles. Amini et al. [29] FSW joints of AA 5083-O aluminium alloys successfully with different tool pin shapes. They examined that the concentric tool pin produced the more temperature relate to the offset pin due to the friction between the tool and work piece surfaces, whereas the temperature of the half pin and arched pin tools significantly increased with increase in rotational speed owing to the other tools. From the Figure 4, the maximum tensile strength was observed at the welding speed of 100 mm/min and rotational speed of 1120 rpm for the offset pin tool compared to other tools.

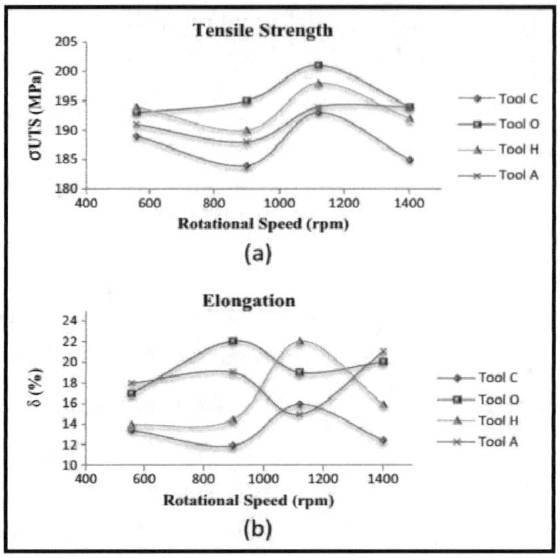

Figure 4 Results of tensile tests with a constant translational speed of 100 mm/min and the change of rotational speed (a). Ultimate tensile strength (b) Percentage of elongation of the tools: Concentric pin with tool shoulder (tool C), offset pin relative to tool shoulder (tool O), tool with half pin (tool H), and tool with arched pin (tool A)

The offset cone pin (Tool O) exhibited superior tensile strength due to the sufficient material flow which produces the fine equiaxed grains in the welded joints compared with other profiled tools, viz., half cone with concentric pin (Tool H), and arched concentric pin (Tool A), whereas concentric cone pin (Tool C) produced the lower tensile strength due to the transferring flow of material was impossible in the direction from front side to the back side of the tool. Dehabadi et al. [30] made an investigation on the microhardness of friction stir welded AA 6061 aluminium alloys using the two different pin profiles, such as triangular pin and tapered cylindrical pin profiles and employed the effect of the pin profiles on the welded joints in the stir zone. They reported that both the tools increased the tensile strength and yield strength of the welded joints due to the generation of higher heat input between the contact surfaces and also found that the quality of the welded joints on the strength was achieved by using conical tool.

Dressler et al. [31] investigated the feasibility study on the weld properties of the dissimilar friction stir welding of TiAl6V4 titanium alloy and AA2024-T3 aluminium alloy with 2 mm sheets. Due to the variations in properties of these alloys, the profiled tool pin was positioned to an offset distance from the weld centre towards the Al alloy side. They observed that the maximum tensile strength was found at the weld interface which was 73 % more than the strength of the Al alloy, whereas minimum harness and maximum local elongation was presented in Al alloy side. Due to the swirl-like structure formation at the nearer to the weld interface, the fracture was occurred at the interface between the alloys. Wang et al. [32] investigated the effect of tool rotational speed on the microstructure and mechanical properties of bobbin tool friction stir welding of Al–Li alloy joints. The asymmetric material flow occurred around the tool in the weld SZ due to the different frictional heat of the rotational speed of the tool which leads heterogeneous plastic deformation in the SZ. The increasing rotational speeds produce the low tensile elongation in SZ.

4.2 Welding Speed

Guo et al. [33] reported that the effects of materials position and welding speed of the tool on the flow behavior, microstructure and microhardness of the dissimilar aluminium AA6061 and AA7075 alloy joints. They reported that the AA6061 aluminium alloy side is more effective for proper material mixing than the AA7075 aluminium alloy side at the nugget zone of the welded joints. Also they observed the onion rings formation and decrease in hardness due to increase in welding speed of the friction stir welding tool. Lee et al. [34] studied the microstructure and mechanical properties of the friction stir welded AA6061 aluminium alloys and observed that the fine equiaxed grains, elongated grains and coarse grains were examined in the stir zone (SZ), Thermo-mechanically affected zone (TMAZ) and Heat affected zone (HAZ) respectively in the weld zone of the joints. They reported that the decrease in width of the strain and maximum strain was found nearer to the retreating side due to increment of welding speed. They realized that the increment of the welding speed decreases the ultimate tensile strength significantly.

Sabari et al. [35] made an investigation of the effect of tool pin profiles on the tensile properties in the stir zone of underwater friction stir welding of AA2519-T87 aluminium alloy joints. They used the four different tool pin profiles, such as straight cylindrical pin (SC), straight threaded cylindrical pin (STC), taper cylindrical pin (TC) and taper threaded cylindrical (TTC) for the welded joints. The defect-free welded joints were found for the straight threaded cylindrical and taper threaded cylindrical pin profiles compared to other pin profiles. The dissimilar lap welding between TC1 Titanium alloy and LF6 Aluminium alloy using FSW process by Yu-hua et al. [36]. The titanium alloy particles are stirred with aluminium alloys in stir zone by the FSW process conditions, such as increment of welding speed and/or decrement of rotational speed with minimum axial force of the tool.

4.3 Rotational Speed

Suri [37] reported that the increase in hardness was found by increasing the rotational speed of the tool on dissimilar friction stir welded aluminium alloy joints. The generation of more heat was found due to the fine grains at the stir zone which increases the hardness. Ghorbanzade et al. [38] investigated the effect of tool rotating speed on the microstructural evolution and mechanical properties of friction stir welded AA 2024-T3 aluminium alloys. They reported that 10 % increase in tensile strength and yield strength at the rotational speed range of 400 rpm to 800 rpm due to precipitates distributed evenly at the stir zone and the lower hardness was found at the speed range 400 rpm to 630 rpm at the stir zone compared to the hardness of the base material.

Ilangovan et al. [39] investigated the effects of tool pin profile on microstructure and tensile properties of dissimilar hot worked AA 6061 aluminium alloy with cold worked AA 5086 aluminium alloy materials. They reported that the threaded cylindrical tool gave the superior performance of the joints over the tapered cylindrical tool followed with straight cylindrical tool due to the formation of finer and uniformly distributed precipitates, circular onion rings and fine equiaxed grains at the Stir Zone.

Saravanan et al. [40] investigated the effects of the shoulder diameter to pin diameter ratio on tensile strength, microhardness and microstructure of the dissimilar FSWed aluminium alloys of AA2024-T6 and AA7075-T6 joints. They examined that the maximum hardness was occurred in weld nugget zone due to the formation of fine grains whereas the minimum hardness was observed in HAZ region on advancing side. Also they found that the superior tensile strength and microhardness were confirmed in NZ due the presence of fine recrystallized structures. Taveres et al. [41] investigated successfully the friction stir welding of AA2198 aluminium alloy plates and studied the microhardness distribution and tensile behavior of the joints. The Figure 5 showed that the hardness values in the SZ was more than the TMAZ and lower than the base metal. Also, they reported that the 90% of the tensile strength and yield strength was achieved in the stir zone whereas only 40% elongation was found at the zone for all the welded joints.

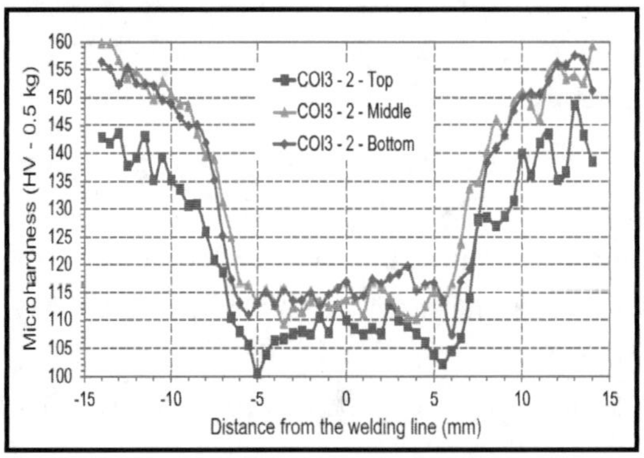

Figure 5 Microhardness measurements along cross section of the weld

4.4 Axial Force

Liu et al. [42] observed that the effect of solid solution strengthening was weaker than the effect of precipitation strengthening in TMAZ and SZ region due to the formation of β" and β'grains at the peak temperature. The appearance of these grains near the centre of the weld reduced the hardness gradually at the stir zone. From the Figure 6, the β" precipitates were increased by increasing the axial force which initially decrease the microhardness and then increased due to the formation of the solid solution strengthening in the SZ and TMAZ and "W" shaped microhardness profiles were observed from the retreating side .

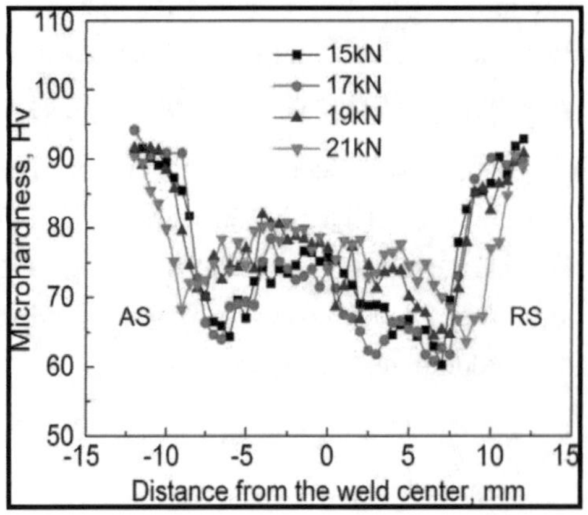

Figure 6 Effect of axial force on the microhardness distributions of the HSFSW joints

Song et al. [43] made an investigation of the effects of probe offset distance on the microstructure and mechanical properties at the interface of the dissimilar 2 mm thickness titanium alloy Ti-6Al-4V and aluminium alloy AA6061-T6 joints. They reported that the sound defect free joints occurred at the 750 rpm and 1000 rpm rotational speeds and 120 mm/min welding speed with proper probe edge offset distance. Also they found that the higher joining performance and formation of intermetallic compounds at the interface between the two dissimilar alloys were due to the proper selection of probe offset distance. Li et al. [44] investigated the dissimilar butt joints of Ti-6Al-4V and Al-6Mg alloys using FSW process. The reduced shoulder attrition of the FSW tool with no back flow of materials was benefited with the use of modified joint configuration and pin plunge setup of the joints. They examined the continuous intermetallic layers at the interface of the dissimilar joints which improve the properties of the joints and also the joint strength was 92% more than the strength of the aluminium base alloys.

5. Effect of Process Parameters on the Microstructure of the joints

The microstructure of the welded joints plays a major role in deciding the good quality of both fabricated welded joints. From this analysis, the presence of defects, nature of the defects and corresponding justification has obtained on the responses of the joints. Xue et al. [45] discussed the effect of interfacial microstructure evolution on mechanical properties and fracture properties of FSW 1060 aluminium alloy and pure copper plates of 5 mm thickness. They examined that the presence of intermetallic compounds (IMCs) enhances the mechanical properties in the welded joints. Due to the cooling time of the welded joints, the increasing size of the IMC layer in the welded SZ was inferior the strength properties.

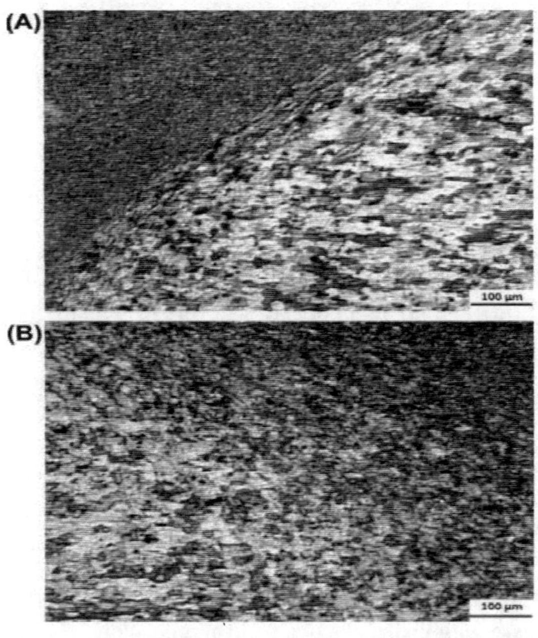

Figure 7 Microstructures of the nugget zone: (A) advancing side and (B) retreating side

The microstructural study was made in different zones, namely thermo-mechanically affected zone (TMAZ) and Stir zone (SZ) of the friction stir welded AA2024-T3 aluminium alloy joints by Franchim et al. [46]. The fine small grains sizes were observed in weld stir zone due to the proper material mixing by the frictional heat. From the Figure 2.5, the grains were arranged parallel to the weld interface in the advancing side of the weld transition zone which was contrasted grains appearance in base Al alloys, while the formation of highly deformed grains were presented in the weld interface. Barenji [47] was made the investigation on the mechanical properties and microstructure of the welded joints based on the process parameters of the friction stir welding process and also he investigated that the effects of the welding speed on the microstructure of the AA 7020 aluminium alloy welded butt joints. Also, they discussed the relationship between the heat input and the welding speed with rotational speed of the tool.

The stationary shoulder friction stir welded Ti-6Al-4V titanium alloy with the 7 mm thickness was successfully carried out by varying welding speeds and tool rotation speeds by Jiang et al. [48]. The formation inherited α-phase texture and reconstructed β-phase texture at the weldment were found by the variant selection analysis. They reported that the decrement in heat input and smooth surface appearance were found at the welded joints due to increase in welding speed and decrease in tool rotational speed. Komerla et al. [49] focused on the dissimilar welding of AA6016 aluminium alloy with DC04 deep drawing mild steel using friction stir welding process and investigated the microstructure, microhardness, surface texture and Grain size of the welded joints. The fine Fe grains were observed in the weld nugget zone (WNZ) whereas grain growth was examined in the thermo-mechanically affected zone (TMAZ) owing to the interaction between the alloy material and tool surface. Mao et al. [50] investigation was made on the microstructure and mechanical properties of the nugget zone on the friction stir welded AA7075-T6 aluminium alloy plates. The smallest grains at the top side and coarse grains at the root of the welded joints were observed by EBSD analysis. They examined that the maximum values of 415 MPa tensile strength, 255 MPa yield strength and 8.1% elongation were found for the AA7075-T6 aluminium alloy plates.

Palani et al. [51] studied the effects of the process parameters on the microstructure and microhardness of the dissimilar friction stir welded/processed AA5083-H321 with AA8011-H24 aluminium alloys. They showed that the defect-free sound welded joints were found compared to other joints from the macrostructural in addition to Al2O3 and SiC nanoparticles. Also, they reported that the maximum hardness was observed at the weld SZ in addition to Al2O3 nanoparticles in dissimilar AA8011-H24 and AA5083-H321 welded joints due to sufficient material flow occurred at the welded region at the SZ whereas, the variations of hardness occurred at the TMAZ which produces the lower hardness at the stir zone. The influence of FSP parameters on the tensile properties and microstructure of dissimilar AA8011-H24 and AA6061-T6 aluminium alloy joints in SZ were discussed by Palani et al. [52]. They reported that the better microstructure was occurred at SZ in addition of SiC nanoparticles than the other zones. Also they examined that the tensile strength of 98.58 % and 90.08 % were established at the weld nugget zone of the similar AA6061 aluminium alloys in addition to SiC nanoparticles and dissimilar friction stir welding of AA 8011 with AA 6061 aluminium alloys without nano additions respectively.

6. Conclusions

The friction stir welding and friction stir processing play a critical role in joining of the various combinations of dissimilar alloys due to their difference in mechanical properties and chemical composition in fabrication industries to decide the quality of the joints. The effects of process parameters on the weld properties and material characterization of the welded joints are effectively studied and also the performance of the welded joints is improved by applying the various optimization techniques. From the review, it can be realized that, the various studies are discussed in joining of the dissimilar alloys with the application of the different optimization techniques using friction stir welding and friction stir processing techniques. Also, this study is highly helpful to solve the complex problems which may occur in joining of dissimilar alloys with energy efficient and cost effective process.

7. References

[1] Amini K 2016 *Met Mater* ***54*** 71-75
[2] Aonuma M and Nakata K 2011 *Mater Trans* ***52*** (5) 948-52
[3] Barbini A, Carstensen J and dos Santos JF 2018 *J Mater Sci Tech* ***34*** 119-27
[4] Azizi A, Barenji RV, Barenji AV and Hashemipour M 2016 *Int J Adv Manuf Tech* ***86*** (5-8) 1985-95
[5] Charit I, and Mishra RS 2018 *J Mater Sci Tech* ***34*** 214-18
[6] Dehghani K, Ghorbani R and Soltanipoor AR 2015 *Int J Adv Manuf Tech* ***77*** (9-12) 1671-79
[7] Delijaicov S, Yakabu DY, De Macedo B, Resende HB and Batalha MHF 2017 *Int J Adv Manuf Tech* ***95*** (1-4) 1339-55
[8] Zhang CQ, Robson JD and Prangnell PB 2016 *J Mater Proc Tech* ***231*** 382-88
[9] Cao X and Jahazi M 2011 *Mater and Des* ***32*** (1)1-11
[10] Dorbane A, Mansoor B, Ayoub G, Shunmugasamy VC and Imad A 2016 *Mater Sci Engg: A* ***651*** 720-33
[11] Palani K and Elanchezhian C 2015 *Appl. Mech. Mater* **813** 451-55
[12] Palani K and Elanchezhian C 2015 *Appl. Mech. Mater* **766** 921-27
[13] Palani K and Elanchezhian C 2015 *Appl. Mech. Mater* **813** 446-450
[14] Farzadi A, Bahmani M and Haghshenas DF 2017 *Arab J Sci Engg* ***42*** 4905-16
[15] Sundaram NS and Murugan N 2010 *Mater and Des* ***31***(9) 4184-93
[16] Palani K and Elanchezhian C 2018 *IOP Conf. Ser.: Mater. Sci. Eng.* ***390*** 012061
[17] Winiczenko R 2016 *Int J Adv Manuf Tech 84* (5-8) 941-55
[18] Okuyucu H, Kurt A and Arcaklioglu E 2007 *Mater and Des* ***28*** (1) 78-84
[19] Sahu PK and Pal S 2015 *J Mag Allo* ***3*** (1) 36-46
[20] Palani K, Elanchezhian C, Vijaya Ramnath B and Ramadoss R 2020 *Mater Today: Proc* ***23*** (2) 573-582
[21] Shojaeefard MH, Khalkhali A, Akbari M and Tahani M 2013 *Mater and Des* ***52*** 587-92
[22] Dadashpour M, Mostafapour A, Yeşildal R and Rouhi S 2016 *Mater Sci Engg: A* ***655*** 379-87
[23] Khodabakhshi F, Gerlich AP, Simchi A and Kokabi AH 2015 *Mater Sci Engg: A* ***626*** 458-66
[24] Cioffi F, Ibáñez J, Fernández R and González-doncel G 2015 *Mater and Des* ***65*** 438-46
[25] Bodaghi M and Dehghani K 2017 *Int J Adv Manuf Tech* ***88*** (9-12) 2651-60
[26] Banik A, Roy BS, Barma JD and Saha SC 2018 *J Manuf Proc* ***31*** 395-04
[27] Ramachandran KK, Murugan N and Shashi Kumar S 2015 *Mater Sci Engg: A* ***639*** 219-33
[28] Palani K, Elanchezhian C, Vijaya Ramnath B, Bhaskar GB and Naveen E 2018 *Mater Today: Proc* ***5*** 24515–24524

[29] Amini S, Amiri MR and Barani A 2015 *Int J Adv Manuf Tech* ***76*** (1-4) 255-61
[30] Dehabadi VM, Ghorbanpour S and Azimi G 2016 *J Cen Sou Uni* ***23*** (9) 2146-55
[31] Dressler U, Biallas G and Mercado UA 2009 *Mater Sci Engg A* ***526*** (1-2) 113-17
[32] Wang T, Zou Y and Matsuda K 2016 *Mater & Des* ***90*** 13-21
[33] Guo JF, Chen HC, Sun CN, Bi G, Sun Z and Wei J 2014 *Mater and Des* ***56*** 185-92
[34] Lee WB, Yeon YM and Jung SB 2004 *Mater Trans* ***45*** (5) 1700-05
[35] Sabari SS, Malarvizhi S and Balasubramanian V 2016 *Int J Mech Mater Engg* ***11*** (5) 1-14
[36] Yu-hua C, Quan NI and Li-ming KE 2012 *Trans Nonfer Met Soc China* ***22*** 299-04
[37] Suri GS, Kaur G and Luthra BS 2016 *Ind J Sci Tech* ***9*** (36) 9-11
[38] Ghorbanzade T, Soltanipour A, Dehghani K and Chabok A 2014 *Proc Instit Mech Eng, Part L: J Mater: Des and Appl* ***230*** (1) 75-87
[39] Ilangovan M, Boopathy SR and Balasubramanian V 2015 *Def Tech* ***11*** (2) 174-84
[40] Saravanan V, Rajakumar S and Muruganandam A 2016 *Metal Micro and Anal* ***5*** (6) 476-85
[41] Tavares SMO, dos Santos JF and de Castro PMST 2013 *Theo and Appl Fract Mecha* ***65*** 8-13
[42] Liu CY, Qu B, Xue P, Ma ZY, Luo K, Ma MZ and Liu RP 2018 *J Mater Sci & Tech* ***34*** 112-18
[43] Song Z, Nakata K, Wu A, Liao J and Zhou L 2014 *Mater and Des* ***57*** 269-78
[44] Li B, Zhang Z, Shen Y, Hu W and Luo L 2014 *Mater and Des* ***53*** 838-48
[45] Xue P, Xiao BL and Ma ZY 2015 *Metal Mater Trans A* ***46*** (7) 3091-103
[46] Franchim AS, Fernandez FF and Travessa DN 2011 *Mater and Des* ***32*** (10) 4684-88
[47] Barenji RV 2015 *Proceed Instit Mech Eng, Part L: J of Mater: Des and Appl* ***230*** (2) 663-73
[48] Jiang X, Wynne BP and Martin J 2018 *J Mater Sci & Tech* ***34*** 198-08
[49] Komerla K, Naumov A, Mertin C, Prahl U and Bleck W 2018 *Int J Adv Manu Tech* ***94*** (9-12) 4209-19
[50] Mao Y, Ke L, Chen Y, Liu F and Xing L 2018 *J Mater Sci & Tech* ***34*** 228-36
[51] Palani K, Elanchezhian C, Sai Prakash K H V, Deepak Kumar, Dayanand, Keshav Kumar and Sreekanth K 2018 *IOP Conf. Series: Mater Sci Engg* **390** 012072
[52] Palani K, Elanchezhian C, Avinash K, Karthik C, Chaitanya K, Yugandar K and Karthik S 2018 *IOP Conf. Series: Mater Sci Engg* ***390*** 012108

IOP Conf. Series: Materials Science and Engineering **954** (2020) 012047 doi:10.1088/1757-899X/954/1/012047

Tribological and Mechanical behaviour of reinforced polymer composite material for clutch facing in automobile application

JGK Kumar[1*], R.Venkatesh Babu [2]

[1]Department of Mechanical Engineering, Research Scholar, Bharath Institute of Higher Education and Research, Chennai-600078

[2]Professor, Department of Mechanical Engineering, Bharath Institute of Higher Education and Research, Chennai-600078

[1*]Corresponding Author: kumarseries@yahoo.com

Abstract. In current applications, the conductive composite polymers are investigated by filling up the polymer matrixes along with different Carbon blacks. The carbon black is considered as particulate filler in all the polymer industry. The most available carbon blacks are used from thermal cracking of natural gas and furnace black which are generated by incomplete combustion of oil filled stocks. Hence, it is necessary to implement and develop the alternative best source of the fillers from the renewable resources like wastes in agriculture units, oil palms, bamboo stem and coconut shells. The coconut shell which consists of 70.5% Carbon, 0.99% ash, 31.75% lignin and 19.5% cellulose and 70% hemicelluloses are considered compare to the agricultural waste. In this work, the investigation is made on mechanical properties and tribological behaviour of polymer matrix composite with coconut shell char.

1. Introduction

Abrasive The natural available fibers and the utilization of the fibers as composite material for automobile industry for clutch facing gives the good reinforced polymer composite with the low cost and high end mechanical properties since they are good renewable and biodegradable to the commonly used synthetic reinforce like glass fiber [1-3]. The preparation from polymer matrix composite replacing with the help of synthetic fiber with natural fiber like sisal, jute, accordingly extensive studies on preparation and properties of polymer matrix composite (PMC) replacing the synthetic fiber with natural fiber like Jute, Sisal were carried out [4-6]. These kinds of plants fibers have much more benefits against carbon fiber in form of they are renewable, eco friendly, very low cost and some times freely available, light in weight and also they have high specific mechanical performance over others. The synthetic filler materials like thermoplastic polymers were used in the previous days in epoxy resin composites to function as filler materials [7-9]. In our present research work the main effectiveness of the coconut shell particles as a source of natural material were reinforcing epoxy resins and their mechanical, impact, flexural and wear behaviour were investigated.

2. Materials used for Experimental work

Materials used in this experimental work are listed below:

1. Araldite Epoxy resin LY 556
2. Hardener HY951
3. Coconut shell
4. Jute

2.1 Epoxy resin

The Epoxy resin which is called by grade LY 556 which is an unmodified epoxy resin based on Bisphenyl-A purchased from Roto polymers and chemicals , Chennai.

2.2 Hardener

Hardener HY951is and aliphatic Primary amines which has a viscosity of around 11-25 MPa at 30^0 c which was used along with the matrix material was purchased from S&S Polymers, Bangalore.

2.3 Coconut shell powder

The shell powder of coconuts was prepared locally. They were first cut into small pieces with the help of hammer. Then these pieces were grinded with the help of crushed and made in the form of fine powder. The size of the particle was around -80 to + 50 microns selected from the different mesh sizes from 50 to 70.

3. Experimental work

Figure.1 Procedure of making raw coconut shell powder

3.1 Preparation of coconut shell char

The coconut powder is set up by the carburization temperature chose was 650˚c till 800˚c. The required amounts of coconut shell powder were taken in pots and were set in the heater. It took around three to four hours to achieve the required temperature. At this temperature one hour drenching time was permitted [10]. After this 24 hrs cooling was permitted with the goal that the heater comes to room temperature. At that point the carburized coconut shell powders were taken out from the heater for reuse.

IOP Conf. Series: Materials Science and Engineering **954** (2020) 012047 doi:10.1088/1757-899X/954/1/012047

3.2 Jute fiber

The Jute filaments were acquired from the Linum usitatissimum stems and they are utilized essentially to make cloth, the specific plant utilized for fiber generation. Be that as it may, the structure is increasingly crystalline in nature and make more grounded and stiffer to deal with at any conditions [11]. The Jute strands are in the range long up to 90cm and the normal measurement is around 13 to 15 microns. Since they assimilate and discharge water immediately they are valuable and have no erosion.

A wood made mould was used for preparation of composite sheet for casting purpose. For the fast removal of composite sheet a separate sheet called mould release sheet was placed above the glass plate. And also the spray for mold release was also used at the inner surfaces. The different weight percentage of coconut shell powder, jute and char and flax are about 5, 15, 20 and 25% in weight were mixed with the matrix material which consisting of epoxy resin and also the hardener HY951 in the proportion of 10:1. Proper care was taken in order to avoid the formation of air bubbles during the process of pouring is shown in Figure 1.

Now the pressure was gradually applied from the upside and the mold was allowed to cure about 3 to 4 days at 300C. During the application of pressure some polymer squeezes out from the mould. For this, care has already been taken during pouring. After 72 hrs the samples were taken out of the mold, after curing the laminate was cut into required size of erosion and other mechanical tests by diamond cutter [12]. In the present case the composites prepared for raw coconut shell and the carbonized char at 650°C and 800°C. The density of epoxy resin, coconut shell fiber and char is found respectively. Tables 1 and 2 show the particle size and density of the shell powder and tensile test specimens are represented in Figure 2.

Table 1. Particle size

Sample	Mesh number	Size in Microns (Min)	Size in Micron (Max)	Weight (Grms)	Spacing inches	Weight (%)
1	40	-	410	129	0.0172	20%
2	**65**	**400**	**218**	**156**	**0.0056**	**35%**
3	110	212	150	44	0.0034	10%
4	140	145	105	31	0.0031	7%
5	190	101	75	34	0.0024	12%

(a)

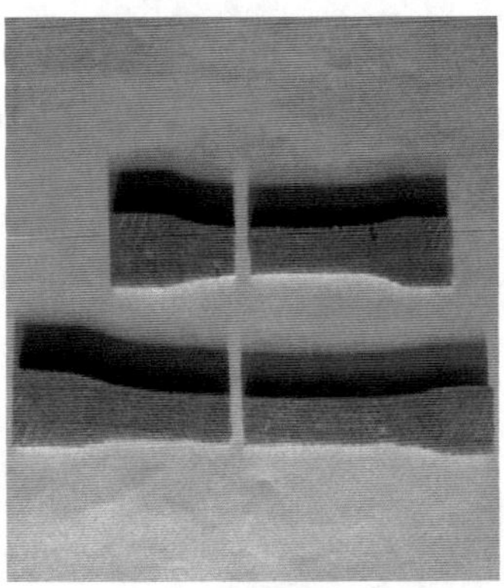

(b)

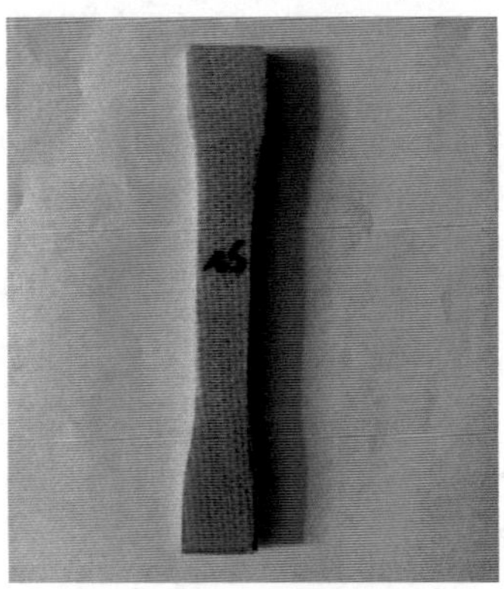

Figure 2 (a) Specimen for Tensile test and (b) Flexural Test

IOP Conf. Series: Materials Science and Engineering **954** (2020) 012047 doi:10.1088/1757-899X/954/1/012047

Table 2 Density of Various Samples under different conditions

Sample	Density(gm/cm^3) For Raw Fiber	Density(gm/cm^3) For 650^{0}c carbonized char	Density(gm/cm^3) For 950^{0}c carbonized char
5%	0.500	1.940	1.20
15%	0.621	1.234	1.00
20%	0.645	1.243	1.00
25%	1.621	1.050	1.04

4. Micro-Hardness test

The micro-hardness test was done with the help of a Lecco Vickers Hardness -LV 700 equipment. A indenter called diamond which is in the shape of a right pyramid and base is in the form of square and at an angle 1350 between the two opposite faces, load (P) was applied on material is shown in Figure 3. The both the X and Y diagonals of the indentation left on the on the above surface area of the sample were measured and noted and further their mean of X was calculated and noted.

Table 3 Hardness of different Samples under various conditions

Sample	Hardness (Hv) for Raw Fiber	Hardness (Hv) for 650^{0}c carbonized char	Hardness (Hv) For 950^{0}c carbonized char
5%	191.9	230.4	249.14
15%	198.2	213.7	286.36
20%	199.7	212.7	211.72
25%	169.1	205.9	289.44

5. Mechanical Properties of Composite Material

The different mechanical tests like tensile, flexural and hardness of the prepared composite material were conducted and results were noted with the help of ASTM Standard.

5.1 Tensile Strength

The tensile test was conducted for 5 different composite samples as per ASTM D 3039-76. The total length of the sample piece was 120 mm and the test was performed with the help of UTM TFUN-400 at the speed of 10mm/min is shown in Figure 3 and the mechanical properties of raw coconut shell powder fiber with flax and epoxy composite is shown in Tables 3 to 6 respectively.

Figure 3. TFUN-400 Testing Machine

5.2 Flexural testing

The Flexural test was also performed on the above machine as per the ASTM standard. The standard size was taken 150x20mm as length by width were tested in the 3-point bending with a span depth ratio 15:1. The tensile test was also performed with the help of same machine with 2mm/min at a load of 15kN. The flexural stress was determined with the help of following relation is shown in Figure 4.

Table 4 Mechanical properties of raw coconut shell powder fiber with flax and epoxy composite

Fiber volume %	Tensile Strength in MPa	Flexural Strength in MPa
Neat epoxy	13.49	13.46
5	6.8	20.34
15	24.32	65.53
20	10.31	51.34
25	9.46	46.68

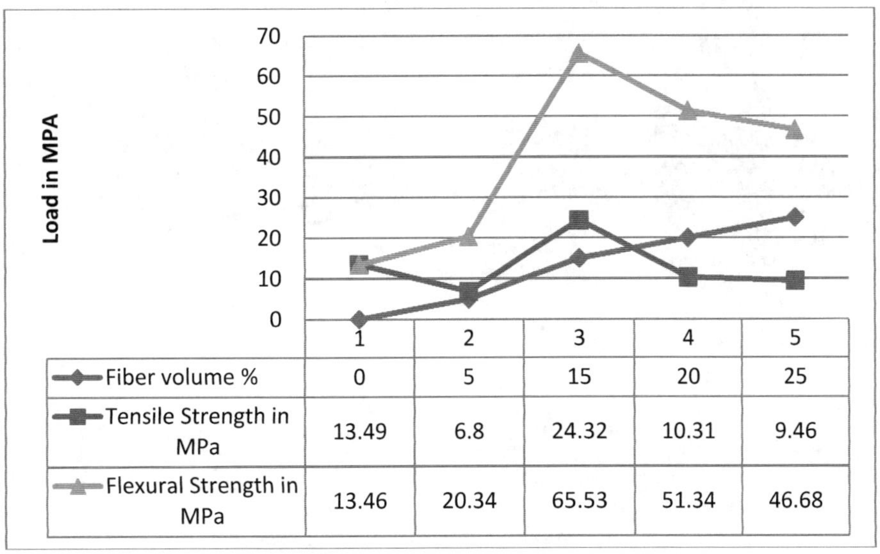

Figure 4. Strength Vs Load

Table 5 Carbonized coconut shell char mechanical properties of epoxy composite at 650^{0}C.

Fiber volume %	Tensile Strength in MPa	Flexural Strength in MPa
Neat epoxy	14.73	19.26
5	15.75	45.54
15	17.25	48.21
20	30.73	59.28

Table 6 Carbonized coconut shell char mechanical properties of epoxy composite at 800^{0}C.

Fiber volume %	Flexural Strength in MPa	Tensile Strength in MPa
Neat epoxy	19.46	14.25
5	69.25	23.76
15	35.26	25.72
20	31.14	17.48

6. Results and Discussions

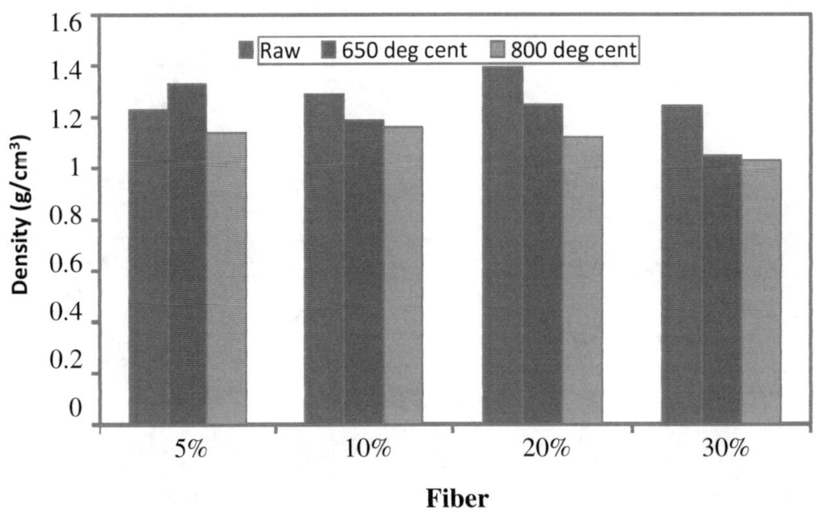

Figure 5. Histogram showing the density of all composites at different carburized temperature

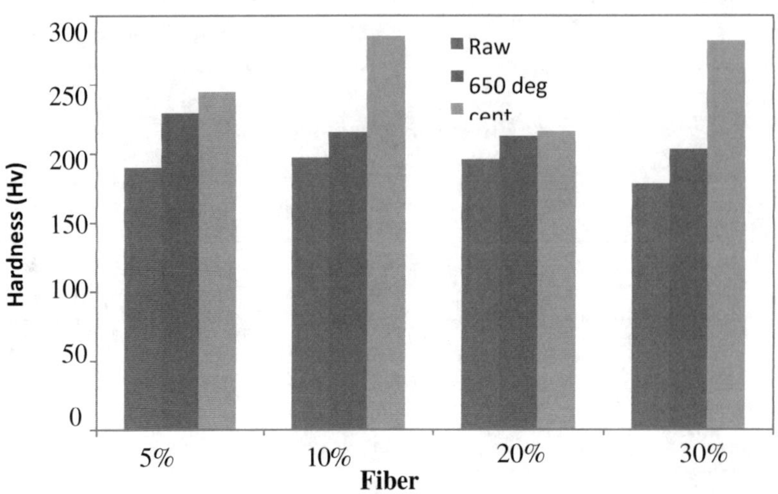

Figure 6 Histogram showing the hardness of all composites at different carburized temperature

IOP Conf. Series: Materials Science and Engineering **954** (2020) 012047 doi:10.1088/1757-899X/954/1/012047

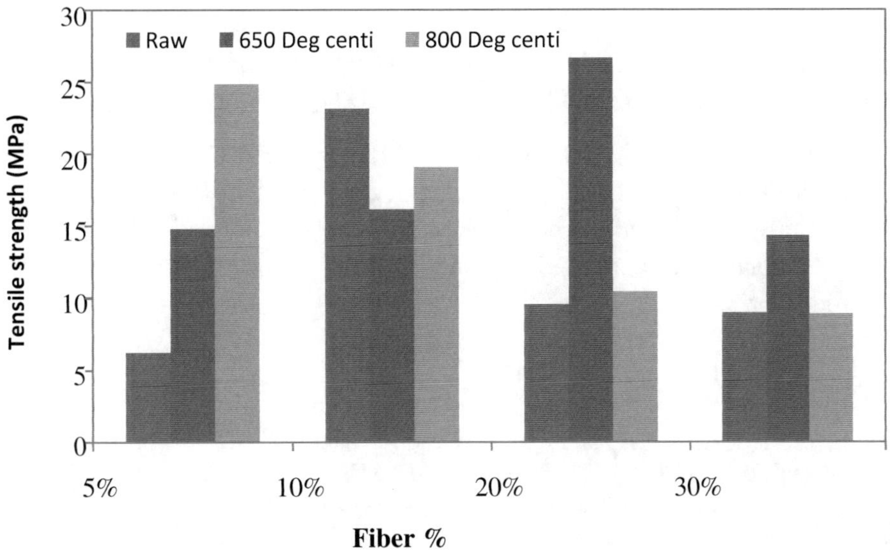

Figure 7 Histogram showing the tensile properties at different carburized temperature

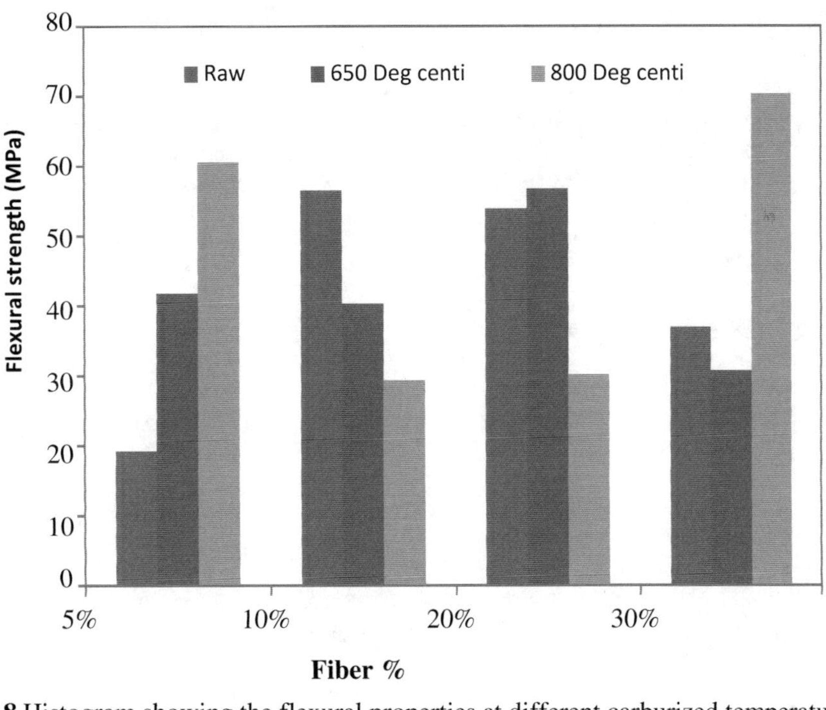

Figure 8 Histogram showing the flexural properties at different carburized temperature

The density test results for various specimens which were prepared with raw coconut and char powder particles with different volume fraction were plotted in figure5. The plot shows that, the density of the composite prepared with char is less when compared to the raw coconut particulate composite. It is also noticed that with increase of fiber concentration the density goes on increasing and samples with 20% fiber volume fraction of fibers and suddenly decreases to some extent because void formation.

7. Conclusions

The effectiveness of the coconut shell particles with reinforcement of epoxy resins are effectively studied on their mechanical, impact, flexural and wear behaviour of the composites.

- The density of the prepared composite with char is less when compared to the raw coconut particulate composite.
- The increase of fiber concentration, the density goes on increasing and samples with 20% fiber volume fraction of fibers and suddenly decreases to some extent because void formation.

8. References

[1] Shadrach Jeya Sekaran A, Palani Kumar K and Pitchandi K 2015 *Bul. Mater Sci* ***38*** 1183
[2] Saheb D and Jog J P 1999 *Adv Polym Technol* ***18*** 351
[3] Hristozov D, Wroblewski L and Sadeghian P 2016 *Compo Part B* ***95*** 82
[4] Ahmad F, Choi H S and Park M K 2015 *Macromo. Mater Eng* **300** 10
[5] Rwawiire S, Tomkova B, Militky J, Jabbar A and Kale B M 2015 *Compo Part B* ***81*** 149
[6] Mangal R, Saxena N S, Sreekala M S, Thomas S and Singh K 2003 *Mate. Sci Eng A* ***339*** 281
[7] Yousif B F, Lau S T W and McWilliam S 2010 *Tribol Int* ***43*** 503
[8] Hadjadj A, Jbara O, Tara A, Gilliot M, Malek F, Maafi E M and Tighzert L 2016 *Compo Struct* ***135*** 217
[9] Anbukarasi K and Kalaiselvam S 2015 *Mater Des* ***66*** 321
[10] Ridzuan MJM, Abdul Majid MS, Afendi M, Aqmariah Kanafiah SN, Zahri JM and Gibson AG 2016 *Mater Des* ***89*** 839
[11] Garkhail S K, Heijenrath R W H and Peijs T 2000 *Appl Compos Mater* ***7*** 351
[12] Pothan L A, Oommen Z and Thomas S 2003 *Compos Sci Technol* ***63*** 283

IOP Conf. Series: Materials Science and Engineering **954** (2021) 012050 doi:10.1088/1757-899X/954/1/012050

Corrigendum: Tribological characteristics of natural fiber composite - Review (*IOP Conf. Ser.: Mater. Sci. Eng.* 954 (2020) 012048)

K.Mohan[1] *, T.Rajmohan [2]
[1] Sri Chandrasekharendra Saraswathi Viswa Mahavidyalaya, Enathur, Kanchipuram - 631561, Tamil Nadu, India

Page 1:

The figure 1 should be read as,

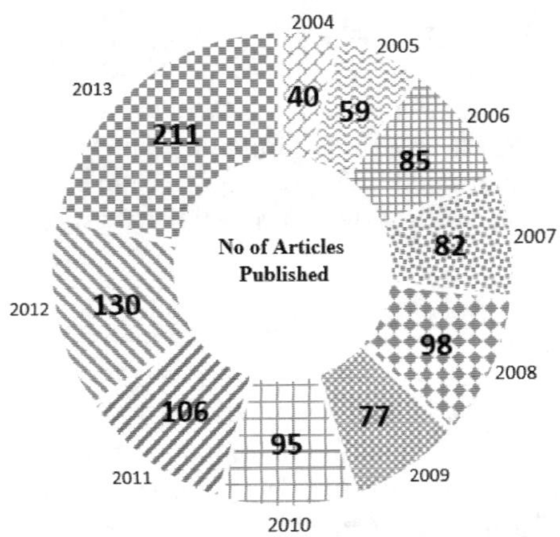

In the Introduction section, the following text appears in below the graph:

" Figure 1: Number of articles published based on tribological evaluation of synthetic and natural fibers [4]."

This should read:

" Figure 1 - Number of articles published based on tribological evaluation of natural fibers [4]."

Tribological characteristics of natural fiber composite - Review

K.Mohan[1 *], T.Rajmohan[1]

[1]Department of Mechanical Engineering, Sri Chandrasekharendra Saraswathi Viswa Mahavidyalaya, Enathur, Kanchipuram - 631561, Tamil Nadu, India.

[1 *]mohank84@gmail.com

Abstract. The continuous research into green environment for engineering based applications the present work tasks to dedicate a gathering of improvements in the research of tribological behaviour of natural fibre reinforced polymeric composites. Information on natural fibres, composites and the wear characteristics such as specific wear rate and co efficient of friction were presented with respect to change of fiber weight or volume fraction, addition of nano fillers and fiber treatments. Various operating parameters and wear mechanism for various types of composites and their performance will be discussed.

1. Introduction

In tribological materials with low noise and self-lubrication capabilities, fiber-reinforced polymer composites (FRP) play an important role. FRP offers more possibilities to optimize the tribological performance of frictional parts through various possible material combinations. Because of this, FRP composites are widely used in different applications and most of the manufacturing parts are subjected to tribological testing based on their applications like clutch pad, brakes lining and brakes pad, including gears, bearings, and prosthetic joints, etc. [1]. The use of natural fibers as reinforcing materials in both thermoplastic and thermoset matrix composite provides optimistic environmental support with consideration for ultimate disposability and best use of raw materials[2-3].

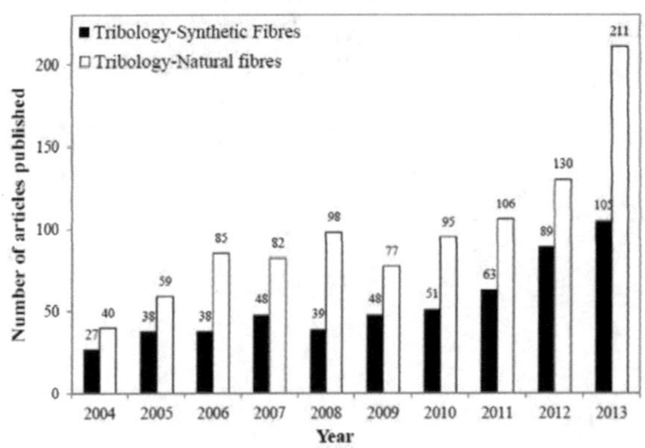

Figure 1 Number of articles published based on tribological evaluation of synthetic and natural fibers [4]

Tribo materials have a unique property profile and have a wear, friction, lubrication and mechanical properties spectrum that are used to design materials for specific tribological applications. The researchers have focused more on the tribological properties of natural fiber-reinforced composites since the last few decades. The Figure 1 shows natural fiber enhancements over synthetic fibers based on the published articles. In the field of tribological applications, a gradual increase can be found in natural fiber research.

Figure 2. Tribological application of Composites [5]

2. Effect of fiber wt% on tribological behavior of natural fiber composites

The application of the spectrum of natural fiber reinforced polymer composites is growing rapidly in various engineering fields. Using hot compression technique, the different types of natural fibers (i.e. nettle, glewia, optical, and sisal) were incorporated into PLA polymer to fabricate composites laminates. Different operating parameters such as applied load (10-30N), sliding speed (1-3 m/s) and sliding distance (1000-3000 m) under dry contact condition tribological characteristics were analyzed. The results indicates that the incorporation of natural fiber mats into the PLA matrix significantly improves the neat polymer's wear behavior. Compared to neat PLA, there was 10-44% reduction in friction coefficient and more than 70% reduction in specific wear rate reported by pajbai et al [6]. The effects of zeolite particles on the mechanical and tribological properties of the Ultra-high molecular weight polyethene (UHMWPE) composites. The tribological properties are carried out under various applied loads of 10, 20, and 30 N with various sliding speeds of 0.209 m/s, 0.419 m/s and 0.838 m/s. Using compression molding technique the composites were fabricated. They found that the lowest average COF was also obtained with the 20% zeolite/UHMWPE composite. With increasing applied loads for all samples, the wear volume loss was found to increase[7].

The wear behavior of polyphenylene sulfide composites has been investigated by Besnea [8]. The composites fabricated with 10 wt% carbon fiber, 10 wt% graphite and 10 wt% polytetrafluoroethylene and another composites with 40 wt% glass fiber. Coefficient of friction and wear rate were analyzed using a pin-on-disc device. The polymer composite with polyphenylene sulfide matrix, carbon fibers, graphite, and polytetrafluorethylene exhibits enhanced wear resistance. The wear behavior of jute reinforced polypropylene composites at various loads and varying sliding velocities. The composites were prepared by using an injection molding machine at different fiber weight percentages 5%, 7.5, 10%, 12.5%, and 15%. At different sliding speeds and loads are 50cm / sec, 75cm / sec and 100cm / sec and 4, 6 and 8 kg the composites wear behavior has been studied. The wear and friction coefficient of jute is improved with the increase in normal loads. Decrease in friction coefficient at 50cm / sec and 100cm with the incorporated fillers[9].

Nordin et al. [10] investigated the kenaf polyester composites tribological behavior. The lengthy kenaf fiber is used to fabricate the composite. The epoxy resin and unsaturated polyester were mixed with hardener at weight ratios of 3:1 and 10:1, respectively. The wear test was carried out using an Abrasion Resistance Tester (TR-600) in dry sliding conditions at a constant velocity of 1.4 m/s. With different applied loads ranging from 5 to 30N the test was conducted at room temperature. The results indicates that the wear characteristic of KEC is higher than the KPC composite material. The centre of tribological consideration such as friction and wear of tribo-composites depends on various issues such as the class of polymers, surface characteristics, type (chemical or mechanical), and degree of adhesion between the transfer layer and counterface material reported by Yousif et al. [11]. Among the various effects, the transfer layer was found to be the critical factor deciding the tribological performance of polymer-based composite materials. Generally, the depletion of soft polymer arising from the wear process may result in the formation of a thin film, which is often referred to as the transfer layer. The existence of the transfer layer is beneficial by maintaining low friction and reduced wear[12].

Yousif et al. investigated the frictional and wear behavior of the composites using a block on disk machine at different applied loads and sliding distances at 2.8 m/s sliding velocity under dry/wet contact conditions. They found that composite polyester reinforced by betelnutfiber (BFRP) had better wear and frictional performance compared to dry under wet contact condition. The wear mechanism of the BFRP composite was predominated in the polyester regions by micro and micro cracks and fiber debonding. They used water to improve BFRP composite wear and friction performance by about 50% and 94% respectively[13].

The Sugarcane fiber/polyester (SCRP) and glass fiber/polyester (GRP) composites (i.e. with chopped fibers of 1, 5, 10mm length randomly distributed and unidirectional mat fibers) using compression mold and hand-lay-up techniques. The friction coefficients and wear rates of the composites SCRP and GRP were determined by applying the load at 2.8 m/s in the range between 5 -30N sliding speed. They found the composite of SCRP that can compete with the composite of GRP. When tested in AP-O and tested in P-O, the U-SCRP composite displayed the lowest friction coefficient. [14] Nomex fabric/phenolic composite, and its tribological properties and evaluated under dry and water-bathed sliding conditions by a pin-on-disc tribometer. The Nomex fabric was cleaned by a Soxhlet extractor in petroleum ether and ethanol in turn, and then dried in the oven. It was found that the penetration of distilled water/sea water induces damage to the mechanical properties. Owing to the NaCl, Mg $(OH)_2$ and $CaCO_3$ deposited on the pin surface, the wear results show that Nomex fabric / phenolic composite was much higher in a water-borne sliding condition than in a distilled water-borne sliding condition. [15]. The tribological behavior of novolac phenolic resin matrix composites reinforced with three kinds of carbonaceous fibers in sliding contact against cast iron. Samples with different type and volume fraction of carbonaceous fibers were prepared and tested in sliding contact against cast iron in a pin-on-disc wear testing machine. The worn surfaces and the debris were analyzed by SEM analogy. They found a reduction in wear rate of composites was observed with the increase in carbonaceous fiber volume fraction. Friction coefficient also showed a trend to reduce with the increase of fiber volume fraction[16]. Natural fibers are used in varied applications of wear and friction. The advantage of natural fibers over ancient reinforcing materials includes low cost, low density, Biodegradability, and recyclability. Abrasive wear rate is tested by employing a two-body abrasion wear tester and also the bagasse fiber reinforced epoxy composite powerfully depends upon load and abrasive grit size. With a rise of load and grit size, the wear and tear rate will increase. The orientation of fiber in composites contains an important influence on the wear and tear rate of composite reported by Punyapriya et al. [17]. The importance of natural fiber with the initiation of sustainable development is narrated, and also the composite materials have currently become a lot of prominent in several applications. Several natural fibers reinforced hybrid composites are being introduced in the aviation business, construction, industrial applications, automotive components, bearing and plenty of others, creating tribo testing a lot of demanding reported by Nirmal et al. [18]

The wear behavior of lantana fiber reinforced polymer epoxy matrix composites. The composites were prepared using hand layup method then the respective size of wear specimens has been prepared. The tribological test was carried out on a pin on disc machine against the 400 grit size abrasive paper with a load of 5,10,15,20 and 25N and a test speed of 0.314m/s. The composites having 40% of fiber content indicates the optimum wear reduction also it is observed with increase in normal load the wear losses are increased. [19]. Tribological behavior of short glass fiber reinforced thermosetting polyester composite. The wear rate of polyester is much lesser than the unreinforced polyester. At 10 wt% of glass fiber content, the wear rate and coefficient of friction are minimal. Further increasing the proportion of glass fiber, wear rate and coefficient of friction is increased. The wear rate is increased

while increasing the sliding speed and load but there is no significant change in the coefficient of friction. [20] Glass fiber based epoxy composite using hand layup method by varying their weight percentage of fiber such as 40%, 60%, and 80%. The fillers are added with and without 10wt%. The wear rate is decreased by increasing the fiber content. The filled composites exhibit superior wear performances and mechanical behavior than the unfilled composites. [21]. the tribological performances of short PALF reinforced Bisphenol-A composites. The composites have been fabricated by changing their fiber length such as 2, 4, 6, 8, 12,14mm. Using pin on disc machine the wear behavior of the composite is tested by varying a load of 5N, 10N and 15N at constant sliding speed, velocity, and speed. They have reported, when increasing the load the wear rate is increased due to the less adhesive bond between fiber and matrix. [22]
Nirmal et al. made a review on tribological performances of natural fiber reinforced polymeric composites. Various operating input parameters were considered such as load, speed, sliding distance for different natural fibers and concluded, significant improvement of wear and friction is attained in polymeric composites reinforced with natural fibers. The cellulose content of natural fibers mainly attributing to reduce or less abrasive damage i.e fibers generally soft by nature. Due to the well adhesive of natural fibers its tends to reduce the impact of thermo mechanical loading and the formation of transfer film or back film protect the surfaces from severe wearing or improve the resistance of wear rate and friction. Fiber treatments and fiber loading and nano fillers can further enhance the tribological properties. [23]. The wear properties of jute fiber reinforced polypropylene composites and reported that the wear resistance is increased while addition of woven jute fabric into polypropylene matrix and the coefficient of friction is reduced to 45% and the specific wear rate also decreased by 65% for further addition of Jute fiber.[24] The effect of unidirectional cotton fiber reinforcement on friction and wear characteristics of polyester. They established that the friction coefficient increases and the wear rate decreases with increasing fiber volume fraction up to 40%. [25]

3. Effect of Nano fillers on tribological behavior of natural fiber composites

Wang et al. [26-28] have reported filled PEEK with various weight fractions of nano fillers such as ZrO_2, SiC, SiO_2, Si_3N_4, and MWCNT. The wear resistance is improved and the friction coefficient is reduced by the addition of nano fillers less than 10wt% infractions. Two factors attribute enhanced wear and friction coefficient are transfer film formation and steel counterface smoothing. Investigated the effect of varying weight percentages of sustainable kenaf particles as specific fillers and reinforcement in epoxy thermoset matrix on the wear and friction properties. They demonstrated that the connotation of slight wear occurs at advanced loading condition with mild plastic deformation. There was more substantial origination of transfer films at higher loads, which would defend the composite from additional wear and thus dip the wear rate. [29]

Due to their superior qualities such as large aspect ratio, good mechanical strength, thermal conductivity and a higher thermal degradation limit, the MWCNTs are established among the numerous types of nano materials available. In the case of modified MWCNTs at high temperature, hybrid composites of PTFE / cotton fabrics in the phenolic matrix with the addition of MWCNTs show better wear resistance. The well-dispersed MWCNTs in a matrix and the transfer layer adhesion can be credited to enhance wear resistance and it is shown that adding MWCNTs shows improved COF wear performance. The well-distributed MWCNTs in a matrix can be ascribed to better wear resistance due to the thermal insulation behavior and rolling effect of MWCNT reported by Zhang et al. [30] The wear behavior of PEEK composites under dry sliding conditions. The presence of constantly decreasing coefficient of friction due to counter face transfer layer formation. Even in the case of higher loading due to the disintegration of agglomeration and improved interfacial adhesion between filler and matrix, the superior friction properties were found.[31] Prepared the short carbon fiber reinforced polyimide (PI) composites filled with particulates of graphite by a heat molding technique. The friction and wear behaviours of PI composites sliding against GCr15 steel rings were evaluated on an M-2000 model ring-on-block test rig. The anti-wear and friction-reducing abilities of the short carbon fiber reinforced PI composites are significantly improved by incorporation particulates of graphite. The differences in the friction coefficient and wear properties of the short carbon fiber reinforced PI composites filled with a solid lubricant are closely related with the sliding condition such as sliding rate and the applied load. [32]

Various types of nano- filling materials, such as Al_2O_3, TiO_2, ZrO_2, ZnO, Si_3N_4, $CaCO_3$, SiC, SiO_2, and nano-CuO, were used in different types of polymers like Epoxy, Polyesters, PMMA, PEEK and PTFE. Nano scale fillers are exhibiting good tribological properties rather than micro scale fillers. The wear resistance and friction coefficient of nano scale filled composites show an increase in values while increasing the filler concentration. Wear behavior of alumina (Al_2O_3) filled glass fabric reinforced epoxy (G-E) composites. Alumina filled G-E composites containing 0, 5, 7.5 and 10 wt% were prepared using the hand lay-up technique by compression molding. Studies of two-body abrasive wear against waterproof silicon carbide abrasive paper were conducted. It was found that the loading of the filler increases the volume of wear and increases as the abrading distance increases. For Al_2O_3 filled G-E composites, excellent wear resistance was obtained. In addition, filler loading of 10wt% resulted in a much lower wear loss.[33]

Tribological and mechanical properties of silicon carbide filled a hybrid glass and basalt fiber reinforced composites. The composites were prepared using a compression molding technique. Increased flexural strength and modulus while increasing silicon carbide's wt%. The specific wear rate of the composite is decreased when the sliding distance is increased. The optimum wear resistance obtained in 6% of silicon carbide filled hybrid composites.[34] The friction and wear behavior Cu-Ni-TiC composites at four different normal loads of 5, 10, 15, and 20 N. A constant sliding speed of 1.25 m/s was maintained while sliding against a hardened counter face made of EN31 steel (HRC 60) under ambient conditions using a pin-on-disc test rig. The addition of 4 wt% TiC was showed the optimum friction and wear performance due to its higher hardness and ability to hold a transfer layer. The composite wear rate and the Cu4Ni matrix alloy increased with the load linearly. [35]

4. Effect of fiber treatment on tribological behavior of natural fiber composites

Treated and untreated carbon fiber reinforced composites and investigated wear behavior of the developed composites sliding against GCr15 steel rings using on an M-2000 model ring on a block test rig. Following the rare earth treatment, the treated carbon fiber enhanced composite surfaces become rougher and form many active groups. In the treated carbon fiber reinforced composites, the wear rate and friction coefficient were reduced compared to the untreated carbon fiber reinforced composites. The optimum content of composites enhanced by 15wt% carbon fiber produces the continuous film transfer in the counterpart surfaces. [36].The tribological properties of rice husk (RH) reinforced composites. They have prepared composites laminates in two different sets such as modified rice husk and unmodified rice husk with polymer epoxy matrix. Due to the unique properties of natural fibers, it has been used by increasing their strength through surface treatments. The better tribological properties were obtained in modified RH composites than the unmodified RH composites. [37] The wear behavior of coir fiber reinforced composites under the condition such as treated and untreated, changing the filler loading and sliding distance. From the results, wear volume increases as the sliding distance increases. The composites reinforced with untreated coir fiber show the higher specific wear rate than the composites treated with alkali and silane. [38] The wear performances of natural fiber hybrid composites. The result reveals that the behavior of wear and friction on normal oriented fibers exhibited better than the fibers treated. The best orientation against the direction of sliding that affects wear and friction is identified. Here is the fibers orientation that dominates the one being treated. Performance varies depending on fiber volume fraction or weight fraction, applied load selection and fiber matrix selection. Normally the wear rate will be increased while the applied rate increases. [39] BF Yousif et al analyzed the wear behavior of untreated oil palm fibre-reinforced polyester and treated oil palm fibre-reinforced polyester composites. The results revealed that, the treated oil palm fiber reinforced composites are enhancing the wear properties [40]. The wear behavior of the composites are determined by the film generated on the counter surface of the composites. Generally the composites are having the frictional characteristic like, the high coefficient of friction at the running stage then followed by steady state. It shows the stability of the rubbed surface. On the other hand the friction coefficient is decreased in the steady state than the running in condition. This is due to the smooth film transfer is generated on the counter surface and its more stability [41-43].

5.Conclusion

In this article, the tribological characteristics of natural fiber reinforced composites in associated with various polymers and improvements have been discussed. In addition to that the factors like weight %of fibers, addition of nano fillers and treatment of fibers are effecting on the wear behaviour of the composites such as specific wear rate and co efficient of friction also been reviewed. The various input parameters like load, sliding speed, time, temperature and fibers are effecting on the wear behaviour have also been reviewed. Still many types of natural fiber and plants which have not been discovered. More opportunities in the new research, especially tribological behaviour of natural fiber reinforced composites would lead to new discoveries that would bring green environment.

6. References

[1] Wang J.X., ShirongGe, Zhang X.Y. (2000) 'Progress of research on tribology of fiber reinforced polymer composites', Tribology, Vol 20, pp76–80.

[2] Seena Joseph, Sreekala M.S.,Oommen Z., Koshy P.,SabuThomas(2002) 'A comparison of the mechanical properties of phenol formaldehyde composites reinforced with banana fibres and glass fibres', Composites Science and Technology, Vol 62(14), pp 1857–1868. https://doi.org/10.1016/S0266-3538(02)00098-2.

[3] Dahlke B., Larbig H., Scherzer H.D.,Poltrock R.(1998) 'Natural fiber reinforced foams based on renewable resources for automotive interior applications', Journalof Cellular Plastics, Vol34(4), pp361–379.doi.org/10.1177/0021955X9803400406.

[4] Umar Nirmal, JamilHashim, Megat Ahmad M.M.H.(2015) 'A review on tribological performance of natural fibrepolymeric composites',Tribology International, Vol 83, pp 77-104.http://dx.doi.org/10.1016/j.triboint.2014.11.003.

[5] Marius Alexandrescu, Adriana Cotetiu, DinuDarabâ (2016) 'Correlations Regarding Lubricant Film Thickness to the Narrow Sliding Radial Bearing Working under Hard Shocks', Scientific Bulletin, Series C, Fascicle: Mechanics, Tribology, Machine Manufacturing Technology, ISSN 1224-3264, VolXXX,Part 1,pp 2-5.

[6] Pramendra Kumar Bajpai, Inderdeep Singh and JitendraMadaan (2013) 'Tribological behaviour of natural fibre reinforced PLA composites', Wear, Vol 297(1-2), pp 829–840.doi.org/10.1016/j.wear.2012.10.019.

[7] Boon PengChanga, HazizanMdAkila, Ramdziah Md. Nasir (2013) 'Mechanical and Tribological Properties of Zeolite-reinforced UHMWPE Composite for Implant Application, Procedia Engineering,Vol 68, pp 88-94.doi.org/10.1016/j.proeng.2013.12.152.

[8] Besnea M.A.C., Trufasu D.C., Gabriel Andrei, Marian Bastiurea, Rodeanu M.S. (2015)'Estimation of Wear Behaviour of Polyphenylene Sulphide Composites Reinforced with Glass/Carbon Fibres, Graphite and Polytetrafluoroethylene, by Pin-on-disc Test', Tribology in industry, Vol. 37(1), pp 88-96.

[9] Vijayakumar N., Gowthamipriyanka V., and Srikanth K. (2014) 'Evaluation of Wear properties of Jute Composites', International Journal of Engineering Research and Technology,Vol 3(11), pp 1212-1215.

[10] Nor AmalinaNordin, FauziahMdYussof, SalmiahKasolang, ZuraidahSalleh and Mohamad Ali Ahmad (2013) 'Wear Rate of Natural Fibre: Long Kenaf Composite', Procedia Engineering, Vol 68, pp 145–151.

[11] Yousif B.F., Nirmal U., Wong K.J. (2010) 'Three-body abrasion on wear and frictional performance of treated betelnut fibre reinforced epoxy (T-BFRE) composite', Materials & Design, Vol 31(9), pp 4514–4521.

[12] Basavarajappa S., Ellangovan S. (2012) 'Dry sliding wear characteristics of glass-epoxy composite filled with silicon carbide and graphite particles', Wear, Vol 296(1-2), pp 491–496.doi.org/10.1016/j.wear.2012.08.001.
[13] Yousif B.F., Saijod Lau T.W., McWilliam S. (2010) 'Polyester composite based on betelnutfibre', Tribology International, Vol 43, pp 503–511. doi.org/10.1016/j.triboint.2009.08.006.
[14] El-Tayeb N.S.M. (2008) 'A study on the potential of sugarcane fibers/polyester composite for tribological applications', Wear, Vol 265(1-2), pp 223–235.doi.org/10.1016/j.wear.2007.10.006.
[15] Guina Ren, Zhaozhu Zhang, Xiaotao Zhu, Xuehu Men, Wei Jiang, Weimin Liu (2014) 'Sliding wear behaviours of Nomex fabric/phenolic composite under dry and water-bathed sliding conditions',Friction,Vol2(3), pp 264–271.
[16] Betancourt S., Cruz L.J., Toro A.(2011) 'Effect of the addition of carbonaceous fibers on the tribological behaviour of a phenolic resin sliding against cast iron', Wear, Vol 272(1), pp 43– 49. doi.org/10.1016/j.wear.2011.07.014.
[17] Punyapriya Mishra and Acharya S. K. (2010) 'Anisotropy abrasive wear behavior of bagasse fiber reinforced polymer composite', International Journal of Engineering, Science and Technology, Vol 2(11), pp 104-112.
[18] Nirmal U., Hashim J and Lau S.T.W. (2011) 'Testing Methods in Tribology of Polymeric Composites', International Journal of Mechanical and Materials Engineering (IJMME), Vol 6(3), pp 367-373.
[19] CharanjitDeo and Acharya S.K. (2010) 'Effect of fiber content on abrasive wear of lantana camarafiber reinforced polymer matrix composites', Indian Journal of Engineering And Material Sciences, Vol 17(3), pp 219-233.
[20] Bahadur S. and Zheng Y. (1990) 'Mechanical and tribologicalbehavior of polyester reinforced with short glass fibers', Wear, Vol 137(2), pp 251-266. doi.org/10.1016/0043-1648(90)90138-Z.
[21] Sandeep Kumar, Amit Joshi and Brijesh Gangi (2013) 'Physico-Mechanical and Tribological Properties of Glass Fiber based Epoxy Hybrid Natural Composite', Proceedings of International Conference on Emerging Trends in Engineering and Technology, ACEEE, pp 378-383.
[22] Supreeth S., Vinod B., Dr.Sudev L.J. (2014) 'Influence of Fiber Length on the Tribological Behaviour of Short PALF Reinforced Bisphenol-A Composite', International Journal of Engineering Research and General Science, Vol 2(4),pp 825-830.
[23] Umar Nirmal, JamilHashim and Megat Ahmad M.M.H. (2015) 'A review on tribological performance of natural fibre polymeric composites', Tribology International,. Vol 83, pp 77–104.
[24] TemesgenBerhanuYallew, Pradeep Kumar, Inderdeep Singh (2014) 'Sliding wear properties of jute fabric reinforced polypropylene composites', Procedia Engineering, Vol 97, pp 402 – 411. doi.org/10.1016/j.proeng.2014.12.264.
[25] Eleiche A.M., and Amin G.M. (1986) 'The effect of unidirectional cotton fibre reinforcement on the friction and wear characteristics of polyester',Wear, Vol 112(1), pp 67–78. doi.org/10.1016/0043-1648 (86) 90201-2.
[26] Qi-Hua Wang, QunjiXueHuiwenLiu,WeichangShen, JinfenXu (1996) 'The Effect of Particle Size of Nanometer ZrO2 on the Tribological Behaviour of PEEK', Wear, Vol 198(1-2), pp. 216-219. doi:10.1016/0043-1648(96)07201-8.
[27] Qi-Hua Wang, JinfenXu,WeichangShenQunjiXue (1997) 'The Effect of NanometerSiC Filler on the TribologicalBehavior of PEEK',Wear, Vol 209(1-2), pp 316-321. doi:10.1016/S0043-1648(97)00015-X.
[28] Qi-Hua Wang, Qun-JiXue, Wei-Min Liu and Jian-Min Chen (2000) 'Effect of NanometerSiC Filler on the TribologicalBehavior of PEEK under Distilled Water

Lubrication', Journal of Applied Polymer Science, Vol 78(3), pp 609-614. doi:10.1002/1097-4628(20001017)78:33.0.CO;2-D.

[29] Umar Nirmal, Saijod TW Lau, JamilHashim, Alvin Devadas, Yuhazri MY (2015) 'Effect of kenaf particulate fillers in polymeric composite for tribological applications', Textile Research Journal, Vol 85(15), pp 1602–1619.

[30] Hui-juan Zhang, Zhao-Zhu Zhang, Fang Guo, Kun Wang (2009) 'Enhanced wear properties of hybrid PTFE/cotton fabric composites filled with functionalized multi-walled carbon nanotubes', Mater ChemPhys,Vol 116(1), pp 183–190. DOI: 10.1016/j.matchemphys.2009.03.008.

[31] Friedrich K., Karger-Kocsis J., Lu (1991) 'Effect of steel counter face roughness and temperature on the friction and wear of PEEK composites under dry sliding conditions', Wear, Vol148(2), pp 235–247. doi.org/10.1016/0043-1648(91)90287-5.

[32] Xinrui Zhang, Xian-Qiang Pei and Qihua Wang (2008) 'Effect of Solid Lubricant on the Tribological Properties of Polyimide Composites reinforced with carbon fibres', Journal of Reinforced Plastics and Composites, Vol 27(18), pp 2005-2012.doi: 10.1177/0731684408090718.

[33] Ayman A. Aly, El-Shafei B. Zeidan, AbdAllah A. Alshennawy, Aly A. El-Masry, Wahid A. Wasel (2012) 'Friction and Wear of Polymer Composites Filled by Nano-Particles: A Review', World Journal of Nano Science and Engineering, Vol 2, pp 32-39.DOI: 10.4236/wjnse.2012.21006.

[34] BhadrabasolRevappaRaju, BheemappaSuresha, RageraParameshwarappaSwamy, Bannanga diSwamyGowdaKanthraju (2013) 'Investigations on Mechanical and Tribological Behaviour of Particulate Filled Glass Fabric Reinforced Epoxy Composites', Journal of Minerals and Materials Characterization and Engineering, Vol 1(4), pp 160-167.DOI: 10.4236/jmmce.2013.14027.

[35] Prasanna S. M., Dr.Vitala H. R., Madhusudhan T., Dr.Raju B. R. (2016) 'Evaluation of Mechanical and Tribological Characterization of Glass-Basalt Hybrid Composites', International Journal of Engineering Research and Advanced Technology (IJERAT), Special Vol 02(1), pp 27-34.

[36] PushkarJha, Rakesh Kumar Gautam, RajneshTyagi, (2017) 'Friction and wear behavior of Cu–4 wt.%Ni–TiC composites under dry sliding', Friction,Vol5(4),pp 437–446.

[37] Xinrui Zhang (2007) 'The effect of fiber oxidation on the friction and wear behaviors of short-cut carbon fiber/polyimide composites', eXPRESS Polymer Letters, Vol.1(5), pp 318–325. DOI: 10.3144/expresspolymlett.2007.45.

[38] SudhakarMajhi, Samantarai S.P., Acharya S.K. (2012) 'TribologicalBehavior of Modified Rice Husk Filled Epoxy Composite', International Journal of Scientific & Engineering Research,Vol 3(6), pp 1-6.

[39] Divya G. S., AnandKakhandaki, Suresha B. (2014) 'Wear Behavior of Coir Reinforced Treated and Untreated Hybrid Composites', International Journal of Innovative Research & Development. Vol 3(5), pp 631-639.

[40] B F Yousif1and NSMEl-Tayeb. Wet adhesive wear characteristics of untreated oil palm fibre-reinforced polyester and treated oil palm fibre-reinforced polyester composites using the pin-on-disc and block-on-ring techniques. Proc. IMechE Vol. 224 Part J: J. Engineering Tribology. 2009.

[41] Chand N, Dwivedi UK. Sliding wear and friction characteristics of sisal fibre reinforced polyester composites: effect of silane coupling agent and applied load. Polym Compos 2008;29:280–4

[42] Nirmal U, Hashim J, Low KO. Adhesive wear and frictional performance of bamboo fibres reinforced epoxy composite. Tribol Int 2012;47:122–33.

[43] Yousif BF. Frictional and wear performance of polyester composites based on coir fibres. Proc Inst Mech Eng, Part J: J Eng Tribol 2009;223:51–9.

IOP Conf. Series: Materials Science and Engineering **954** (2020) 012049 doi:10.1088/1757-899X/954/1/012049

Analysis of Inertance Pulse Tube Refrigerator using CFD

M. Sai baba[1], Pankaj kumar[2]

[1]M. Tech Scholar, Department of Mechanical Engineering, GMRIT, Rajam, Srikakulam-532127, Andhra Pradesh, India.

[2]Assistant Professor, Department of Mechanical Engineering, GMRIT, Rajam, Srikakulam - 532127, Andhra Pradesh, India.

Email: mandalemulasaibaba@gmail.com, pankajcryo@gmail.com

Abstract. The current research focuses on the study of the Inertance pulse tube refrigerator [IPTR] using ANSYS FLUENT code. In the workbench with suitable dimensions, an axis symmetric model is developed, which could then be modeled to determine the validity of the Inertance pulse tube refrigerator. The additional Inertance pulse tube refrigerator is updated, i.e. 2 further reservoirs were attached to above system and they are respectively connected by valves. The efficiency of this new model of pulse tube refrigerator is analyzed and connected with the refrigerator of the Inertance pulse tube.

Keywords: PT, IPTR, CFD, 3RPTR.

1. Introduction

The pulse tube cooler has become a closed mechanism where there will be no mass contact in between device and the atmosphere [1]. The only moving component in PTR is the compressor that oscillates back and forth creating pressure waves (and rotary valve if the PTR is G-M type) [2]. The mostly used working fluid in PTR is the helium [3]. Helium has the lowest critical temperature than other gases which can be used as working fluid. It also has thermal conductivity [4]. Because of the periodic flow, the analysis of the PTRs is difficult and accurate analysis is required for optimum design of the PTR [5]. In order to design the PTR, the thermos-fluid processes in the system must be thoroughly analyzed. One-way of doing that is solving the spectrum is numerically governed based on the universal theory without having any conditional conclusions [6]. While using efficient computational fluid dynamics (CFD) software that is capable of modeling and addressing dynamic and multidimensional flow and temperature difference in complicated geometrical, its other approach analyses the problems [7].

2. Objective

In the present work the continuum governing equations are solved numerically using CFD software FLUENT for analysis of the IPTR and 3RPTR. The objectives of the current work are as follows:

- The simulation of the IPTR by using 3D axisymmetric model by using FLUENT software.
- To develop the geometric model for new type of the PTR i.e. 3 RPTR using commercial software analysis of 3RPTR.
- To study the performance of the 3RPTR by simulating it by using FLUENT software.
- To compare the results of the IPTR and 3RPTR to evaluate the performance the 3RPTR

IOP Conf. Series: Materials Science and Engineering **954** (2020) 012049 doi:10.1088/1757-899X/954/1/012049

3. Modeling and mesh generation of IPTR and 3RPTR

The modeling of IPTR and 3 buffers PTR is done using ANSYS software. The schematic geometry is shown for both in following figures. The system is composed of a transfer pipe, a cooler and a regenerator, a heat exchanger with a cold end, a connecting tube and a reservoir. Each part of the IPTR system is actually circular in shape and all the parameters are arranged in series to create an axis-symmetric system in Figure 1. Consequently, in a three dimensional axis-symmetric reference method, the IPTR is defined.

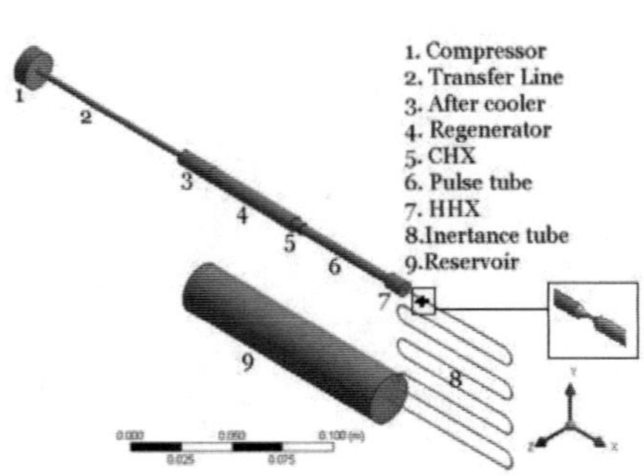

Figure 1. Geometry 3D modeling of Inertance

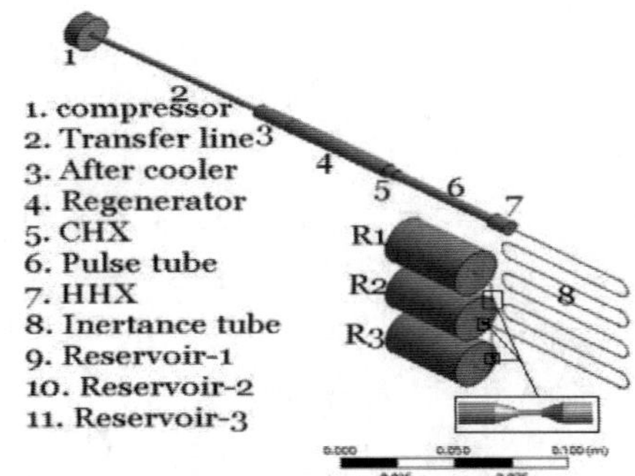

Figure 2. Schematic diagram of 3D Model of 3 RPTR pulse tube refrigerator with single reservoir

For 3 buffer type dimensions are similar only 2 more buffers are added to this geometry. The system consists of a transfer line, a cooler, a regenerator, a heat exchanger with a cold end, a PT, a hot end heat exchanger, a connecting tube and a tank. Figure 2 indicates that each 3-buffer form PTR part is then classified on an IPTR-like 3-dimensional axis-symmetric co-ordinate structure.

A dimensions of various components IPTR and 3RPTR are shown Table 1. The dimensions are same both IPTR and 3RPTR has extra dimensions of two reservoirs [last two dimensions].

Table 1. Dimension of components of IPTR and 3 RPTR

Sl. No	Components	Radius[m]	Length[m]
1	Compressor	0.00954	0.0075
2	Transfer line	0.00155	0.101
3	After cooler	0.004	0.02
4	Regenerator	0.004	0.058
5	Cold end heat exchanger	0.003	0.0057
6	Pulse tube	0.0025	0.06
7	Hot end heat exchanger	0.004	0.01
8	Inertance tube	0.000425	0.0684
9	Reservoir 1	0.013	0.13
10	Reservoir 2	0.013	0.0433

Pressure inlet is given at inlet of transfer line by defining the pressure profile in UDF as given below:

```
#include "udf.h"
DEFINE_PROFILE (unsteady_pressure, thread, position)
{
face_t f;
real t = CURRENT_TIME;
begin_f_loop (f, thread)
{
F_PROFILE (f, thread, position) = 101325.0*(5.0 * sin(2*3.141592654*34*t));
}
end_f_loop (f, thread)
}
```

Mesh of IPTR and 3 RPTR is shown in following Figure 3.

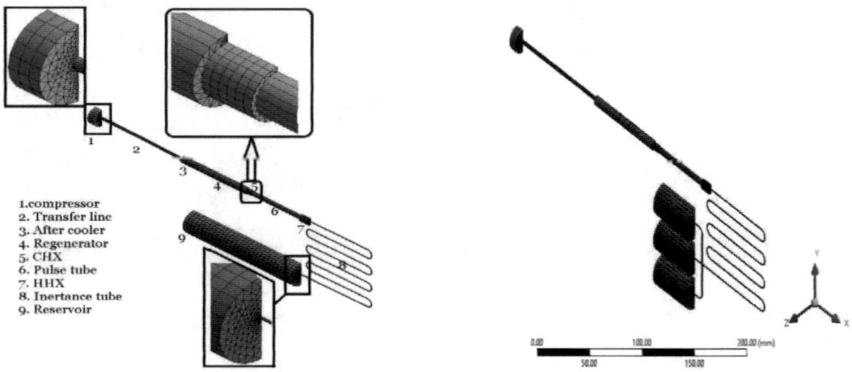

Figure 3. Mesh of different components of IPTR and 3RPTR

4. Results and discussion

Inertance pulse tube refrigerator has been simulated using FLUENT software and results are shown below. Figure 5 shows the temperature cool down with respect to time. Since the simulation is initiated 100k it shows increase in temperature for initial period then starts to cool off. The temperature appears in Figure 6 variation with time when steady periodic state is reached.

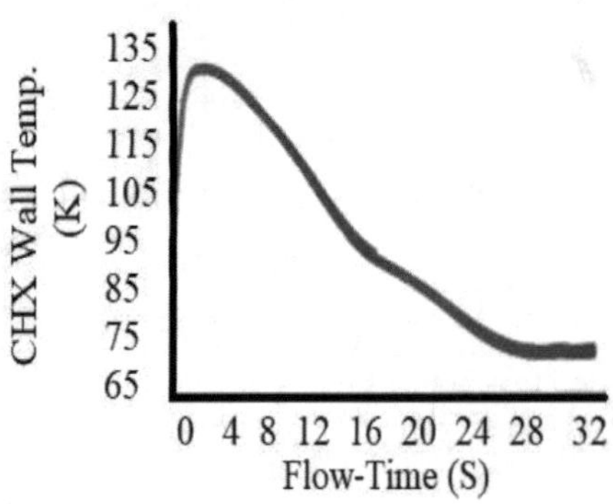

Figure 4. Temperature cool down of IPTR

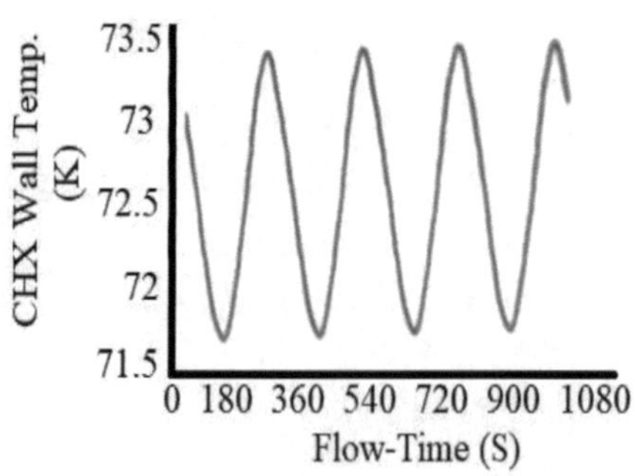

Figure 5. Cyclic variation of temperature in IPTR

IOP Conf. Series: Materials Science and Engineering **954** (2020) 012049 doi:10.1088/1757-899X/954/1/012049

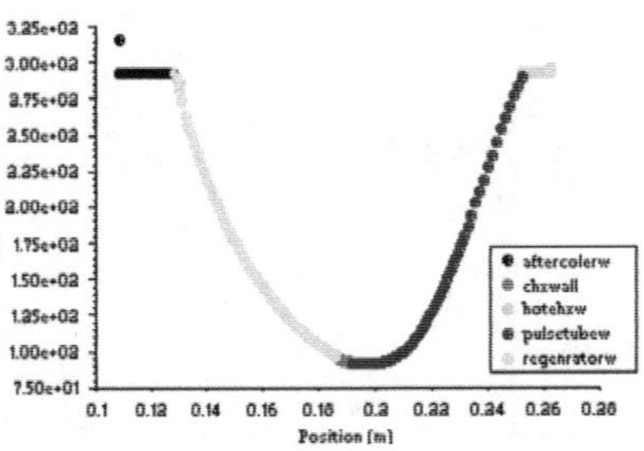

Figure 6. Temperature variation along length

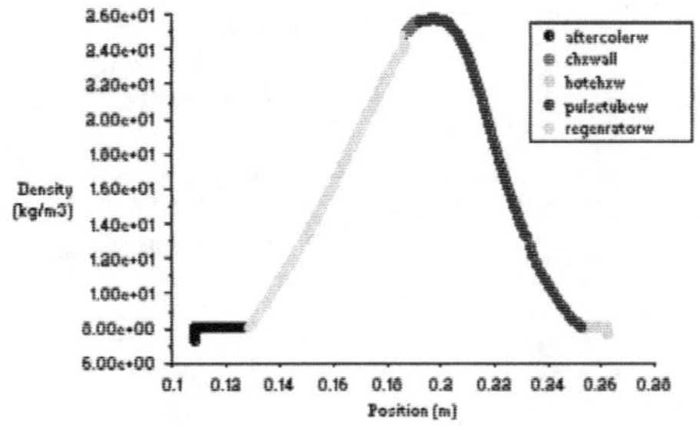

Figure 7. Density contour of IPTR

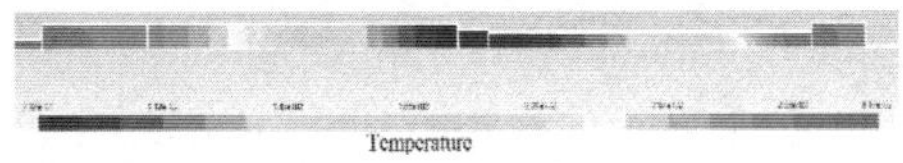

Figure 8. Temperature contour of IPTR

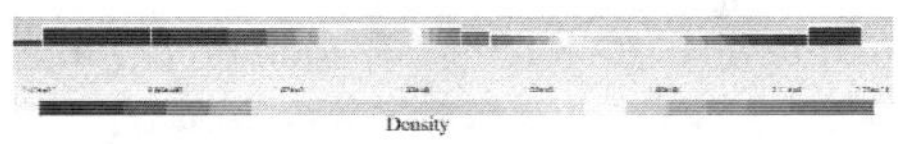

Figure 9. Density contour of IPTR

As shown in above figures the lowest temperature of CHX wall achieved in IPTR is 76.7k, this temperature is reached after 24 seconds of flow time since the simulation is initiated from 100k. FIGURE 6 and 7 shows the Density and temperature over the duration of the IPTR. As seen from these figures in CHX while density is highest in CHX.Figure 8 and 9 shows the contours of the temperature and density for regenerator, CHX and pulse tube of IPTR.

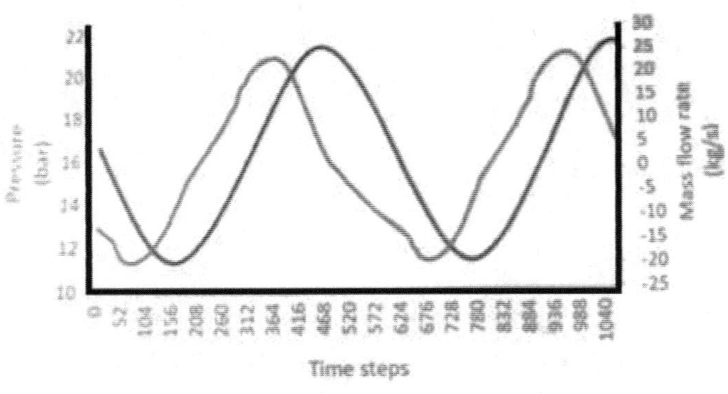

Figure 10. Phase relation between pressure and mass flow rate of IPTR

The Figure 10 shows the graph showing Pressure and intensity of mass transfer versus the time step. That could be clearly seen that there is phase difference of 74o.The 3-buffer type pulse tube refrigerator [3RPTR] is simulated using fluent software with initial guess of 100 K. Initial temperature profile is shown in fig. 11 this can be done by using patch option provided by FUENT.

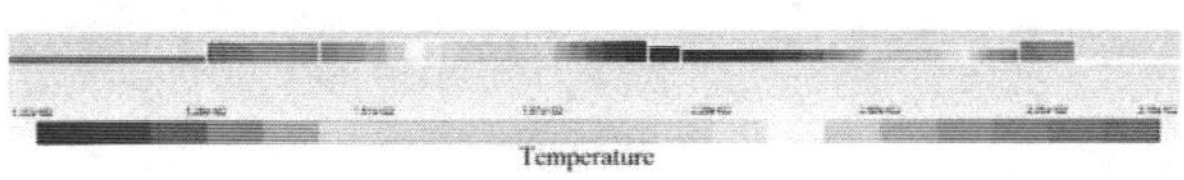

Figure 11. Initial temperature guess in 3RPTR

The simulation of 3 RPTR is done with two cases: one with only valve operating and one with all three valves operating figure 12 and 13 shows the temperature cool down for case with 1 valve and case with all valves working respectively.

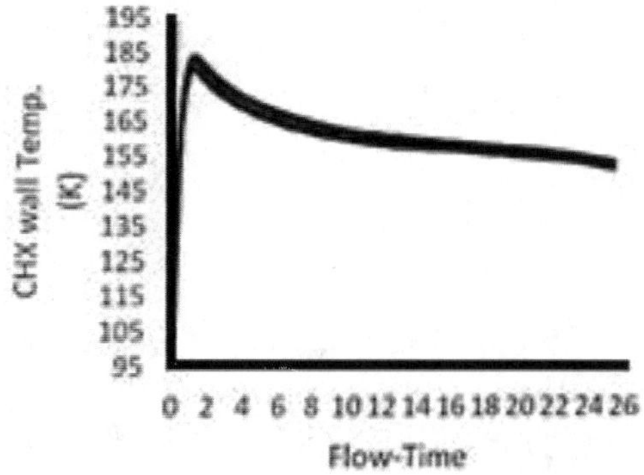

Figure 12. Temperature cool down of case 1

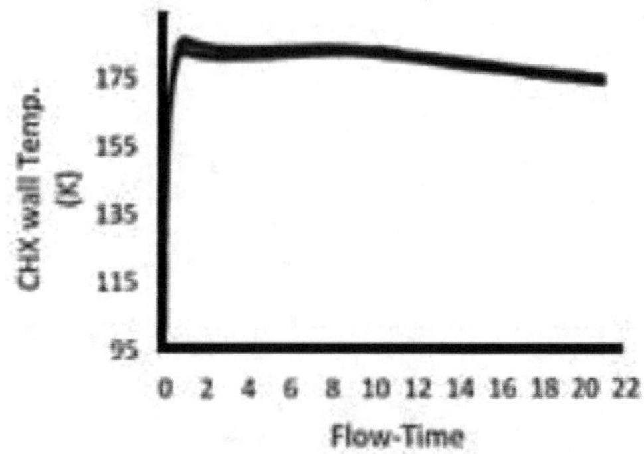

Figure 13. Temperature cool down of case 2

Figure 14 and 15 shows the cyclic temperature of case 1 and case 2 respectively. As shown from above figures, it shows that in case 1 minimum temperature of 153.24k is reached in 25 seconds and in case 2 minimum temperatures of 173.86k is reached in 22 seconds. Figure 16 and 17 shows the temperature and density variation for case 1 and figure 18 and 19 shows the same case 2.

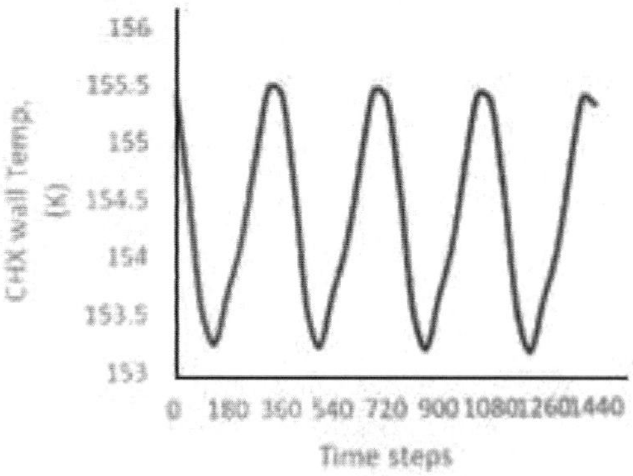

Figure 14. Cyclic temperature for case 1

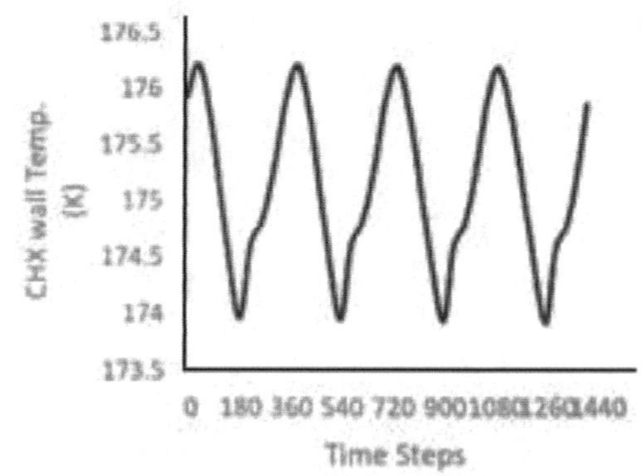

Figure 15. Fig 15 Cyclic temperature for case 2

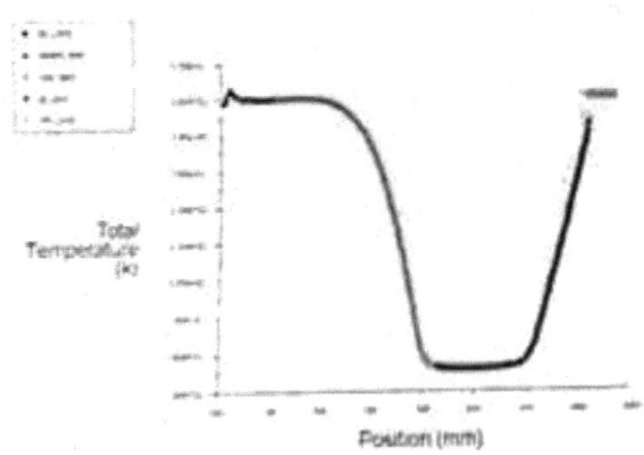

Figure 16 Temperature variation for case 1

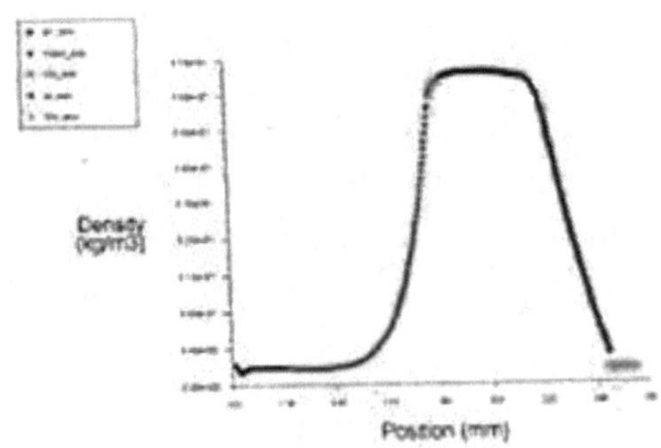

Figure 17. Density variation for case 1

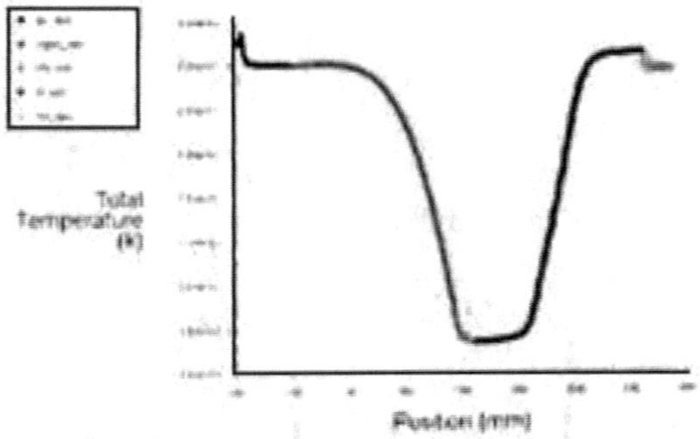

Figure 18. Temperature variation for case 2

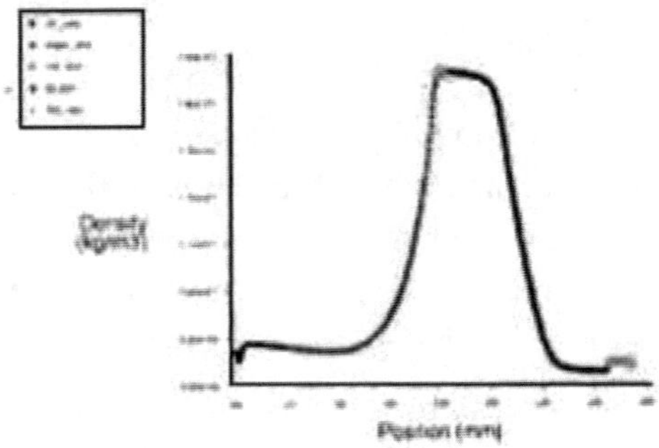

Figure 19. Density variation for case 2

Following figures shows the contours of temperature and density for Case 1 and case 2, respectively.

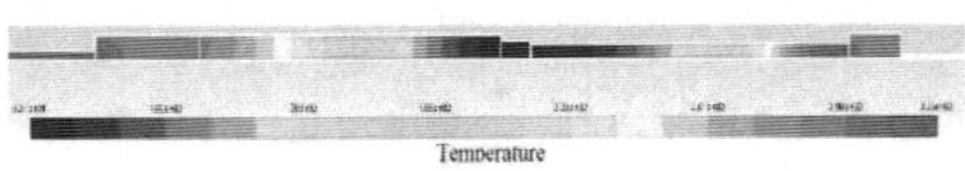

Figure 20. Temperature contour for case 1

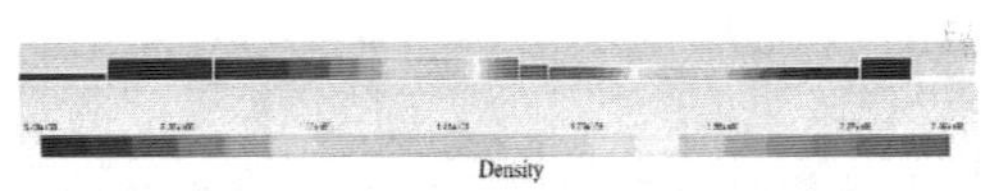

Figure 21 Density contour for case 1

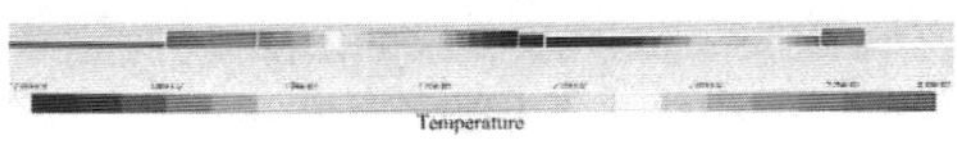

Figure 22. Temperature contour for case 2

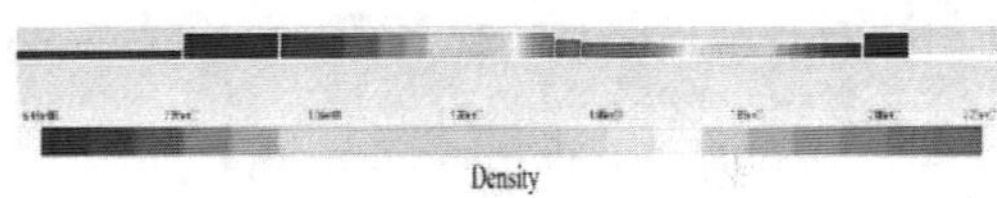

Figure 23. Density contour for case 2

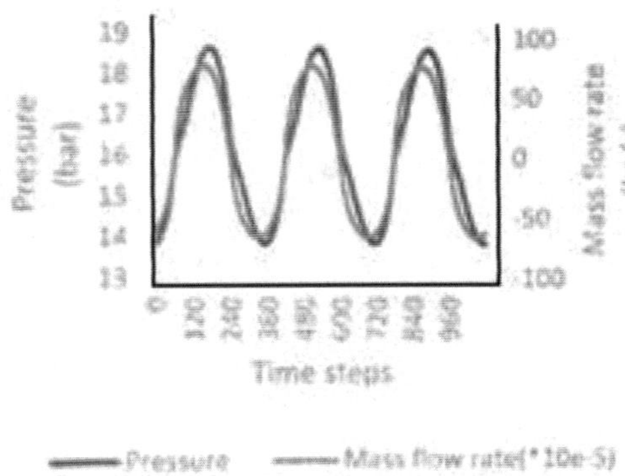

Figure 24. Phase relation for case 1

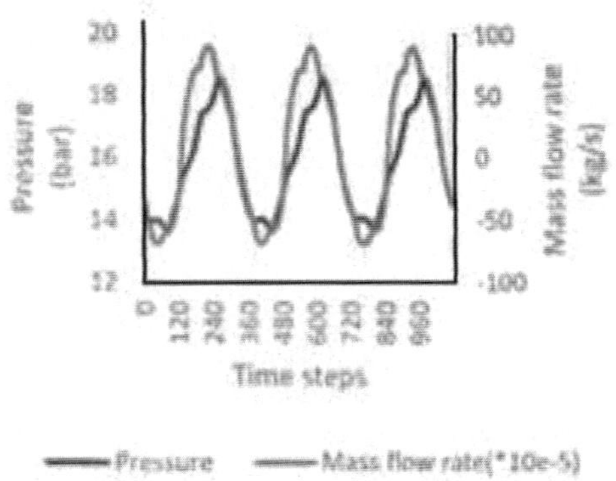

Figure 25. Phase relation for case 2

Figure 24 and 25 shows the phase relation between pressure and mass flow rate for case 1 and case 2 respectively. As seen from these graphs the phase difference between pressure and mass flow rate is almost same for both cases (250 for case 1 and 310 for case 2) and lower than that of IPTR.

IOP Conf. Series: Materials Science and Engineering **954** (2020) 012049 doi:10.1088/1757-899X/954/1/012049

5 Conclusions

The simulation of IPTR and two cases of 3RPTR is done by using ANYSYS FLUENT software. The lowest temperature observed in IPTR is 76.7 k after 24 seconds flow time. For IPTRR, its phase difference in pressure and mass flow rate is 74o. Two cases of 3 RPTR which are case 1where only one valve is operating and other two are closed and case 2 where all valves are operating. The lowest temperature is achieved in case 1 which is 152.24k in 25 seconds of flow rate and in case 2 is 178.5k in 22 seconds.

6 Future work

The simulation of 3RPTR can be extended by using different valve operating timings and different valve openings. As can be seen from results, by adjusting the parameter to achieve the higher output, there is potential for improving 3RPTR efficiency.

7 References

[1] Gifford, W. E., &Longsworth, R. C. (1964). Pulse-tube refrigeration.
[2] Gifford, W. E., &Longsworth, R. C. (1966). Surface heat pumping. In *Advances in cryogenic engineering* (pp. 171-179). Springer, Boston, MA.
[3]Mikulin, E. I., Tarasov, A. A., &Shkrebyonock, M. P. (1984). Low-temperature expansion pulse tubes. In *Advances in cryogenic engineering* (pp. 629-637). Springer, Boston, MA.
[4] Cha, J. S., Ghiaasiaan, S. M., Desai, P. V., Harvey, J. P., & Kirkconnell, C. S. (2006). Multi-dimensional flow effects in pulse tube refrigerators. *Cryogenics*, *46*(9), 658-665.
[5] Chen, L., Zhang, Y., Luo, E., Li, T., & Wei, X. (2010). CFD analysis of thermodynamic cycles in a pulse tube refrigerator. *Cryogenics*, *50*(11-12), 743-749.
[6]Ghahremani, A. R., Saidi, M. H., Jahanbakhshi, R., &Roshanghalb, F. (2011). Performance analysis and optimization of high capacity pulse tube refrigerator. *Cryogenics*, *51*(4), 173-179.
[7] Hu, J. Y., Ren, J., Luo, E. C., & Dai, W. (2011). Study on the inertance tube and double-inlet phase shifting modes in pulse tube refrigerators. *Energy conversion and management*, *52*(2), 1077-1085.

Corrigendum: Tribological characteristics of natural fiber composite - Review (*IOP Conf. Ser.: Mater. Sci. Eng.* 954 (2020) 012048)

K.Mohan[1] *, T.Rajmohan [2]
[1] Sri Chandrasekharendra Saraswathi Viswa Mahavidyalaya, Enathur, Kanchipuram - 631561, Tamil Nadu, India

Page 1:

The figure 1 should be read as,

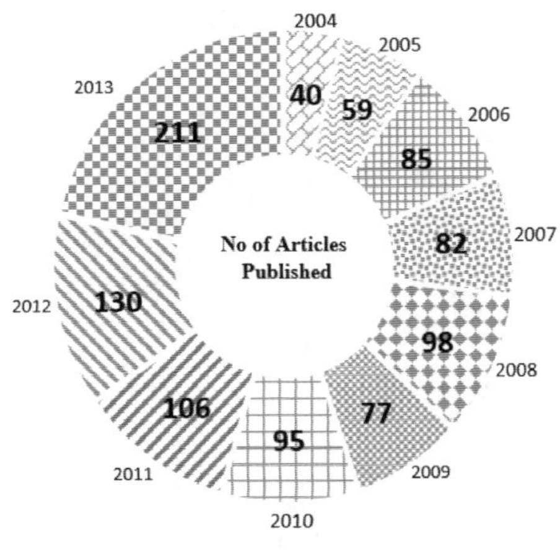

In the Introduction section, the following text appears in below the graph:

" Figure 1: Number of articles published based on tribological evaluation of synthetic and natural fibers [4]."

This should read:

" Figure 1 - Number of articles published based on tribological evaluation of natural fibers [4]."

AUTHOR INDEX